Advanced Nanomaterials for Wastewater Remediation

Advances in Water and Wastewater Transport and Treatment

A SERIES

Series Editor

Amy J. Forsgren

Xylem, Sweden

Advanced Nanomaterials for Wastewater Remediation
Ravindra Kumar Gautam and Mahesh Chandra Chattopadhyaya

Membrane Bioreactor Processes: Principles and Applications
Seong-Hoon Yoon

Wastewater Treatment: Occurrence and Fate of Polycyclic Aromatic Hydrocarbons (PAHs)
Amy J. Forsgren

Harmful Algae Blooms in Drinking Water: Removal of Cyanobacterial Cells and Toxins
Harold W. Walker

ADDITIONAL VOLUMES IN PREPARATION

Advanced Nanomaterials for Wastewater Remediation

Edited by
Ravindra Kumar Gautam
Mahesh Chandra Chattopadhyaya

CRC Press
Taylor & Francis Group
Boca Raton London New York

CRC Press is an imprint of the
Taylor & Francis Group, an **informa** business

CRC Press
Taylor & Francis Group
6000 Broken Sound Parkway NW, Suite 300
Boca Raton, FL 33487-2742

First issued in paperback 2019

© 2017 by Taylor & Francis Group, LLC
CRC Press is an imprint of Taylor & Francis Group, an Informa business

No claim to original U.S. Government works

ISBN-13: 978-1-4987-5333-3 (hbk)
ISBN-13: 978-0-367-87629-6 (pbk)

Library of Congress Cataloging-in-Publication Data

Names: Gautam, Ravindra Kumar, editor. | Chattopadhyaya, Mahesh Chandra, editor.
Title: Advanced nanomaterials for wastewater remediation / editors, Ravindra Kumar Gautam and Mahesh Chandra Chattopadhyaya.
Description: Boca Raton : Taylor & Francis Group, a CRC title, part of the Taylor & Francis imprint, a member of the Taylor & Francis Group, the academic division of T&F Informa, plc, [2016] | Series: Advances in water and wastewater transport and treatment ; 4 | Includes bibliographical references and index.
Identifiers: LCCN 2016004617 | ISBN 9781498753333 (acid-free paper)
Subjects: LCSH: Water--Purification--Materials. | Nanostructured materials--Industrial applications. | Nanotechnology.
Classification: LCC TD477 .A38 2016 | DDC 628.3--dc23
LC record available at https://lccn.loc.gov/2016004617

Visit the Taylor & Francis Web site at
http://www.taylorandfrancis.com

and the CRC Press Web site at
http://www.crcpress.com

Dedicated to

My loving sister, the late Pratiksha Gautam

(R.K. Gautam)

And to our

Maa & Baba

Contents

Foreword

Water is life. The human body contains, on average, 60% water by weight. The hydrolysis reaction, which is one of the most common reactions that takes place in the human body, requires water for its completion. Thus, of necessity, without pure and nascent water, life will end on our planet. Although it is reported that water was purified in ancient Egypt by using charcoal, it is still used today in different forms to purify water. With the advent of physical and chemical characterization techniques at the atomic level, and the understanding of them through theoretical study in angstrom to nanometer scale, there is an improvement in the understanding of the use of nanotechnology, which can play a pivotal role in improving the standard of life many times over through its application in water, health care, food, and electronics. This book attempts to show the use of nanotechnology in the purification of water. Water is plentifully available on the planet and, thus, we tend to ignore the importance of water treatment and purification and its requirement in daily life. Since water distribution across land masses is quite erratic, it is important that we treat water, recycle, and reuse it as far as possible. It is in this context that the topic chosen by the authors is very relevant in today's world scenario, due to the environmental concern toward both water and air pollution and their nexus to the food chain. I must congratulate the editors and authors who brought together this bouquet of chapters, which encompasses different aspects of wastewater treatment using nanotechnology concepts, and I wish them great success in disseminating the knowledge!

Professor Suddhasatwa Basu, FNASc, FICS
(Former Head of Chemical Engineering Department, I.I.T. Delhi)
(Former Chairman of IIChE NRC)
Chemical Engineering Department
Indian Institute of Technology Delhi

Preface

Water is the driving force in nature.

Leonardo da Vinci (1452–1519)

Contamination of aqueous environments by hazardous chemical compounds has increased tremendously during the last couple of decades, due to enhanced industrialization, burgeoning urbanization, and the haphazard growth of the world population. Thus, the supply of clean water has declined rapidly throughout the globe to meet the recommended safe limits. Pollution by metals, metalloids, dyes, volatile organic compounds, and other inorganic and organic salts is one of the most widespread concerns of aquatic ecosystems. Many of them are difficult to biodegrade, possess high solubility in water, and facilitate increased mobilization in abiotic and biotic environments, thus threatening human health and aquatic organisms. As a consequence of growing pressure on the water supply in drought-prone areas, and for industrial and civil purposes, the use of unconventional water sources such as treated wastewater will be a new norm. Hence, various remediation measures such as membrane filtration, coagulation–flocculation and oxidation–reduction processes, bioremediation, and adsorption have been examined regarding the separation and purification of contaminated aqueous environments. All these technological innovations have their own benefits and drawbacks.

The scientific community and environmental engineers expect a promising technological breakthrough from emerging nanotechnological innovations, which hold great potential for wastewater remediation with improved treatment efficiency and lower energy requirements. Some nanotechnological applications use smart nanomaterials of inorganic and organic origin, which display high surface area, increased functionality, enhanced dispersibility, strong sorption capacity, superparamagnetism, and an excellent energy band gap.

In recent years, various types of nanomaterials, such as advanced oxidation catalysts, carbon nanotubes, nanocatalysts based on transition metal ions, ion-exchange nanocomposites, layered double hydroxides, magnetic nanoparticles, aluminosilicates, and graphene-supported nanosorbents have been widely investigated for the remediation of wastewater. The nanomaterial-based remediation of water and wastewater possesses the marvelous prospect of minimizing the need for treatment and disposal of contaminated water, by removing or transforming organic or inorganic contaminants into harmless forms from highly concentrated to the parts per billion level in treated wastewater. However, in the current scenario, a large number of publications and studies are available in the literature; however, these publications do not cover advanced nanomaterials completely and certain gaps are visible that need to be filled. For the past five years, several research papers dealing with adsorption, separation, and preconcentration or transformation of pollutants using various nanomaterial-based technologies have been reported. However, these results are still new and scattered in various journals, magazines, and proceedings of conferences.

Therefore, it was thought worthwhile and opportune to prepare a reference book that described the synthesis, fabrication, and application of advanced nanomaterials in water treatment processes. Their adsorption, transformation into low toxic forms, or degradation phenomena, and the adsorption and separation of hazardous dyes, organic pollutants, heavy metals and metalloids from aqueous solutions have been highlighted. The book also contains consistent explanations for adsorption kinetics, equilibrium isotherm modeling, and thermodynamic analysis of various pollutants with nanomaterials of various categories. Since the book contains both background and up-to-date information, it will serve not only as a reference source for researchers, but also as an introductory work for graduate and advanced undergraduate students in water, wastewater remediation, and environmental sciences.

Ravindra Kumar Gautam
Mahesh Chandra Chattopadhyaya

Acknowledgments

First of all, we are thankful to Irma Britton, senior acquisitions editor, at CRC Press/Taylor & Francis Group for encouraging us to write this book and accepting our proposal. Without the kind e-mails, guidance, and cooperation from Irma, it would not have been possible for us to complete this task. Thanks Irma! We are also grateful to all the editors and technical staff, especially Ariel Crockett and Laurie Oknowsky of CRC Press, for their kind e-mails and messages. Thanks are due to all the contributors for writing excellent chapters for this book. We wish to thank Professor Suddhasatwa Basu for accepting our request to write a few words for the foreword. We are indebted to all the reviewers for their fruitful comments, which greatly improved the quality of the chapters. The support and encouragement of Professor V.S. Tripathi from the Department of Chemistry, University of Allahabad is also appreciated. Ravindra Gautam is thankful to all those in the laboratory and at home—particularly Dr. Sushmita Banerjee—whose encouragement has helped him to bring the manuscript of this book to completion. We are also thankful to Kokila Banerjee, Ajit K. Banerjee, and Tuhina Banerjee for their kind support and cooperation. Mahesh Chattopadhyaya is thankful to his wife, Alpana Chattopadhyaya, and his sons, Shushant and Shashank Chattopadhayaya, for making this journey beautiful and memorable. Finally, no words can express the feelings toward our family members who have contributed and sacrificed a lot for us to accomplish this task and will always remain the sole source of inspiration in our life to achieve higher goals.

Last, but certainly not least, we thank every reader of this book, and solicit your comments to our e-mails: ravindragautam1987@gmail.com and mcchattopadhyaya@gmail.com. Please let us know what you think of this edition; we will earnestly try to incorporate your suggestions to strengthen future editions. Enjoy this book!

Ravindra Kumar Gautam
Mahesh Chandra Chattopadhyaya

Editors

Ravindra Kumar Gautam completed his postgraduate studies in environmental science at the University of Allahabad, India, in 2009. He did a postgraduate diploma in disaster management at Indira Gandhi National Open University, New Delhi in 2010. Thereafter, he worked for one year in the National Environmental Engineering Research Institute, Council of Scientific and Industrial Research (NEERI-CSIR), Nagpur, India. He qualified for the CSIR-UGC national eligibility test for a junior research fellowship. He has published 66 research papers, including original research articles, reviews, books, book chapters, and conference proceedings. He has written a book entitled *Environmental Magnetism: Fundamentals and Applications* (ISBN-10: 3659209090 | ISBN-13: 978-3659209093), which was published by LAP Lambert Academic Publishing, Saarbrücken, Germany. Five research articles and one book are in the pipeline. He was selected as a fellow of the Indian Chemical Society, and as a life member of the Indian Science Congress Association in 2013. He is a member of the editorial boards of the *International Journal of Nanoscience and Nanoengineering, American Journal of Environmental Engineering and Science, International Journal of Environmental Monitoring and Protection*, and *International Journal of Industrial Chemistry and Biotechnology*. He also serves as a reviewer for more than 20 journals of international repute. Currently, he is engaged in doctoral work in the Department of Chemistry, University of Allahabad, India. His areas of interests are adsorption and nanomaterials, and their analogs for water/wastewater remediation.

Mahesh Chandra Chattopadhyaya is professor in the Department of Chemistry at the University of Allahabad, India. He carried out his postgraduate studies in chemistry with a specialization in inorganic chemistry at Gorakhpur University in 1967. He was selected for the postgraduate course in radiological physics by Bhabha Atomic Research Centre. He obtained a PhD degree from the Indian Institute of Technology Bombay, Mumbai. He completed a short course at the American Chemical Society on the interpretation of infrared spectra. Thereafter, he joined the University of Allahabad as lecturer in 1974, and subsequently, he became a reader and then professor of chemistry. He served the university as head of the Department of Chemistry during the period 2008–2010. Besides teaching inorganic and analytical chemistry, he also taught environmental chemistry at the university. He was elected as a fellow of Cambridge Philosophical Society; the Chemical Society, London; the Indian Chemical Society; and the Institution of Chemists, India. He was president of the Indian Chemical Society during the years 2011–2013. He has published more than 150 research papers in different national and international journals. Twenty-eight research scholars have been awarded doctoral degrees under his supervision. Currently, he is working on nanomaterials and nanocomposites and their application for environmental remediation, solid oxide fuel cells, synthesis of inorganic materials, and development of sensors for estimation of surfactants.

Contributors

Vishnu Agarwal
Department of Biotechnology
Motilal Nehru National Institute of
  Technology
Allahabad, India

Nida Alam
Analytical and Polymer Research Laboratory
Department of Applied Chemistry, Faculty of
  Engineering and Technology
Aligarh Muslim University

and

Environmental Research Laboratory
Department of Applied Chemistry, Faculty of
  Engineering and Technology
Aligarh Muslim University,
Aligarh, India

Sushmita Banerjee
Environmental Chemistry Research
  Laboratory
Department of Chemistry
University of Allahabad
Allahabad, India

Ramasamy Boopathy
Environment and Sustainability Department
CSIR-Institute of Minerals and Materials
  Technology
Bhubaneshwar, India

Dhevagoti Manjula Dhevi
Department of Chemistry
SRM University
Chennai, India

Vinod Kumar Garg
Centre for Environmental Sciences and
  Technology
Central University of Punjab
Bathinda, India

Ravindra Kumar Gautam
Environmental Chemistry Research Laboratory
Department of Chemistry
University of Allahabad
Allahabad, India

Farshid Ghanbari
Department of Environmental Health
  Engineering, School of Public Health
Ahvaz Jundishapur University of Medical
  Sciences
Ahvaz, Iran

Debasis Ghosh
Department of Chemical Engineering
Durgapur Institute of Advanced Technology
  and Management
Durgapur, India

Tang Shu Hui
Centre of Lipids Engineering and Applied
  Research (CLEAR)
Universiti Teknologi Malaysia
Johor, Malaysia

Navish Kataria
Department of Environmental Science and
  Engineering
Guru Jambheshwar University of Science and
  Technology
Hisar, India

Anupreet Kaur
Basic and Applied Sciences Department
Punjabi University, Patiala
Punjab, India

Asif Ali Khan
Analytical and Polymer Research Laboratory
Department of Applied Chemistry, Faculty of
  Engineering and Technology
Aligarh Muslim University
Aligarh, India

Heecheul Kim
Department of Architectural Engineering
Kyung Hee University
Suwon, South Korea

Kap Jin Kim
Department of Advanced Materials
  Engineering for Information and
  Electronics
Kyung Hee University
Suwon, South Korea

Anamika Kushwaha
Department of Biotechnology
Motilal Nehru National Institute of Technology
Allahabad, India

Xubiao Luo
Key Laboratory of Jiangxi Province for
  Persistent Pollutants Control and Resources
  Recycle
Nanchang Hangkong University
Nanchang, People's Republic of China

Mahsa Moradi
Department of Environmental Health
  Engineering
School of Medical Sciences
Tarbiat Modares University
Tehran, Iran

Barun Kumar Nandi
Department of Fuel and Mineral Engineering
Indian School of Mines Dhanbad
Dhanbad, India

Marta Pazos
Department of Chemical Engineering
University of Vigo
Vigo, Spain

Arun Anand Prabu
Department of Chemistry
VIT University
Vellore, India

Mihir Kumar Purkait
Department of Chemical Engineering
Indian Institute of Technology Guwahati
Guwahati, India

Mehabub Rahaman
Department of Chemical Engineering
Jadavpur University
Kolkata, India

Puja Rai
Environmental Chemistry Research Laboratory
Department of Chemistry
University of Allahabad
Allahabad, India

Radha Rani
Department of Biotechnology
Motilal Nehru National Institute of Technology
Allahabad, India

Vandani Rawat
Environmental Chemistry Research Laboratory
Department of Chemistry
University of Allahabad
Allahabad, India

Emilio Rosales
Department of Chemical Engineering
University of Vigo
Vigo, Spain

Nalini Sankararamakrishnan
Centre for Environmental Science and
  Engineering
Indian Institute of Technology Kanpur
Kanpur, India

M. Ángeles Sanromán
Department of Chemical Engineering
University of Vigo
Vigo, Spain

Shakeeba Shaheen
Analytical and Polymer Research Laboratory
Department of Applied Chemistry
Aligarh Muslim University
Aligarh, India

Geoffrey S. Simate
School of Chemical and Metallurgical
  Engineering
University of the Witwatersrand
Johannesburg, South Africa

Annamalai Sivaraman
Department of Chemistry
VIT University
Vellore, India

Lubinda F. Walubita
Texas A&M Transportation Institute (TTI)
The Texas A&M University System
Vellore College Station, Texas

Muhammad Abbas Ahmad Zaini
Centre of Lipids Engineering and Applied
    Research (CLEAR)
Universiti Teknologi Malaysia
Johor, Malaysia

Lee Lin Zhi
Centre of Lipids Engineering and Applied
    Research (CLEAR)
Universiti Teknologi Malaysia
Johor, Malaysia

Abbreviations

σ_s	saturation magnetization
AA	acrylic acid
AAS	atomic absorption spectrometry
AC	activated carbon
ACCs	activated carbon cloths
ACF	activated carbon fiber
AES	atomic emission spectrometer
AFM	atomic force microscopy
AOPs	advanced oxidation processes
APTES	γ-aminopropyltriethoxysilane
APTMS	γ-aminopropyltrimethoxysilane
As	arsenic
ASC	amino sulfonic-Cu-(4,4′-bipy)$_2$
ATR-IR	attenuated total reflectance infrared
BDD	boron-doped diamond
BET	Brunauer–Emmett–Teller
BIS	Bureau of Indian standards
BJH	Barrett–Joyner–Halanda
CA	crotonic acid
CAC	commercial activated carbon
CD	cyclodextrin
CDTA	trans 1,2-diaminocyclohexane– N,N,N′,N′ tetra acetic acid monohydrate
CL	cellulose
CMCS	carbon disulfide modified chitosan
CMMC	cross-linked magnetic modified chitosan
CNT	carbon nanotube
COD	chemical oxygen demand
CPs	coordination polymers
CS	chitosan
CR	Congo red
CUSs	coordinative unsaturated sites
CVAAS	cold vapor atomic absorption spectrometry
CVD	chemical vapor deposition
d_w	mass mean particle diameter
DI	deionized
DRS	diffuse reflection spectroscopy
DTA	differential thermal analysis
DTG	differential gravimetric analysis
EC	electrocoagulation
ECPs	electrochemical processes
ED	electrodialysis
ED	ethylene diamine
EDX	energy-dispersive x-ray spectroscopy
EGDMA	ethylene glycol dimethacrylate
EGMA	ethylene glycol methacrylate
ENPs	engineered nanoparticles

EPMs	electrophoretic mobilities
FASS	flame atomic absorption spectrometry
Fe–BTC	iron-1,3,5-benzenetricarboxylic
Fe$_3$O$_4$	magnetite
γ-Fe$_2$O$_3$	maghemite
FESEM	field emission scanning electron microscopy
FSP	Fe$_3$O$_4$–SiO$_2$-poly(1,2-diaminobenzene)
FTIR	Fourier transform infrared spectroscopy
GAC	granular activated carbon
GFAAS	graphite furnace atomic absorption spectrometry
GLA	glutaraldehyde
GMA	glycidyl methacrylate
GNs	graphene nanosheets
Gn-MNP	dendrimer-conjugated magnetic nanoparticle
GO	graphene oxide
H_c	coercivity
HF	hydrofluoric acid
HKUST	Hong Kong University of Science and Technology
HPLC	high-performance liquid chromatography
HR-TEM	high resolution transmission electron microscope
HSAB	hard and soft acids and bases
ICMR	Indian Council of Medical Research
ICP	inductively coupled plasma
ICP-AES	inductively coupled plasma atomic emission spectrometry
ICP-MS	inductively coupled plasma mass spectrometry
ICP-OES	inductively coupled plasma optical emission
IDA	iminodiacetic acid
IL	ionic liquid
K_F	Freundlich constant related to intensity
K_L	Langmuir constant
LDHs	layered double hydroxides
M_R	remanent magnetization
M_S	saturation magnetization
MA	maleic anhydride
MAC	magnetic activated carbon
MAPA	magnetically active polymeric adsorbent
MB	methylene blue
MCM	magnetic composite microsphere
MCNCs	magnetic chitosan nanocomposites
MDA	malondialdehyde
MF	microfiltration
MIL	materials of Institut Lavoisier
MMA	methyl methacrylate
MMO	mixed metal oxide
MNC	magnetic nanocomposite
MNPs	magnetic nanoparticles
MOF	metal organic frameworks
MPA	magnetic polymer adsorbent
M-PVA	magnetite-polyvinyl alcohol
M-PVAc	magnetite-polyvinyl acetate

M-PVEP	magnetite-polyvinyl propenepoxide
MST	magnetic separation technology
MW	molecular weights
MWCNT	multiwalled carbon nanotube
n	Freundlich constant related to adsorption capacity
NF	nanofiltration
NHMRC	National Health and Medical Research Committee
NMCM	nanoporous magnetic composite microspheres
NOM	natural organic matter
NPs	nanoparticles
nZVI	zero-valent iron
PAA	poly(acrylic acid)
PAC	powdered activated carbon
PAMAM	polyamidoamine
PAN-CFB	polyacrylonitrile-based carbon fiber brush
PCN	porous coordination network
PEF	photoelectro-Fenton
pH_{pzc}	pH of zero point charge
PPMS	physical property measurement system
PPY	protease peptone yeast extract medium
PPy	polypyrrole
PR	polyrhodanine
PSD	pore size distribution
PVAc	polyvinyl acetate
Q_m	maximum adsorption capacity
Re	removal efficiency
RO	reverse osmosis
ROS	reactive oxygen species
RS	Raman spectroscopy
RT	room temperature
S_E	specific area of external surface
SAED	selected area electron diffraction
SAXS	small angle x-ray diffraction
SBUs	secondary building units
SEM	scanning electron microscopy
SOD	superoxide dismutase
SPEF	solar photoelectro-Fenton
SQUID	superconducting quantum interference device
STEM	scanning transmission electron microscope
TATAB	4,4′,4″-s-triazine-1,3,5-triyltri-p-amino benzoate
TDS	total dissolved solids
TEM	transmission electron microscopy
TEPA	tetraethylenepentamine
TGA	thermogravimetric analysis
TG-DSC	thermogravimetry differential scanning calorimetry
TPA	terephthalic acid
UF	ultrafiltration
USEPA	United States Environmental Protection Agency
UV	ultraviolet radiation
VIM	vinyl imidazole

VSM	vibrating sample magnetometer
WHO	World Health Organization
XAFS	x-ray absorption fine structure
XPS	x-ray photoelectron spectroscopy
XRD	x-ray powder diffraction

1 Nanomaterials Applications for Environmental Remediation

Anupreet Kaur

CONTENTS

ABSTRACT

Nanoremediation has the potential to clean up large contaminated sites *in situ*, reducing cleanup time, eliminating the need for removal of contaminants, and hence reducing contaminant concentration to near zero. Advances in nanoscale science and engineering suggest that many of the current problems involving water quality could be resolved or greatly diminished by using nonadsorbent, nanocatalyst, bioactive nanoparticles, nanostructured catalytic membranes, submicrons, nanopowder, nanotubes, magnetic nanoparticles, nanosensors, and fullerenes. Nanomaterials are the key players that promise many benefits through their nanoenabled applications in multiple sectors. The aim of this chapter is to give an overall perspective of the use of nanomaterials to solve potential issues, such as the treatment of contaminated water for drinking and more effective reuse than through conventional means. The explosive growth in nanotechnology research has opened the doors to new strategies in environmental remediation.

Keywords: Nanomaterials, Nanoadsorbents, Nanosensors, Nanostructured catalytic membranes

1.1 INTRODUCTION

Nanomaterials are typically defined as materials smaller than 100 nm in at least one dimension. At this scale, materials often possess novel size-dependent properties different from their large counterparts, many of which have been explored for applications in water and wastewater treatment. Some of these applications utilize the smoothly scalable size-dependent properties of nanomaterials that relate to their high specific surface area, such as fast dissolution, high reactivity, and strong sorption. Others take advantage of their discontinuous properties, such as superparamagnetism, localized *surface plasmon resonance* (SPR), and quantum confinement effect. Most applications discussed in this chapter are still in the stage of laboratory research. Pilot-tested or field-tested exceptions will be noted in the text. According to Bhattacharyyal et al. (2010), the word "nano" is developed from the Greek word meaning "dwarf." In more technical terms, the word "nano" means 10^{-9}, or one billionth of something. For example, a virus is roughly 100 nm in size (Cao 2004; World Health Organization and UNICEF 2013; U.S. Bureau of Remediation and Sandia National Laboratories 2003; Mara 2003). Naturally, the word nanotechnology evolved due to the use of nanometer-sized particles (sizes of 1–100 nm). The potential uses and benefits of nanotechnology are enormous. Environmental pollution is a serious day-to-day problem faced by the developing and developed nations in the world. Air, water, and solid waste pollution due to anthropogenic sources contribute a major share to the overall imbalance of the ecosystem. Water is the most essential substance for all life on Earth and a precious resource for human civilization. Reliable access to clean and affordable water is considered one of the most basic humanitarian goals, and remains a major global challenge for the twenty-first century. Water has a broad impact on all aspects of human life including, but not limited to, health, food, energy, and the economy. In addition to the environmental, economic, and social impacts of poor water supply and sanitation, the supply of fresh water is essential for the safety of children and the poor (Cloete et al. 2010; Elimelech and Phillip 2011; Mara 2003). It is estimated that 10–20 million people die every year due to waterborne diseases, and nonfatal infections cause the death of more than 200 million people every year. Every day, about 5000–6000 children die due to the water-related problem of diarrhea. There are currently more than 0.78 billion people around the world who do not have access to safe water resources (Moore et al. 2003; Montgomery and Elimelech 2007; Eshel 2007), resulting in major health problems. It is estimated that more than one billion people in the world lack access to safe water, and within a couple of decades the current water supply will decrease by one-third. The increasingly stringent water quality standards, compounded by emerging contaminants, have brought new scrutiny to existing water treatment and distribution systems widely established in developed countries. The rapidly growing global population and the improvement in living standards continuously drive up demand. Moreover, global climate change accentuates the already uneven distribution of fresh water, destabilizing the supply. Growing pressure on water supplies makes the use of unconventional water sources (e.g., stormwater, contaminated fresh water, brackish water, wastewater, and seawater) a new norm, especially in historically water-stressed regions. Furthermore, current water and wastewater treatment technologies and infrastructure are reaching their limit for the provision of adequate water quality to meet human and environmental needs.

Recent advances in nanotechnology offer leapfrogging opportunities to develop next-generation water supply systems. Our current water treatment, distribution, and discharge practices, which rely heavily on conveyance and centralized systems, are no longer sustainable. The highly efficient, modular, and multifunctional processes enabled by nanotechnology are envisaged to provide high-performance and affordable water and wastewater treatment solutions that rely less on large infrastructures (Hutton et al. 2007; Guzman et al. 2006; Kaur and Gupta 2009a; Chun et al. 2006; Qu et al. 2013; WHO 2012). Nanotechnology-enabled water and wastewater treatment promises not only to overcome major challenges faced by existing treatment technologies, but also to provide new treatment capabilities that could allow economic utilization of unconventional water sources to expand the water supply. Nanotechnology holds great potential in advancing water and wastewater treatment to improve treatment efficiency, as well as to augment the water supply through the safe use of unconventional water sources. Here, we review recent developments in nanotechnology for water and wastewater treatment. The discussion covers candidate nanomaterials, properties, and mechanisms that enable the applications, advantages, and limitations as compared with existing processes, and the barriers to and research needs for commercialization. By tracing these technological advances to the physicochemical properties of nanomaterials, the present review outlines the opportunities and limitations to further capitalize on these unique properties for sustainable water management.

Nanoremediation has the potential not only to reduce the overall costs of cleaning up large-scale contaminated sites, but also to reduce cleanup time, eliminate the need for treatment and the disposal of contaminated soil, and reduce some contaminant concentrations to near zero—all *in situ*. A proper evaluation of nanoremediation—particularly via full-scale ecosystem-wide studies—needs to be conducted to prevent any potential adverse environmental impacts. In addition to remediating pollution, nanoparticles can be used as sensors to monitor toxins, heavy metals, and organic contaminants in land, air, and water environments and have been found to be more sensitive and selective than conventional sensors. In this chapter, we present a background and overview of current practice, research findings related to nanotechnology, issues surrounding the use of nanotechnology for environmental remediation, and future directions.

1.2 NANOREMEDIATION USING TiO$_2$ NANOPARTICLES

Tiny particles of titanium dioxide are found as key ingredients in wall paints, sunscreens, and toothpaste; they act as reflectors of light or as abrasives. However, with decreasing particle size and a corresponding change in their surface-to-volume ratio, their properties change so that they acquire catalytic ability. Activated by the ultraviolet (UV) component in sunlight, they break down toxins or catalyze other relevant reactions.

Titanium oxide photocatalysts have been widely studied for solar energy conversion and environmental applications in the past several decades, because of their high chemical stability, good photoactivity, relatively low cost, and nontoxicity. In the photocatalytic oxidation process, organic pollutants are destroyed in the presence of semiconductor photocatalysts, an energetic light source, or an oxidizing agent such as oxygen or air. Only photons with energies greater than the band gap energy (ΔE) can result in the excitation of valence band (VB) electrons that then promote possible reactions. The absorption of photons with energy lower than ΔE or a longer wavelength usually causes energy dissipation in the form of heat (Figure 1.1).

The illumination of the photocatalytic surface with surface energy leads to the formation of a positive hole (h$^+$) in the VB and an electron (e$^-$) in the conduction band. The positive hole oxidizes either the pollutant directly or water to produce OH radicals, whereas the electron in the conduction band reduces the oxygen adsorbed on the photocatalyst (TiO$_2$) (Figure 1.2). The activation of TiO$_2$ by UV light may be as follows:

Overall reaction:

$$TiO_2 + h\nu \rightarrow e^- + h^+ \tag{1.1}$$

$$e^- + O_2 \rightarrow O_2^-$$

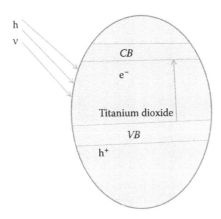

FIGURE 1.1 Photocatalytic activity of TiO$_2$.

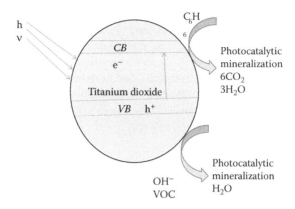

FIGURE 1.2 Decomposition of pollutants by photocatalytic activity of TiO_2.

Oxidative reaction:

$$h^+ + \text{organic moiety} \rightarrow CO_2 \qquad (1.2)$$

$$h^+ + H_2O \rightarrow OH + H^+ \qquad (1.3)$$

Reductive reaction:

$$OH + \text{organic moiety} \rightarrow CO_2 \qquad (1.4)$$

In recent years, advanced oxidation processes (AOPs) using titanium dioxide (TiO_2) have been used effectively to detoxify recalcitrant pollutants present in industrial wastewater. TiO_2 has singular characteristics that make it an extremely attractive photocatalyst: high photochemical reactivity, high photocatalytic activity, low cost, stability in aquatic systems, and low environmental toxicity. The general detailed mechanism of dye degradation upon irradiation is described as follows:

$$\text{Dye} + h\nu \rightarrow \text{Dye} * \qquad (1.5)$$

$$\text{Dye} * + TiO_2 \rightarrow \text{Dye} \cdot + + TiO_2(e) \qquad (1.6)$$

$$TiO_2(e) + O_2 \rightarrow TiO_2 + O_2^{\cdot-} \qquad (1.7)$$

$$O_2^{\cdot-} + TiO_2(e) + 2H^+ \rightarrow H_2O_2 \qquad (1.8)$$

$$H_2O_2 + TiO_2(e) \rightarrow \bullet OH + OH^- \qquad (1.9)$$

$$\text{Dye} \bullet^+ + O_2 \left(\text{or } O_2^{\cdot-} \text{ or } \bullet OH \right) \rightarrow \text{peroxylated or hydroxylated intermediates}$$
$$\rightarrow \text{degraded or mineralized products} \qquad (1.10)$$

However, the photocatalytic capability of TiO_2 is limited to ultraviolet light only. To overcome this problem, both chemical and physical modification approaches were developed to extend the absorption band-edge of TiO_2 into the visible region (Hoffmann et al. 1995; Toma et al. 2004; Sin et al. 2012; Tsai et al. 2009; Katsumata et al. 2009). Photocatalytic degradation of rhodamine

6G (R-6G), methyl red, malachite, 4-nitrophenol, yellow 27, yellow 50, violet 51, and bisphenol-A has been described. Photocatalytic degradation of Diuron by TiO_2 and by Pt/TiO_2, also Au/TiO_2, for Diuron as well as its didemethylated product, 3,4-dichlorophenyl urea, has also been reported. The photocatalytic degradation of s-triazine herbicides (atrazine, simazine, trietazine, prometon, and prometryn) was first studied by Pelizzetti et al. (1990). Several studies were published about the solar TiO_2 photocatalyzed oxidation of s-triazines (2-chloro-, 2-methoxy-, and 2-methylthio-s-triazines) and the mechanistic pathways of the observed photoproducts (Muszkat et al. 1995; Konstantinou et al. 2001; Pelizzetti et al. 1992; Borio et al. 1998; Sanlaville et al. 1996; Textier et al. 1999). Recently, herbicides belonging to atrazine, simazine, cyanazine, cyromazine (N-cyclopropyl-1,3,5-triazine-2,4,6-triamine), and metamitron {4-amino-6-phenyl-3-methyl-1,2,4-triazin-5(4H)-one} were studied and similar results were observed.

The photocatalytic degradation of four representative compounds of anilide and amide herbicides (3,4-dichloropropioamide, propanil, alachlor, propachlor) have been studied (Konstantinou et al. 2001; Pathirana and Maithreepala 1997; Muszkat et al. 1992; Peñuela and Barceló 1996). The principal intermediates in the photocatalytic destruction of p-chlorophenol (PCP) in the presence of TiO_2 were identified as *p*-chloraniline, tetrachlorohydroquinone, H_2O_2, tetrachlorocatechol, and *o*-chloraniline (Vulliet et al. 2002; Minero et al. 1993, 1996; Mills and Hoffmann 1993). The major PCP intermediates detected were 2,3,5,6-tetrachloro-1,4-benzoquinone, 2,3,5,6-tetrachloro-1,4-hydroquinone, and 2,3,5,6-tetrachlorophenol (Jardim et al. 1997). The pesticide permethrin can be easily photodecomposed into CO_2 and Cl^- ions in a fluorosurfactant/TiO_2 aqueous dispersion (Hidaka et al. 1992). Atrazine (2-chloro-4-ethyl-amino-6-isopropylamino-1,3,5-triazine) is one of the most common pesticides found in groundwater sources and drinking water supplies. TiO_2 nanoparticles were used for the preconcentration of trace arsenite and arsenate in natural water. TiO_2 has also been used for the preconcentration of metal ions such as Cu, Cr, Zn, Cd, Se, Ho, Au, Nd, Tm, La, Y, Tb, Eu, and Dy from various environmental samples such as wastewater, sediments, coal ash, vehicle exhaust particulates, and geological samples, and also physical modification with diethyldithiocarbamate (DDTC), 1-(2-pyridylazo)-2-naphthol (PAN), dithizone, and 8-hydroxyquinoline used for the separation and extraction of Cr, Cu, Pb, Zn, Fe, Al, Y, and Yb from natural, waste, and environmental water samples, biological samples, and also food samples (Li et al. 2004; Liang et al. 2001, 2003; Shunxin et al. 1999; Hang et al. 2003; Zheng et al. 2004).

Since the expression of superhydrophilicity with a TiO_2 photocatalyst, it has been increasingly applied to fogproofing and self-cleaning applications for mirrors, including road mirrors (curved mirrors) and door mirrors on cars, as well as glass window panels. Since the hydrophilicity of TiO_2 photocatalysts is more positively maintained by the addition of SiO_2 or a more porous structure of TiO_2 particles, improvement in the composition and layer-forming method for TiO_2 photocatalysts is now underway. Air purification is one example of the most advanced applications of TiO_2 photocatalysts. For example, photocatalysts are used in deodorizing filters in air purifiers incorporating UV lamps to eliminate aldehydes or volatile organic compounds (VOCs) in indoor air. A TiO_2 photocatalyst oxidizes NO into NO_2 and eventually into NO_3^-, hence removing NO from the air (Yang et al. 2004). Nanogold is supported on TiO_2-coated glass fiber for the removal of toxic CO gas from the air. Outstanding catalytic activities of nanogold for oxidizing CO at low temperatures and various reactions over nanogold catalysts have been studied. These include CO oxidation, preferential oxidation of CO in the presence of excess hydrogen (PROX), water gas shift reaction (WGSR), hydrogenation, and oxidation.

Photocatalytic oxidation is an AOP for removal of trace contaminants and microbial pathogens. It is a useful pretreatment for hazardous and nonbiodegradable contaminants to enhance their biodegradability. Photocatalysis can also be used as a polishing step to treat recalcitrant organic compounds. The major barrier for its wide application is slow kinetics, due to limited light fluence and photocatalytic activity. Current research focuses on increasing photocatalytic reaction kinetics and photoactivity range.

1.3 GOLD NANOPARTICLES FOR NANOREMEDIATION

Gold nanoparticles (AuNPs), one of the wide varieties of core materials available, coupled with tunable surface properties in the form of inorganics or inorganic–organic hybrids, have been reported as an excellent platform for a broad range of analytical methods. The modification of the Au surface with appropriate chemical species can improve separation and preconcentration efficiency, analytical selectivity, and method reliability. Because of their high surface-to-volume ratio, easy surface modification, and simple synthesis methods, AuNPs are becoming an attractive material as an alternative to conventional solvent extraction and solid-phase extraction. Through covalent bond formation (Au–S bonds), electrostatic attraction, hydrophobic adsorption, and molecular recognition, AuNPs have been applied successfully to the extraction/removal of a variety of compounds, peptides, proteins, heavy metal ions, and polycyclic aromatic hydrocarbons (PAHs). Nanocomposites of gold and aluminum nanoparticles have been used for the preconcentration of Hg(II) from natural water.

Gold nanoparticles loaded onto activated carbon (AuNP-AC) with 1-(((6-(-(2,4-dihydroxyben-zylideneamino))hexylimino) methyl)benzene-2,4-diol (DHBAHMB) have been applied for the enrichment and preconcentration of trace amounts of Cu(II), Fe(III), and Zn(II) ions in real samples (Marahel et al. 2011). Modified citrate-stabilized AuNPs have been used for the enrichment and preconcentration of endocrine disrupters in real samples (Noh et al. 2010). The determination of picogram Hg(II) has been carried out using 2,5-dimercapto-1,3,4-thiadiazole stabilized gold nanoparticles (DMT-AuNPs) in environmental samples.

N-1-(2-mercaptoethyl)thymine modification of gold nanoparticles—a highly selective and sensitive colorimetric chemosensor for Hg(II)—has been undertaken (Vasimalai and John 2012; Chen et al. 2011). A novel alternative approach using the so-called solid-phase nanoextraction (SPNE) for the preconcentration of PAHs from drinking water has been proposed by Wang et al. (Wang and Campiglia 2008; Wang et al. 2009). Alkanethiol-modified AuNPs coated on silica gel were used for the SPNE of steroids (progesterone and testosterone propionate) by Liu (2008). Qu et al. (2008) used AuNPs for the SPNE of nine compounds (ethanol, benzene, 1-butanol, chlorobenzene, 1-pentanol, anisole, phenol, methyl benzoate, benzyl alcohol). Aromatic compounds (benzene, naphthalene, phenanthrene, anthracene) were preconcentrated by the immobilization of n-octadecanethiol-modified Au-coated polystyrene particles on capillary by Kobayashi et al. (2006). Immobilization of BSA-modified AuNPs on capillary was used for the SPNE of dansyl-norvaline by Liu et al. (2003).

1.4 ZEROVALENT IRON NANOPARTICLES

Iron is one of the most abundant elements on Earth. Elemental iron has been used as an ideal candidate for remediation, because it is inexpensive, extremely easy to prepare and apply to a variety of systems, and devoid of any known toxicity induced by its usage. The concept of using metals such as iron as remediation agents is based on reduction–oxidation or "redox" reactions, in which a neutral electron donor (a metal) chemically reduces an electron acceptor (a contaminant). Nanoscale iron particles have surface areas significantly greater than larger-sized powders or granular iron, which leads to enhanced reactivity for the redox process. As a result, iron nanoparticles have been extensively investigated for the decomposition of halogenated hydrocarbons to benign hydrocarbons and the remediation of many other contaminants including anions and heavy metals (Yantasee et al. 2007). Zerovalent iron nanoparticles are highly reactive and react rapidly with surrounding media in the subsurface (Maity and Agrawal 2007). A significant loss of reactivity can occur before the particles are able to reach the target contaminant. In addition, zerovalent iron nanoparticles tend to flocculate when added to water, resulting in a reduction in the effective surface area of the metal. Therefore, the effectiveness of a remediation depends on the accessibility of the contaminants to the nanoparticles, and the maximum efficiency of remediation will be achieved only if the metal nanoparticles can effectively migrate without oxidation to the contaminant or the water–contaminant interface. To overcome such difficulties, a commonly used strategy is to incorporate iron nanoparticles within support materials, such as polymers, porous carbon, and polyelectrolytes (Laurent et al. 2008; Zhang and Elliot 2006). Zerovalent iron removes aqueous contaminants by reductive dechlorination in the case of chlorinated solvents, or by reduction to an insoluble form in the case of aqueous metal ions (Quinn et al. 2005; Schrick et al. 2002). Increasing the surface area of zerovalent iron nanoparticles results in an increased rate of remediation. In general, chlorinated organics ($C_xH_yCl_z$) and iron in aqueous solutions can be expressed by the equation:

$$CxHyClz + zH^+ + zFe^0 \rightarrow CxHy + z + zFe^{2+} + zCl^- \tag{1.11}$$

Iron undergoes classical electrochemical/corrosion redox reactions, in which iron is oxidized from exposure to oxygen and water:

$$2Fe^0(s) + O_2(g) + 2H_2O \rightarrow 2Fe^{2+}(aq) + 4OH - (aq) \tag{1.12}$$

$$Fe^0(s) + 2H_2O(g) \rightarrow Fe^{2+}(aq) + H_2(g) + 2OH - (aq) \tag{1.13}$$

Fe(II) reacts to give magnetite (Fe_3O_4), ferrous oxide ($Fe(OH)_2$), and ferric hydroxide ($Fe(OH)_3$) depending on redox conditions and pH. For example, chromium(VI) can be reduced by Fe(II) to the generic reaction scheme shown in Ponder et al. (2000):

$$Cr^{6+} + 3Fe^{2+} \rightarrow Cr^{3+} + 3Fe^{3+} \tag{1.14}$$

Iron nanoparticles have been used for the separation of As(III) (Savina et al. 2011). Phosphates are a growth nutrient for microorganisms in water. As a result of increased phosphorus concentration, an excessive growth of photosynthetic aquatic micro- and macroorganisms occurs and ultimately becomes a major cause of eutrophication, or extensive algae growth. All parameters being equal, hydrated iron oxide nanoparticles (HAIX), containing hydrated iron oxide nanoparticles, was compared with a granular ferric hydroxide (GFH) without any ion exchange material. HAIX provided significantly greater phosphate removal capacity (Zelmanov and Semiat 2011). Phosphate breakthrough with HAIX occurred after nearly 4000 bed volumes, while the commercially available GFH column from the U.S. Filter Corporation showed a breakthrough after 1000 bed volumes. Iron nanoparticles have been encapsulated with silica to increase stability and prevent aggregation. Iron oxide has been shown to retard the proliferation of bacteria. The incorporation of iron oxide–catalyzed ozonation technology increases the retention of bacteria to the surface of membranes, resulting in improved remediation of water. Iron oxide–catalyzed ozonation and membrane filtration will combine to improve inactivation and/or the removal of bacteria (Mak and Chen 2004). Fast adsorption of methylene blue on polyacrylic acid-bound iron oxide magnetic nanoparticles has been obtained (Huang et al. 2010). Figures 1.3 and 1.4 show the dechlorination and nitrate reduction pathways by zerovalent iron nanoparticles, respectively.

1.5 SILICON OXIDE NANOPARTICLES FOR NANOREMEDIATION

Silica nanoparticles are a promising material as a solid-phase extractant, because of their large surface area, high adsorption capacity, low temperature modification, lesser degree of unsaturation, and low electrophilicity (Figures 1.5 and 1.6). The sequence of reactivity is expressed as follows:

$$Zr(OR)_4, Al(OR)_4 > Ti(OR)_4 > Sn(OR)_4 \gg Si(OR)_4$$

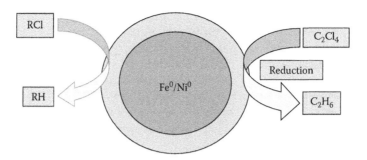

FIGURE 1.3   Dechlorination by zerovalent iron nanoparticles.

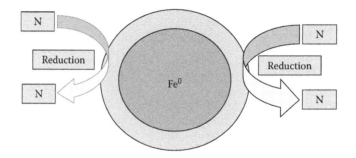

FIGURE 1.4   Nanoremediation of nitrate by zerovalent iron nanoparticles.

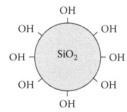

FIGURE 1.5   Silica nanoparticle.

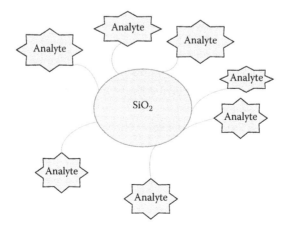

FIGURE 1.6   Nanoremediation of analyte by silica nanoparticles.

Some pollutants are poorly adsorbed on nanoparticles. To overcome this problem, physical or chemical modification of the surface of these nanoparticles with certain functional groups containing some donor atoms such as oxygen, nitrogen, sulfur, and phosphorus is necessary. The method used most often is to load a kind of specific chelating reagent by physical or chemical procedure. The formal method is simple, but the loaded reagent is prone to leaking from the sorbent, while the chemically bonded material is more stable and can be used repeatedly. The modification of nanometer-sized materials is usually required to prevent a conglomeration of particles and to improve their consistency in relation to other materials. Also, the modification of nanometer-sized materials can improve their selectivity toward pollutants. Selectivity of suitable specific functional groups toward metal ions depends on certain factors such as (1) the size of modifiers; (2) the activity of the loaded group; and also (3) the basis of the concept of hard–soft acids and bases. Chemisorption of nanoparticles provides immobility, mechanical stability, and water insolubility, thereby increasing efficiency, sensitivity, and selectivity.

Chemical modification is a process that leads to change in the chemical characteristics of the surface of nanoparticles. By this modification, adsorption properties are significantly affected. Chemisorption of chelating molecules on nanoparticle surfaces provides immobility, mechanical stability, and water insolubility, thereby increasing the efficiency, sensitivity, and selectivity of nanoparticles for analytical application. The chemical modification of nanoparticles by silylation, using different silylating agents such as 3-aminopropyltriethoxysilane, 3-chloropropyltriethoxysilane, and 3-mercaptopropyltriethoxysilane provides immobility, mechanical stability, and water insolubility. N-(3-(trimethoxysilyl)propyl)ethylenediamine–modified SiO_2 nanoparticles have been used for the preconcentration of some toxic heavy metal ions such as Hg(II), Cu(II), and Zn(II) (Shu et al. 2011). Modified silica nanoparticles have also been used for the preconcentration of drugs and pesticides. Silylation of silica nanoparticles followed by their chemical modification has been carried out using 4-(2-pyridylazo)-resorcinol (Lin and Lung 2012) and these modified SiO_2 nanoparticles have been used for the selective preconcentration of Hg(II). Also, SiO_2 nanoparticles modified with acetylsalicylic acid, p-dimethylaminobenzaldehyde, and 5-sulfonylsalicylic acid have been used for the preconcentration of Cr(III), Fe(III), Pb(II), and Cu(II) (Yantasee et al. 2007; Maity and Agrawal 2007; Laurent et al. 2008).

Preconcentration and separation of ferbam has been done by modified SiO_2-PAN nanoparticles. Nano-organo-composites were used for the nanoextraction as well as the separation of various toxic metal ions such as Hg(II), Pb(II), Sb(III), Cd(II), and Ni(II); also, other metal ions were studied by Kaur and Gupta (2009a). Silica-1,8-dihydroxyanthraquinone nanoparticles (SiO_2-DHAQ) nanoparticles as well as silica-resacetophenone nanoparticles (SiO_2-RATP) were used for the preconcentration of Co(III), Ni(II), and other toxic metal ions from various environmental as well as food samples (Kaur and Gupta 2008; 2009a–e, 2010a–c; 2015; 2016).

1.6 OTHER MATERIALS FOR NANOREMEDIATION

A major challenge for water/wastewater treatment is water quality monitoring, due to the extremely low concentration of certain contaminants and the lack of fast pathogen detection, as well as the high complexity of water/wastewater matrices.

There is a great need for innovative sensors with high sensitivity and selectivity and, also, we know that fast responses are strongly required. Biosensors and affinity sensor devices have been shown to have the ability to provide rapid, cost-effective, specific, and reliable quantitative and qualitative analysis. To date, the developments in nanomaterials and biosensor fabrication technology are moving rapidly, with new and novel nanobiorecognition materials being developed, which can be applied as sensing receptors for mycotoxin analysis. Biosensors, as tools, have proved to be able to provide rapid, sensitive, robust, and cost-effective quantitative methods for on-site testing. Developing biosensor devices for different mycotoxins has attracted much research interest in recent years, with a range of devices being developed and reported in the scientific literature. However, with the advent of nanotechnology and its impact on developing ultrasensitive devices, mycotoxin analysis is benefiting also from the advances taking place in applying nanomaterials in sensor development. The application of nanotechnology in biosensors can range from the transducer device, the recognition ligand, the label, and the running systems (e.g., instruments). Their application in sensor development has been due to the excellent advantages offered by these materials in the miniaturization of the devices, signal enhancements that result in high precision and accuracy, and also amplification of signal by the use of nanoparticles as labels. The high surface-to-volume ratio offered by nanomaterials makes these devices very sensitive and can allow the detection of a single molecule, which is very attractive in contaminant monitoring such as toxins. The development of micro/nanosensor devices for toxins analysis is increasing, due to their extremely attractive characteristics for this application (Saini and Kaur 2012a,b, 2013). In principle, these devices are miniature transducers fabricated using conventional thin and thick film technology. Their novel electron transport properties make them highly sensitive for low-level detection. Multitoxin detection (e.g., of mycotoxins) in foods can be conducted using a single micro/nanoelectrode array chip with high sensitivity and rapid analysis time. The use of micro/nanoarrays for analysis applications in foods can produce highly sensitive sensors. Multimycotoxin detection has also been reported in the literature, using different sensor platforms combined with enzyme-linked immunosorbent assay (ELISA). Therefore, multiple toxins can be detected on a single microelectrode array chip with a multiarray working electrode, where a different antibody is immobilized to detect a specific mycotoxin. Micro/nanoelectrode arrays have unique properties, which include a small capacitive charging current and faster diffusion of electroactive species, which will result in an improved response time and greater sensitivity. The use of a lab-on-a-chip is expanding in all areas of analysis due to the advantages of using small samples to analyze several markers/toxins, that is, to offer high throughput analysis. These types of devices will be attractive for mycotoxin analysis, since several toxins may exist in the same food or feed sample.

A range of sensors are being developed for mycotoxins based on these technologies, which can be applied on the farm or in the factory and be operated by unskilled personnel. Current trends to produce chip-based micro/nanoarrays for multimycotoxin analysis are challenging but possible, and will have significant impact on risk assessment testing. The use of nanoparticles such as gold, silver, metal oxides, and quantum dots (QDs) in assay developments will enhance the capability of the biosensor technology for mycotoxin analysis. Early and sensitive detection will aid in eliminating these toxins from entering the food chain and preventing ill health and protecting life.

The development of biosensors for the rapid, reliable, and low-cost determination of mycotoxins in foodstuffs has received considerable attention in recent years, and various types of assays have already been devised for several of the major groups of mycotoxins. One format uses the phenomenon of surface plasmon resonance (SPR) to detect the change in mass that occurs when mycotoxin-specific antibodies attach to a mycotoxin that has been covalently bonded to the surface of a sensor chip (Moskovits 2005). A recent application, developed and optimized for measuring deoxynivalenol in wheat extracts, gave results that were in good agreement with liquid chromatography-mass spectrometry data. Moreover, SPR sensor chips with immobilized deoxynivalenol may be reused more than 500 times without significant loss of activity. Because the instrumentation is now commercially available, this format could find widespread application to future mycotoxin analysis. A second format, using *fiber-optic probes*, may be adapted for continuous monitoring of mycotoxin levels. This sensor uses the evanescent wave of light that can form around the surface of an optical fiber. Antibodies attached to the surface of the fiber trap fluorescent mycotoxins (e.g., aflatoxins) or fluorescent analogs of mycotoxins (e.g., derivatized fumonisins) within the evanescent zone, permitting their detection. Two different bench-top devices have been designed for fumonisins and aflatoxins. Unfortunately, most of the SPR and fiber-optic biosensor procedures for mycotoxin analysis still require some form of sample cleanup/preconcentration to be truly effective in the analysis of real samples and to achieve adequate sensitivity. Moreover, the majority of these devices lack the ability to perform simultaneous analyses of multiple samples. Recently, array biosensors have been developed and demonstrated for a variety of applications. The ability of array biosensors to analyze multiple samples simultaneously for multiple analytes offers a significant advantage over other types of biosensors. In particular, a rapid multianalyte array biosensor, developed by Ngundi et al. (Chen et al. 2011) at the Naval Research Laboratory, Washington D.C., has demonstrated its potential to be used as a screening and monitoring device for clinical, food, and environmental samples. The device, which is portable and fully automated, can be used with different immunoassay formats. One interesting application is the development of a competitive immunoassay for the detection and quantification of Ochratoxin A in a variety of spiked food and beverage samples. A simple extraction procedure was employed with no need for cleanup or preconcentration of the sample extract. This is the first demonstration that a rapid biosensor can be used in a competitive assay format to detect a mycotoxin in extracts of relevant foods. However, further work aimed at developing a dual-analyte assay for deoxynivalenol and Ochratoxin A showed that improvements are still necessary to reduce analysis time and increase sensitivity.

Carbon nanotubes were used for the determination of zearalenone in urine samples by Andres et al. (Wang and Campiglia 2008). Multiwalled carbon nanotubes (MWCNTs) were modified with an enzyme, aflatoxin-detoxifizyme (APTZ). MWCNTs were used for enzyme immobilization of APTZ, for the determination of sterigmatocystin (Wang et al. 2009), and also carbon nanotubes field effect transistors (FET) that had been functionalized with proteins G and IgG to detect *Aspergillus flavus* in contaminated milled rice (Liu 2008). Optical sensors based on nanomaterials have been applied much less to the detection of analytes of interest in the food industry. QDs are practically the only nanomaterial used. QDs are nanocrystals of inorganic semiconductors that are somewhat restricted to a spherical shape of around 2–8 nm diameter (Qu et al. 2008). Their fluorescent properties are size dependent and, therefore, they can be tuned to emit at desired wavelengths (between 400 and 2000 nm) if synthesized in different compositions and sizes. In this way, QDs of different sizes can be excited with a single wavelength and emission controlled at different wavelengths, thus providing for simultaneous detection. These, together with their highly robust emission properties, make them more advantageous for labeling and optical detection than conventional organic dyes (Kobayashi et al. 2006). Their high quantum yields and narrow emission bands produce sharper colors and lead to higher sensitivity and the possibility of multiplexing of the analysis (Liu et al. 2003; O'Mahony et al. 2003). Costa et al. (Yang et al. 2005a) have reviewed the progress in exploiting these novel probes in optical sensing, as well as their still unexploited sensing capabilities. In the analytical chemistry field, their major application has been as fluorescent labels, while an application to food analysis is, up to now, unexploited. Goldman et al. (Yang et al. 2005b) have used QDs for fluoroimmunoassays of toxins. They detected four toxins simultaneously, three of which are naturally responsible for food or waterborne sickness. The CdSe-ZnS core-shell QDs were capped with dihydrolipoic acid and bioconjugates with the appropriate antibodies were prepared. A sandwich immunoassay was performed in microtiter plates where the toxins and different QDs were incubated for an hour. Fluorescence was measured at adequate wavelengths and, although there was spectral overlap, deconvolution of spectra revealed the fluorescence contribution of all toxins. Signals increased with toxin concentration in different ranges according to the particular toxin. No limit of detection were reported. Although authors treated the bioconjugate QDs as fluororeagents, they can be considered as "chemosensing devices." Ultrasensitive densitometry detection of cytokines was carried out with nanoparticle-modified aptamers (Abhijith and Thakur 2012; Xiulan 2006). Gold nanoparticles were also employed for an immunoassay for the detection of aflatoxin B1 (AFB1) in foods (Tertykh et al. 2000).

In the last few years, metal-organic-frameworks (MOFs), obtained by linking metal cations (or cationic metal clusters) with organic linkers, have attracted significant interest in the last few years, mainly due to the advantage of showing a large variety of structural types and chemical compositions, high surface area and permanent nanoscale porosity (Jiang and Xu 2011; Li et al. 2011). MOFs have been widely studied as materials for catalysis (Surblé et al. 2006; Sumida et al. 2011), gas storage and separation (Li et al. 2012; Chae et al. 2004), and sensing and drug delivery (Shekhah et al. 2011); and, more recently, analytical applications of MOFs have emerged (Bagheri et al. 2012).

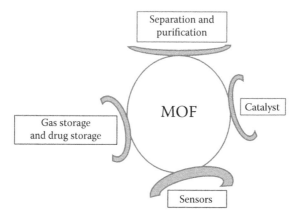

FIGURE 1.7   Different potential applications of MOFs.

In this field, MOFs have been shown to be promising materials as sorbents for sample preparation (Schwarzenbach et al. 2010; Harada 1995) as chromatographic stationary phases, as well as for the development of improved detection systems and sensors (Liu et al. 2011). However, MOFs' crystalline powders generally possess a random crystal size and shape, which makes their direct application troublesome and has led to the engineering of hybrid materials to contain them, such as through supports, magnetic beads, beads coated with a MOF shell, or MOF crystals trapped in a porous monolith (Furukawa et al. 2013). Porous materials are defined as solids containing empty voids that can host other molecules. The fundamental features of these materials are their porosity, the ratio between the total occupied and empty space, the (average) size of the pores, and the surface area. Typical surface area values for the porous materials applied in technological processes range between 2000 and 8000 $m^2\,g^{-1}$. The most important applications of such materials are the storage of small molecules and filtering. The metal organic frameworks are defined as a nanocomposite material that can be composed of either inorganic or organic materials. MOFs have shown high potential in gas storage, separation, chemical sensing, drug delivery, and heterogeneous catalysis applications (Figure 1.7). In general, the flexible and highly porous structure of MOFs allows guest species such as metal ions to diffuse into their bulk structure. The shape and size of the pores lead to selectivity over the guests that may be adsorbed. These features make MOFs an ideal sorbent in solid-phase extraction of heavy metals. However, there is little information about MOFs as an adsorbent (Meek et al. 2011).

For many decades, the high intrinsic reactivity of nanoscale zerovalent iron (nZVI) has justified its widespread use in the environmental remediation of organic pollutants and heavy metals. However, the formation of oxide layers on the nanoparticle surface during the reaction with surrounding media may limit its reaction rate (O'Carroll et al. 2013). To enhance the reactivity and functionality of nZVI, it can be impregnated with a second metal such as Pd, Cu, or Ni to form bimetallic nanoparticles (Nagpal et al. 2010). This is because the bimetallic nanoparticles not only make the nanoparticles more stable in air by inhibiting oxidation, but also increase reactivity by forming a catalyst (Su et al. 2011). A new strategy for stabilizing Fe/Pd nanoparticles with sodium carboxymethyl cellulose (CMC) has also been developed. In addition, the tendency is for nZVI to be aggregated because of the high surface energy resulting in decreased reactivity and reduction efficiency. To overcome these difficult issues, bimetallic nanoparticles have been immobilized on polymers or copolymers for the dechlorination of trichloroethylene (TCE), PCBs, and trichloroacetic acid (Xu and Bhattacharyya 2008). Furthermore, clays have been used as a support matrix in our previous studies involving bentonite and kaolinite, which enhanced the reactivity and stability of nZVI in removing Pb(II), Cr(VI), p-chlorophenol, and methyl orange. These activities suggest that this is a promising method for the removal of both organic compounds and metal ions. It is possible to simultaneously catalytically remove dyes and metal ions in effluents from the textile industry. However, the removal of cocontaminants of dyes and metal ions by ion-based nanoparticles is still unclear, because of the competition between dyes and metal ions for reactive sites and electrons in the passivation of the nZVI surface.

1.8 CONCLUSION

Nanotechnology for water and wastewater treatment is gaining momentum globally. The unique properties of nanomaterials and their convergence with current treatment technologies present great opportunities to revolutionize water and wastewater treatment. Decontamination is the reduction or removal of chemical and biological agents by means of physical, chemical neutralization, or detoxification techniques. Nanotechnology has shown huge potential in areas as diverse as drug development, water decontamination, information and communication technologies, the production of stronger, lighter materials, and human health care. Water and air are two vital components of life on Earth; the existence of life on Earth is made possible largely because of their importance to metabolic processes within the body. Clean and fresh water and air are essential for the existence for life. The recent development of nanotechnology has raised the possibility of environmental decontamination through several nanomaterial cut the process and tools. This chapter summarizes the expertise of various approaches of decontamination for the successful realization of remediation in the environment.

REFERENCES

Abhijith, K.S., and Thakur, M.S. *Analytical Methods*, 2012, 4, 4250–4256.

Anupreet, K., and Gupta, U. A review on applications of nanoparticles for the preconcentration of environmental pollutants. *Journal of Material Chemistry*, 2009, 19, 8279.

Bagheri, A., Taghizadeh, M., Behbahani, M., Asgharinezhad, A.A., Salarian, M., Dehghani, A., Ebrahimzadeh, H., and Amini, M.M. Synthesis and characterization of magnetic metal-organic framework (MOF) as a novel sorbent, and its optimization by experimental design methodology for determination of palladium in environmental samples. *Talanta*, 2012, 99, 132–139.

Borio, O., Gawlik, B.M., Bellobono, I.R., Muntau, H. Photooxidation of prometryn and prometon in aqueous solution by hydrogen peroxide on photocatalytic membranes immobilising titanium dioxide. *Chemosphere*, 1998, 37, 975–989.

Cao, G.Z. *Nanostructures and Nanomaterials, Synthesis, Properties and Application*, Imperial College Press, London, 329, 2004.

Chae, H.K., Perez, D.Y.S., Kim, J., Go, Y., Eddaoudi, M., Matzger, A.J., O'Keeffe, M., and Yaghi, O.M. A route to high surface area, porosity and inclusion of large molecules in crystals. *Nature*, 2004, 427, 523–525.

Chen, L., Lou, T., Yu, C., and Kang, Q. N-1-(2-mercaptoethyl)thymine modification of gold nanoparticles: A highly selective and sensitive colorimetric chemosensor for Hg^{2+}. *Analyst*, 2011, 136, 4770–4773.

Chun, C.L., Penn, R.L., and Arnold, W.A. *Environmental Science and Technology* 2006, 40, 3299–3304.

Cloete, T.E., Kwaadsteniet, M.D., Botes, M., and Lopez-Romero, J.M., *Nanotechnology in Water Treatment Applications*. Caister Academic Press, Wymondham, UK, 2010.

Elimelech, M., and Phillip, W.A. The future of seawater desalination: Energy, technology, and the environment. *Science*, 2011, 333, 712.

Eshel, K. *British Medical Journal*, 2007, 334, 610–616.

Furukawa, H., Cordova, K.E., O'Keeffe, M., and Yaghi, O.M. The chemistry and applications of metal-organic frameworks. *Science*, 2013, 341, 1230–1234.

Guzman, K.A.D., Taylor, M.R., and Banfield, J.F. *Environmental Science and Technology*, 2006, 40, 1401–1407.

Hang, Y., Qin, Y., and Shen, J. Separation and microcolumn preconcentration of traces of rare earth elements on nanoscale TiO_2 and their determination in geological samples by ICP-AES, *Journal of Separation Science*, 2003, 26, 957–960.

Harada, M. Minamata disease: Methylmercury poisoning in Japan caused by environmental pollution. *Critical Reviews in Toxicology*, 1995, 25, 1–24.

Hidaka, Il., Jou, H., Nohara, K., and Zhao, J. Photocatalytic degradation of the hydrophobic pesticide Permethrin in fluoro surfactant/TiO_2 aqueous dispersions. *Chemosphere*, 1992, 25, 1589–1597.

Hoffmann, M.R., Martin, S.T., Choi, W., and Bahnemann, D.W. Environmental applications of semiconductor photocatalysis. *Chemical Reviews*, 1995, 95, 69–95.

Huang, Y.F., Wang, Y.F., and Yan, X.P. Amine-functionalized magnetic nanoparticles for rapid capture and removal of bacterial pathogens. *Environmental Science and Technology*, 2010, 44, 7908–7913.

Hutton, G., Haller, L., and Bartram, J. *Economic and Health Effects of Increasing Coverage of Low Cost Household Drinking Water Supply and Sanitation Interventions*, World Health Organization, Geneva, Switzerland, 2007.

Jardim, W.F., Moraes, S.G., and Takiyama, M.M.K. Photocatalytic degradation of aromatic chlorinated compounds using TiO_2: Toxicity of intermediates. *Water Research*, 1997, 31, 1728–1732.

Jiang, H., and Xu, Q.L. Porous metal–organic frameworks as platforms for functional applications. *Chemical Communications*, 2011, 47, 3351–3370.

Katsumata, H., Sada, M., Nakaoka, Y., Kaneco, S., Suzuki, T., and Ohta, K. Photocatalytic degradation of diuron in aqueous solution by platinized TiO_2. *Journal of Hazardous Materials*, 2009, 171, 1081–1087.

Kaur, A. Applications of organo-silica nanocomposites for SPNE of Hg(II). *Applied Nanoscience*, 2016, 6, 183–190.

Kaur, A., and Gupta, U. A preconcentration procedure using 1-(2-pyridylazo)-2-napthol anchored to silica nanoparticle for the analysis of cadmium in different samples. *E-Journal of Chemistry*, 2008, 5, 930–939.

Kaur, A., and Gupta, U. A review on applications of nanoparticles for the preconcentration of environmental pollutants. *Journal of Material Chemistry*, 2009a, 19, 8279–8289.

Kaur, A., and Gupta, U. Sorption and preconcentration of lead on silica nanoparticles modified with resacetophenone. *E-Journal of Chemistry*, 2009b, 6, 633–638.

Kaur, A., and Gupta, U. *Chinese Journal of Chemistry*, 2009c, 27, 1833–1838.

Kaur, A., and Gupta, U. Preconcentration of zinc and manganese using 1-(2-pyridylazo)-2-naphthol anchored SiO_2 nanoparticles. *Eurasian Journal of Analytical Chemistry*, 2009d, 4, 175–183.

Kaur, A., and Gupta, U. *Eurasian Journal of Analytical Chemistry*, 2009e, 4, 234–244.

Kaur, A., and Gupta, U. Solid-phase extraction of antimony using chemically modified SiO_2-PAN nanoparticles. *Journal of AOAC International*, 2010a, 93, 1302–1307.

Kaur, A., and Gupta, U. Solid phase extraction of Cd(II) using mesoporous organosilica nanoparticles modified with resacetophenone. *Separation Science*, 2010b, 2, 11–16.

Kaur, A., and Gupta, U. *Electronic Journal of Environment, Agriculture, Food Chemistry*, 2010c, 9, 1334–1342.

Kaur, A., and Gupta, U. Preparation of silica-PAN functionalized nanoextractants for the extraction of ferbam from various samples. *Separation Science and Technology*, 2015, 50, 661–669.

Kobayashi, K., Kitagawa, S., and Ohtani, H. Development of capillary column packed with thiol-modified gold-coated polystyrene particles and its selectivity for aromatic compounds. *Journal of Chromatography A*, 2006, 1110, 95–101.

Konstantinou, I.K., Sakkas, V.A., and Albanis, T.A. Metolachlor photocatalytic degradation using TiO_2 photocatalysts. *Applied Catalysis B: Environmental*, 2001, 34, 227–239.

Konstantinou, I.K., Sakellarides, T.M., Sakkas, V.A., and Albanis, T.A. Photocatalytic degradation of selected s-triazine herbicides and organophosphorus insecticides over aqueous TiO_2 suspensions. *Environmental Science and Technology*, 2001, 35, 398–405.

Laurent, S., Forge, D., Port, M., Roch, A., Robic, C., Vander Elst, L., and Muller, R.N. Magnetic iron oxide nanoparticles: Synthesis, stabilization, vectorization, physicochemical characterizations, and biological applications. *Chemical Reviews*, 2008, 108, 2064–2110.

Li, J.R., Ma, Y., McCarthy, M.C., Sculley, J., Yu, J., Jeong, H.K., Balbuena, P.B., and Zhou, H.C. Carbon dioxide capture-related gas adsorption and separation in metal-organic frameworks. *Coordination Chemistry Reviews*, 2011, 255, 1791–1823.

Li, J.R., Sculley, J., and Zhou, H.C. Metal–organic frameworks for separations. *Chemical Reviews*, 2012, 112, 869–932.

Li, P., Liu, Y., and Guo, L. Determination of molybdenum in steel samples by ICP-AES after separation and preconcentration using nanometre-sized titanium dioxide. *Journal of Analytical Atomic Spectrometry*, 2004, 19, 1006–1009.

Liang, P., Hu, B., Jiang, Z., Qin, Y., and Peng, T. Nanometer-sized titanium dioxide micro-column on-line preconcentration of La, Y, Yb, Eu, Dy and their determination by inductively coupled plasma atomic emission spectrometry. *Journal of Analytical Atomic Spectrometry*, 2001, 16, 863–866.

Liang, P., Qin, Y., Hu, B., and Peng, T. Speciation of chromium by selective separation and preconcentration of Cr(III) on an immobilized nanometer titanium dioxide microcolumn. *Analytica Chimica Acta*, 2001, 440, 207–213.

Liang, P., Shi, T., Lu, H., Jiang, Z., and Hu, B. Speciation of Cr(III) and Cr(VI) by nanometer titanium dioxide micro-column and inductively coupled plasma atomic emission spectrometry. *Spectrochimica Acta Part B: Atomic Spectroscopy*, 2003, 58, 1709–1714.

Liang, P., Yang, L., Hu, B., and Jiang, Z. Simultaneous determination of hydroxyanthraquinones in rhubarb and experimental animal bodies by high-performance liquid chromatography. *Analytical Science*, 2003, 19, 1163–1167.

Lin, J.H., and Lung, T.W. The effect of nanoparticle size, shape, and surface chemistry on biological systems. *Reviews in Analytical Chemistry*, 2012, 31, 153–159.

Liu, F.K., Solid phase extraction of neutral analytes through silica gel coated with layers of Au nanoparticles self-assembled with alkanethiols. *Journal of the Chinese Chemical Society*, 2008, 55, 69–78.

Liu, F.K., Wei, G.T., and Cheng, F.C. *Journal of Chinese Chemical Society*, 2003, 50, 931–937.

Liu, Q., Shi, J., Sun, J., Wang, T., Zeng, L., and Jiang, G. Graphene and graphene oxide sheets supported on silica as versatile and high-performance adsorbents for solid-phase extraction. *Angewandte Chemie*, 2011, 50, 5913–5917.

Maity, D., and Agrawal, D.C. Synthesis of iron oxide nanoparticles under oxidizing environment and their stabilization in aqueous and non-aqueous media. *Journal of Magnetism and Magnetic Materials*, 2007, 308, 46–55.

Mak, S.Y., and Chen, D. H. Fast adsorption of methylene blue on polyacrylic acid-bound iron oxide magnetic nanoparticles. *Dyes and Pigments*, 2004, 61, 93–98.

Mara, D.D. Water, sanitation and hygiene for the health of developing nations. *Public Health*, 2003, 117, 452.

Marahel, F., Ghaedi, M., Montazerozohori, M., and Khodadoust, S. Chemical functionalization of silica gel with 2-((3-silylpropylimino) methyl) phenol (SPIMP) and its application for solid phase extraction and preconcentration of Fe (III), Pb (II), Cu (II), Ni (II), Co (II) and Zn (II) ions. *Indian Journal of Science and Technology*, 2011, 4, 1234–1240.

Mauter, M.S., and Elimelech, M. Environmental applications of carbon-based nanomaterials. *Environmental Science and Technology*, 2008, 42, 5843–5859.

Meek, S.T., Greathouse, J.A., and Allendorf, M.D. Metal-organic frameworks: A rapidly growing class of versatile nanoporous materials. *Advanced Materials*, 2011, 23, 249–267.

Mills, G., and Hoffmann, M.R. Photocatalytic degradation of pentachlorophenol on titanium dioxide particles: Identification of intermediates and mechanism of reaction. *Environmental Science and Technology*, 1993, 27, 1681–1689.

Minero, C., Pelizzetti, E., Malato, S., and Blanco, J. Large solar plant photocatalytic water decontamination: Degradation of pentachlorophenol. *Chemosphere,* 1993, 26, 2103–2119.

Minero, C., Pelizzetti, E., Malato, S., and Blanco, J. Solar photocatalytic processes for the purification of water: State of development and barriers to commercialization. *Solar Energy*, 1996, 56, 421–428.

Montgomery, M.A., and Elimelech, M. Water and sanitation in developing countries: Including health in the equation. *Environmental Science and Technology*, 2007, 41, 17–24.

Moore, M., Gould, P., and Keary, B.S. Global urbanization and impact on health. *International Journal of Hygiene and Environmental Health*, 2003, 206, 269–278.

Moskovits, M. Surface-enhanced Raman spectroscopy: A brief retrospective. *Journal of Raman Spectroscopy*, 2005, 36, 485–496.

Muszkat, L., Bir, L., and Feigelson, L. *Journal of Photochemistry and Photobiology A: Chemistry*, 1995, 87, 85–88.

Muszkat, L., Halmann, M., Raucher, D., and Bir, L. Reaction pathways and mechanisms of photodegradation of pesticides. *Journal of Photochemistry and Photobiology A: Chemistry*, 1992, 65, 109–115.

Nagpal, V., Bokare, A.D., Chikate, R.C., Rode, C.V., and Paknikar, K.M. Reductive dechlorination of γ-hexachlorocyclohexane using Fe–Pd bimetallic nanoparticles. *Journal of Hazardous Materials*, 2010, 175, 680–687.

Noh, H.B., Lee, K.S., Lim, B.S., Kim, S.J., and Shim, Y.B. Total analysis of endocrine disruptors in a microchip with gold nanoparticles. *Electrophoresis*, 2010, 31, 3053–3060.

Noh, H.B., Lee, K.S., Lim, B.S., Kim, S.J., and Shim, Y.B. Assembly of polymer–gold nanostructures with high reproducibility into a monolayer film SERS substrate with 5 nm gaps for pesticide trace detection. *Analyst*, 2012, 137, 3349–3335.

O'Carroll, D., Sleep, B., Krol, M., Boparai, H., and Kocur, C. Nanoscale zero valent iron and bimetallic particles for contaminated site remediation. *Advances in Water Resources*, 2013, 5, 104–122.

O'Mahony, T., Owens, V.P., Murrihy, J.P., Guihen, E., Holmes, J.D., and Glennon, J.D. Alkylthiol gold nanoparticles in open-tubular capillary electrochromatography. *Journal of Chromatography A,* 2003, 1004, 181–193.

Pathirana, H.M.K.K., and Maithreepala, R.A. Photodegradation of 3,4-dichloropropionamide in aqueous TiO$_2$ suspensions. *Journal of Photochemistry and Photobiology A: Chemistry*, 1997, 102, 273–277.

Pelizzetti, E., Carlin, V., Minero, C., Pramauro, E., and Vincenti, M. Degradation pathways of atrazine under solar light and in the presence of TiO$_2$ colloidal particles. *Science of the Total Environment*, 1992, 123, 161–169.

Pelizzetti, E., Maurino, V., Minero, C., Carlin, V., Pramauro, E., Zerbinati, O., Tosato, M.L. Photocatalytic degradation of selected s-triazine herbicides and organophosphorus insecticides over aqueous TiO$_2$ suspensions. *Environmental Science and Technology*, 1990, 24, 1559–1565.

Pelizzetti, E., Minero, C., Carlin, V., Vincenti, M., and Pramauro, E. Degradation pathways of atrazine under solar light and in the presence of TiO$_2$ colloidal particles. *Chemosphere*, 1992, 24, 891–910.

Peñuela, G.A., and Barceló, D. Comparative degradation kinetics of alachlor in water by photocatalysis with FeCl$_3$, TiO$_2$ and photolysis, studied by solid-phase disk extraction followed by gas chromatographic techniques. *Journal of Chromatography A*, 1996, 754, 187–195.

Ponder, S., Darab, J., and Mallouk, T. Remediation of Cr(VI) and Pb(II) aqueous solutions using supported, nanoscale zero-valent iron. *Environmental Science and Technology*, 2000, 34, 2564–2569.

Qu, Q.S., Shen, F., Shen, M., Hu, X.Y., Yang, G.J., Wang, C.Y., Yan, C., and Zhang, Y.K. Rapid probing of photocatalytic activity on titania-based self-cleaning materials using 7-hydroxycoumarin fluorescent probe. *Analytical Chimica Acta*, 2008, 609, 76–81.

Qu, X.L., Brame, J., and Li, Q., and Alvarez, J.J.P. Fabrication and characterization of heparin-grafted poly-L-lactic acid–chitosan core–shell nanofibers scaffold for vascular gasket. *Accounts of Chemical Research*, 2013, 46, 834–843.

Quinn, J., Geiger, C., Clausen, C., Brooks, C., and Coon, K. Field demonstration of DNAPL dehalogenation using emulsified zero-valent iron. *Environmental Science and Technology*, 2005, 39, 1309–1318.

Saini, S.S., and Kaur, A. The analysis of aflatoxin M$_1$ in dairy products. *Separation Science*, 2012a, 4, 13–17.

Saini, S.S., and Kaur, A. *Global Advanced Research Journals*, 2012b, 14, 63–70.

Saini, S.S., and Kaur, A. Aflatoxin B1: Toxicity, characteristics and analysis. Mini review: Molecularly imprinted polymers for the detection of food toxins. *Advances in Nanoparticles*, 2013, 2, 60–65.

Sanlaville, Y., Guittonneau, S., Mansour, M., Feicht, E.A., Meallier, P., and Kettrup, A. Effect of a dichlorophenol-adapted consortium on the dechlorination of 2,4,6-trichlorophenol and pentachlorophenol in soil. *Chemosphere*, 1996, 33, 353–362.

Savina, I.N., English, C.J., Whitby, R.L.D., Zheng, Y., Leistner, A., Mikhalovsky, S.V., and Cundy, A.B. High efficiency removal of dissolved As(III) using iron nanoparticle-embedded macroporous polymer composites. *Journal of Hazardous Materials*, 2011, 192, 1002–1008.

Schrick, B., Blough, J., Jones, A., and Mallouk, T.E. Room temperature negative differential resistance in molecular nanowires. *Chemistry of Materials*, 2002, 14, 5140–5147.

Schwarzenbach, R.P., Egli, T., Hofstetter, T.B., Gunten, U.V., and Wehrli, B. In situ synthesis and excellent photocatalytic activity of tiny Bi decorated bismuth tungstate nanorods. *Annual Review of Environment and Resources*, 2010, 35, 109–136.

Shekhah, O., Liu, J., Fischer, R.A., and Wöll, C. MOF thin films: Existing and future applications. *Chemical Society Reviews*, 2011, 40, 1081–1106.

Shu, W.C., Li, F.K., and Ko, F.H. Nanoadsorbents: Classification, preparation, and applications. *Analytical and Bioanalytical Chemistry*, 2011, 399, 103–118.

Shunxin, L., Shahua, Q., Ganquan, H., and Fei, H. Preconcentration of selenium by living bacteria immobilized on silica for microwave induced plasma optical emission spectrometry with continuous powder introduction. *Fresenius Journal of Analytical Chemistry*, 1999, 365, 469–471.

Sin, J.C., Lam, S.M., Mohamed, A.R., and Lee, K.T. Efficient photodegradation of resorcinol with Ag$_2$O/ZnO nanorods heterostructure under a compact fluorescent lamp irradiation. *International Journal of Photoenergy*, 2012, 67, 1277–1284.

Su, J., Lin, S., Chen, Z.L., Megharaj, M., Naidu, M.R. Dechlorination of p-chlorophenol from aqueous solution using bentonite supported Fe/Pd nanoparticles: Synthesis, characterization and kinetics. *Desalination*, 2011, 280, 167–173.

Sumida, K., Rogow, D.L., Mason, J.A., McDonald, T.M., Bloch, E.D., Herm, Z.R., Bae, T.H., and Long, J.R. Carbon dioxide capture in metal-organic frameworks. *Chemical Reviews*, 2011, 112, 724–781.

Surblé, S., Millange, F., Serre, C., Düren, T., Latroche, M., Bourrelly, S., Llewellyn, P.L., and Férey, G. The development of new materials such MOFs for CO$_2$ capture and alkylation of aromatic compounds. *Journal of the American Chemical Society*, 2006, 128, 14896–14889.

Tertykh, V.A., Yanishpolskii, V.V., and Panova, O.Y. Immobilization of optically active olefins on the silica surface by combined hydrosilylation and sol-gel technology. *Journal of Thermal Analysis and Calorimetry*, 2000, 62, 545–549.

Textier, I., Giannotti, C., Malato, S., Richter, C., and Delaire, J. *Catalysis Today*, 1999, 54, 297–307.

Toma, F.L., Bertrand, G., Klein, D., and Coddet, C. *Environmental Chemistry Letters*, 2004, 2, 117–121.

Tsai, W.T., Lee, M.K., Su, T.Y., and Chang, Y.M. *Journal of Hazardous Materials*, 2009, 168, 269–275.

U.S. Bureau of Remediation and Sandia National Laboratories, Desalination and Water Purification technology roadmap: A report of executive committee water purification, Sandia National Laboratories, New Mexico/California, 2003.

Vasimalai, N., and John, S.A. Mercaptothiadiazole capped gold nanoparticles as fluorophore for the determination of nanomolar mercury(II) in aqueous solution in the presence of 50000-fold major interferents. *Analyst*, 2012, 37, 3349–3354.

Vulliet, E., Emmelin, C., Chovelon, J.M., Guillard, C., and Herrmann, J.M. Physicochemical properties and photocatalytic activities of TiO_2-films prepared by sol-gel methods. *Applied Catalysis B: Environment*, 2002, 38, 127–137.

Wang, H.Y., and Campiglia, A.D. Determination of polycyclic aromatic hydrocarbons in drinking water samples by solid-phase nanoextraction and high-performance liquid chromatography. *Analytical Chemistry*, 2008, 80, 8202–8209.

Wang, H.Y., Yu, S.J., and Campiglia, A.D. Potential role of gold nanoparticles for improved analytical methods: An introduction to characterizations and applications. *Analytical Biochemistry*, 2009, 385, 249–256.

World Health Organization and UNICEF, Progress on drinking-water and sanitation, World Health Organization, Geneva, Switzerland, 2012.

World Health Organization and UNICEF, Progress on sanitation and drinking-water, World Health Organization, Geneva, Switzerland, 2013.

Wu, C.S., Liu, F.K., Ko, F.H., Potential role of gold nanoparticles for improved analytical methods: An introduction to characterizations and applications. *Analytical and Bioanalytical Chemistry*, 2011, 399, 103–118.

Xiulan, S. *Food Control*, 2006, 17, 256–262.

Xu, J., and Bhattacharyya, D. Modeling of Fe/Pd nanoparticle-based functionalized membrane reactor for PCB dechlorination at room temperature. *The Journal of Physical Chemistry C*, 2008, 112, 9133–9144.

Yang, L., Guihen, E., and Glennon, J.D. Use of cyclodextrin-modified gold nanoparticles for enantioseparations of drugs and amino acids based on pseudostationary phase-capillary electrochromatography. *Journal of Separation Science*, 2005, 28, 757–766.

Yang, L., Guihen, E., Holmes, J.D., Loughran, M., O'Sullivan, G.P., and Glennon, J.D. Gold nanoparticle-modified etched capillaries for open-tubular capillary electrochromatography. *Analytical Chemistry*, 2005, 77, 1840–1846.

Yang, L., Hu, B., Jiang, Z., and Pan, H. On-line separation and preconcentration of trace metals in biological samples using a microcolumn loaded with PAN-modified nanometer-sized titanium dioxide, and their determination by ICP-AES. *Microchimica Acta*, 2004, 144, 227–231.

Yantasee, W., Warner, C.L., Sangvanich, T., Addleman, R.S., Carter, T.G., Wiacek, R.J., Fryxell, G.E., Timchalk, C., and Warner, M.G. Removal of heavy metals from aqueous systems with thiol functionalized superparamagnetic nanoparticles. *Environmental Science and Technology*, 2007, 41, 5114–5119.

Zelmanov, G., and Semiat, R. Phosphate removal from water and its recovery by using iron (3) oxide-based nano-adsorbent. *Journal of Environmental Engineering and Management*, 2011, 10, 1923–1933.

Zhang, W.X., and Elliot, D.W. Applications of iron nanoparticles for groundwater remediation. *Remediation J.*, 2006, 16, 7–21.

Zheng, H., Chang, X., Lian, N., Wang, S., Cui, Y., and Zhai, Y. *International Journal of Environmental Analytical Chemistry*, 2004, 86, 431–439.

2 Treatment of Fluoride-Contaminated Water by Electrocoagulation Followed by Microfiltration Technique

Debasis Ghosh, Barun Kumar Nandi,
Mehabub Rahaman, and Mihir Kumar Purkait

CONTENTS

ABSTRACT

The contamination of drinking water has been a major issue due to the presence of various organic, inorganic, and pathogenic ingredients. Fluoride contamination is a serious problem in several parts of India, as well as in different parts of the world, causing serious damage to health. Different techniques like adsorption, precipitation, membrane separation, ion exchange, and various hybrid techniques have been reported for the removal of fluoride from drinking water. This chapter deals with the effective removal of fluoride from drinking water using electrocoagulation (EC) followed by the microfiltration (MF) technique. During EC, several parameters such as the initial fluoride concentration, current density, electrode connection (monopolar and bipolar), pH level, and interelectrode distance affect the removal of fluoride contaminants from drinking water. An aluminum electrode is considered for the batch mode of the EC operation. The corrosion of electrodes, as well as the sludge formed during the process, is estimated. By-products obtained from the EC bath are analyzed using scanning electron microscopy (SEM), elemental analysis (EDAX), Fourier transform infrared spectroscopy (FTIR), and x-ray diffraction (XRD) and explained. A comparative cost estimation for both electrode connections is also presented. It is found that drinking water contamination caused by the significant presence of fluoride is successfully monitored by bipolar EC for 45 min at a current density of 625 A m^{-2} and an interelectrode distance of 0.005 m. EC performance is estimated in terms of percentage removal of fluoride, and up to 93.2% of fluoride removal can be achieved. However, the electrocoagulated solution is not suitable for drinking purposes as the solution has an alkaline pH, along with agglomerated suspended sludge with a size range of 10–100 μm. To make the electrocoagulated solution drinkable, it has to be filtered using suitable membrane separation techniques such as MF. The MF membranes are efficient in retaining the suspended particulates formed during the EC process. In recent times, ceramic MF membranes have been widely used for their better mechanical, thermal, and chemical strength compared with the commercial flat-sheet type of polymeric membranes. The hybrid techniques of EC followed by MF can successfully remove fluoride from contaminated drinking water. The content of this chapter may be useful for the further advancement of the hybrid technique to design a drinking water treatment system in continuous mode.

Keywords: Drinking water, Fluoride, Electrocoagulation, Microfiltration membranes

2.1 INTRODUCTION

2.1.1 BACKGROUND

Industrialization, along with urbanization, results in the rapid deterioration of drinking water quality. Scientific evidence proves that the effluents released from various process industries, namely textile, leather, paint, and so on, comprise different hazardous and toxic compounds, some of which are known carcinogens and others probable carcinogens. Over the past few decades, the ever-growing population, urbanization, industrialization, and the inexpert utilization of water resources have led to the degradation of water quality and reduction in its per capita availability in various developing countries. Due to various ecological factors, either natural or anthropogenic, groundwater is becoming polluted because of deep percolation from intensively cultivated fields, disposal of hazardous wastes, liquid and solid wastes from industries, sewage disposal, surface impoundments, and so on. During its complex flow history, groundwater passes through various geological formations leading to consequent contamination in shallow aquifers (Anwar 2003). The presence of various hazardous contaminants such as fluoride, arsenic, nitrates, sulfates, pesticides, other heavy metals, and so on in underground water has been reported from different parts of India. In many cases, water sources have been rendered unsafe, not only for human consumption, but also for other activities such as irrigation and industrial needs (Mulligan et al. 2001; Liu et al. 2005).

2.1.2 SOURCES OF FLUORIDE IN DRINKING WATER AND ITS TOXICITY

Fluorine is highly reactive and is found naturally as CaF_2. It is an essential constituent in minerals such as topaz, fluorite, fluorapatite, cryolite, phosphorite, theorapatite, and so on (Singh and Maheshwari 2001). Fluoride is found in the atmosphere, soil, and water. It enters the soil through weathering of rocks, precipitation, or waste runoff. Surface waters generally do not contain more than 0.3 mg L^{-1} of fluoride unless they are polluted from external sources. Though drinking water is the major contributor (75%–90% of daily intake), other sources of fluoride poisoning are food, industrial exposure, drugs, cosmetics, and so on (Garg and Malik 2004). The fluoride content of some major food products is given in Table 2.1.

Fluoride—in minute quantities—is an essential component for the normal mineralization of bones and formation of dental enamel (Bell and Ludwig 1970). However, its excessive intake may result in a slow, progressively crippling affliction known as fluorosis. There are more than 20 developed and developing nations in which fluorosis is endemic. These are Argentina, the United States, Morocco, Algeria, Libya, Egypt, Jordan, Turkey, Iran, Iraq, Kenya, Tanzania, South Africa, China, Australia, New Zealand, Japan, Thailand, Canada, Saudi Arabia, the Persian Gulf, Sri Lanka, Syria, India, and so on (Mameri et al. 1998). In India, it was first detected in the Nellore district of Andhra Pradesh in 1937 (Shortt et al. 1937). Since then, considerable work has been done in different parts of India to explore the fluoride-laden water sources and their impacts on humans

TABLE 2.1

Fluoride Concentration in Agricultural Crops and Other Edible Items

Food Item	Fluoride Concentration (mg kg)$^{-1}$	Food Item	Fluoride Concentration (mg kg^{-1})
Cereals		**Nuts and oil seeds**	
Wheat	4.6	Almond	4.0
Rice	5.9	Coconut	4.4
Maize		Mustard seeds	5.7
Pulses and legumes	5.6	Groundnut	5.1
Green gram dal	2.5	**Beverages**	
Red gram dal	3.7	Tea	60–112
Soybean	4.0	Aerated drinks	0.77–1.44
Vegetables		**Spices and condiments**	
Cabbage	3.3	Coriander	2.3
Tomato	3.4	Garlic	5.0
Cucumber	4.1	Turmeric	3.3
Kidney vetch	4.0	**Food from animal sources**	
Spinach	2.0	Mutton	3.0–3.5
Lettuce	5.7	Beef	4.0–5.0
Mint	4.8	Pork	3.0–4.5
Potato	2.8	Fish	1.0–6.5
Carrot	4.1	**Others**	
Fruits		Rock salt	200.0–250.0
Mango	3.7	Areca nut (*supari*)	3.8–12.0
Apple	5.7	Betel leaf (*paan*)	7.8–12.0
Guava	5.1	Tobacco	3.2–38

Source: Sengupta, S. R., and Pal, B. *Ind. J. Nutr. Dicter.*, 8, 66–71, 1971.

as well as on animals. At present, it has been estimated that fluorosis is prevalent in 17 states of India. Table 2.2 depicts the fluoride concentrations in different states of India.

2.1.3 MAXIMUM CONTAMINATION LEVEL (MCL) AND HEALTH EFFECTS OF FLUORIDE

The MCL of fluoride in drinking water is 1.5 mg L^{-1} (WHO 1984). The endemic fluorosis in India is largely of hydrogeochemical origin. It has been observed that low levels of calcium and high bicarbonate alkalinity favor a high fluoride content in groundwater. Water with high fluoride content is generally soft, has a high pH value, and contains large amounts of silica. In groundwater, the natural concentration of fluoride depends on the geological, chemical, and physical characteristics of the aquifer, the porosity and acidity of the soil and rocks, temperature, the action of other chemicals, and the depth of wells. Due to the large number of variables, the fluoride concentrations in groundwater range from well under 1.0 mg L^{-1} to more than 35.0 mg L^{-1}. As the amount of water consumed and consequently the amount of fluoride ingested are influenced primarily by air temperature, the U.S. Public Health Service (USPHS 1962) drinking water standards have set a range of concentrations for the maximum allowable fluoride in drinking water for communities based on climatic conditions, as shown in Table 2.3. As fluorine is a highly electronegative element, it has an extraordinary tendency to be attracted by positively charged ions such as calcium. Hence, the effect of fluoride on mineralized tissues such as bones and teeth, leading to developmental alterations, is of clinical significance, as these tissues have the highest amounts of calcium and thus attract the maximum amount of fluoride, which is deposited as calcium–fluorapatite crystals. Tooth enamel

TABLE 2.2

Fluoride Concentrations in Different States of India

States	Districts	Range of Fluoride Concentration (mg L⁻¹)
Assam	Karbianglong, Nagaon	0.2–18.1
Andhra Pradesh	All districts except Adilabad, Nizamabad, West Godhavari, Visakhapattnam, Vijzianagaram, Srikakulam	0.11–20.0
Bihar	Palamu, Daltonganj, Gridh, Gaya, Rohtas, Gopalganj, Paschim, Champaran	0.6–8.0
Delhi	Kanjhwala, Najafgarh, Alipur	0.4–10.0
Gujarat	All districts except Dang	1.58–31.0
Haryana	Rewari, Faridabad, Karnal, Sonipat, Jind, Gurgaon, Mohindergarh, Rohtak, Kurukshetra, Kaithal, Bhiwani, Sirsa, Hisar	0.17–24.7
Jammu and Kashmir	Doda	0.05–4.2
Karnataka	Dharwad, Gadag, Bellary, Belgam, Raichur, Bijapur, Gulbarga, Chitradurga, Tumkur, Chikmagalur, Manya, Banglore, Mysore	0.2–18.0
Kerala	Palghat, Allepy, Vamanapuram, Alappuzha	0.2–2.5
Maharashtra	Chandrapur, Bhandara, Nagpur, Jalgaon, Bulduna, Amravati, Akola, Yavatmal, Nanded, Sholapur	0.11–10.2
Madhya Pradesh	Shivpuri, Jabua, Mandla, Dindori, Chhindwara, Dhar, Vidhisha, Seoni, Sehore, Raisen, Bhopal	0.08–4.2
Orrissa	Phulbani, Koraput, Dhenkanal	0.6–5.7
Punjab	Mansa, Faridcot, Bhatinda, Muktsar, Moga, Sangrur, Ferozpur, Ludhiana, Amritsar, Patila, Ropar, Jallandhar, Fatehgarh sahib	0.44–6.0
Rajasthan	All 32 districts	0.2–37.0
Tamilnadu	Salem, Periyar, Dharampuri, Coimbatore, Tiruchirapalli, Vellore, Madurai, Virudunagar	1.5–5.0
Uttar Pradesh	Unnao, Agra, Meerut, Mathura, Aligarh, Raibareli, Allahabad	0.12–8.9
West Bengal	Birbhum, Bhardaman, Bankura	1.5–13.0

Source: Susheela, A. K. *Curr. Sci.* 77, 1250–1256, 1999.

TABLE 2.3

USPHS Recommendations for Maximum Allowable Fluoride in Drinking Water

Annual Average of Maximum Daily Air Temperature (°C)	Recommended Fluoride Concentration (mg L⁻¹)			Maximum Allowable Fluoride Concentration (mg L⁻¹)
	Lower	Optimum	Upper	
10–12	0.9	1.2	1.7	2.4
12.1–14.6	0.8	1.1	1.5	2.2
14.7–17.7	0.8	1.0	1.3	2.0
17.8–21.4	0.7	0.9	1.2	1.8
21.5–26.2	0.7	0.8	1.0	1.6
26.3–32.5	0.6	0.7	0.8	1.4

is composed principally of crystalline hydroxylapatite. Under normal conditions, when fluoride is present in the water supply, most of the ingested fluoride ions are incorporated into the apatite crystal lattice of calciferous tissue enamel during its formation.

The hydroxyl ion is substituted with a fluoride ion, since fluorapatite is more stable than hydroxylapatite. Thus, a large amount of fluoride is bound into these tissues and only a small amount is

TABLE 2.4

Effects of Fluoride in Water on Human Health

Fluoride Concentration (mg L^{-1})	Effects
<1.5	Safe limit
1.5–3.0	Dental fluorosis (discoloration, mottling, and pitting of teeth)
3.0–4.0	Stiffened and brittle bones and joints
4.0–6.0 and above	Deformities in knee and hip bones, and finally paralysis, making the person unable to walk or stand with a straight posture, crippling fluorosis

Source: Choubisa, S. L., and Sompura, K. *Poll. Res.*, 15, 45–47, 1996.

excreted through sweat, urine, and stools. The intensity of fluorosis is not merely dependent on the fluoride content in water, but also on fluoride from other sources, physical activity, and dietary habits. The various forms of fluorosis arising due to excessive intake of fluoride are briefly presented in Table 2.4.

2.1.3.1   Dental Fluorosis

Due to excessive fluoride intake, tooth enamel loses its luster. In its mild form, dental fluorosis is characterized by white, opaque areas on the tooth surface, and in its severe form, it manifests as yellowish-brown to black stains and severe pitting of the teeth. This discoloration may be in the form of spots or horizontal streaks.

2.1.3.2   Skeletal Fluorosis

Skeletal fluorosis affects children as well as adults. It does not easily manifest until the disease attains an advanced stage. Fluoride is mainly deposited in the joints of the neck, knee, pelvis, and shoulder bones and makes it difficult to move or walk. The symptoms of skeletal fluorosis are similar to spondylitis or arthritis. Besides skeletal and dental fluorosis, excessive consumption of fluoride may lead to muscle fiber degeneration and low hemoglobin.

2.1.4   EXISTING PROCESSES FOR THE SEPARATION OF FLUORIDE

Defluoridation is the process of removal of fluoride ions from drinking water. The different methods tried so far for the removal of excess fluoride from water can be broadly classified into four categories:

1. Adsorption methods
2. Ion exchange methods
3. Precipitation methods
4. Membrane separation (reverse osmosis [RO])

Adsorption processes using different adsorbents for the removal of fluoride from aqueous media such as trimetal oxide (Wu et al. 2007), waste carbon slurry (Gupta et al. 2007), and many low-cost materials (Srimurali et al. 1998) have been investigated. Membrane separation techniques have also been investigated for the effective separation of fluoride using electrodialysis (Amor et al. 2001), Donnan dialysis (Hichour et al. 1999), nanofiltration (Hu and Dickson 2006), and anion-exchange membranes (Tor 2007). Garmes et al. (2002) has performed defluoridation of groundwater by a hybrid process combining adsorption and Donnan dialysis. Fluoride distribution in the electrocoagulation (EC) defluoridation process was investigated by Zhu et al. (2007). The kinetics were developed empirically in the fluoride removal process, which uses a monopolar electrode connection

TABLE 2.5

Advantages and Disadvantages of Different Fluoride Removal Techniques

Technique	Adsorption	Ion Exchange	Coagulation-Precipitation	Membrane Process
Remarks	*Adsorbents:* Activated alumina, activated carbon, calcite, activated sawdust, activated coconut shell carbon and activated fly ash, groundnut shell, coffee husk, rice husk, bone charcoal, activated soil sorbent, etc.	Strongly basic anion-exchange resin containing quaternary ammonium functional groups.	*Nalgonda technique:* In the first step, precipitation occurs by lime dosing, which is followed by a second step, in which alum is added to cause coagulation.	NF and RO are generally used for fluoride removal.
Advantage	Process can remove fluoride up to 90%. Treatment is cost-effective.	Removes fluoride up to 90%–95%. Retains the taste and color of water intact.	Two-step process has been claimed as the most effective technique by the National Environmental Engineering Research Institute (NEERI) under *Rajib Gandhi Drinking Water Mission*; several fill and draw (F&D) type and hand pump attached (HPA) plants based on the Nalgonda technique have been established in rural areas, for which design and technology has been developed by NEERI.	Process is highly effective for fluoride removal. Membranes also provide an effective barrier to suspended solids, all inorganic pollutants, organic micropollutants, pesticides, etc. No chemicals are required. Process works in a wide pH range. No interference by other ions is observed.
Disadvantage	Process is highly dependent on pH level, presence of sulfate, phosphate, or carbonate; results in ionic competition.	Efficiency is reduced in presence of other ions. The technique is expensive because of the cost of resin.	Process removes only a smaller portion of fluoride (18%–33%) in the form of precipitates and converts a greater portion of ionic fluoride (67%–82%) into soluble aluminum fluoride complex ion. Therefore, this technology is unsuitable. Silicates have adverse effect on defluoridation by Nalgonda technique.	Process is expensive in comparison to other options.

(Emamjomeh and Sivakumar 2006; Wu et al. 2007). The advantages and disadvantages of different fluoride removal techniques are summarized in Table 2.5.

2.1.5 ELECTROCOAGULATION

2.1.5.1 Applications of Electrocoagulation

EC uses an electrochemical cell to treat polluted water. Sacrificial anodes corrode to release active coagulant cations, usually aluminum or iron, to the solution. Accompanying electrochemical reactions are dependent on the species present and usually evolve electrolytic gases. The coagulant's

TABLE 2.6

Various Pollutants Removed by Different Electrocoagulation Arrangements

References	Pollutant	Current/Cell Voltage	Electrodes Anode/Cathode
Suspended Solids			
(Abuzaid et al. 1998)	Bentonite	0.2, 0.5, 1 A	Stainless steel
(Belongia et al. 1999)	Silica (SiO_2) and alumina	2.5–10.0 V cm^{-1}	304 stainless steel
(Matteson et al. 1995)	Kaolinite	0.01 A m^{-2}	Stainless steel
Color			
(Nandi and Patel 2013)	Dye	27.8–138.9 A m^{-2}	Fe
(Nandi and Patel 2014)	Dye	69.4–416.7 A m^{-2}	Al
(Kashefialasl et al. 2006)	Dye	127.8 A m^{-2}	Fe
(Golder et al. 2005)	Dye	1.5 KWh	Fe
(Daneshvar et al. 2006)	Dye	60–80 A m^{-2}	Fe
Organics			
(Baklan and Kolesnikova 1996)	Sewage	120 A m^{-2}	Fe and Al
(Pouet and Grasmick 1995)	Urban wastewater	3.9 A	Al/Al
(Pouet and Grasmick 1994)	Municipal wastewater	4–10 A	Al/Al
(Vik et al. 1984)	Aquatic humus	6–12 V	Al/Al
(Kobya et al. 2006)	Potato chips wastewater	20–300 A m^{-2}	Al/Fe
Fats and Oils			
(Balmer and Foulds 1986)	Oil	200–781 mA	Fe/Pt
(Rubach and Saur 1997)	Oil, salt, and chemicals	40–220 A	Al
(Woytowich et al. 1993)	Hydrocarbons		Al and steel tubes
Ions			
(Mameri et al. 1998)	Fluorides	75 A m^{-2}	Al/Al
(Kumar et al. 2004)	Arsenic	0.65–1.53 mA cm^{-1}	Ai/Fe/Ti

delivery and its nature influence the coagulation and separation processes. In recent years, smaller-scale EC processes have advanced to the point where they are seen as a reliable and effective technology. Numerous examples of pollutant removal from water by EC techniques have been reported in the recent literature, as summarized in Table 2.6. The pollutant's physicochemical properties influence its interactions within the system and eventual removal path. For example, ions are most likely electroprecipitated, while charged suspended solids are adsorbed on to the charged coagulant. EC's ability to remove a wide range of pollutants is the reason for its ongoing attractiveness to industry.

2.1.5.2 Electrochemistry

EC reactors are electrochemical cells. All such reactors consist of an electrode arrangement in contact with the polluted water, with coagulant production *in situ* being their distinguishing feature. To release the coagulant, an applied potential difference is required across the electrodes. Potential requirements for the electrodes can be deduced from the electrochemical half-cell reactions occurring at each electrode, which will vary according to the operational pH and the species present in the system. Reported electrode designs are numerous, including aluminum pellets in a fluidized bed reactor, bipolar aluminum electrodes, mesh electrodes, and bipolar steel Raschig rings, as well as simple plate electrodes. Various electrode materials have also been reported, including aluminum, iron, stainless steel, and platinum. The electrode material used determines the coagulant type.

Thus, regardless of the electrode design employed, the electrode material determines the electrochemical reactions that occur and, hence, the coagulant cation. For this reason, electrochemistry is one of the foundations for EC. Aluminum, the most commonly used anode material, is used here as an example. Equation 2.1 shows the dissolution of aluminum in the anode.

$$Al^{+3} + 3e^- \leftrightarrow Al \tag{2.1}$$

Oxygen evolution is also possible at the anode:

$$4OH^- \rightarrow O_2 + 2H_2O + 4e^- \tag{2.2}$$

Simultaneously, an associated cathodic reaction, usually the evolution of hydrogen, occurs. The reaction that occurs at the cathode is dependent on the pH value. At a neutral or alkaline pH, hydrogen is produced via Equation 2.3,

$$2H_2O + 2e^- \rightarrow 2OH^- + H_2 \tag{2.3}$$

while under acidic conditions, Equation 2.4 best describes hydrogen evolution at the cathode.

$$2H^+ + 2e^- \rightarrow H_2 \tag{2.4}$$

2.1.5.3 Electrocoagulation and Chemical Coagulation

Coagulation is a key feature of all EC reactors, describing the interaction between the coagulant and any pollutant material. The coagulant's role here is to destabilize the colloidal suspension by reducing any attractive forces, thereby lowering the energy barrier and enabling particles to aggregate. Depending on the physical and chemical properties of the solution, pollutant, and coagulant, a number of coagulation mechanisms (e.g., charge neutralization, double layer compression, bridging, and sweep) have been postulated in Letterman et al. (1999). For any given EC reactor, the dominant coagulation mechanism will vary with the reactor's operating conditions, the pollutant type and its concentration, and the coagulant concentration.

EC has been compared with chemical coagulation to assess its efficiency and advantages. Chemical dosing delivers the coagulant as a salt that dissociates in solution, with hydrolysis of the aluminum cation (and associated anions) determining solution speciation and pH value. Alum (i.e., aluminum sulfate) addition, for example, acidifies the water. By contrast, aluminum added via EC does not bring with it any associated salt anions, with the result that the pH value typically stabilizes in the alkaline range. However, Donini et al. (1994) claimed that the coagulation mechanism for electrochemical and chemical dosing are very similar, yet neither author supports their claim with rigorous experimental evidence.

In EC, a pollutant's stability is determined by its physicochemical properties. Pollutants composed of similarly charged particles repel each other, with the repulsive forces creating a stable, colloidal system with oppositely charged ions, typically hydroxyl (OH^-) or hydrogen ions (H^+), being attracted to the charged pollutant particles. The attraction of counterions to a charged pollutant forms an electric double layer—referred to as the Stern and diffuse layers (Letterman et al. 1999). Electrostatic repulsion between electric double layers drives particles apart, while van der Waals forces act to bring them together. The energy is such that attraction dominates at small separations. However, to reach a small separation, a repulsive energy barrier must first be overcome. The zeta potential is generally used as an experimental measure of the particle's effective charge as it moves through the solution, thus providing a direct indicator of solution stability (Letterman et al. 1999). Hence, zeta potential measurement provides an important characterization for any EC system, providing an indication of stability and of possible coagulation mechanisms.

2.1.5.4 Electrocoagulation Mechanism for the Removal of Fluoride

The EC process operates on the principle that the cations produced electrolytically from iron or aluminum anodes enhance the coagulation of contaminants from an aqueous medium. Electrophoretic motion tends to concentrate negatively charged particles in the region of the anode and positively charged ions in the region of the cathode. The consumable, or sacrificial, metal anodes are used to continuously produce polyvalent metal cations in the vicinity of the anode. These cations neutralize the negative charge of the particles carried toward the anodes by electrophoretic motion, thereby facilitating coagulation. In the following EC techniques, the production of polyvalent cations from the oxidation of the sacrificial anodes (Fe and Al) and the electrolysis gases (H_2 and O_2) work in combination to flocculate the coagulant materials. Even inert electrodes, such as titanium, and the passage of an alternating current have also been observed to remove metal ions from solutions and to initiate the coagulation of suspended solids. As previously mentioned, gas bubbles produced by the electrolysis carry the pollutant to the top of the solution where it is concentrated, collected, and removed. The removal mechanisms in EC may involve oxidation, reduction, decomposition, deposition, coagulation, absorption, adsorption, precipitation, and flotation.

Different electrodes have been reported in the literature such as carbon, mild steel, graphite, titanium, iron, and aluminum. But, iron and aluminum have been reported to be very effective and successful in removing pollutants at favorable operating conditions. The electrode reactions are summarized as follows:

$$\text{Anode: } Al \Rightarrow Al^{3+} + 3e \text{ (same as Equation 2.1)}$$

$$\text{Cathode: } 3H_2O + 3e \Rightarrow \tfrac{3}{2}H_2 \uparrow + 3OH^- \text{ (same as Equation 2.3)}$$

During the final stages, coagulated aggregates interact with bubbles and float to the surface or settle to the bottom of the EC bath. Flotation is the dominant pollutant removal path for high operating currents, while sedimentation is dominant at lower currents. The shift is because the number of bubbles concentrated at low currents is insufficient to remove the aggregated material, allowing sedimentation to dominate. Al(III) and OH^- ions generated by Equations 2.1 and 2.3 react to form various monomeric species such as $Al(OH)^{+2}$, $Al(OH)^+_2$, $Al_2(OH)_2^{4+}$, and $Al(OH)_4^-$ and polymeric species such as $Al_6(OH)_{15}^{3+}$, $Al_7(OH)_{17}^{4+}$, $Al_8(OH)_{20}^{4+}$, $Al_{13}O_4(OH)^{7+}_{24}$, and $Al_{13}(OH)_{34}^{5+}$, which transform finally into $Al(OH)_{3(S)}$ according to complex precipitation kinetics:

$$Al^{3+} + 3H_2O \Rightarrow Al(OH)_3 + 3H^+ \tag{2.5}$$

Freshly formed amorphous $Al(OH)_{3(S)}$ occurs as "sweep flocs" having large surface areas. These flocs are active in rapid adsorption of soluble organic compounds and trapping of colloidal particles, and are easily separated from aqueous media by sedimentation or H_2 flotation. These flocs polymerize as $nAl(OH)_3 \Rightarrow Al_n(OH)_{3n}$.

This $Al(OH)_3$ complex is believed to have strong fluoride adsorption capacity as Equation 2.6 (Ghosh et al. 2008):

$$Al(OH)_3 + xF^- \Leftrightarrow Al(OH)_{3-x}F_x + xOH^- \tag{2.6}$$

2.1.5.5 Flotation

The production of electrolytic gases is an inevitable by-product of EC. These gases lift pollutant particles and coagulant aggregates to the surface by a flotation-like process, while encouraging contact between pollutant particles and coagulant by providing a certain amount of mixing action. The main difference between this "electrolytic flotation" and more conventional flotation techniques is the method of bubble production and resultant bubble size. Expertise from other flotation techniques, including electroflotation, dissolved air flotation, and airlift reactors, can be employed to understand the flotation process in EC reactors. Electroflotation describes the production of electrolytic gases for the sole purpose of pollutant removal. One of the main advantages of flotation by electrolytic gases is the small size of the bubbles produced. For a given gas volume, a smaller bubble diameter results in both a greater surface area and more bubbles, thereby increasing the probability of collision and the ability to remove fine pollutant particles (Matteson et al. 1995). Also, as noted, electrolytic bubbles enhance mixing in the bulk solution via their overall upward momentum flux, increasing the likelihood of effective contact between coagulant and pollutant particles.

Bubble movement within a reactor is a function of the bubble density, bubble path, and bubble residence time. Current density determines the production rate of electrolytic gas and, thus, the bubble density, while reactor geometry (size, height, electrode positioning, effective electrode surface area-to-volume ratio) determines the bubble path. The average time a bubble spends in the reactor is referred to as its residence time, which is a function of bubble size and path length. It should be noted that shear forces from any mixing source affect the growth of aggregates. Operation at a low current density produces relatively few bubbles, resulting in gentle agitation—conditions that are ideal for aggregate growth and flocculation. As the current density increases, however, bubble density and the net upward momentum flux increase. These increases change the reactor's hydrodynamic behavior and the degree of mixing. High shear forces induced by mixing can damage and break flocs apart, reducing the effectiveness of pollutant removal. Electrochemistry, coagulation, and flotation thus form the three foundation stones for EC. Each component is a well-studied technology in its own right. However, it is clear from the published literature that what is lacking is a quantitative appreciation of the way in which these technologies interact to provide an EC system.

2.1.6 APPLICATION OF MEMBRANE TECHNOLOGY FOR THE TREATMENT OF DRINKING WATER

The membrane separation process in water treatment has gradually gained popularity because it effectively removes a variety of contaminants from raw water. While microfiltration (MF) and ultrafiltration (UF) membranes can mainly remove suspended particles, nanofiltration (NF) membranes are an effective technology to remove dissolved organic contaminants with molecular weights (MW) of larger than 200 Da and about 70% of monovalent ions by electrostatic repulsion (charge effect), size exclusion (sieving effect), and a combination of the rejection mechanisms (Petersen 1993). NF membranes offer an attractive approach to meeting multiple objectives of advanced drinking water treatment, such as the removal of disinfection by-product precursors, natural organic matter (NOM), endocrine-disrupting chemicals, and pesticides (Escobar et al. 2000). However, the decrease of permeate flux (i.e., membrane fouling) is a major obstacle to the application of NF membranes to drinking water treatment. Fouling worsens membrane performance and ultimately shortens membrane life, resulting in the increase of operational cost. Efficient control of membrane fouling is, therefore, required for the successful application of NF technology. Since a broad spectrum of constituents in feed waters can cause membrane fouling, it is important to investigate what types of materials (i.e., foulants) should be removed from feed waters by pretreatments prior to the NF process. NF membranes are subject to fouling by dissolved and macromolecular organic substances, sparingly soluble inorganic compounds, colloidal and particulate matter, and microorganisms (Potts et al. 1981; Li and Elimelech 2006; Hörsch et al. 2005), which are not primarily removed by the pretreatment process, such as coagulation–flocculation. To date, conventional pretreatment (coagulation followed by filtration) and MF are usually used in drinking water treatment (Speth et al. 2000; Van der Bruggen et al. 2004). The NF process may only be applied directly without pretreatment to groundwaters containing very low turbidity. Since surface waters contain higher turbidity and the water quality characteristics of surface waters vary significantly, the selection of pretreatment processes is one of the most important factors that determine the success or failure of NF process in drinking water treatment, because membrane fouling is caused by a combined effect of the membrane and feed water properties. Better understanding of the properties of tested waters and the efficiency of different pretreatments to remove foulants from feed waters may help us to prevent membrane fouling and design the best pretreatment process for NF membranes. However, there is little information on the effect of pretreatment processes on NF membrane fouling in the treatment of actual surface water.

2.1.7 Advantages and Disadvantages of Membrane Process

Although various conventional techniques of water purification described earlier are currently used to solve the problem of groundwater pollution, none of them is a user-friendly and cost-effective technique, due to some or other limitation, and has either no or a very long payback period. In recent years, the RO membrane process has emerged as a preferred alternative to provide safe drinking water without posing the problems associated with other conventional methods. RO is a physical process, in which the contaminants are removed by applying pressure on the feed water to direct it through a semipermeable membrane. The process is the reverse of natural osmosis, as a result of the pressure applied to the concentrated side of the membrane, which overcomes the natural osmotic pressure. The RO membrane rejects ions based on size and electrical charge. The factors influencing membrane selection are cost, recovery, rejection, raw water characteristics, and pretreatment. Efficiency of the process is governed by different factors such as raw water characteristics, pressure, temperature, regular monitoring and maintenance, and so on. There are two types of membranes that can remove fluoride from water: NF and RO. NF is a relatively low-pressure process that primarily removes the larger dissolved solids as compared with RO. Conversely, RO operates at higher pressures with greater rejection of all dissolved solids. Fluoride removal efficiency—up to 98%—by membrane processes has been documented by many researchers. In the past, the use of membrane technology for water treatment, particularly for drinking water production, was considered uneconomical in comparison with conventional means; but, in recent years, increased demand and contamination of water, the rise in water quality standards, and the problems associated with other methods have led to reconsideration of membrane technology for water purification. The progressive technical improvements in the design and materials of the membranes have made the water treatment process economically competitive and highly reliable. Also, the capital and operational costs of RO plants continue to decrease with increasing plant capacity (Barba et al. 1997). Thus, with improved management, this new technology for drinking water production may be the best option. Furthermore, membrane processes present several advantages as compared with other treatment methods (Arora et al. 2004).

2.1.7.1 Advantages

- The process is highly effective for fluoride removal. Membranes also provide an effective barrier to suspended solids, all inorganic pollutants, organic micropollutants, pesticides and microorganisms, and so on.
- The process permits the treatment and disinfection of water in one step.
- It ensures constant water quality.
- No chemicals are required and very little maintenance is needed.
- The life of the membrane is sufficiently long, so the problem of regeneration or replacement is encountered less frequently.
- It works in a wide pH range.
- No interference by other ions is observed.
- The process works in a simple, reliable, automated operating regime, with minimal manpower, using a compact modular model.

2.1.7.2 Disadvantages

- It removes all the ions present in the water, though some minerals are essential for proper growth; remineralization is required after treatment.
- The process is expensive in comparison to other options.

- The water becomes acidic and needs pH correction.
- A lot of water is wasted as brine.
- Disposal of brine is a problem.

2.1.8 Ceramic Membrane: An Overview

Existing and ongoing policies in process industries enforce the replacement of conventional process technology with membrane technology. Some of the aspirant features that membrane technology confers include compact design, high product quality, and lower operating cost. Alumina- (Van Gestel et al. 2002; DeFriend et al. 2003), zirconia- (Falamaki et al. 2004), titania- (Wang et al. 2006), and silica-based (Yoshino et al. 2005) ceramic membranes with high chemical, mechanical, and thermal stability and longer lifetimes are found to be useful in different membrane-based applications (Sourirajan 1970). However, the cost of these membranes is very high compared with polymeric membranes. This is primarily due to the higher cost of the inorganic precursors such as alumina, zirconia, titania, and silica, as well as the very high sintering temperatures (greater than 1100°C) required for the fabrication of the membranes. Due to these two key issues, the economic competitiveness of inorganic membranes to drive their industrial sustainability has not been appreciable to date.

To circumvent the higher costs of inorganic membranes, existing and ongoing research in the preparation of low-cost inorganic membranes is dovetailed toward the use of low-cost inorganic precursors and lower sintering temperatures (below 1000°C). However, these variants in ceramic membrane research need to guarantee cheaper membranes that have the inherent ability to provide a stable, on-time performance and lifetime, similarly to the existing expensive ceramic membranes.

Recently, much work has been reported on the fabrication of inorganic membranes using cheaper raw materials such as apatite powder (Masmoudi et al. 2007), fly ash (Saffaj et al. 2004), natural raw clay (Saffaj et al. 2006), dolomite (Bouzerara et al. 2006), and kaolin (Almandoz et al. 2004; Belouatek et al. 2005). However, the sintering temperature used in these works was greater than 1100°C and the average pore size of the membranes was more than 1 µm. In this regard, it is well known that MF membranes with a pore size in the submicron range (0.1–0.5 µm) are preferable for industrial application to obtain excellent solute separation efficiency. Currently, the surface modification technique is used to prepare a submicron-sized membrane from a micron-sized MF membrane. In this method, various inorganic precursors such as zeolite (Potdar et al. 2002; Workneh and Shukla 2008), zirconia, alumina (Workneh and Shukla 2008), and titania (Das and Dutta 1999) are coated over the membrane surface separately. However, the fabrication of such multilayer ceramic membranes involves a tedious cycle of sintering processes, which further contributes to the cost of the membrane (Potdar et al. 2002; Workneh and Shukla 2008). Therefore, there exists a necessity to develop alternate formulations using low-cost precursors and low sintering temperatures for the preparation of submicron-range ceramic membranes with an average pore size of around 100–500 nm to further the industrialization of ceramic membranes.

Ceramic membranes are found to be suitable in different process applications such as desalination (Saffaj et al. 2006), high temperature membrane reactors (Vivanpatarakij et al. 2009), membrane bioreactors, food processing, colored effluent treatment, drinking water purification, and treatment of industrial wastewater (Luque et al. 2008). Among all of these, the treatment of wastewater using membrane technology in industrial systems appears more relevant and to be the greatest current need, due to stricter and tighter environmental and health legislation.

2.1.8.1 Advantages of Ceramic Membranes over Polymeric Membranes

The advantages of inorganic membranes have been recognized for a long time. In fact, studies of the use of some inorganic membranes such as platinum and porous glass were evident even in the last century. Thermal and pH resistance characteristics of inorganic and organic membranes are compared in Table 2.7 with approximate ranges of operating temperature and pH value. Although inorganic polymers have not yet been used commercially, polyphosphazene, which is under development,

TABLE 2.7

Commonly Used Membrane Materials and Their Properties

Material	Application(s)	Approximate Maximum Working Temperature (°C)	pH Range
Cellulose acetates	RO, UF, MF	50	3–7
Aromatics	RO, UF	60–80	3–11
Fluorocarbon	RO, UF, MF	130–150	1–14
Polyimides	RO, UF	40	2–8
Polysulfone	UF, MF	80–100	1–13
Nylon	UF, MF	150–180	—
Polycarbonate	UF, MF	60–70	—
PVDF	UF	130–150	1–13
Alumina (gamma)	UF	300	5–8
Alumina (alpha)	MF	>900	0–14
Glass	RO, UF	700	1–9
Zirconia	UF, MF	400	1–14
Zirconia (hydrous)	RO, UF	80–90	4–11
Silver	MF	370	1–14
Stainless steel (316)	MF	>400	4–11

Source: Hsieh, H. P. 1996. *Inorganic Membranes for Separation and Reaction.* Membrane Science and Technology Series, 3. Amsterdam, Netherlands: Elsevier.

is included in the table. It can be seen that the thermal stabilities of organic polymers, inorganic polymers, and inorganic materials as membrane materials can be conveniently classified as <100°C, 100°C–350°C, and >350°C, respectively. It should be stressed that the use of pH stability is only a crude indication of the chemical stability of the membrane material. An example is silver, which can withstand strong bases and certain strong acids. Although silver is resistant to strong hydrochloric or hydrofluoric acid, it is subject to attack by nitric and sulfuric acids and cyanide solutions. Therefore, it cannot be deduced from the wide operating pH range of silver that it is resistant to all acids.

The operating temperature limits of inorganic membranes are obviously much higher than those of organic polymeric membranes. The majority of organic membranes begin to deteriorate structurally at around 100°C. The thermal stability of membranes is becoming not only a technical problem but also an economic issue. Inorganic membranes can generally withstand organic solvents, chlorine, and other chemicals better than organic membranes. This also permits the use of more effective, yet corrosive, cleaning procedures and chemicals. Many organic membranes are susceptible to microbial attack during applications. This is not the case with inorganic ones, particularly ceramic membranes. In addition, inorganic membranes, in general, do not suffer from the mechanical instability of many organic membranes, where the porous support structure can undergo compaction under high pressures and cause a decrease in permeability.

2.1.9 COMBINATION OF VARIOUS SEPARATION PROCESSES

In some studies, better separation was achieved by combining two or more processes. Dhale and Mahajani (2000) used an integrated process including NF and wet oxidation to treat a disperse dye bath waste. The dye bath waste stream underwent NF for the recovery of reusable water from the permeate stream, thus reducing fresh water consumption and minimizing effluent discharge. The concentrate obtained from the membrane unit had a high chemical oxygen demand (COD) content, and was then treated by the wet oxidation (WO) process. In another study by Abdessemed and Nezzal (2003), a combination of coagulation, adsorption, and UF was used for the treatment of a primary effluent.

This primary effluent contained organic mineral and dissolved and suspended matter (colloids). MF or UF was adequate to produce disinfected clear water suited for different applications. However, direct filtration onto the membrane was limited by the fouling phenomena, which led to a substantial and continuous decrease of the permeate flux during filtration at constant pressure. On the other hand, coagulation and adsorption made it possible to remove the colloidal fraction, which played a significant role in membrane fouling. It was observed that coagulation significantly improved the UF performances. In Lin and Wang's (2002) study, treatment of high-strength phenolic wastewater was investigated by a novel two-step method. This two-step treatment method consisted of chemical coagulation of the wastewater by metal chloride, followed by further phenol reduction by resin adsorption. The combined treatment was found to be highly efficient in removing the phenol concentration from the aqueous solution and proved to be capable of lowering the initial phenol concentration considerably. The effectiveness of a combined reduction–biological treatment system for the decolorization of nonbiodegradable textile dyeing wastewater was investigated by Ghoreishi and Haghighi (2003). In this treatment system, a bisulfite-catalyzed sodium borohydride reduction followed by an activated sludge technique was used to remove the color at ambient temperature and pressure. Combinations of adsorption and membrane-based processes in the treatment of process wastewater were reported in the literature. Baudin et al. (1997) used a combination of adsorption and UF for the treatment of surface water in 12 full-scale drinking water treatment plants in Europe. Later, Meier et al. (2002) used a combination of adsorption and NF for the treatment of severely contaminated wastewater. Powdered adsorbent was injected into the feed of an NF unit and subsequently removed from the concentrate by a thickener. The adsorbent in the feed had a positive effect on permeate quality, permeate flux, and the fouling layer in the NF unit. In comparison to RO, the combination process had a higher maximum recovery rate, lower operating pressure, and energy consumption, which resulted in lower treatment costs. In another recent work by Métivier-Pignon et al. (2003), the combination of adsorption and UF was used in the treatment of colored wastewaters. In this process, adsorption was carried out in batch reactors using activated carbon cloths (ACCs). Adsorption was carried out to uptake low molecular weight compounds (mainly dyes), while UF was used to remove large-sized compounds (macromolecules, colloids, and turbidity). The membrane filtration step allowed a significant removal of turbidity (about 98%), whereas adsorption onto ACCs resulted in the decolorization of the stream with a high adsorption capacity. The continuous process of adsorption onto ACC and UF resulted in successful discoloration with a high permeate flow rate. During EC, dissolution of electrodes in the form of metallic hydroxides creates flocs, which, in turn, increase the alkalinity and turbidity. Therefore, further treatment is strongly recommended. In view of this, combinations of other techniques such as electroflotation, NF, and RO were investigated along with EC for the treatment of drinking water. Zuo et al. (2008) studied a combination of EC and electroflotation for the removal of fluoride from drinking water. Aouni et al. (2009) examined a hybrid EC/NF process for the treatment of textile wastewater. Removal of silica from brackish water by EC followed by RO has been investigated by Den and Wang (2008). The particle size of the flocs generated during EC would definitely determine the type of filtration process to be combined with the existing process. It was found that the particle size of the electrocoagulated products was in the range of 10–100 μm. It was therefore realized that MF may be a suitable process rather than UF, NF, or RO. MF membranes were prepared according to the requirement. EC followed by MF was adopted to remove fluoride from drinking water. Detailed discussions on EC, the preparation of the inorganic MF membrane, and its application in the separation of electrocoagulated by-products have been carried out.

2.2 TREATMENT OF FLUORIDE USING A COMBINATION OF ELECTROCOAGULATION AND MICROFILTRATION

Many processes have been developed for the removal of fluoride from drinking water. To implement any process scheme in industrial application, a detailed parametric study is required along with cost estimation. However, little literature is available on such detailed experimental study together with an estimation of costs. Moreover, literature on the posttreatment process of the EC technique is scant. This section deals with a combinational process for the removal of fluoride from drinking water involving both EC and MF. Aqueous synthetic solutions of fluoride are selected separately for EC. Later on, the process scheme is applied to the treatment of mixtures of fluoride. Ceramic membranes are prepared for MF experiments. Both paste and uniaxial methods are adopted to prepare MF membranes. After EC, the treated water is passed through the MF unit, where the unsettled electrocoagulated by-products are separated and the water from the MF unit can safely be used for drinking.

2.3 ELECTROCOAGULATION: EXPERIMENTAL

2.3.1 ELECTROCOAGULATION BATH

EC in batch mode exhibits time-dependent behavior, as the coagulant is continuously generated in the reactor due to anodic oxidation. An EC bath operating in batch mode is always best suited to laboratory-scale applications. Nonetheless, a batch EC bath is very difficult to model mathematically, due to its inherently dynamic behavior coupled with the interplay between thermodynamic considerations. An EC bath operating in batch mode consists of a constant volume, where the contents are well mixed and internal concentrations change with time. The performance of such a bath depends mostly on the reaction time, that is, the time in the reactor. The discussion here uses a perspex tank of working volume of around 2.8 L and dimensions of $0.14 \times 0.147 \times 0.137$ m for the EC bath (Figure 2.1). The performance of the process depends on the interactions between species, which must come into contact with each other and be well mixed. Mixing is a rudimentary measure and depicts the homogeneity within a reactor. Therefore, mixing strongly influences the performance and effectiveness of the EC bath. Mixing in an EC bath operating in batch mode is primarily a function of fluid flow and agitation. A magnetic stirrer is used for mixing purposes.

2.3.2 ELECTRODE

In the EC process, an applied potential generates the coagulant species *in situ* as the sacrificial metal anode (aluminum or iron) dissolves, while hydrogen is simultaneously evolved at the cathode. Coagulant species aggregate the suspended particles and adsorb the dissolved contaminants. The choice of electrode material depends on various criteria such as low cost, low oxidation potential, inertness toward the system under consideration, and so on. Different electrodes were reported in the literature such as carbon (Alverez-Gallegos and Pletcher 1999), mild steel (Golder et al. 2005),

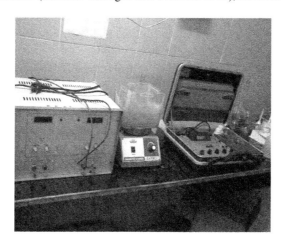

FIGURE 2.1   (See color insert) Picture of electrocoagulation setup.

iron (Yildiz et al. 2007), graphite titanium (Linares-Hernández et al. 2007), and aluminum (Wang et al. 2004). Aluminum was reported to be very effective and successful in removing pollutants in favorable operating conditions (Ghosh et al. 2008). In the present case, aluminum sheets of $0.15 \times 0.05 \times 0.002$ m are used as electrodes and the effective surface area of each electrode is 40×10^{-4} m^2. Two types of electrode connection (monopolar and bipolar) are considered during the EC of fluoride present in drinking water.

2.3.3 MONOPOLAR ELECTRODE CONNECTION

In a monopolar electrode connection, one end of the electrode is connected to the positive terminal of the power source, while the other end of the electrode is connected to the negative terminal of the same power source, thus completing the formation of a primary cell, which, under the application of potential in contact with the contaminant solution, initiates the EC process.

2.3.4 BIPOLAR ELECTRODE CONNECTION

In a bipolar connection, more than two electrodes are used to make such an arrangement to reduce the effective corrosion of the electrode and improve the process performance. Only two ends of the electrodes are connected to the direct current (dc) power source, whereas the other two electrodes have no connection to it. In such conditions, induced polarization takes place when voltage is applied to the end electrodes, so that the inner electrodes start to act as a secondary cell. Therefore, the total assembly is bipolarized with a primary and a secondary cell acting together. The entire electrode assembly is fitted on nonconducting wedges and hung from the top of the EC tank. The assembly is connected to the dc power source (Textronics 36D, Agarwal Electronics, Mumbai, India) to constitute an electrochemical cell with galvanostatic mode for constant current supply. A schematic diagram of monopolar and bipolar electrochemical cells is shown in Figure 2.2. The electrode assembly is placed in the cell, the electrodes are connected to the respective terminals of the dc power supply, and a constant current is supplied for a given time.

2.3.5 SOLUTION PREPARATION

Sodium fluoride (NaF, supplied by Aldrich Chemical Company, United States) is used in this study for the preparation of a fluoride solution. In all the cases, a measured quantity of salts was added to tap water (1 L) for the preparation of fluoride-contaminated drinking water. The conductivity and pH value of the tap water were 12 S m^{-1} and 7.5, respectively. The initial fluoride concentration varied from 4 to 10 mg L^{-1}. The range of concentration was selected based on the availability of fluoride concentrations in drinking water.

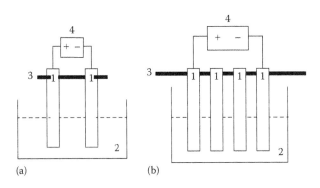

(a)   (b)

FIGURE 2.2 (a,b) Schematic diagram of electrochemical cell for EC experiments.

2.3.6 Procedure

In each EC experiment, 1 L of water solution was placed into the electrolytic cell (EC bath). The current density was adjusted to a desired value and the operation started. After the experiment, the power was switched off and the electrodes were dismantled. The treated sample collected at different time intervals was filtered before analysis. Before each run, the electrodes were washed with acetone to remove surface grease, and the impurities on the aluminum electrode surfaces were removed by dipping them in an acetone solution for 5 min, followed by rubbing with an abrasive paper (C220), to ensure the complete removal of impurities before every experiment. After each experiment, the used anode and cathode plates were interchanged for effective electrode utilization. All the runs were performed at a constant temperature of 25°C. The parameters chosen in these experiments are reported in subsequent sections. A fluoride ion meter was used to determine the fluoride concentration, whereas the conductivity and pH were measured using a conductivity meter and pH meter, respectively.

2.3.6.1 Fluoride Ion Meter

A fluoride ion meter (make: Eutech Instruments, Singapore; model: ECFOO301BEU) was used to determine the fluoride concentration after calibration using a total ionic strength adjuster buffer (TISAB). TISAB is a solution containing CH_3COOH (acetic acid), NaCl (sodium chloride), CDTA (trans-1,2-diaminocyclohexane-N,N,N′,N′-tetraacetic acid monohydrate), and NaOH (sodium hydroxide).

2.3.6.2 pH Meter

A pH meter (make: Century Instruments (P) Ltd., Chandigarh, India) was used to determine the pH of the solution. The meter was first calibrated using two different buffers of pH 4 and pH 9.2. Each time, the meter was thoroughly washed with deionized water before and after dipping into the buffer solutions.

2.3.6.3 Conductivity Meter

The conductivity of solution in the EC bath was measured using a digital conductivity meter (make: Electronics Pvt. Ltd., India; model: VS1) before and after the treatment to match the recommended limit of 0.2 mmhos (Weber 1972). The conductivity meter was also calibrated by matching the known conductivity of 0.1 N KCl solution given in the manual supplied by the manufacturer. After calibration, the conductivity meter was thoroughly washed with deionized water and soaked with tissue paper without touching the coated detector inside the meter.

2.3.6.4 Operating Conditions

Other important parameters such as initial concentration, interelectrode distance, electrode connection, current density, and time of treatment were investigated to discover sufficient information about the EC process, applied mainly to the purification of drinking water contaminated with fluoride. All the experimental conditions considered in this study are shown in Table 2.8.

2.3.7 Characterization of By-Products

A whitish precipitate was formed inside the EC bath at the end of the EC process for the treatment of drinking water contaminated with fluoride.

2.3.7.1 Scanning Electron Microscopy and Elemental Analysis (SEM and EDAX)

The morphology of the by-products obtained from the EC bath was analyzed by SEM, whereas elemental information about the by-products were recorded using EDAX. Furthermore, it is worth mentioning that before the SEM analysis the sample was finely ground and coated with Au inside a

TABLE 2.8

Experimental Conditions for the Removal of Fluoride Using EC

Parameters	Conditions Maintained
Initial concentration (mg L^{-1})	4, 6, 8, 10
Interelectrode distance (m)	0.005, 0.01, 0.015
Current density (A m^{-2})	250, 375, 500, 625
Experiment time (min)	45
Temperature (°C)	25
Electrode connection	Monopolar and bipolar

plasma chamber operating under vacuum for 135 s with a leakage current of 4 mA. This was done to ensure a lower possibility of charging during the analysis. The sample was analyzed under the application of the probe current of 94 pA and at different magnifications with the primary electron hitting the sample with energy of 10–15 kV. EDAX was an integral part of the SEM, where energy dispersion according to element was calibrated with a standard Co (cobalt) sample before analysis.

2.3.7.2   X-Ray Diffraction (XRD)

The phase of the by-products was recorded using XRD. Powder XRD of the by-product was recorded by a diffractometer operating with a Cu K α radiation source in nano-Bragg mode. The XRD analysis result is recorded from 5° to 90° with a step increase rate of 0.05° s^{-1}. The spectrum of the analyzed by-products showed broad and diffuse peaks. Therefore, the identification of peaks with such broad humps is a well-known confirmed characteristic of phases that are amorphous or poorly crystalline in nature.

2.3.7.3   Fourier Transform Infrared Spectroscopy (FTIR)

The finely ground by-product was mixed well with potassium bromide and pressed under a weight of 5 t for 10 s to prepare the pellet necessary for the analysis. Furthermore, the pellet was subjected to the infrared spectra, and scans were recorded with wave numbers ranging from 4000 to 450 cm^{-1}. Different peaks at different wave numbers were leveled and matched with the standard data available for certain identified bond stretching. Some peaks were not identified due to the lack of standard data related to the work. Nonetheless, the available data matched well with the experimental results presented here.

2.4   ELECTROCOAGULATION: RESULTS AND DISCUSSION

This section discusses the results obtained for the removal of fluoride from drinking water using the EC technique. By-products obtained from the EC bath were characterized by SEM, EDAX, FTIR, and XRD analysis. Operating costs for the removal of fluoride were calculated and presented as well. In the calculation of the operating costs, only material and energy costs were considered. Other cost items such as labor, maintenance, and solid/liquid separation costs were not taken into account. A simplified cost equation was used to evaluate the operating costs of EC for the removal of fluoride (Ghosh et al. 2008). In this work, two different electrode connections (monopolar and bipolar) were investigated to choose the better alternative to intensify the performance of the process. Different initial concentrations (4–10 mg L^{-1}) of fluoride were considered for the experiment, which had a duration of 45 min. Experiments were carried out with different current densities. Corrosion of electrodes, as well as sludge formation, was estimated during the experiments for both electrode connections. Variations of film thickness deposited over the electrode surfaces with changes in initial fluoride concentrations and current densities were determined.

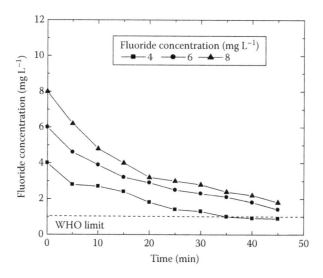

FIGURE 2.3 Variation of fluoride concentration in the EC bath over time. Interelectrode distance: 0.005 m; current density: 250 A m^{-2}; temperature: 25°C; electrode connection: monopolar.

2.4.1 Effect of Initial Fluoride Concentration

In the fluoride removal process by EC, the initial fluoride concentration is an important parameter for a particular mode of electrode connection. Figure 2.3 reveals the variations of fluoride concentration in the EC bath during the experiment using a monopolar electrode connection. It was observed that, after the treatment, the final fluoride concentration did not drop below the recommended upper limit suggested by the World Health Organization (WHO) for all initial fluoride concentrations. It was also observed (see Figure 2.3) that around 35 min were required to attain the final fluoride concentration of 1 mg L^{-1} for an initial fluoride concentration of 4 mg L^{-1} and current density of 250 A m^{-2}. Again, for an initial fluoride concentration of 6 mg L^{-1} with a monopolar connection, 45 min were needed to attain a final fluoride concentration of 1.4 mg L^{-1}. It was very clear that, with an increase in the initial fluoride concentration, the treatment time increased to reach a final fluoride concentration of the recommended limit (1 mg L^{-1}). In EC, aluminum cations initially contribute to charge neutralization of the pollutant particles as the isoelectric point is attained. Here, a sorption coagulation mechanism occurs, resulting in the formation of loose aggregates. As time progresses, further aluminum cation addition results in amorphous aluminum hydroxide precipitation that promotes pollutant aggregation. As the current density was kept constant, so the production of the aluminum cation remained fixed and, therefore, with an increase in initial fluoride concentration, the complex formation process between the amorphous aluminum hydroxide and fluoride was insufficient at the applied current density mentioned above. This could be the reason that the drinking water, with initial fluoride concentrations of 6 and 8 mg L^{-1}, had final fluoride concentrations of 1.8 and 2.2 mg L^{-1} after 45 min treatment by EC with a monopolar electrode connection and an initial fluoride concentration of 10 mg L^{-1}. Other parameters such as current density, interelectrode distance, and duration of the experiment were maintained at 250 A m^{-2}, 0.005 m, and 45 min, respectively. It was observed that, with the passage of time, the fluoride concentration inside the EC bath decreased for both electrode connections. It is also seen from Figure 2.4 that, after 45 min with the bipolar connection, the final fluoride concentration was 1.7 mg L^{-1}, whereas for the monopolar connection, the fluoride concentration dropped down to a value of 2.2 mg L^{-1}, which was far above the recommended limit. It is necessary to mention that, with the bipolar connection, two pairs of electrodes were used, of which only the end electrodes were connected to the respective anode and cathode connections of the dc source. With the bipolar connection, two electrochemical cells acted

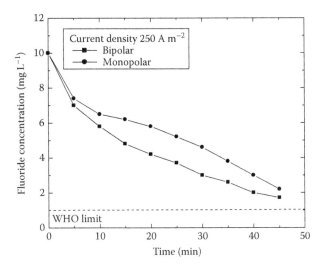

FIGURE 2.4 Variation of fluoride concentration in the EC bath over time. Interelectrode distance: 0.005 m; current density: 250 A m^{-2}; temperature: 25°C; initial fluoride concentration: 10 mg L^{-1}; electrode connection: bipolar and monopolar.

together, which favored adequate anodic oxidation with a higher surface area compared with that of the monopolar connection. As a result, with the same current density applied to both kinds of connection, the intensity was higher with the bipolar connection. Therefore, the final fluoride concentration in the solution was found to be lower than that observed with the monopolar electrode connection. However, the fluoride concentration in the bath after 45 min did not attain the permissible limit with either type of electrode connection.

2.4.2 Effect of Bipolar Connection of Electrodes

To improve fluoride removal efficiently from contaminated drinking water with a higher initial fluoride concentration, electrode connections can have a significant effect on sludge formation as well as on the corrosion of the electrode. The effect of electrode connections (monopolar and bipolar) for fluoride removal by EC is shown in the inset of Figure 2.4. An investigation was carried out into the treatment of drinking water.

2.4.3 Effect of Interelectrode Distance

The setup of the electrode assembly is very important to obtain the required effective surface area of the electrode and interelectrode distance. Variations of fluoride concentrations in the EC bath and interelectrode distances are shown in Figure 2.5. It is observed from the figure that, at any given point in time, fluoride concentration in the EC bath was lower with a smaller interelectrode distance. For example, the final fluoride concentration in the EC bath decreased from around 3 to 0.8 mg L^{-1} after 45 min of EC when the interelectrode distance was decreased from 0.015 m to 0.005 m for an initial fluoride concentration of 10 mg L^{-1}. It is well known that, in the EC bath, anodic oxidation begins as the potential is initially applied to the electrodes. Over time, a very fine film of metal hydroxides is formed on the anode, generating an extra resistance that even increases with a greater interelectrode distance. Hence, after some time of the EC process, the current drops. To maintain a constant current level, applied potential has to be increased. As the rate of anodic oxidation becomes lower, the number of cations at the anode also decreases. These cations are responsible for the formation of the coagulant. Therefore, at a higher interelectrode distance, the rate of aggregation

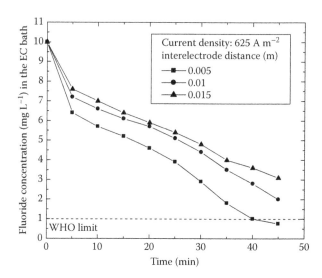

FIGURE 2.5 Effect of interelectrode distance on fluoride concentration in the EC bath over time. Initial fluoride concentration 10 mg L^{-1}; current density: 625 A m^{-2}; interelectrode distance: 0.005 m; temperature: 25°C; electrode connection: monopolar.

of suspended particles, as well as adsorption of contaminants, would be low. This may be the reason behind the finding that there is lower removal efficiency at a greater interelectrode distance.

2.4.4 FLUORIDE REMOVAL WITH VARYING CURRENT DENSITY

2.4.4.1 Monopolar

In any EC process, current density (A m^{-2}) and time of electrolysis are important operational parameters, ultimately establishing fluoride removal and defining electrical energy and power consumption; so, eventually, deciding the ultimate operating cost for the process. Current density was estimated as the amount of current passing through the electrode divided by the effective surface area of the electrode submerged inside the solution. A sufficient number of electrons passing through the unit area of the electrode surface corresponded to a favorable occurrence of anodic oxidation. This further enhanced the formation of an adequate amount of metal hydroxide species, followed by hydrolysis and polymerization. Some investigators reported that in EC, the current density can influence the efficiency of treatment, while others pointed out that current density has no significant role in pollutant removal. Therefore, it remained unclear whether current density affects treatment efficiency or not. The choice of electrode material is also vital, affecting cell voltage (different oxidation potential for different electrode materials), which is responsible for initiating anodic oxidation. Since the metallic hydroxides cannot be formed without anodic oxidation, the formation rate of such species is therefore dependent on the voltage applied as well as the electrode material. Further, these species take part in the agglomeration of flocs followed by adsorption, through which separation is achieved. Figure 2.6 shows the removal of fluoride from fluoride-rich water for four different current densities with time. The current density varied from 250 to 625 A m^{-2}. It is seen from the figure that for all the current densities, a sharp decrease in fluoride concentration occurred at a point up to 5 min of operation. Formation of a thin film of gelatinous aluminum hydroxide on the anode surface inhibits the cation formation rate as well as the agglomeration of the contaminant particles from the solution in the form of flocs. Therefore, after a certain period of time (5 min in this case), the decrease in fluoride concentration became gradual. It is also observed from the figure that, with an increase in current density, defluoridation is improved. Generally, an increase

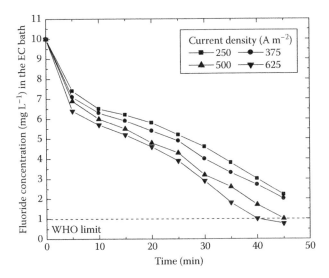

FIGURE 2.6 Effect of current density on the fluoride concentration in the EC bath over time. Initial fluoride concentration: 10 mg L^{-1}; interelectrode distance: 0.005 m; temperature: 25°C; electrode connection: monopolar.

in current density causes anodic oxidation to take place more readily, which, in turn, favors the formation of amorphous aluminum hydroxides species in the vicinity of the electrode as well as in the bulk. Formation of such hydroxides at the vicinity of the anode–water interface manifests as the accumulation of a fine gelatinous film on the electrode surface. It is well known that in a system with coexisting anions, defluoridation occurs in the bulk of the solution, and without coexisting anions, it occurs on the surface of the anode. Therefore, fluoride ions present in the solution may be allowed to interact with such species and form a complex. It was found that the final fluoride concentration of 1 mg L^{-1} was achieved after 40 min of EC with an application of 625 A m^{-2} current density.

2.4.4.2 Bipolar

Figure 2.7 represents the effect of current density on the fluoride concentration in the EC bath during the experiment. Current density varied from 375 to 625 A m^{-2} using a bipolar electrode connection. It is seen from Figure 2.7 that defluoridation improved with an increase in current density. This is due to the formation of adequate amorphous aluminum hydroxides species in the vicinity of the electrode as discussed for the monopolar case. It was found that the final fluoride concentration of 1 mg L^{-1} was achieved after 30 min of EC using a bipolar connection with an application of 625 A m^{-2}. Furthermore, this was also supported by the fact that with an increase in current density and initial fluoride concentration, the corrosion of the electrodes increased, as well as sludge formation.

2.4.5 Variation of pH in Electrocoagulation Bath

The removal of fluoride from the EC bath is believed to be a favorable technology, due to the formation of more OH^- ions in the electrolysis of water. It was observed that aluminum electrodes produced $Al(OH)_3$ precipitation in a slightly basic medium and followed the sweep-floc mechanism (Sujana et al. 1998). Hence, a change in pH of the solution in the EC bath would be a measure of $Al(OH)_3$ formation, influencing fluoride removal. The duration of EC was 45 min, and after every 5 min, the pH of the solution was checked using a digital pH meter (CONSORT, Belgium; model: C863). The variations in pH in the EC bath are shown in Figure 2.8 for different current densities

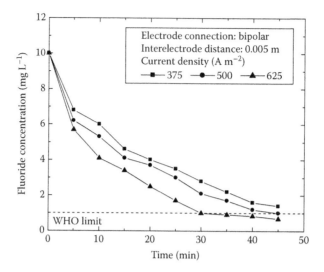

FIGURE 2.7 Effect of current density on the fluoride concentration in the EC bath over time. Initial fluoride concentration: 10 mg L^{-1}; interelectrode distance: 0.005 m; electrode connection: bipolar.

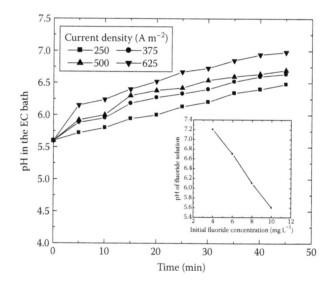

FIGURE 2.8 Variation of pH of drinking water containing fluoride in the EC bath over time at different current densities. Interelectrode distance: 0.005 m; initial fluoride concentration: 10 mg L^{-1}; electrode connection: monopolar.

(250, 375, 500, and 625 A m^{-2}) for the initial fluoride concentration of 10 mg L^{-1}. For all cases, the initial pH of the solution was taken as similar to that of synthetic fluoride containing water and shown in the inset of Figure 2.8. It was seen that the pH of the solution increased over time for all the current densities. At the end of the EC process, the pH was seen to reach values of around 6.5, 6.6, 6.7, and 6.9 for the current densities of 250, 375, 500, and 625 A m^{-2}, respectively. The increase in pH value with time and current density is due to the formation of a greater concentration of OH^{-} ions in the EC bath due to electrolysis. This increases anodic oxidation as well as the formation of aluminum hydroxide complexes. A whitish sludge, due to the formation of fluoroaluminum complexes, was observed at the end of the experiment. This may have been due to the generation of more aluminum hydroxides, which, in turn, adsorbed fluoride from the solution and remained in the solution as flocs.

2.4.6 VARIATION OF ELECTRODES CORROSION

Anodic oxidation gives rise to the corrosion of the electrode material (aluminum, in the present case) due to aluminum hydroxide formation during EC. The corrosion of the electrodes gives us an overview of the lifetime as well as the experimental cost of the electrode material. Therefore, electrode corrosion during EC can be considered to be an important parameter that reflects the viability of the process. Corrosion of electrodes can be defined as the weight loss (mg) of the electrodes due to anodic oxidation. This loss was calculated by subtracting the weight of the electrodes at the end of the experiment from the weight of the same electrodes taken before the experiment. Figure 2.9 describes the extent of electrode corrosion at different current densities while using bipolar as well as monopolar electrode connections. The investigation was carried out using different initial fluoride concentrations. Parameters such as interelectrode distance and duration of the experiment were kept constant at the values mentioned in Figure 2.3. It was observed that with an increase in initial fluoride concentration and current density, electrode corrosion was enhanced due to the increased dissolution of the electrodes. In addition to this, it was seen that electrode corrosion was greater for the bipolar connection. It can be seen from Figure 2.9 that electrode corrosion with the monopolar connection was 14.5 mg whereas with the bipolar connection it was 22.8 mg at an applied current density of 250 A m^{-2} for the initial fluoride concentration of 4 mg L^{-1}. This was an obvious result, since anodic oxidation with the bipolar connection was higher compared with the monopolar connection. Sludge generated in the EC bath was estimated for the bipolar and monopolar electrode connections. It was found that there was more sludge formation with the bipolar connection, whereas in both cases it increased with greater current density. In the treatment of drinking water with an initial fluoride concentration of 4 mg L^{-1} by EC at a current density of 250 A m^{-2}, the amount of sludge produced was 68.6 and 42 mg with bipolar and monopolar connections, respectively. It was noticed that when removing fluoride from drinking water at an initial fluoride concentration of 4 mg L^{-1} using the bipolar connection, the amount of sludge generated was equal to 90.2, 148.5, and 190.3 mg at 375, 500, and 625 A m^{-2}, respectively. It was also observed that with an increase in initial fluoride concentration, more sludge was produced. Figure 2.10 represents the pH variation with time in the

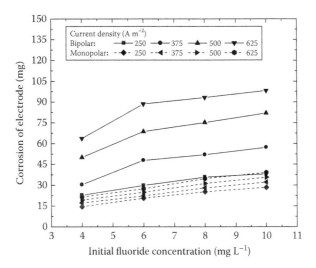

FIGURE 2.9 Variation of electrode corrosion with bipolar and monopolar connections at different current densities. Interelectrode distance: 0.005 m; duration of the experiment: 45 min; temperature: 25°C.

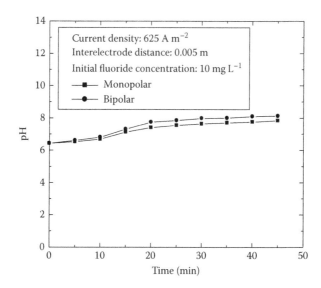

FIGURE 2.10 Variation of pH over time in the EC bath for bipolar and monopolar connections. Initial fluoride concentration: 10 mg L^{-1}; interelectrode distance: 0.005 m; duration of the experiment: 45 min; temperature: 25°C.

EC bath during the experiment using bipolar and monopolar connections. The pH increased with time in both cases (varying from 6.4 to 8.1). It was also observed that for the bipolar connection, the pH of the solution was marginally higher than the monopolar connection. For example, at a current density of 625 A m^{-2} and an initial fluoride concentration of 10 mg L^{-1}, the pH of the solution always had a value of around 8 at the end of all trials. According to the complex precipitation kinetics, most of the charged aluminum species transformed into Al(OH)$_3$, which exists in equilibrium with some of the soluble species (Duan and Gregory 2003). It can be also seen from Figure 2.10 that after attaining a pH close to 8, it does not increase significantly irrespective of the connection mode. This is probably because of an equilibrium between the Al(OH)$_3$ and the soluble aluminum species at a pH value of about 8 for both types of electrode connections. This observation also supported the reason behind electrode dissolution, as discussed in the preceding section.

2.4.7 VARIATION OF FILM THICKNESS

In the EC process, anodic oxidation favors the formation of gelatinous hydroxide species with an interaction between positive metal ions and negative OH$^-$ ions. This behavior allows such hydroxides to adhere to the electrode surface, growing like a film with the progress of the process. Therefore, an extra resistance is created, which, in turn, affects the performance of the process. Hence, it is very clear that the film generated during the EC process plays a major role in the operation of the technique for the removal of pollutants. Film thickness was calculated using the following simple mass balance equation:

$$\delta \cong \frac{(m_1 - m_2) \times 10^{-6}}{\rho \times A} \tag{2.7}$$

where:
   δ is the film thickness (nm)
   m_1 is the weight of the electrodes (mg) after the experiment without cleaning
   m_2 is the weight of the electrodes (mg) after the experiment after cleaning

ρ is the density of the film (g L^{-1})

A is the area of the electrodes (m^2)

Figure 2.11 depicts the variation of film thickness over the electrode surface with different current densities and initial concentrations of fluoride. It was seen that with an increase in current density and initial fluoride concentration of the drinking water, film thickness increased for both monopolar (Figure 2.11) and bipolar connections. It can be observed from Figure 2.11 that the film thickness increased from around 131 to 470 nm when the initial fluoride concentration increased from 4 to 10 mg L^{-1} at an applied current density of 375 A m^{-2} for the bipolar connection. This parameter also increased from 147 to 180 nm when the current density was increased from 500 to 625 A m^{-2} for the initial fluoride concentration of 6 mg L^{-1}. Variations of film thickness using the monopolar connection have been estimated under the same conditions as those of the bipolar connection and are shown in the inset of Figure 2.11. It was observed that film formation followed the same trend, but it was much lower than in the case of bipolar connection. For example, film thickness increased from 23.5 to 57 nm when the initial fluoride concentration varied from 4 to 10 mg L^{-1} at an applied current density of 375 A m^{-2} for 45 min of operation. Furthermore, it was also found that with the increase in current density from 500 to 625 A m^{-2}, film thickness increased from 69 to 76.4 nm for an initial fluoride concentration of 10 mg L^{-1}. An enhancement to the current density favored anodic oxidation and, hence, it increased the possibility of the formation of more metal ions as well as interactions between the metal hydroxides and the fluoride ions. Therefore, the generation of the film formed during the experiment became thicker.

2.4.8   VARIATION OF CONDUCTIVITY OF TREATED WATER

To examine the quality of water treated by EC, a measurement of conductivity was made so that it would be acceptable for drinking purposes. Before each experiment using the bipolar electrode connection, the conductivity of the contaminated water was measured as 0.4, 0.6, and 0.9 mmhos

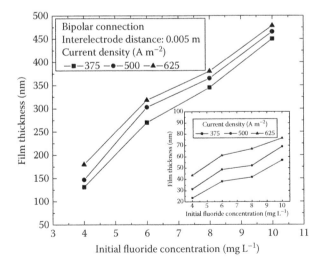

FIGURE 2.11   Variation of film thickness over electrode surface with different current densities and initial fluoride ion concentrations. Current density: 375 A m^{-2}; interelectrode distance: 0.005 m; duration of the experiment: 45 min; electrode connection: bipolar; temperature: 25°C. Inset: Film thickness variation over electrode surface with different current densities and initial fluoride ion concentrations for monopolar electrode connection.

for initial fluoride concentrations of 6, 8, and 10 mg L^{-1}, respectively. After the treatment, the sludge was filtered out and the clear solution was collected. The conductivity of the treated water was around 0.2 mmhos for all cases, that is, within the range of the permissible limits for drinking water (Weber 1972).

2.4.9 ESTIMATION OF OPERATING COSTS

In any electrical process, cost is incurred due to electrical energy demand, which affects the operating cost. For the EC process, the operating cost includes material (mainly electrodes) and electrical energy costs, as well as labor, maintenance, sludge dewatering and disposal, and fixed costs. In this preliminary economic investigation, energy and electrode material costs were taken into account as major cost items in the calculation of the operating cost (US$ m^{-3} of fluoride solution) in the form:

$$\text{Operating cost} = aC_{\text{energy}} + bC_{\text{electrode}} \tag{2.8}$$

where:

C_{energy} and $C_{\text{electrode}}$ (kg Al m^{-3} of fluoride solution) are consumption quantities for fluoride removal, which are obtained experimentally
a and b are given for the Indian market in February 2008, and are as follows:
a is the electrical energy price (0.0065 US$ kWh^{-1})
b is the electrode material price (0.3 US$ kg Al^{-1})

Costs due to electrical energy were calculated from the expression:

$$C_{\text{energy}} = \frac{U \times I \times t_{\text{EC}}}{V} \tag{2.9}$$

where:

U is the cell voltage (V)
I is the current (A)
t_{EC} is the time of electrolysis (s)
V is the volume (m^3) of the fluoride solution

The cost of the electrode was calculated from Faraday's law:

$$C_{\text{electrode}} = \frac{I \times t \times M_{\text{w}}}{z \times F \times V} \tag{2.10}$$

where:

I is the current (A)
t is the time of electrolysis (s)
M_{w} is the molecular mass of aluminum (26.98 g mol^{-1})
z is the number of electrons transferred ($z = 3$)
F is Faraday's constant (96,487 C mol^{-1})
V is the volume (m^3) of the fluoride solution (Daneshvar et al. 2006)

Total operating costs for the treatment of different initial fluoride concentrations using the monopolar and bipolar electrode connections were calculated and are shown in Figure 2.12. It is seen from Figure 2.12 that the operating cost increased with the initial fluoride concentration. This is due to the increase in electrode cost, as dissolution of electrodes increases with initial fluoride

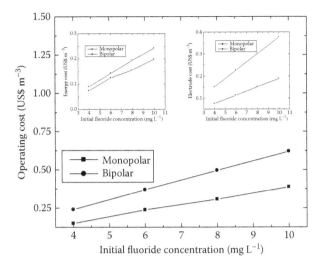

FIGURE 2.12 Cost for treatment of drinking water containing different concentrations of fluoride. Current density: 625 A m^{-2}; interelectrode distance 0.005 m; duration of the experiment: 45 min; temperature: 25°C. Inset: Cost for electrode and energy required for monopolar and bipolar connections.

concentration, as described in the preceding section. Figure 2.12 shows that the theoretical electrode cost varied from 0.075 to 0.12 US\$ m^{-3} and from 0.15 to 0.38 US\$ m^{-3} for the monopolar and bipolar electrode connections, respectively. Similarly, energy cost increased from 0.07 to 0.19 US\$ m^{-3} and from 0.09 to 0.24 US\$ m^{-3} for the monopolar and bipolar connections, respectively, while the initial fluoride concentration varied from 4 to 10 mg L^{-1}. In addition to this, it is necessary to mention that the energy cost was 21% higher for the bipolar connection but the electrode cost dominated energy costs due to the greater dissolution of electrodes compared with the monopolar connection. Figure 2.12 also demonstrates that the operating costs for the treatment of drinking water containing an initial fluoride concentration of 10 mg L^{-1} with the bipolar and monopolar connections under the same operating conditions (current density of 625 A m^{-2}, interelectrode distance of 0.005 m, 45 min of operation, 25°C, and pH 8) were 0.62 and 0.38 US\$ m^{-3}, respectively. The higher operation cost for the bipolar connection may be related to its greater surface area compared with the monopolar one.

2.4.10 CHARACTERIZATION OF THE BY-PRODUCTS OBTAINED FROM THE EC BATH

A whitish precipitate was formed at the bottom of the EC bath at the end of the EC process. The precipitate was removed after filtration using a filter paper (HM2, Indiachem, India) and dried inside a hot-air oven for 3–4 h. It was then ground into a fine powder and prepared for SEM, EDAX, FTIR, and XRD analysis to characterize the by-product obtained from the EC bath.

2.4.11 SCANNING ELECTRON MICROSCOPY AND ENERGY-DISPERSIVE X-RAY

The morphology of the by-products obtained from the EC bath using the bipolar connection is shown in Figure 2.13a. It can be seen from the figure that the by-products formed during the process were whitish and the amounts of aluminum hydroxides and dissolved electrode were much higher than that of the fluoride present in the EC bath. It is seen from Figure 2.13b that the by-products formed during EC were composed of elements such as Al, O, and F. This analysis confirmed that fluoride was captured within the aluminum hydroxide complex and formed sludge.

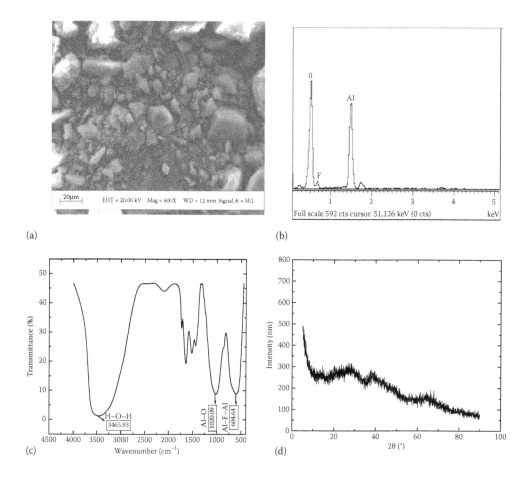

FIGURE 2.13 Characterization of by-products obtained from EC bath. Current density: 625 A m^{-2}; initial fluoride concentration: 10 mg L^{-1}; duration of the experiment: 45 min; temperature: 25°C; electrode connection: bipolar. (a) SEM image, (b) elemental analysis, (c) FTIR analysis, (d) XRD analysis.

2.4.12 Fourier Transform Infrared Spectroscopy

FTIR was performed using potassium bromide pellets and the result is shown in Figure 2.13c. The wave numbers ranged from 4000–450 cm^{-1}. Peaks at 3466, 1020, and 605 cm^{-1} corresponding to H–O–H, Al–O, and Al–F–Al bond stretching, respectively, can be observed in Figure 2.13d. Al–F–Al bond stretching was matched to the analysis made by Gross et al. (2007) for various amorphous trifluoride complexes. From this analysis, it was confirmed that fluoride was linked with aluminum hydroxide complexes and precipitated at the bottom of the EC bath.

2.4.13 Powder X-Ray Diffraction

The XRD of the by-product was recorded by a diffractometer operating with a Cu K α radiation source and in nano-Bragg mode. The XRD analysis result is shown in Figure 2.13d. The 2θ scans were recorded from 5° to 90°. The spectrum of the analyzed by-products showed broad and diffuse peaks. Therefore, the identification of peaks with such broad humps is a well-known confirmed characteristic of phases that are amorphous or poorly crystalline in nature (Gomes et al. 2007).

2.5 SUMMARY OF ELECTROCOAGULATION EXPERIMENTS

EC is solely dependent on the extent of anodic oxidation occurring with an application of a certain voltage to the electrodes connected through an electrical dc source. During EC treatment, the pH value is an important indicator that depicts the overall effect of different ion–ion interactions. Hence, several parameters such as initial fluoride concentration, duration of the experiment, interelectrode distance, electrode connection, current density, and pH value were investigated during EC treatment of contaminated drinking water containing fluoride.

The following are excerpts of the experimental findings:

1. Different concentrations of fluoride in drinking water were treated in an EC bath of 1 L capacity. The highest initial concentration of fluoride to be considered was 10 mg L^{-1}. It was observed that the fluoride concentration decreased from 10 to 1 mg L^{-1} (the WHO limit) in 40 min.

2. The results showed that the efficiency of removal increases with the rise in current density. A current density of 625 A m^{-2} was found to be effective for the efficient removal of fluoride. It was observed that the final fluoride concentrations of 0.76 mg L^{-1}, 0.2 mg L^{-1}, and 6.2 μg L^{-1} in drinking water were attained by EC at an applied current density of 625 A m^{-2} in 45 min, using a monopolar electrode connection.

3. It was observed that, with an increase in interelectrode distance, the efficiency of removal decreased. In terms of their charges from one electrode to another, the mobility of ions carrying currents is mostly dependent on the interelectrode distance. Therefore, EC is also favored when the interelectrode distance decreases. It was observed that the final fluoride concentration decreased from around 3 mg L^{-1} to 0.8 mg L^{-1} when the interelectrode distance decreased from 0.015 to 0.005 m, respectively.

4. The results also confirmed that the bipolar connection can serve the process better than the monopolar connection. However, electrode corrosion, as well as sludge formation, occurred more with the bipolar connection. With the bipolar electrode connection, fluoride concentration was found to have dropped from 10 to 1.7 mg L^{-1} in 45 min. On the other hand, with the monopolar electrode connection, fluoride concentration was found to have declined from 10 to 2.2 mg L^{-1}. The final fluoride concentration of 0.8 mg L^{-1} was recorded after 40 min when using the bipolar connection and at an applied current density of 625 A m^{-2}.

5. The pH of the solution was found to increase insignificantly over time in all cases. During the EC treatment, with the passage of time, a sufficient amount of metallic hydroxide species was generated inside the bath, which compelled the pH to increase over time in all cases (varying from 6.4 to 8.1). It was also observed that for the bipolar connection, the pH of the solution was marginally higher than for the monopolar connection. For example, at a current density of 625 A m^{-2} and an initial fluoride concentration of 10 mg L^{-1}, the pH of the solution always had a value of about 8 at the end of the trials.

6. The operating costs for the monopolar and bipolar connections were 0.38 and 0.62 US\$ m^{-3}, respectively, for the initial fluoride concentration of 10 mg L^{-1}.

7. Finally, the by-products formed inside the EC bath were characterized using SEM, EDAX, FTIR, and XRD, which confirmed the removal of the abovementioned contaminants from drinking water during treatment. It was observed that by-products formed during EC treatment were irregular in shape and size in all cases, which was supported by the XRD data. Elemental analysis, along with the FTIR analysis, clearly supported the credentials of EC using aluminum electrodes in successfully capturing the fluoride from the contaminated drinking water.

These findings indicate that the health risk due to the amount of fluoride in drinking water can be well monitored by EC, and that this investigation may be effective for the future development of an

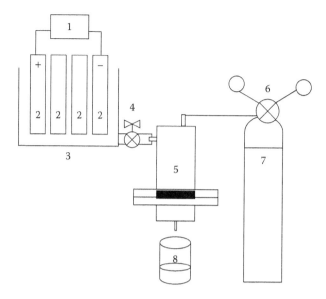

FIGURE 2.14 Schematic diagram of the hybrid process.

economical electrocoagulator. However, the by-products so obtained during EC must be separated using the MF technique, which is elaborated in the next section.

2.5.1 ELECTROCOAGULATION FOLLOWED BY MICROFILTRATION

In this section, the applied aspect of the proposed combination method (EC followed by MF) is verified with a mixture of drinking water containing fluoride. The mixture was first treated in a batch EC process. The treated water, along with the electrocoagulated by-products, was then passed through a MF unit for further treatment. The schematic representation of the hybrid process is shown in Figure 2.14.

2.5.2 EXPERIMENT: ELECTROCOAGULATION FOLLOWED BY MICROFILTRATION

An EC bath (as discussed in the preceding section) with four electrodes made of aluminum sheet, having dimensions of $0.15 \times 0.05 \times 0.002$ m in the bipolar mode of connection, was used for the EC treatment. The working volume of 1 L was considered for the treatment of contaminated water containing fluoride at a concentration of 10 mg L^{-1}. EC was continued for 45 min. Current density and interelectrode distance were maintained at 625 A m^{-2} and 0.005 m, respectively. A membrane cell of 125 mL capacity was used for MF experiments. The outlet of the electrocoagulator was introduced to the membrane cell. The ceramic membrane, prepared by uniaxial methods, sintered at 950°C, was used in the MF cell. The membrane cell was pressurized by N$_2$ gas up to 202 kPa. During the experiments, the pressure was kept constant, and after every 5 min interval, the permeate was collected and the flux (J) was measured using the following equation:

$$J = \frac{Q}{A \times t} \tag{2.11}$$

where:
  Q is the volume of permeate
  A is the effective membrane area
  t is the time of permeation

Permeate conductivity, pH value, and concentration of fluoride were measured to ensure water quality for drinking purposes. At the end of the filtration treatment using the ceramic MF membrane, the retained electrocoagulated sludge was collected from the membrane surface by scrubbing it out and drying it in an air oven at 125°C for 4 h. Thereafter, the dried sludge was ground into powder and made ready for the SEM, EDAX, and XRD analyses to investigate the morphology, qualitative elemental visualization, and phase transformations, respectively. In addition to this, the ceramic MF membrane used for the EC experiments was also dried in the same conditions and was subjected to SEM analysis to visualize the morphological changes that occurred to the membrane surface before and after the treatment due to the blocking of active pores by the electrocoagulated sludge.

2.5.3 Electrocoagulation Followed by Microfiltration: Results and Discussion

2.5.3.1 Particle Size Distribution of Electrocoagulated By-Product

The by-products formed during the EC treatment were analyzed with the help of a particle size analyzer to obtain confirmation about the size of the suspended flocs that were to be separated out of the solution to make it suitable for drinking purposes. Figure 2.15 demonstrates the particle size distribution of the electrocoagulated by-products. From the figure, it can be seen that most of the particles generated during the treatment were in the range of 10 µm with a volume of more than 5%. Some of the larger-sized particles were also formed due to the agglomeration among flocs, so that particle size distribution had two other nodes, one at 100 µm and the other at 1000 µm, with volume percentages of 1.4% and 2%, respectively. This observation undoubtedly supported the agglomeration of particles between the generated electrocoagulants. Comparing the particle size distribution of membranes prepared by the uniaxial method and the particle size distribution of the electrocoagulated by-product, it might be concluded that the MF membrane would be suitable for the separation of the electrocoagulated by-product.

2.5.3.2 Fluoride Removal Performance

This subsection elaborates on the discussion of the performance of the hybrid process for the treatment of drinking water containing fluoride. The performance of EC and MF was analyzed separately in terms of the removal efficiency of fluoride, conductivity, and pH value of the treated

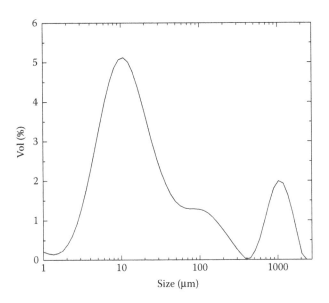

FIGURE 2.15 Particle size distribution of the electrocoagulated by-product before membrane filtration.

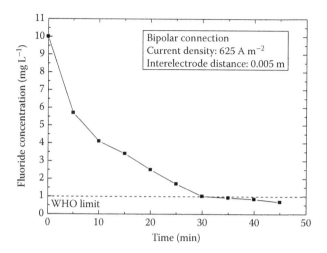

FIGURE 2.16   Concentration of fluoride during the EC process.

water. Figure 2.16 shows the removal performance of fluoride during the EC process in terms of concentrations. It was seen that the concentration decreased with time and the final concentration was observed to be below the recommended upper limit (i.e., 1 mg L^{-1}) 35 min after the experiment. For example, the fluoride concentration decreased from 10 to 0.92 mg L^{-1} after 35 min. In summary, from the figures it may be concluded that the concentrations of fluoride in treated water were below the WHO limit after 35 min of the EC process.

2.5.3.3   Permeate Flux Profile After MF Experiments

The electrocoagulated solution was subjected to MF. Permeate flux and quality were measured and analyzed. Figure 2.17 shows the permeate flux profile during MF. It was observed that the permeate flux declined with time for all the operating pressure differentials. For example, at 101.3 kPa the permeate flux declined from 3.87×10^{-6} to 1.7×10^{-6} m^3 m^{-2} sec^{-1} after 45 min

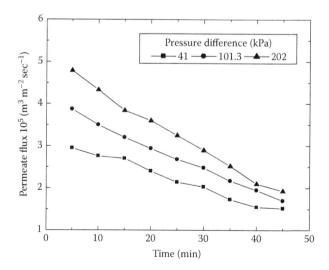

FIGURE 2.17   Decrease in flux of drinking water treated for fluoride contamination by electrocoagulation followed by ceramic microfiltration membrane at different pressures.

TABLE 2.9

Quality of Electrocoagulated Solution and Permeate of MF

Parameters	After Electrocoagulation	After Microfiltration	Drinking Water Specification (Garmes et al. 2002)
pH	8.6	7.8	6.5–8.2
Conductivity (mS cm^{-1})	0.34	0.2	0.2–2.0
TDS (mg L^{-1})	1325	580	500–700
Concentration of fluoride (mg L^{-1})	0.8	0.8	1

of the MF experiment. A similar trend was also noticed for the experiments operating at 41 and 202 kPa. In these cases, the flux declined from 2.95×10^{-5} to 1.52×10^{-5} m^3 m^{-2} sec^{-1} and from 4.79×10^{-6} to 1.93×10^{-6} m^3 m^{-2} sec^{-1} for the transmembrane pressure differentials of 41 and 202 kPa, respectively. It was found that at higher pressure differentials, the drop in flux was greater. The permeate flux was found to have decreased by 48.47% and 56.07% after 45 min of MF at 41 and 101.3 kPa, respectively. The decline in flux with time is because the deposition of suspended particles over the membrane surface restricts the permeate flux by blocking the active pores of the membranes. Over time, the amount of sludge deposition increases, which causes the permeate flux to decrease. Also, the rise in permeate flux with pressure is due to a greater driving force. The quality of the electrocoagulated solution and permeate after MF was measured in terms of conductivity, pH value, total dissolved solids (TDS), and concentrations of fluoride and is shown in Table 2.9. It may be seen that after EC treatment, the final concentration of all the contaminants were at their respective safety limits as recommended by WHO. However, the solution's pH value, TDS, and conductivity were beyond the recommended limits. It was noticed that the pH value was around 8.6, while conductivity and TDS were found to be 0.34 mS cm^{-1} and 1325 mg L^{-1}, respectively, after EC treatment for 45 min. Hence, the treated water could not be used for drinking purposes after EC. Furthermore, the suspended particles generated during the treatment also reflected the unsuitable quality of the treated water for drinking purposes. Therefore, after completion of EC, a further treatment was necessary to retain drinking water quality in terms of pH value, conductivity, and TDS as specified in Table 2.9. In view of this, MF was considered as an alternative in combination with the EC treatment to overcome the situation. It was observed that after the MF treatment, the pH value, conductivity, and TDS were found to be 7.9, 0.2 mS cm^{-1}, and 580 mg L^{-1}, respectively. It was also observed that there were no noticeable changes in the final concentrations of fluoride in the permeate collected after the MF treatment. It was finally confirmed that EC followed by MF for the treatment of water with fluoride contamination could be an effective alternative in the near future to combat the drinking water problem associated with such a type of contamination.

2.5.3.4 Characterization of Membranes and By-Products

2.5.3.4.1 *Characterization of MF Membrane (Before and After Experiment)*

SEM analysis of the MF membranes before and after each experiment was carried out at 2KX magnification, which confirmed the successful retention of the electrocoagulated sludge by the ceramic MF membrane prepared by uniaxial cold pressing at 950°C. Figure 2.18a shows the SEM images of the membrane before use, whose morphological structure consolidated enough to retain particles bigger than the pore size, as can be visualized from the dimensions of the pores. Figure 2.18b clearly shows the morphological view of the used membrane at the same magnification to clarify the successful retention of agglomerated sludge generated during the EC treatment of contaminated drinking water. In addition, it is also clear from the image that almost all the agglomerates were trapped over the membrane surface and, therefore, hardly any active pores were visible.

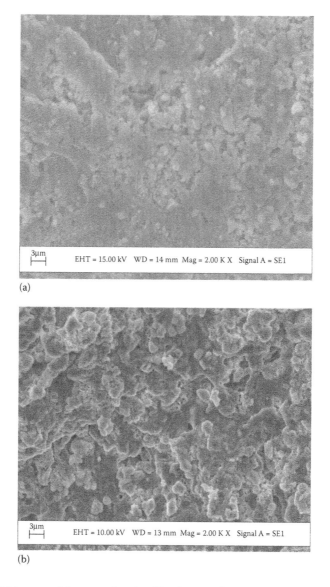

FIGURE 2.18 SEM images of the ceramic microfiltration membrane (a) before and (b) after filtration of electrocoagulated sludge.

2.5.3.4.2   By-Product Characterization

The agglomerated sludge was successfully captured by the ceramic MF membrane and removed from the membrane surface after filtration treatment by scrubbing for its characterization to ensure the efficient removal of the contaminants from the drinking water. The sludge collected from the membrane surface was dried inside a hot-air oven for 3–4 h, mostly to remove the unbounded moisture. After completion of drying, it was observed that several lumps of dried solid were left. These lumps were ground into powder and made ready for the XRD, SEM, and EDAX analyses for their characterization.

XRD analysis showed broad peaks with noise due to the interference of irregular reflection of the reflected monochromatic light beam throughout the prescribed 2θ° range during the analysis, and thus confirmed the amorphous nature of the sludge generated during the EC treatment and retained

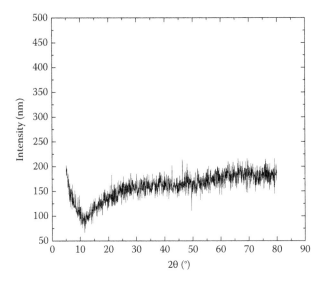

FIGURE 2.19   Characterization images of the electrocoagulated by-product retained by the ceramic micro-filtration membrane by XRD.

FIGURE 2.20   SEM images of the electrocoagulated by-product retained by the ceramic microfiltration membrane.

by the ceramic MF membrane thereafter. Figure 2.19 shows the XRD analysis of the sample. The sample was also subjected to SEM analysis to obtain confirmation of the amorphous nature of the product from its morphological overview. Figure 2.20 supports the amorphous nature, observed during XRD, that no definite shape or size of particles were noticed in the morphological view, even at 2KX magnification, with primary electrons hitting the sample at an energy level of 10 kV without charging the main sample during analysis.

2.5   SUMMARY

Drinking water contaminated with significant amounts of fluoride was treated by EC followed by the MF technique. Aluminum sheet electrodes with a bipolar connection, and a ceramic MF membrane, were used in the EC and membrane filtration processes, respectively. The main findings are as follows:

1. Drinking water contamination caused by the significant presence of fluoride was successfully monitored by bipolar EC for 45 min at 625 A m^{-2} current density and at an interelectrode distance of 0.005 m.
2. The electrocoagulated water solution was unsuitable for drinking purposes unless it was treated further, combining with another technique such as MF. This combination process produced water of drinkable quality with a pH value of 7.9 and conductivity of 0.2 mS cm^{-1}.
3. A ceramic MF membrane was beneficial for the MF treatment of the electrocoagulated water.
4. SEM analysis of the ceramic MF membrane used for the MF treatment of electrocoagulated water confirmed the successful retention of suspended agglomerates generated during EC treatment.
5. XRD and SEM analysis of the by-product removed from the ceramic MF membrane after the treatment clearly supported the successful removal of contaminants from drinking water by EC followed by ceramic MF.

All the abovementioned findings may be useful for the further advancement of the hybrid technique to design a drinking water treatment system in continuous mode.

REFERENCES

Abdessemed, D., and Nezzal, G. 2003. Treatment of primary effluent by coagulation-adsorption-ultrafiltration for reuse. *Desalination.* 152:367–373.

Abuzaid, N. S., Bukhari, A. A., and Al-Hamouz, Z. M. 1998. Removal of bentonite causing turbidity by electrocoagulation. *J. Env. Sci. Health Part A.* 33:1341–1358.

Almandoz, M. C., Marchese, J., Prádanos, P., Palacio, L., and Hernández, A. 2004. Preparation and characterization of non-supported micro-filtration membranes from aluminosilicates. *J. Membr. Sci.* 241:95–103.

Alverez-Gallegos, A., and Pletcher, D. 1999. The removal of low level organics via hydrogen peroxide formed in a reticulated vitreous carbon cathode cell. Part 2: The removal of phenols and related compounds from aqueous effluents. *Electrochimica Acta* 44:2483–2492.

Amor, Z., Bariou, B., Mameri, N., Taky, M., Nicolas, S., and Elmidaoui, A. 2001. Fluoride removal from brackish water by electrodialysis. *Desalination.* 133:215–223.

Anwar, F. 2003. Assessment and analysis of industrial liquid waste and sludge disposal at unlined landfill sites in an arid climate. *Waste Manage.* 23:817–824.

Aouni, A., Fersi, C., Ali, M. B. S., and Dhahbi, M. 2009. Treatment of textile wastewater by a hybrid electro-coagulation/nanofiltration process. *J. Hazard. Mater.* 168:868–874.

Arora, M., Maheshwari, R. C., Jain, S. K., and Gupta, A. 2004. Use of membrane technique for potable water production. *Desalination.* 170:105–112.

Baklan, V. Y., and Kolesnikova, I. P. 1996. Influence of electrode material on the electrocoagulation. *J. Aerosol. Sci.* 27:S209–S210.

Balmer, L. M., and Foulds, A. W. 1986. Separating oil from oil-in-water emulsions by electroflocculation/electroflotation. *Filtrat. Sep.* 23:366–370.

Barba, D., Caputi, P., and Cifoni, D. 1997. Drinking water supply in Italy. *Desalination.* 113:111–117.

Baudin, I., Chevalier, M. R., Anselme, C., Cornu, S., and Laîné, J. M. 1997. L'Apié and Vigneux case studies: First months of operation. *Desalination.* 113:273–275.

Bell, M. C., and Ludwig, T. G. 1970. The supply of fluoride to man: Ingestion from water. In *Fluorides and Human Health*, WHO Monograph Series 59. Geneva, Switzerland: World Health Organization.

Belongia, B. M., Haworth, P. D., Baygents, J. C., and Raghavan, S. 1999. Treatment of alumina and silica chemical mechanical polishing waste by electrodecantation and electrocoagulation. *J. Electrochem. Soc.* 146:4124–4130.

Belouatek, A., Benderdouche, N., Addou, A., Ouagued, A., and Bettahar, N. 2005. Preparation of inorganic supports for liquid waste treatment. *Micropor. Mesopor. Mater.* 85:163–168.

Bouzerara, F., Harabi, A., Achour, S., and Larbot, A. 2006. Porous ceramic supports for membranes prepared from kaolin and doloma mixtures. *J. Eur. Ceram. Soc.* 26:1663–1671.

Choubisa, S. L., and Sompura, K. 1996. Dental fluorosis in tribal villages of Dungerpur district (Rajasthan). *Poll. Res.* 15:45–47.

Daneshvar, N., Oladegaragoze, A., and Djafarzadeh, N. 2006. Decolorization of basic dye solutions by electrocoagulation: An investigation of the effect of operational parameters. *J. Hazard. Mater.* 129:116–122.

Das, R., and Dutta, B. K. 1999. Permeation and separation characteristics of supported alumina and titania membranes. *Sep. Sci. Technol.* 34:609–625.

DeFriend, K. A., Wiesner, M. R., and Barron, A. R. 2003. Alumina and aluminate ultra-filtration membranes derived from alumina nanoparticles. *J. Membr. Sci.* 224:11–28.

Den, W., and Wang, C. J. 2008. Removal of silica from brackish water by electrocoagulation pretreatment to prevent fouling of reverse osmosis membranes. *Sep. Puri. Technol.* 59:318–325.

Dhale, A. D., and Mahajani, V. V. 2000. Studies on treatment of disperse dye waste: Membrane-wet oxidation process. *Waste Manage.* 20:85–92.

Donini, J. C., Kan, J., Szynkarczuk, J., Hassan, T. A., and Kar, K. L. 1994. Operating cost of electrocoagulation. *Can. J. Chem. Eng.* 72:1007–1012.

Duan, J., and Gregory, J. 2003. Coagulation by hydrolyzing metal salts. *Adv. Colloid Interf. Sci.* 10:475–502.

Emamjomeh, M. M., and Sivakumar, M. 2006. An empirical model for defluoridation by batch monopolar electrocoagulation/flotation (ECF) process. *J. Hazard. Mater.* 131:118–125.

Escobar, I. C., Hong, S., and Randall, A. A. 2000. Removal of assimilable and biodegradable dissolved organic carbon by reverse osmosis and nanofiltration membranes. *J. Membr. Sci.* 175:1–17.

Falamaki, C., Afarani, M. S., and Aghaie, A. 2004. Initial sintering stage pore growth mechanism applied to the manufacture of ceramic membrane supports. *J. Eur. Ceram. Soc.* 24:2285–2292.

Garg, V. K., and Malik, A. 2004. Ground water quality in some villages of Haryana, India: Focus on fluoride and fluorosis. *J. Hazard. Mater. B.* 106:85–97.

Garmes, H., Persin, F., Sandeaux, J., Pourcelly, G., and Mountadar, M. 2002. Defluoridation of groundwater by a hybrid process combining adsorption and Donnan dialysis. *Desalination.* 145:287–291.

Ghoreishi, S. M., and Haghighi, R. 2003. Chemical catalytic reaction and biological oxidation for treatment of non-biodegradable textile effluent. *Chem. Eng. J.* 95:163–169.

Ghosh, D., Medhi, C. R., and Purkait, M. K. 2008. Treatment of fluoride contaminated drinking water by electrocoagulation using monopolar and bipolar electrode connection. *Chemosphere.* 73:1393–1400.

Golder, A. K., Hridaya, N., Samanta, A. N., and Ray, S. 2005. Electrocoagulation of methylene blue and eosin yellowish using mild steel electrodes. *J. Hazard. Mater.* 127:134–140.

Gomes, J. A., Daida, P., Kesmez, M., Weir, M., Moreno, H., Parga, J. R., Irwin, G., et al. 2007. Arsenic removal by electrocoagulation using combined Al–Fe electrode system and characterization of products. *J. Hazard. Mater.* 139:220–231.

Gross, U., Rüdiger, S., Kemnitz, E., Brzezinka, K. W., Mukhopadhyay, S., Bailey, C., Wander, A., and Harrison, N. 2007. Vibrational analysis study of aluminum trifluoride phases. *J. Phys. Chem. A.* 111:5813–5819.

Gupta, V. K., Ali, I., and Saini, V. K. 2007. Defluoridation of waste waters using waste carbon slurry. *Water Res.* 41:3307–3316.

Hichour, M., Persin, F., Sandeaux, J., and Gavach, C. 1999. Fluoride removal from waters by Donnan dialysis. *Sep. Puri. Technol.* 18:1–11.

Hörsch, P., Gorenflo, A., Fuder, C., Deleage, A., and Frimmel, F. H. 2005. Biofouling of ultra- and nanofiltration membranes for drinking water treatment characterized by fluorescence *in situ* hybridization (FISH). *Desalination.* 172:41–52.

Hsieh, H. P. 1996. *Inorganic Membranes for Separation and Reaction.* Membrane Science and Technology Series, 3. Amsterdam: Elsevier.

Hu, K., and Dickson, J. M. 2006. Nanofiltration membrane performance on fluoride removal from water. *J. Membr. Sci.* 279:529–538.

Kashefialasl, M., Khosravi, M., Marandi, R., and Seyyedi, K. 2006. Treatment of dye solution containing colored index acid yellow 36 by electrocoagulation using iron electrodes. *Int. J. Environ. Sci. Tech.* 2:365–371.

Kobya, M., Senturk, E., and Bayramoglu, M. 2006. Treatment of poultry slaughterhouse waste waters by electrocoagulation. *J. Hazard. Mater.* 123:172–176.

Kumar, P. R., Chaudhari, S., Khilar, K. C., and Mahajan, S. P. 2004. Removal of arsenic from water by electrocoagulation. *Chemosphere.* 55:1245–1252.

Letterman, R. D., Amirtharajah, A., and O'Melia, C.R. 1999. Coagulation and flocculation. In eds. R. D. Letterman and American Water Works Association, *Water Quality and Treatment: A Handbook of Community Water Supplies*, Chapter 8. New York: McGraw-Hill.

Li, Q., and Elimelech, M. 2006. Synergistic effects in combined fouling of a loose nanofiltration membrane by colloidal materials and natural organic matter. *J. Membr. Sci.* 278:72–82.

Lin, S. H., and Wang, C. S. 2002. Treatment of high-strength phenolic wastewater by a new two-step method. *J. Hazard. Mater.* 90:205–216.

Linares-Hernández, I., Barrera-Díaz, C., Roa-Morales, G., Bilyeu, B., and Ureña-Núñez, F. 2007. A combined electrocoagulation–sorption process applied to mixed industrial wastewater. *J. Hazard. Mater.* 144:240–248.

Liu, A., Ming, J., and Ankumah, R. O. 2005. Nitrate contamination in private wells in rural Alabama, United States. *Sci. Tot. Environ.* 346:112–120.

Luque, S., Gómez, D., and Álvarez, J. R. 2008. Industrial applications of porous ceramic membranes (pressure-driven processes). *Membr. Sci. Technol.* 13:177–216.

Mameri, N., Lounici, H., Belhocine, D., Grib, H., Piron, D. L., and Yahiat, Y. 2001. Defluoridation of Sahara water by small plant electrocoagulation using bipolar aluminum electrodes. *Sep. Puri. Technol.* 24:113–119.

Mameri, N., Yeddou, A. R., Lounici, H., Belhocine, D., Grib, H., and Bariou, B. 1998. Defluoridation of septentrional Sahara water of North Africa by electrocoagulation process using bipolar aluminum electrodes. *Water Res.* 32:1604–1610.

Masmoudi, S., Larbot, A., El Feki, H., and Amar, R. B. 2007. Elaboration and characterisation of apatite based mineral supports for microfiltration and ultrafiltration membranes. *Cer. Inter.* 33:337–344.

Matteson, M. J., Dobson, R. L., Glenn, R. W., Kukunoor, N. S., Waits, W. H., and Clayfield, E. J. 1995. Electrocoagulation and separation of aqueous suspensions of ultrafine particles. *Colloid Surf. A.* 104:101–109.

Meier, J., Melin, T., and Eilers, L. H. 2002. Nanofiltration and adsorption on powdered adsorbent as process combination for the treatment of severely contaminated waste water. *Desalination.* 146:361–366.

Métivier-Pignon, H., Faur-Brasquet, C., and Le Cloirec, P. 2003. Adsorption of dyes onto activated carbon cloths: Approach of adsorption mechanisms and coupling of ACC with ultrafiltration to treat coloured wastewaters. *Sep. Puri. Technol.* 31:3–11.

Mulligan, C. N., Yong, R. N., and Gibbs, B. F. 2001. Remediation technologies for metal contaminated soils and groundwater: An evaluation. *Eng. Geol.* 60:193–200.

Nandi, B. K., and Patel, S. 2013. Removal of pararosaniline hydrochloride dye (basic red 9) from aqueous solution by electrocoagulation: Experimental, kinetics, and modeling. *J. Disper. Sci. Tech.* 34:1713–1724.

Nandi, B. K., and Patel, S. 2014. Removal of brilliant green from aqueous solution by electrocoagulation using aluminum electrodes: Experimental, kinetics, and modeling, *Sep. Sci. Tech.* 49:1–12.

Petersen, R. J. 1993. Composite reverse osmosis and nanofiltration membranes. *J. Membr. Sci.* 83:81–150.

Potdar, A., Shukla, A., and Kumar, A. 2002. Effect of gas phase modification of analcime zeolite composite membrane on separation of surfactant by ultra-filtration. *J. Membr. Sci.* 210:209–225.

Potts, D. E., Ahlert, R. C., and Wang, S. S. 1981. A critical review of fouling of reverse osmosis membranes. *Desalination.* 36:235–264.

Pouet, M. F., and Grasmick, A. 1994. Electrocoagulation and flotation: Applications in cross flow microfiltration. *Filtrat. Sep.* 31:269–272.

Pouet, M. F., and Grasmick, A. 1995. Urban wastewater treatment by electrocoagulation and flotation. *Water Sci. Technol.* 31:275–283.

Rubach, S., and Saur, I. F. 1997. Onshore testing of produced water by electroflocculation. *Filtrat. Sep.* 34:877–882.

Saffaj, N., Persin, M., Younsi, S. A., Albizane, A., Cretin, M., and Larbot, A. 2006. Elaboration and characterization of micro-filtration and ultra-filtration membranes deposited on raw support prepared from natural Moroccan clay: Application to filtration of solution containing dyes and salts. *Appl. Clay Sci.* 31:110–119.

Saffaj, N., Younssi, S. A., Albizane, A., Messouadi, Bouhria, M., Persin, M., Cretin, M., and Larbot, A. 2004. Preparation and characterization of ultra-filtration membranes for toxic removal from wastewater. *Desalination.* 168:259–263.

Sengupta, S. R., and Pal, B. 1971. Iodine and fluoride content of foodstuffs. *Ind. J. Nutr. Dicter.* 8:66–71.

Shortt, H. E., Pandit, C. G., and Raghavachari, T. N. S. 1937. Endemic fluorosis in the Nellore district of South India. *Ind. Med. Gazette.* 72:396–8.

Singh, R., and Maheshwari, R. C. 2001. Defluoridation of drinking water: A review. *Ind. J. Environ. Protec.* 21:983–991.

Sourirajan, S. 1970. *Reverse Osmosis.* London: Logos Press.

Speth, T. F., Gusses, A. M., and Summers, R. S. 2000. Evaluation of nanofiltration pretreatments for flux loss control. *Desalination.* 130:31–44.

Srimurali, M., Pragathi, A., and Karthikeyan, J. 1998. A study on removal of fluorides from drinking water by adsorption onto low-cost materials. *Env. Pollution.* 99:285–289.

Sujana, M. G., Thakur, R. S., and Rao, S. B. 1998. Removal of fluoride from aqueous solution by using alum sludge. *J. Colloid Interf. Sci.* 206:94–101.

Susheela, A. K. 1999. Fluorosis management programme in India. *Curr. Sci.* 77:1250–1256.

Tor, A. 2007. Removal of fluoride from water using anion-exchange membrane under Donnan dialysis condition. *J. Hazard. Mater.* 141:814–818.

U.S. Public Health Service Drinking Water Standards (USPHS). 1962. *US Government Printing Office, Department of Health Education and Welfare.* Washington, DC: National Service Center for Environmental Publications.

Van der Bruggen, B., Kim, J. H., DiGiano, F. A., Geens, J., and Vandecasteele, C. 2004. Influence of MF pretreatment on NF performance for aqueous solutions containing particles and an organic foulant. *Sep. Puri. Technol.* 36:203–213.

Van Gestel, T., Vandecasteele, C., Buekenhoudt, A., Dotremont, C., Luyten, J., Leysen, R., Van der Bruggen, B., and Maes, G. 2002. Alumina and titania multilayer membranes for nanofiltration: preparation, characterization and chemical stability. *J. Membr. Sci.* 207:73–89.

Vik, E. A., Carlson, D. A., Eikum, A. S., and Gjessing, E. T. 1984. Electrocoagulation of potable water. *Water Res.* 18:1355–1360.

Vivanpatarakij, S., Laosiripojana, N., Arpornwichanop, A., and Assabumrungrat, S. 2009. Performance improvement of solid oxide fuel cell system using palladium membrane reactor with different operation modes. *Chem. Eng. J.* 146:112–119.

Wang, Y. H., Tian, T. F., Liu, X. Q., and Meng, G. Y. 2006. Titania membrane preparation with chemical stability for very hash environments applications. *J. Membr. Sci.* 280:261–269.

Weber, W. J. 1972. *Physicochemical Processes for Water Quality Control.* New York: Wiley-Interscience.

Workneh, S., and Shukla, A. 2008. Synthesis of sodalite octahydrate zeolite-clay composite membrane and its use in separation of SDS. *J. Membr. Sci.* 309:189–195.

World Health Organization (WHO). 1984. *Guidelines for Drinking Water Quality (Vol. II): Health Criteria and Supporting Information.* Geneva, Switzerland: World Health Organization.

Woytowich, D. L., Dalrymple, C. W., Gilmore, F. W., and Britton, M. G. 1993. Electrocoagulation (CURE) treatment of ship bilge water for the U.S. coast guards in Alaska. *Mar. Technol. Soc. J.* 27:62–67.

Wu, X., Zhang, Y., Dou, X., and Yang, M. 2007. Fluoride removal performance of a novel Fe-Al-Ce trimetal oxide adsorbent. *Chemosphere.* 69:1758–1764.

Yildiz, Y. Ş., Koparal, A. S., İrdemez, Ş., and Keskinler, B. 2007. Electrocoagulation of synthetically prepared waters containing high concentration of NOM using iron cast electrodes. *J. Hazard. Mater.* 139:373–380.

Yoshino, Y., Suzuki, T., Nair, B. N., Taguchi, H., and Itoh, N. 2005. Development of tubular substrates, silica based membranes and membrane modules for hydrogen separation at high temperature. *J. Membr. Sci.* 267:8–17.

Zhu, J., Zhao, H., and Ni, J. 2007. Fluoride distribution in electrocoagulation defluoridation process. *Sep. Puri. Technol.* 56:184–191.

Zuo, Q., Chen, X., Li, W., and Chen, G. 2008. Combined electrocoagulation and electroflotation for removal of fluoride from drinking water. *J. Hazard. Mater.* 159:452–457.

3 Electrooxidation Processes for Dye Degradation and Colored Wastewater Treatment

Farshid Ghanbari and Mahsa Moradi

CONTENTS

ABSTRACT

Dye-consuming industries, especially textile industries, are among the prime consumers of water, by which a huge proportion of the aquatic environment has been compromised due to the discharge of dyes, which are, most often, persistent and toxic. Conventional physico-chemical processes cannot completely remove dyes from water and wastewater due to their complex structures. Nowadays, electrooxidation (EO) processes are proposed for the degradation of synthetic dyes. Employing electrons as the major reagent, EO processes are well known for being among clean processes. The advantages of electrochemical processes (ECPs) include environmental compatibility, versatility, amenability of automation, and high energy efficiency. These processes are classified into two main categories. First, direct EO, in which hydroxyl radicals are produced on the anode surface. The hydroxyl radical generated is a powerful oxidant that degrades dye structure. Second, indirect EO, where either oxygen- or chlorine-based oxidizing agents are electrochemically produced in the solution. Indirect EO is also divided into two processes, involving the production of an oxidizing agent (H_2O_2) at the cathode and the generation of active chlorine species at the anode. The electrogenerated H_2O_2 in the presence of a transitional metal (Fe^{2+}) can produce the hydroxyl radicals in a process called *electro-Fenton*. This chapter presents (1) direct EO theory, principles, and various studies of dye degradation by direct EO; (2) applications of indirect EO in different conditions; and (3) advances and challenges of electrochemical oxidation in colored wastewater treatment.

Keywords: Electrochemical oxidation, Dye, Textile wastewater, Electro-Fenton, Chlorine-mediated electrooxidation

3.1 INTRODUCTION

An urban population demands high quantities of raw materials, water, and energy, some of which turns into environmental pollution. Water is a vital substance for all creatures and its quality is crucial for the future of humankind. One of the main problems confronted by the human race is to provide clean water to the vast majority of the population around the world (Mollah et al. 2001). When consumers make use of water, it is disposed of as wastewater, most often containing a variety of substances and pollutants. Indeed, the "tragedy of the commons" arises when significant amounts of pollutants are constantly and sometimes ignorantly released into the environment, which should be a safe and sound place for all creatures, imposing extremely devastating effects and problems (Feeny et al. 1990). The use of water by humans in several industrial activities increases the amount of dissolved chemicals in water, consequently reducing water quality. Water polluted with chemicals threatens public health and the environment and increases the costs of water remediation. Among chemicals, dyes have long been utilized in dyeing, paper and pulp, textiles, plastics, leather, cosmetics, and food industries. Dyes are compounds that absorb light within the visible range, between 380 and 700 nm. They are very detectable contaminants, because of their intense color in water bodies (dos Santos et al. 2007). Dyes have complex aromatic molecular structures, making them more stable and resistant to microbial degradation (Yavuz et al. 2012). Dyes are composed of a group of atoms responsible for their color, which is called the *chromophore*, as well as an electron-withdrawing or donating substituent that influences the color of the chromophore, called the *auxochrome* (dos Santos et al. 2007). Dyes are classified based on their chemical structures, physical properties, and applications in various processes. The most important chromophores are azo (–N=N–), carbonyl (–C=O), nitro (–NO$_2$), and quinoid groups. Moreover, the major classification of synthetic dyes normally used includes anthraquinone, indigoide, triphenylmethyl, xanthene, azo derivatives, and so on (Panakoulias et al. 2010; Yavuz et al. 2012). Table 3.1 shows the structures of various dyes with some of their properties. Dyes can also be classified according to their usage, such as reactive, disperse, direct, vat, sulfur, cationic, acid, mordant, ingrain, and solvent dyes that are listed in the color index (C.I.) by the Society of Dyers and Colourists and the American Association of Textile Chemists and Colorists. The C.I. name of a dye indicates how it is used with materials and its hue, and its number specifies the chronological order of its commercial introduction (Salleh et al. 2011; Hao et al. 2000; O'Neill et al. 1999).

Since the amount of dye production in the world is not determined, the amount of discharge of residual dyes is also unknown. However, it is reported that over 700,000 t of dyes are produced per year (Hai et al. 2007). In the dyeing process, the degree of fixation of dyes in matter (especially fibers) is never complete; therefore, about 10%–15% of the dyes are lost during the dyeing process, and enter wastewater streams (Moradi et al. 2016). Hence, a large volume of wastewater containing dye is discharged into the waters that receive it. Most dyes are visible in water at even 1 mg/L concentration. Besides the aesthetic aspect and consumers' concern, dyes reduce sunlight emission into the aquatic environment, disturbing aquatic plant life (Ghanbari et al. 2014b; Rodriguez et al. 2009). The color appearing in water bodies stemming from the arrival of dyes can also interfere with the ultraviolet (UV) disinfection processes in wastewater treatment. Moreover, they consume dissolved oxygen in water bodies, leading to the domination of anaerobic conditions, resulting in the death of fish and producing toxic smells (Sun et al. 2011; Petrucci et al. 2015). Dye precursors, including halogenated, methylated, or nitrated anilines and phenols, are listed as priority toxic pollutants. As dyes are synthesized to be stable in the cases of their chemical and photochemical properties, they are highly persistent in the environment, remaining unchanged for long periods of time (Hamza et al. 2009).

Among different groups of dyes, azo dyes are the most common synthetic dyes that are widely used in the textile industry. In fact, azo dyes make up 60%–70% of all dyes. Azo dyes contain xenobiotic substitutions, including azo, sulfonic acid, nitro, chloro, and bromo functional groups (Petrucci et al. 2015; Ramírez et al. 2013). The main concern related to azo dyes stems from the

TABLE 3.1

Introduction to Some Synthetic Dyes along with Their Properties

Name	Formula	Molecular Weight (g/mol)	CAS No.	Type of Dye	Structure
Reactive Black 5; Remazol Black B	$C_{26}H_{21}N_5Na_4O_{19}S_6$	991.81	17095-24-8	Azo	
Acid Yellow 36; metanil yellow; Acid Gold Yellow G	$C_{18}H_{14}N_3NaO_3S$	375.38	587-98-4	Azo	
Acid red 1; Amido Naphthol Red G; azophloxine	$C_{18}H_{13}N_3Na_2O_8S_2$	509.42	3734-67-6	Azo	

(Continued)

TABLE 3.1 (CONTINUED)
Introduction to Some Synthetic Dyes along with Their Properties

Name	Formula	Molecular Weight (g/mol)	CAS No.	Type of Dye	Structure
Orange G; Acid Orange 10; Wool Orange 2G	$C_{16}H_{10}N_2Na_2O_7S_2$	452.37	1936-15-8	Azo	
MO; Acid Orange 52; helianthin	$C_{14}H_{14}N_3NaO_3S$	327.33	547-58-0	Azo	
Rhodamine B; Basic Violet 10; Brilliant Pink B	$C_{28}H_{31}ClN_2O_3$	479.01	81-88-9	Xanthene	
Malachite green; Basic Green 4	$C_{23}H_{25}ClN_2$	364.91	569-64-2	Triphenylmethane	

Dye name	Molecular formula	Molecular weight	CAS number	Dye class	Structure
Crystal violet; Methyl Violet 10B; Basic Violet 3	$C_{25}H_{30}N_3Cl$	407.98	548-62-9	Triphenylmethane	
Indigo carmine; Acid Blue 74; amacid brilliant blue	$C_{16}H_8N_2Na_2O_8S_2$	466.35	860-22-0	Indigoid	
Remazol Brilliant Blue R; Reactive Blue 19	$C_{22}H_{16}N_2Na_2O_{11}S_3$	626.54	2580-78-1	Anthraquinone	
Alizarin Red S; acid mordant red; Mordant Red 3	$C_{14}H_7NaO_7S$	342.26	130-22-3	Anthraquinone	

types of precursors used in their synthesis, which appear only after the reduction and cleavage of the azo bond to aromatic amines, which are toxic and carcinogenic to humans (Jaafarzadeh et al. 2015; Pinheiro et al. 2004). With this background, it is clear why scientists and researchers have been extensively studying dyes and colored effluents. It should also be noted that the most common pollutants used in most studies are dyes, as simple models of organic pollutants.

Synthetic dyes are widely used in the textile industry, and unused dye remains in water after processing. Textile industries consume large amounts of water, organic chemicals, and salts for wet processing activities, such as dyeing and finishing of products. The average water consumption in the textile industry for the production of 1 kg final product is 60–400 L (Ali et al. 2009). Moreover, depending on the fabric's weight, the chemicals used are from 10% to over 100% (Asghar et al. 2015). Hence, billions of liters of wastewater from the world's textile plants are released every day. These wastewaters can contain up to 20 g of dyes, their intermediates, and salts per liter, which can cause health and environmental impacts (Singh et al. 1998). Due to the variation of products and various stages and processes undergone in a textile plant, textile wastewater is a multicomponent wastewater containing various types of pollutants. Therefore, such wastewater is characterized by strong color, high amounts of total dissolved solids (TDS), high chemical oxygen demand (COD), highly fluctuating pH (2–12), and low biodegradability, with a 5-day biochemical oxygen demand/COD of less than 0.25 (Eslami et al. 2013; Ramírez et al. 2013). The presence of synthetic dyes leads to the recalcitrance of textile effluents (Aquino et al. 2014). Within the textile industry, frequent changes of dyes used in the dyeing process induce remarkable variation in wastewater characteristics, especially in parameters such as pH, color, and COD. With this background, there are three points of importance in the case of textiles, including textile production, application, and disposal. There have been remarkable improvements within the first two sectors, while the latter is probably the least improved. In fact, the need for color-free effluent due to public concern, along with strict regulatory requirements, render dye wastewater treatment labored, considerable, and costly. Different qualities of textile wastewaters are documented in the literature. Table 3.2 illustrates some important characteristics of real textile wastewater studied in various countries being treated by the electrochemical processes.

There are various methods for treating textile wastewater and water polluted with dyes. These methods are divided into three main categories, involving biological, chemical, and physical processes (Forgacs et al. 2004; Hao et al. 2000). Biological processes include aerobic and anaerobic treatments that are economical and environmentally friendly. All researchers attempt to take advantage of biological methods. Biological treatments/processes need high reaction times, requiring huge reactors and large areas. These processes are ineffective in the treatment of dyes, which have complicated structures and low biodegradability. Moreover, biological processes cannot completely treat colored effluent due to the toxicity of dyes to the microorganisms employed in these processes. Hence, effluent treated by biological processes cannot meet discharge standards in terms of color and COD (Rocha et al. 2014). Chemical processes include various processes such as adsorption, coagulation, and advanced oxidation processes (AOPs). Typically, AOPs can be widely defined as aqueous phase oxidation techniques intermediated with highly reactive species, such as hydroxyl radicals, in the processes that result in destruction of the dyes. AOPs include the Fenton process, ozonation, photocatalysis, and wet air oxidation, which have been studied to degrade colored wastewater. With hydroxyl radicals as the main oxidative agents, complete mineralization of pollutants to carbon dioxide, water, and inorganic compounds, or at least a conversion of the organics into highly oxidized, more innocuous products, can be obtained (Asghar et al. 2015; Wang et al. 2011). Physical processes (membrane technologies) have rarely been considered for the removal of dyes and the treatment of textile wastewater, due to the high cost and potential of membrane clogging. The latter issue necessitates the cleaning and substitution of membrane modules, imposing high costs. The advantages and limitations of some processes are presented in Table 3.3. Nowadays, electrochemical oxidation opens a new perspective on the field of wastewater treatment, especially the treatment of refractory organic compounds such as dyes.

TABLE 3.2

Some Characteristics of Real Textile Wastewater in Various Countries

Parameters	Brazil (Martínez-Huitle et al. 2012)	Greece (Tsantaki et al. 2012)	Iran (Ghanbari et al. 2015)	Korea (Kim et al. 2003)	China (Lin et al. 1994)	Brazil (Sales Solano et al. 2013)	India (Bhaskar Raju et al. 2009)	Taiwan (Wang et al. 2010)	France (Zongo et al. 2009)	Taiwan (Wang et al. 2009)	India (Bhatnagar et al. 2014)
COD (mg/L)	650	470	1310	870	460–1500	1018	530.7	1224	1787	1354	347
BOD (mg/L)	N/A	N/A	180	N/A	100–500	N/A	175	323.8	N/A	N/A	N/A
pH	10.2	8.8	6.5	11.0	9–10	12.4	8.04	4.8	7.06	8.84	8.7
Conductivity (mS/cm)	2.70	7.2	1.26	2.63	2.1–2.9	5.9	16.3	2.91	2.8	1.93	12
Color	1204 (Hazen unit)	N/A	2024 ADMI	1340 (Pt. Co)	N/A	7504 (Hazen unit)	Dirty green	N/A	N/A	2175 ADMI	1100 (Pt.Co)
Total organic carbon (TOC) (mg/L)	N/A	120	N/A	N/A	N/A	N/A	N/A	394	N/A	N/A	109

TABLE 3.3

Advantages and Limitations of Some Processes Used for the Treatment of Dye and Colored Wastewater

Process	Advantages	Limitations
Adsorption	Efficient removal of various dyes, simple operation	High costs due to the regeneration of adsorbents; rapid saturation of adsorbents
Coagulation	Effective separation of pollutants, simple operation	High production of sludge
Fenton process	High efficiency in removal of soluble and insoluble dyes; Fenton's reagent is relatively inexpensive	Production of sludge, operation in acidic conditions
Photocatalysis	No sludge production	Inefficient in degradation of dyes in high concentrations, high energy costs
NaOCl	Inexpensive, good performance for azo dyes degradation	Release of chloroaromatic compounds
Ozonation	Applied in gaseous state, no change in volume	Short half-life, production of aldehydes as by-products
Membrane	Removal of all dyes, producing high-quality water suitable for reuse	Formation of concentrated sludge, high pressure requirement, high cost
Biological process	Publicly acceptable treatment, cost benefit	Slow reaction, needs nutrition and maintenance, need for huge reactors, not efficient in degradation of dyes with toxicity and complicated molecular structures
Sulfate radical-based processes	Efficient, application in wide range of pH, persulfate and peroxymonosulfate oxidants are relatively inexpensive	Sulfate ion as by-product increases TDS of effluent, need to an activation agent
Sonolysis	Effective in laboratory scale	Requires high energy
Gamma radiation	Effective in laboratory scale	Requires sufficient quantity of dissolved oxygen, high energy consumption

Source: Nidheesh, P. et al., *Environmental Science and Pollution Research*, 20, 2099–132, 2013a; Salleh, M. A. M. et al., *Desalination*, 280, 1–13, 2011; Singh, K. et al., *Critical Reviews in Environmental Science and Technology*, 41, 807–78, 2011.

3.1.1 ELECTROCHEMICAL PROCESSES

Electrochemistry is a branch of physicochemistry whose importance is well recognized within all sciences and technologies. It deals with the charge transfer at the interface of an electrode (a semiconductor or conductor) and the electrolyte (ionic conductor). In fact, electrochemistry either uses electrical energy to bring about a chemical reaction or generates electrical energy by means of chemical action. Electrochemical processes (ECPs) include two components: reduction, which occurs at the cathode (negative electrode), and oxidation, which takes place at the anode (positive electrode) (Grimm et al. 1998).

In the last two decades, electrochemistry science has been utilized for the remediation of polluted water and wastewater. In environmental remediation, ECPs employ electrical energy to remove pollutants through chemical reactions. In fact, electron transfer is mainly responsible for removing pollutants in electrochemical reactors. Therefore, in most cases, no addition of chemicals is required for the remediation of pollutants, thus introducing the electrochemical process as a green technology. ECPs for water and wastewater treatments have many innate advantages that include:

1. Low or moderate capital cost due to operation in ambient temperature and atmospheric pressure

2. Ease of automation
3. Simple control of the reaction rate by adjusting the electrical current
4. Environmental compatibility
5. Process safety regarding operation in mild conditions (Chen 2004; Rajeshwar et al. 1994; Ghanbari et al. 2014a; Walsh 2001)

Conventional ECPs in environmental remediation are: electrocoagulation, electroreduction, electroflotation, and electrooxidation (EO). The electrocoagulation and electroflotation processes are based on the separation mechanism. In other words, dyes are concentrated in the solid phase as sludge. The electroreduction process has rarely been considered for degradation of dyes and colored effluents, because of its low efficiency and persistence of dyes in electrochemical reduction (Chen 2004; Brillas et al. 2015). EO or electrochemical oxidation processes involve methods in which one or more oxidizing agents are produced, either at the surface area of the electrode or within the bulk solution, decomposing dyes and organic pollutants to more oxidizing organics or even mineral compounds. EO processes are able to oxidize or decompose organic compounds, including dyes, through capture of one or more electrons from an organic pollutant (Anglada et al. 2009; Martínez-Huitle et al. 2006).

In this chapter, EO processes, along with configurations and mechanisms, are comprehensively discussed.

3.2 ELECTROCHEMICAL OXIDATION OR ELECTROOXIDATION

Regarding high efficiency and variety in procedures, electrochemical oxidation is presumably the most applicable electrochemical method for the degradation of dyes and the treatment of textile wastewater. In this method, an oxidizing agent is produced on the anode surface or bulk solution that is responsible for the degradation of dyes (Sirés et al. 2014; Ghanbari et al. 2016; Rodriguez et al. 2009).

Two main approaches of EO that have been widely applied for degrading dyes involve:

1. Anodic oxidation
2. Electrogeneration of oxidizing agents in solution

Figure 3.1 illustrates the classification of the EO processes based on this chapter's perspective.

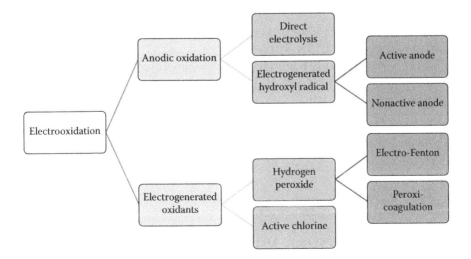

FIGURE 3.1 **(See color insert)** Classification of EO processes.

3.2.1   ANODIC OXIDATION

Anodic oxidation is the oxidation of organic compounds through direct electrolysis (via electron transfer) or electrogenerated hydroxyl radicals on the anode surface. These methods are comprehensively discussed within the following sections.

3.2.1.1   Direct Electrolysis

In this method, electron transfer is responsible for the degradation of organic compounds. In this way, organic compounds are firstly adsorbed on the anode surface, then they are oxidized to other products via electron transfer. Reaction 3.1 indicates the interaction between the anode's surface and the organic compounds. In this reaction, S and C are representatives of the anode's surface and the organic compound, respectively.

$$S + C \xrightarrow{\text{Ads}} S - C + ne^- \to \text{Product} \tag{3.1}$$

This reaction occurs only at potentials lower than that of oxygen evolution, which is highly dependent on the anode material. This process has rarely been applied to degrade dyes, since its removal efficiency is low. Also, a polymer layer is formed on the anode surface, decreasing catalytic activity. This action depends on the type and nature of the electrode and the organic pollutants (Panizza et al. 2009a). Moreover, due to the probable presence of chloride and sulfate ions in the electrolyte, the mechanism identification of direct electron transfer is difficult.

3.2.1.2   Electrogenerated Hydroxyl Radicals: Mediated Anodic Oxidation

Based on the literature, organic pollutants can be mineralized through electrolysis at potentials higher than the thermodynamic potential of oxygen evolution (1.23 V/SHE) by means of some electrodes. Various potentials for the reaction of oxygen evolution on different anodes are summarized in Table 3.4.

In this way, the oxidation of organic compounds at the anode occurs through the transfer of oxygen (originating from water) to the products. This transfer is known as the *electrochemical oxygen transfer reaction* (Reaction 3.2).

$$\text{Organic compounds} + H_2O \to CO_2 + nH^+ + ne^- \tag{3.2}$$

TABLE 3.4
Potentials for Oxygen Evolution Reaction Regarding Electrode Material

Electrode	V/SHE
RuO_2[a]	1.47
IrO_2[a]	1.52
Pt[a]	1.6
Graphite[a]	1.7
SnO_2[a]	1.9
PbO_2[b]	1.9
Boron-doped diamond (BDD)[a]	2.3

[a]   Conditions: 0.5 M sulfuric acid
[b]   Conditions: 1 M sulfuric acid

To activate water for the electrochemical oxygen transfer reaction, two mechanisms have been hypothesized: (1) dissociative adsorption of water; (2) electrolytic discharge of water (Panizza 2010; Panizza et al. 2009a; Fóti et al. 1999). These mechanisms are indicated in the next subsection.

3.2.1.2.1 Dissociative Adsorption of Water

In this mechanism, water is adsorbed on the anode's surface and a hydroxyl radical is formed on the surface of the anode.

$$MO_x + H_2O \rightarrow MO_x(HO^\bullet) + H^+ + e^- \tag{3.3}$$

In the following reaction, the adsorbed hydroxyl radical interacts with the anode in such a way that a higher oxide is formed.

$$MO_x(HO^\bullet) \rightarrow MO_{x+1} + H^+ + e^- \tag{3.4}$$

The redox couple of MO_{x+1}/MO_x in the electrochemical cell acts as a mediator for the oxidation of organic compounds (Reaction 3.5). This oxidation is selective, through which partial oxidation of pollutants occurs in solution. It should be noted that acidic media are necessary for degradation via mediated anodic oxidation.

$$MO_{x+1} + R \rightarrow MO_x + RO \tag{3.5}$$

This species is called chemisorbed *active oxygen*, which is formed using Pt–Ru-based electrodes with catalytic activity. The "active electrode" is defined for this mechanism. In fact, the anode material is of paramount importance in determining the mechanism of the oxidation of organics. This method has limitations in the degradation of organic pollutants, including poor degradation of simple organic pollutants or intermediates, electrode deactivation by chemosorption of carbon monoxide, and the probable occurrence of side reactions (Reaction 3.6) instead of the main reaction (Anglada et al. 2009; Kapałka et al. 2010; Panizza et al. 2009a; Comninellis 1994).

$$MO_{x+1} \rightarrow MO_x + \tfrac{1}{2}O_2 \tag{3.6}$$

3.2.1.2.2 Electrolytic Discharge of Water

In contrast with the first mechanism, in this mechanism, the hydroxyl radical produced has a weak interaction with the anode surface, so that it can directly react with the organic compounds. In fact, physisorbed "active oxygen" is the main agent for the degradation of organic pollutants. A physisorbed hydroxyl radical on the anode surface mineralizes the organic compounds to CO_2 and water. In this mechanism, "nonactive" electrodes are considered, by which no catalytic active site is provided to directly react with the organic compounds. In other words, the anode electrode provides an inert site to remove the electrons and only oxidation of water occurs on the anode surface. Hence, nonselective oxidation of organics takes place and complete mineralization may be carried out based on Reaction 3.7.

$$MO_x(HO^\bullet) + R \rightarrow MO_x + CO_2 + H_2O + H^+ + e^- \tag{3.7}$$

This reaction should compete with preventive side reactions of O_2 evolution or hydrogen peroxide generation, based on Reactions 3.8 and 3.9, respectively:

$$MO_x(HO^\bullet) \rightarrow MO_x + \tfrac{1}{2}O_2 + H^+ + e^- \tag{3.8}$$

$$2MO_x(HO^\bullet) \rightarrow 2MO_x + H_2O_2 \tag{3.9}$$

The nonactive electrodes, thanks to their high efficiency, have been considered for water and wastewater treatment. Among the nonactive electrodes, PbO_2, SnO_2, and boron-doped diamond (BDD) are the most applicable electrodes for the degradation of organics. Looking at Table 3.4, BDD is a promising electrode with high oxidation power; its bond with the hydroxyl radical is weak. Moreover, its high stability and resistance against deactivation have persuaded scientists and researchers to pay attention to the synthesis and application of BBD in different technologies (Panizza et al. 2009a; Simond et al. 1997; Sirés et al. 2014; Comninellis 1994).

In EO processes, two mechanisms may occur simultaneously. Therefore, continued reactions occur on the anode's surface and electrolytes in the degradation of pollutants such as dyes. The mechanism of the electrochemical oxidation of organic compounds on "active" and "nonactive" anodes is schematically presented in Figure 3.2.

It is worthwhile emphasizing that the efficiency of EO in both mechanisms depends on three factors: (1) the electrode material and its properties; (2) the amount of active oxygen produced on the anode's surface (chemically or physically adsorbed); (3) the occurrence of the oxygen evolution reaction instead of the main reaction (hydroxyl radical production reaction) (Martínez-Huitle et al. 2006).

To assess the efficiency of ECPs, two indicators have been frequently considered for a comparison of processes: current efficiency (CE) and electrical energy consumption (EEC).

$$CE(\%) = \frac{(COD_0 - COD_t)FV_e}{zIt} \times 100 \tag{3.10}$$

where:

COD_0 and COD_t	are chemical oxygen demands (g/L) at time $= 0$ and time $= t$, respectively
I	is the applied current (A)
F	is Faraday's constant (96,485 C/mol)
V_e	is the volume of electrolyte (L)
$z = 8$	is the oxygen equivalent mass (g/eq)

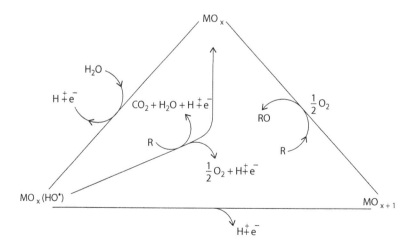

FIGURE 3.2 Scheme of the electrochemical oxidation of organic compounds on "active" and "nonactive" anodes. (Reproduced from Comninellis, C., *Electrochimica Acta*, 39, 1857–62, 1994).

$$\text{EEC}(\text{kWh}) = \frac{I.t.V}{1000} \qquad (3.11)$$

where:
  I is the applied current (A)
  V is the applied voltage (V)

EEC can be calculated based on the COD removed (kWh/COD removed) (Brillas et al. 2015), volume of electrolyte (kWh/m³) (Ghanbari et al. 2014c), and dye removed (kWh/dye removed) (Martínez-Huitle et al. 2009).

3.2.1.2.3 Dye Degradation by Pt Anode

A platinum anode is a conventional electrode for electrochemical degradation with high catalytic activity. Sanromán et al. (2004) studied the decolorization of structurally varied dyes by platinum anode. Azo dye (methyl orange [MO]), indigoide dye (indigo), polymeric (Poly R-478), and triphenylmethane (phenol red, fuchsin, crystal violet, bromophenol blue, and methyl green) were selected for the evaluation of electrochemical oxidation. The decolorization rate principally depended on the chemical structure of the dyes. Complete decolorization was first attained for bromophenol blue, followed by MO and methyl green, while the efficiency of decolorizing phenol red was 30% after 60 min of electrolysis. Accordingly, the rate constant of bromophenol blue decolorization was 30-fold more than that of phenol red.

Jović et al. (2013) used the platinum anode for degradations of Reactive Blue 52, Reactive Black 5, Reactive Green 15, and Reactive Yellow 125. Complete decolorization occurred for all dyes in less than 60 min of electrolysis. EEC for 60 min of electrolysis for all the studied dyes was 0.048–0.090 kWh/m³. The highest COD removal efficiency was obtained for Reactive Blue 52, in such a way that COD reduced from 138.5 to <30 mg/L in 20 min and at 12 V applied voltage.

Hattori et al. (2003) studied the performances of platinum and diamond electrodes in different configurations for the degradation of amaranth dyestuff. The results showed that platinum, in the role of anode, was not able to mineralize amaranth dye, while a diamond anode could significantly mineralize it. In another study, decolorization of amaranth dye was investigated using platinum wire as the anode and 1000 mV potential, by which the removal efficiencies of color, total organic carbon (TOC), and COD were 95.4%, 30%, and 35%, respectively (Fan et al. 2008). Socha et al. (2005) investigated the effect of temperature on anodic oxidation with a platinum anode in the range of 30°C–80°C. Regarding their results, it can be stated that temperature did not markedly influence dye degradation. With an increase in temperature from 30°C to 80°C, decolorization, COD, and TOC removal increased from 76.7% to 78.9%, 35.1% to 36.4%, and 13.8% to 18.7%, respectively. Decolorization of Reactive Orange 4 was also investigated, using a Ti/PtO$_x$ anode. The results of this study indicated that decolorization was dependent on the electrolyte. In this way, chloride salt was more efficient in comparison with sulfate salt, particularly at low current densities. Solar irradiation and UV light significantly improved decolorization compared with dark conditions. The rate constants (k_d) of sunlight, UV light, and dark conditions were 0.084, 0.607, and 0.024 h^{-1} respectively (López-Grimau et al. 2006).

3.2.1.2.4 Dye Degradation by PbO₂ Anode

Lead dioxide (PbO$_2$), thanks to its high oxygen evolution overpotential, can generate hydroxyl radicals as a nonactive electrode. Several studies have been conducted on the decolorization of synthetic dyes and colored wastewater using a PbO$_2$ anode. Awad et al. (2005) investigated decolorization and degradation of two azo dyes (Acid Blue 120 and Basic Brown 4). Based on the results, the electrolyte type was remarkable in influencing the degradation and decolorization of the dyes. An NaOH electrolyte performed better compared with an H$_2$SO$_4$ electrolyte. Using an H$_2$SO$_4$ electrolyte, electrode poisoning occurred as a consequence of the formation of an adherent film on the anode surface. The COD removals from Acid Blue 120 using NaOH and H_2SO_4 electrolytes were 76% and 37.5%, respectively, while the values for Basic Brown 4 were 77.5% and 45%.

Zhou and He studied the efficiency of EO on Basic Red 46 via a β-PbO_2 electrode modified with fluorine resin. The color and COD removals obtained by EO were 20.1% and 15.8%, respectively, under these conditions: 0.5 A applied current, pH 5.0, temperature 25°C, and 3 g/L Na_2SO_4. Combining wet oxidation (high temperature [160°C] and pressure [0.5 MPa]) and EO, removal efficiency increased considerably to 98.5% and 43.2% for color and COD, respectively. This combination depicted that no decay took place when the initial concentration rose from 500 to 2000 mg/L. The authors expressed that the combination of wet oxidation and EO is a suitable configuration for a wider concentration range (Zhou et al., 2007).

Ghalwa et al. (2012) studied decolorization of Reactive Orange 7 (RO7) by three lead oxide–based anodes. A carbon rod/PbO_2 anode could decolorize 97.6% of RO7 after 15 min electrolysis, while complete COD removal was achieved within 300 and 350 min of electrolysis for RO7 and real textile wastewater, respectively. Within 15 min reaction time, the decolorization efficiency of RO7 using Pb + Sn/PbO_2 + SnO_2 and Pb/PbO_2 anodes was 95.3% and 94.6%, respectively. Using Pb + Sn/PbO_2 + SnO_2 and Pb/PbO_2 anodes, further reaction time was required to achieve complete COD removal of both RO7 and real textile wastewater in comparison with the carbon rod/PbO_2 anode. Hence, the carbon rod-modified PbO_2 anode was more efficient than the other modifications.

In another study, Yahiaoui et al. (2014) used central composite design (CCD) for the optimization of Basic Yellow 28 (BY28) removal by Pb/PbO_2 anodes at pH 2. The effects of temperature, current density, initial BY28 concentration, and agitation speed on BY28 removal were examined. The optimum conditions obtained were a temperature of 60°C, current density of 8.125 mA/cm^2, agitation speed of 720 rpm, and initial BY28 concentration of 134 mg/L; under these conditions, 93% decolorization was attained. They also showed that after 3 h of electrolysis, biodegradability in terms of the BOD_5/COD index increased from 0.076 to 0.3, revealing that anodic oxidation with a PbO_2 anode can be used as a pretreatment before the biological processes.

Panizza and Cerisola studied methyl red degradation by anodic oxidation using a PbO_2 anode. The results indicated that to achieve 90% COD removal at 0.5, 1, 1.5, and 2 A applied current, the required reaction times were 7.6, 6.4, 5.6, and 4.3 h, respectively. The authors stated that a pH value in the range of 3.0–7.0 did not significantly affect the degradation rate. Methyl red degradation followed first-order kinetics, and the rate constant increased with an increase in applied current (Panizza et al. 2007).

3.3.1.2.5 Dye Degradation by DSA Anode

In 1966, Beer discovered metal oxides of titanium that were coated by the thermochemical coprecipitation method. These metal oxides include platinum-group metals (ruthenium, rhodium, palladium, osmium, iridium, and platinum). These anodes were registered under the trademark DSA®. Since that time, many applications of DSA have been observed in different industries, as some industries, such as the chlor-alkali industry, need a nonconsumable anode in the electrolyte. In the EO process, this property is an important factor, because release of metal ions into water resources can be an issue of concern from the public health point of view (Panizza 2010). Nowadays, various metal oxides and combinations of them are coated on titanium and have been used for the degradation of textile dyes. Several studies have been carried out on the decolorization of dyes by Ti/MO_n anodes. DSA electrode applications are more related to chlorine evolution for indirect EO. For instance, many studies have explored the performance of DSA electrodes on the decolorization of Reactive Blue 4 (Carneiro et al. 2002), Reactive Red 120 (Panakoulias et al. 2010), Reactive Red 198 (Catanho et al. 2006), Reactive Blue 81 (Kusmierek et al. 2011), Orange II (Li et al. 2011), and methyl red (Morais et al. 2013).

Ciríaco et al. (2011) studied the degradation of Acid Orange 7 (AO7) by Ti/SnO$_2$_Sb$_2$O$_4$ anode. They used the following equation to determine the degree of combustion of organic pollutants in terms of combustion efficiency:

$$\eta_c = \frac{32}{12}\left(\frac{n}{4x}\right)\frac{dTOC}{dCOD} \tag{3.12}$$

where:

TOC and COD   are given in mg/L

n   is the number of electrons transferred to the electrode in the process for complete combustion of the AO7 molecule

x   is the number of carbon atoms of the organic compound

The electrochemical combustion reaction of AO7 is

$$C_{16}H_{11}N_2SO_4Na + 32H_2O \rightarrow 16CO_2 + 2NH_4^+ + SO_4^{2-} + Na^+ + 67H^+ + 68e^- \tag{3.13}$$

Based on these equations, the value of η_c for AO7 was 0.39 at 10 mA/cm^2. The combustion efficiency of AO7 was low, because the azo bond was oxidized more easily, which resulted in the reduction of COD without any reduction of TOC. The reductions of the studied parameters were as follows: TOC < COD < absorbance (230 nm) < absorbance (486 nm).

Chaiyont et al. (2013) synthesized a mixed metal oxide film coating of IrO$_2$–SnO$_2$–Sb$_2$O$_5$ supported on a titanium substrate. The performance of Ti/IrO$_2$–SnO$_2$–Sb$_2$O$_5$ was evaluated by decolorization of MO. Applying electrode potentials of 1.5 and 1.75 V, MO decolorization efficiencies were similar, achieving over 98% decolorization after 50 min of electrolysis. At a higher potential (2.0 V), decolorization decreased to 89% after 300 min of electrolysis. This reduction in decolorization can be attributed to the fact that the oxidation of water to evolve oxygen competes with hydroxyl radical generation, which is the main agent that oxidizes organic matter.

An undivided mode of the filter press reactor for electrochemical oxidation was applied by del Río et al. (2011). They used stainless steel and Ti/SnO$_2$–Sb–Pt electrodes as the cathode and anodes respectively. Reactive Orange 4 was selected to be degraded in this system under conditions of 0.8 g/L Reactive Orange 4, 0.1 M Na$_2$SO$_4$, and 125 mA/cm^2. The results demonstrated that TOC removal was higher than COD removal. This higher mineralization rate could be attributed to the previous oxidation of the intermediates, making them more easily mineralized. This reveals that, although more TOC removal efficiency was obtained, the remaining TOC was more difficult to oxidize because a greater number of moles of O$_2$ were required.

Shestakova et al. (2015) combined anodic oxidation (Ti/Ta$_2$O$_5$–SnO$_2$) with the sonochemical process for the decolorization of methylene blue. Using electrolysis alone, complete decolorization occurred, applying 20 mA and 3 h of electrolysis with an energy consumption of 8.46 kWh/m^3; while, using the sonochemical process, complete decolorization was obtained, requiring 162 kWh/m^3 energy in similar conditions. Combining the sonochemical and electrolysis processes, the reaction time and energy consumption were reduced to 1 h and 56.43 kWh/m^3, respectively. TOC removal efficiency of 38.4% was achieved after 2 h of applying a combination of sonochemical and electrolysis processes, which was 1.5 times higher compared with EO only and 10.5 times higher than the sonochemical process alone.

Duan et al. (2015) fabricated Ti/Ru–Sb–SnO$_2$ successfully by the selective potential pulse electrodeposition method in a single electrolytic solution. Dye decolorization was evaluated for various deposition cycles of Ti/Ru–Sb–SnO$_2$. Ti/Ru–Sb–SnO$_2$ with 200 deposition cycles showed the best performance in the decolorization of methylene blue, Orange II, and MO, as the rate constants of first-order kinetics were 23.0×10^{-3}, 25.1×10^{-3}, and 25.2×10^{-3} min^{-1} respectively. Moreover, the

TABLE 3.5

Comparison of Anodic Oxidation of Various Reactive Dyes on Ti/BDD and Ti/SnO₂-Sb2O₅ Electrodes

Dyes	Charge (Ah/L)	Initial COD (mg/L)	Residual COD (mg/L)		Current Efficiency (%)	
			Ti/BDD	Ti/SnO₂-Sb₂O₅	Ti/BDD	Ti/SnO₂-Sb₂O₅
Cibacron Yellow HW200	3.52	737	28	526	67.5	20.1
Cycafix Yellow FLN250	3.02	610	31	386	64.2	24.8
Cycafix Navy-Blue F2B	2.77	659	64	453	72.0	24.9
Monozol Black SGRN	2.52	710	78	451	84	34.4
Monozol Blue BRF-150	2.52	634	71	466	74.8	22.3
Monozol Red F3B150	2.39	654	72	425	81.6	32.1
Monozol T-blue HFG	4.03	980	93	712	73.7	22.3
Monozol Yellow F3R150	2.52	667	74	423	78.8	32.4
Procion Blue HE-RD	3.02	902	89	551	90.2	38.9
Reactive Blue R	2.90	803	61	390	85.7	47.7
Reactive Red HE-7B	2.52	402	19	216	51.0	24.7
Samafix Red S-3B	2.52	607	38	370	75.6	31.5
Samafix Yellow S-3R	2.52	440	50	277	51.8	21.7
Unicion Green S6B	4.03	711	8	484	58.4	18.9
Unicion Red S-3BF80	2.27	589	66	381	77.2	30.7

Source: Chen, X. et al., *Journal of Applied Electrochemistry*, 35, 185–91, 2005.

Note: Current density: 100 A/M²; temperature: 30°C; volume 25 mL; initial dyes concentration: 1000 mg/L; initial pH: 4.70–6.73.

service life of Ti/Ru−Sb−SnO₂ with 200 deposition cycles was 18.3 h, which was 1.54 times more than that of the control electrode.

Chen et al. (2005) compared the performances of Ti/BDD with Ti/SnO₂–Sb₂O₅ for COD removal of various reactive dyes. Table 3.5 shows the corresponding results. As can be seen, not only are the COD removal efficiencies of the Ti/BDD anode higher than those of Ti/SnO₂–Sb₂O₅, but a Ti/BDD anode also provides higher CE compared with Ti/SnO₂–Sb₂O₅.

In general, DSA, as the active electrode, exhibited poor efficiency in the production of hydroxyl radicals for oxidation of dyes in a nonchlorine electrolyte.

3.2.1.2.6 Dye Degradation by BDD Anode

BDD makes an excellent electrode with a very wide potential window in aqueous solution, low capacitance, and low background current. Recently, BDD electrodes have been attracting much attention in many sciences and technologies such as electrochemistry, analytical chemistry, environmental engineering, medical sciences, and so on. Nowadays, BDD has been stabilized as the best anode for EO of organic compounds (Brillas et al. 2015; Panizza et al. 2009a, Alfaro et al. 2006). The application of anodic oxidation with BDD for oxidation of various compounds is ever increasing. Several research groups have focused on BDD application in pollutant remediation.

The effect of current densities (30, 60, and 90 mA/cm²) on AlphazurineA (AZA) removal has been studied by Bensalah and coworkers (Bensalah et al. 2009). The results indicated that complete COD removal was achieved after 10 h of electrolysis, which was independent of the current density. In contrast, CE was highly dependent on current density. Obviously, an increase in current density reduces CE. Bensalah et al. (2009) declared that the oxidation of AZA could be carried out both by direct EO and mediated oxidation. Due to the presence of sodium sulfate as the supporting electrolyte, it was also possible for persulfate to be formed during the electrolysis with the BDD anode. Persulfate is a strong oxidant that can efficiently oxidize organic compounds.

$$2SO_4^{2-} \rightarrow S_2O_8^{2-} + 2e^- \tag{3.14}$$

The authors also indicated that EEC decreased with increases in temperature and agitation speed. The EEC (kW/kg COD) values for 95% COD removal were 102.3, 92.4, and 89.6 for temperatures of 25°C, 40°C, and 60°C, respectively, under conditions of 60 mA/cm² current density, initial COD of 500 mg/L, and 0.5 M Na_2SO_4 as the supporting electrolyte.

Abdessamad et al. (2013) studied the effect of electrode arrangement on Alizarin Blue Black B (ABB) removal by anodic oxidation with BDD. The results demonstrated that the removal efficiency of a bipolar arrangement was 1.2 times higher than that of a monopolar arrangement. In addition, the electrochemical oxidation of ABB in the bipolar arrangement resulted in almost direct oxidation of ABB, while large amounts of intermediates were not accumulated. In the monopolar arrangement, the electrolysis led to the formation of different intermediates. The phytotoxicity test, in terms of the germination index, indicated that anodic oxidation can significantly detoxify ABB.

The effect of the reactor type (divided and undivided) on Orange G azo dye was studied by El-Ghenymy and coworkers (El-Ghenymy et al. 2014). They concluded that the divided reactor had a superior oxidation ability in comparison with the undivided one at constant current density. Complete color removal was observed in both systems. However, the color removal process was always more rapid in the divided reactor. At similar current densities, complete color removal was achieved in a shorter time in the divided reactor. TOC decay was also faster using the divided reactor, in such a way that total mineralization was obtained in an electrolysis time of less than 330 min at 66.7 mA/cm², while total mineralization in the undivided reactor was achieved after 420 min of electrolysis at 150 mA/cm². The dye degradation consistently followed a pseudo-first-order kinetic, and its apparent rate constant rose with an increase in the current density.

Chen and coworkers (Chen et al. 2006) investigated Orange II degradation by anodic oxidation with a Ti/BDD electrode. The results demonstrated that alkaline conditions, a high temperature, and low current density were suitable for Orange II degradation. Nevertheless, the pH value did not significantly influence system efficacy. The concentration of Na_2SO_4, as the supporting electrolyte, in the range of 1500–3000 mg/L did not influence oxidation efficiency.

Araújo et al. (2015) indicated that the type of electrolyte (Na_2SO_4, $HClO_4$, H_3PO_4, or NaCl) did not significantly affect Rhodamine B degradation by anodic oxidation using conductive-diamond electrodes (p-Si–BDD). However, applying these electrolytes, the rate constants obtained differed in such a way that NaCl had the lowest rate constant value. This result was in contrast to the results gained in other studies. This phenomenon can be attributed to Cl_2 generation at the BDD surface, which decreased the generation of active chlorine species. Thus, the oxidation efficiency of the dye was decreased. In the case of chloride media, chlorinated organics were formed as intermediates by the attack of ClO^- on organic compounds.

Tsantaki and colleagues (Tsantaki et al. 2012) selected an actual textile wastewater sample to evaluate anodic oxidation by BDD. At optimum conditions (current density 8 mA/cm², inherent temperature, and acidic conditions), TOC removal was about 70% after 180 min of electrolysis. Compared with synthetic dyes (a sample of 17 commercial dyes), COD removal was higher in the case of the actual textile wastewater. This behavior was associated with the presence of other organic compounds in the textile wastewater that may have been oxidized more easily than the synthetic dyes. Moreover, the energy consumed was lower for the textile waste water in comparison with the synthetic effluent (75 vs. 135 kWh/kg COD).

Another study on real textile wastewater was conducted by Martínez-Huitle et al. (2012). It was observed that an increase in temperature had a strong influence on COD removal at all current densities. Complete COD removal was achieved after 9 or 10 h of electrolysis. It was reported that adding only 5 g/L of sodium sulfate to the textile wastewater increased removal efficiency remarkably. In addition, in the presence of sodium sulfate, less time was needed to remove COD and, also, the cell potential required was lower. Hence, EEC was notably reduced. Table 3.6 presents the studies of dye removal using a BDD anode.

TABLE 3.6
Color and COD (or TOC) Removals for the Electrochemical Oxidation Using BDD Anode

Electrode	Dye	Color Removal (%)	COD or TOC Removal (%)	Conditions	References
p-Si/BDD	Alizarin Red S	100	Complete TOC removal at 28 Ah/L	Dye = 5 mM, T = 25°C, time = 300 min, j = 30 mA/cm²	Faouzi et al. 2007
BDD	Orange G	100	99 (TOC), time = 270 min	Dye = 0.52 mmol, pH = 3.0, T = 35°C, j = 100 mA/cm², time = 60 min	El-Ghenymy et al. 2014
BDD	Novacron Yellow C-RG	>95	86 (TOC)	Dye = 200 mg/L, 0.25 M Na$_2$SO$_4$, T = 25°C, agitation speed: 400 rpm, time = 480 min, j = 30 mA/cm²	Rocha et al. 2014
BDD	Methyl violet	100	100 (TOC), time = 360 min	TOC = 100 mg/L, pH = 3, 0.05 M Na$_2$SO$_4$, T = 35°C, time = 60 min, j = 100 mA/cm²	Hamza et al. 2009
BDD	MO	94	60.3 (TOC)	Dye = 100 mg/L, pH = 3, 0.05 M Na$_2$SO$_4$, T = ambient, time = 138 min, j = 31 mA/cm²	Ramírez et al. 2013
BDD	Reactive Green 19	20	50 (TOC), time = 3 h	Dye = 100 mg/L, pH = 7, 0.05 M Na$_2$SO$_4$, T = 25°C, time = 30 min, j = 300 mA/cm²	Petrucci et al. 2015
BDD	MO	>90	35.5 (COD), 25.1 (TOC)	Dye = 50 mg/L, 0.1 M Na$_2$SO$_4$, time = 120 min, j = 50 mA/cm²	Zhou et al. 2011
BDD	Reactive Black 5	97	51 (COD), 29.3 (TOC), time = 25 min	Dye = 100 mg/L, pH = neutral, 0.02 M Na$_2$SO$_4$, Q = 100 mL/min, time = 50 min, j = 1 mA/cm²	Yavuz et al. 2012
BDD	Eriochrome Black T	100	77 (TOC) at 17 Ah/L	Dye = 100 mg/L, pH = 4.6, 5000 mg/L Na$_2$SO$_4$, time = 120 min, j = 30 mA/cm²	Bedoui et al. 2009

Anode	Dye	Conditions	Color removal (%)	COD/TOC removal (%)	References
BDD	Basic Blue 3	Dye = 40 mg/L, pH = 4.6, 0.02 M Na_2SO_4, T = 30°C, Q = 109.5 mL/min, time = 60 min, j = 0.875 mA/cm²	99	86.7 (COD)	Yavuz et al. 2011
BDD	Bromoamine acid	Dye = 125 mg/L, pH = natural, 0.001 M Na_2SO_4, T = 30°C, Q = 550 mL/min, time = 120 min, j = 15.5 mA/cm²	100	85.29 (TOC)	Liu et al. 2015
p-Si/BDD	Methylene blue	Dye = 58 mg/L, pH = 3, time = 240 min, j = 40 mA/cm²	100	100 (COD) at time = 6 h	Akrout et al. 2015
BDD	Malachite green	Dye = 20 mg/L, pH = 3, 0.1 M Na_2SO_4, time = 60 min, j = 32 mA/cm²	100	91 (COD) at time = 3 h	Guenfoud et al. 2014
BDD	Auramine-O	COD = 150 mg/L, pH = 6, 0.05 M Na_2SO_4, T = 30°C, time = 90 min, j = 50 mA/cm²	N/A	95 (COD)	Hmani et al. 2012
Si/BDD	Acid Black 210	Dye = 500 mg/L, pH = 6.8, 0.2 M buffer phosphate, time = 120 min, j = 25 mA/cm²	100	64.6 (TOC) at time = 5 h	Costa et al. 2009
BDD	Amido black	Dye = 0.05 mM, pH = 7, 0.01 M Na_2SO_4, time = 90 min, j = 40 mA/cm²	77.7	40 (COD)	Akrout et al. 2013
Nb/BDD	Vat Blue 1	Dye = 0.1 mM, pH = 11.3, 1.0 mM sodium dithionite, time = 90 min, I = 500 mA	100	90 (TOC) at time = 6 h	Diagne et al. 2014
BDD	Acid Yellow 1	Dye = 1 g/L, pH = 11.3, T = 30°C, 0.5 M $HClO_4$, time = 120 min, I = 1500 mA	100	Nearly 85 (COD)	Rodriguez et al. 2009

Note: j = current density; T = temperature; I = applied current.

3.2.1.2.7   Degradation of Dye by Carbon-Based Electrodes

Carbon-based electrodes have scarcely been utilized for dye degradation through anodic oxidation. They include graphite, activated carbon fiber (ACF), graphite felt, carbon pellet, and carbon fiber. However, thanks to large surface areas and the low price of carbon electrodes, both EO and adsorption mechanisms for dye removal from aqueous solutions may be fulfilled (Panizza et al. 2009a).

Yi and Chen (Yi et al. 2008) investigated Alizarin Red S (ARS) dye wastewater using ACF as anode material. They used ACF (with a specific surface area of about 1000 m^2/g) as both adsorbent and anode. The experimental results showed that the amount of decolorization was greater than 98% after 60 min of electrolysis, while it was 84.7% when ACF was used as an adsorbent. They compared the ACF and carbon fiber (with a specific surface area of about 4.4 m^2/g) on decolorization. The decolorization of ARS was much higher with the ACF anode than with the carbon fiber anode. The reason for this was the high specific surface area of the ACF anode compared with the carbon fiber anode.

The effect of the specific surface area of ACF was also studied for AO7 degradation. Three ACFs were used: ACF$_a$ (823.4 m^2/g), ACF$_b$ (1292.5 m^2/g), and ACF$_c$ (1558.1 m^2/g), and their performances were compared with the performances of Ti/Pt and graphite. The decolorization efficiency of AO7 followed this order: ACF$_c$ > ACF$_b$ > ACF$_a$ > Pt/Ti > graphite. These results proved the influence of the specific surface area on EO with ACF electrodes. It was reported that the sodium sulfate concentration did not influence the decolorization of AO7 (Zhao et al. 2010).

Kariyajjanavar et al. (2013) showed that an increase of sodium sulfate from 5 to 20 g/L did not affect the degradation of Vat Black 27, while with an increase to more than 20 g/L, decolorization was significantly enhanced, in a way that decolorization efficiencies were <5% and >90% for 20 and 35 g/L, respectively.

3.2.1.2.8   Comparison of Anode Materials

The performance of an anode depends on its material. Not only does the anode material determine the efficiency of the dye removal system, but it also identifies the degradation mechanism of organic compounds. Panizza and Cerisola (Panizza et al. 2007) compared the efficiencies of Ti–Ru–Sn ternary oxide, platinum, lead dioxide, and BDD anodes for the oxidation of methyl red. A decrease in absorbance at 520 nm is illustrated in Figure 3.3. As can be seen, as expected, BDD performed best

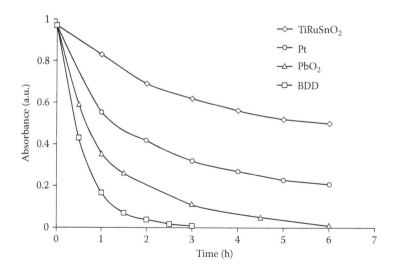

FIGURE 3.3 Comparison of the decrease of the absorbance band at 520 nm during the oxidation of 200 mg/L methyl red in 0.5 M Na$_2$SO$_4$ using different anodes. Conditions: I = 500 mA; flow rate = 180 L/h. (Reproduced from Panizza, M. et al., *Applied Catalysis B: Environmental*, 75, 95–101, 2007).

in the decolorization of methyl red; it could completely decolorize after 3 h, while PbO_2 provided the same level of decolorization after 6 h. Pt and $TiRuSnO_2$ could decolorize only about 80% and 50%, respectively, after 6 h of electrolysis. The decolorization and COD removal efficiencies for methyl red performed in the following order: BDD > PbO_2 > Pt > $TiRuSnO_2$.

Panakoulias et al. (2010) noted that decolorization and mineralization rates of Reactive Red 120 were very fast using BDD in comparison with Ti/IrO_2–RuO_2, which is a DSA electrode. Moreover, intermediates produced utilizing a BDD anode were aliphatic organics, while aromatic organics were formed during electrolysis with Ti/IrO_2–RuO_2. The BDD anode demonstrated high efficiency in degrading dyes through anodic oxidation. This point was also observed in Table 3.5, where a Ti/BDD anode was compared with Ti/SnO_2–Sb_2O_5 (DSA anode). In all cases, the BDD anode was more successful than the DSA anode in removing COD. In addition, BDD had a higher CE.

Ammar et al. (2012) degraded alizarin red using three anodes (BDD, PbO_2, and Pt). Anodic oxidation was comparatively examined under these conditions: 100 mg/L dye solution (containing 70 mg/LTOC), 0.05 M Na_2SO_4, 100 mA/cm², pH 7.0. The results demonstrated that Pt reduced only 28% of TOC at the end of electrolysis, while TOC was reduced by 98% and 90% using BDD and PbO_2 anodes, respectively, at 18 Ah/L. The apparent rate constant for decolorization followed this sequence: Pt (2.5×10^{-3} s⁻¹) < PbO_2 (4.5×10^{-3} s⁻¹) < BDD (8.66×10^{-3} s⁻¹), which indicates the order of oxidation power of these three anode materials in anodic oxidation of the dye.

Regarding the literature and experts' points of view, Table 3.7 summarizes the advantages and limitations of the anode materials.

3.2.2 ELECTROGENERATION OF OXIDIZING AGENTS IN SOLUTION

Electrogeneration of oxidizing agents is usually divided into two categories; (1) electrogeneration of active chlorine species and (2) electrogeneration of hydrogen peroxide. In the electrogeneration of active chlorine species, these are generated at anodes such as DSA, BDD, PbO_2, and Pt, while the electrogeneration of hydrogen peroxide occurs at carbon-based cathodes. The electrogeneration of

---

TABLE 3.7
Advantages and Limitations of Anode Materials in Electrochemical Oxidation Process

Electrode	Advantages	Limitations
Platinum	Low oxygen evolution overpotential, stable and fairly available, genuine, without requiring further processing	Low efficiency in anodic oxidation, expensive
Boron-doped diamond (BDD)	High oxygen evolution overpotential, high stability, very effective in production of oxidants, high corrosion steadfastness, inert surface with low adsorption properties and strong tendency to resist deactivation	Extremely expensive
Dimensionally stable anode (DSA)	High oxygen evolution overpotential, fairly inexpensive	Fast fading of efficiency, short life service
PbO_2	Easy preparation, high oxygen evolution overpotential, cheap and available	Lead ions may be released into the solution
Graphite	Inert, very cheap and abundant, low oxygen evolution overpotential,	Low efficiency in anodic oxidation, surface corrosion occurs at high potentials

Source: Panizza, M. et al., *Chemical Reviews*, 109, 6541–69, 2009a; Särkkä, H. et al., *Journal of Electroanalytical Chemistry*, 754, 46–56, 2015.

active chlorine species is performed using an electrolyte containing chloride ion, while a chloride-free electrolyte is usually used for the electrogeneration of hydrogen peroxide.

3.2.2.1 Chlorine-Mediated Electrooxidation

Chloride is one of the most abundant ions in polluted waters and wastewaters. Chlorine-mediated oxidation is claimed to be the most commonplace *in situ* oxidant generation process. In fact, electrochlorination is counted among promising treatment processes for wastewaters containing high amounts of chloride. In wastewaters containing chloride ions, active chlorine species such as Cl_2, $HClO$ and/or ClO^-, ClO^{2-}, ClO^{3-}, and ClO^{4-} are formed via EO. In electrochlorination, regarding the oxidation power of the active chlorine species, either partial oxidation or complete mineralization may occur. Applying chloride-mediated EO favors textile wastewater treatment, due to the presence of chloride in such wastewater resulting from dye fixation on fibers (Palma-Goyes et al. 2015). The active chlorine species generated during electrochlorination have a great affinity to chromophoric groups, in such a way that a high rate and efficient decolorization of dyes are fulfilled. Based on a study conducted by Petrucci et al. (2015), in which BDD was used as the anode material and the target dye to be degraded was Reactive Green 19 (RG19), the reaction time needed to achieve more than 90% dye removal was significantly reduced merely in the presence of chloride. Accordingly, electrochlorination is undoubtedly a viable option for the treatment of textile wastewaters, unless the formation of chloroderivatives is ignored. In other words, the main drawback of chlorine-based treatment processes is the formation of toxic organic chloroderivatives. Numerous studies have been conducted to trace and monitor these compounds. However, it has been claimed that the concentrations of these derivatives are negligible or zero if the process is comprehensively optimized, taking these derivatives into consideration. In fact, it has been stated that even in the case of the formation of such derivatives, further electrolysis of wastewater most often fulfills degradation and destruction of these intermediates. For instance, based on the results achieved in the above mentioned study, optimum values of chloride lead to further degradation of RG19 without the formation of persistent intermediates.

Referring back to the introduction of the process, ambiguities and complexities are still there in the exact nature of the chemical and electrochemical reactions occurring during electrochlorination. An indirect transfer of oxygen occurs both on the anode's surface and in the bulk solution via chlorine, hypochlorous acid, or hypochlorite, which is strongly dependent on the pH value. Initially, chlorine is electrogenerated as a result of chloride oxidation at the anode, based on the following reactions:

$$Cl^- \leftrightarrow Cl^{\bullet}_{ads} + e^- \tag{3.15}$$

$$Cl^- + Cl^{\bullet}_{ads} \leftrightarrow Cl_2 + e^- \tag{3.16}$$

With pH values less than 3.3, the primary active chloro-species is Cl_2 (Anglada et al. 2009). When the point concentration of Cl_2 exceeds its solubility, the electrogenerated Cl_2 forms bubbles. Likewise, it may be partially diffused away from the anode to react with chloride, forming trichloride as a consequence (Reaction 3.17), or else it quickly becomes hydrolyzed to hypochlorous acid as well as chloride ions (Reaction 3.18). Depending on the pH value, the hypochlorous acid can be further dissociated toward the production of hypochlorite ions (Reaction 3.19). Further on, ClO^- may take part in an oxidation process based on Reaction 3.20.

$$Cl_2(aq) + Cl^- \leftrightarrow Cl_3^- \tag{3.17}$$

$$Cl_2(aq) + H_2O \leftrightarrow HOCl + H^+ + Cl^- \tag{3.18}$$

$$HClO \leftrightarrow ClO^- + H^+ \tag{3.19}$$

TABLE 3.8

Domination of Chlorine Species at Various pH Levels

Active Chlorine Species	pH of Abundance	Species E^0 (V vs. SHE)
Cl_2	About 3	1.36
HOCl	3–8 (pH < 7.5)	1.49
ClO^-	Higher than 8	0.89

$$ClO^- + H_2O + 2e^- \leftrightarrow Cl^- + 2OH^- \tag{3.20}$$

In an alkaline medium, the hypochlorite ions can oxidize the organics in the vicinity of the anode and/or in the bulk of the solution (Scialdone et al. 2014).

$$Organics + OCl^- \rightarrow Intermediates \rightarrow CO_2 + Cl^- + H_2O \tag{3.21}$$

HClO and ClO^- are strong oxidants; nevertheless, their actions to degrade organics in volume reactions are highly dependent on the pH value and chlorine concentration in the solution. The domination of chlorine species is strongly dependent on the pH value (Table 3.8).

Anodically generated chlorine and hypochlorite are the major agents responsible for oxidation of organic pollutants. Regarding Table 3.8, the active chlorine species–mediated oxidation is supposed to occur at higher rates in acidic conditions, due to the higher E^0 values allocated to HOCl and Cl_2. Nevertheless, the EO of the wastewater causes desorption of the chlorine, inhibiting it from acting as an oxidizer. In this way, higher pH levels may theoretically promote electrochemical oxidation, since HOCl and OCl^- are not highly affected by the desorption of gases and are efficient oxidizers acting throughout the reactor (Anglada et al. 2009). Likewise, hypochlorite ion concentration can be diminished through its anodic oxidation to chlorite, chlorate, and perchlorate:

$$ClO^- + H_2O \rightarrow ClO_2^- + 2H^+ + 2e^- \tag{3.22}$$

$$ClO_2^- + H_2O \rightarrow ClO_3^- + 2H^+ + 2e^- \tag{3.23}$$

$$ClO_3^- + H_2O \rightarrow ClO_4^- + 2H^+ + 2e^- \tag{3.24}$$

Within active chlorine species–mediated systems, oxychloro species take the responsibility for the transfer of oxygen, based on Figure 3.4. Bonfattiet et al. (2000) have described the mediating role of chloride, illustrating the electrochemical oxygen transfer reaction (EOTR) via the adsorbed oxychloro species that are intermediates of the chlorine evolution reaction. In fact, chloro and oxychloro radicals coadsorbed at the electrode surface are of quite high importance in determining the mechanism of electrochemical incineration. Likewise, coelectroadsorption of hydroxyl radicals should also be noted. Within EOTR, $\cdot OH_{ads}$ plays a critical role which its reactivity as a function of metal oxide electrode nature has been declared by Comninelis while in chlorine mediated EO systems, oxychloro species are also in charge of oxygen transfer.

Based on Figure 3.4, chloride ions inhibit the oxygen evolution reaction, causing a rise in anode potential and higher reactivity of adsorbed oxidized chloride species.

Based on the results reported within the literature, chloride concentration, as an influential parameter in electrochlorination, has a bilateral effect on the process's efficiency that should be optimized carefully. In a study carried out by Bonfatti et al. (2000), it was found that increasing the chloride concentration up to 5 g/dm^3 brought about more efficient electrochemical mineralization with a higher rate, while a further increase of chloride to 10 g/dm^3 revealed adverse effects

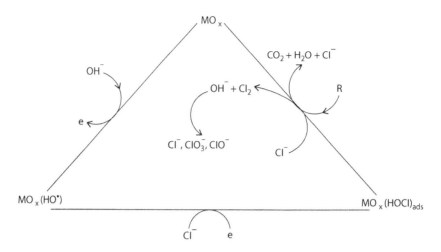

FIGURE 3.4 Electrochemical incineration of organics by active chlorine species-mediated process. (Reproduced from Bonfatti, F. et al., *Journal of the Electrochemical Society*, 147, 592–96, 2000).

on the efficiency of the process. But, generally, it has been well established that dye degradation is strongly enhanced in the presence of inorganic mediators such as active chlorine species. It has been reported that chloride affects the behavior of the anodes in an efficient manner (Sala et al. 2012). In the presence of chloride, EO can be conducted within lower potentials in comparison with the potentials needed in direct anodic oxidation. Sala et al. stated that the amount of chlorine species generated electrochemically is highly influenced by the amounts of hydroxyl radicals. Based on a comprehensive review conducted by Trasatti and coworkers (Trasatti 1987), adsorption of oxy-chloro radicals has been reported to be the most probable mechanism. Accordingly, the reaction mechanism of chlorine electrogeneration is supposed to be related to the acid base equilibrium at the anode surface, which is actually the source and start point of chlorine electrogeneration. The corresponding reactions are as follows:

$$AS-OH_2^+ \leftrightarrow AS-(OH)_{ads}+H^+ \tag{3.25}$$

$$AS-(OH)_{ads} \leftrightarrow AS-O_{ads}+H^++e^- \tag{3.26}$$

$$AS-O_{ads}+Cl^- \rightarrow AS-(OCl)_{ads}+e^- \tag{3.27}$$

$$AS-(OCl)_{ads}+Cl^-+H^+ \leftrightarrow AS-(OH)_{ads}+Cl_2 \tag{3.28}$$

$$AS-(OCl)_{ads}+Cl^- \leftrightarrow AS-(O)_{ads}+Cl_2+e^- \tag{3.29}$$

where AS represents active sites on the electrode surface.

Sales Solano et al. (2013) investigated the decontamination of real textile wastewater using a BDD anode. They found that NaCl, in comparison with Na_2SO_4, provided higher degradation efficiency regarding the electrogeneration of strong chlorine active species on the BDD anode surface, attacking the chromophore group of dyes, thereby eliminating the wastewater color. Based on their work, by increasing NaCl concentration, it was observed that degradation efficiency in terms of COD removal was enhanced up to 100% temperature is also of importance, although less so in comparison with the pH level and chloride concentration. With temperatures lower than 100°C, there is a clear evolution of hydrogen, oxygen, and chlorine, while as the temperature increases, their generation is strongly suppressed. Nevertheless, marginal amounts of hydrogen can be found even at higher temperatures, although these amounts are much lower than those expected to be generated based on Faraday's law.

Although electrochlorination is reported to be highly dependent on chloride concentration and pH, various results have been gained in electrochlorination studies applying active and nonactive anodes, indicating that in such processes, the anode material is also of importance, since the EO of chloride to chlorine should occur at high rates. It should also be noted that it is essential for the anode material to be inactive, due to the need to hinder active sites being blocked through adsorption (physio-sorption). Based on the literature, platinum or mixtures of metal oxides have been employed in the electrochlorination process, as they have suitable electrocatalytic features along with good stability.

DSA electrodes are mainly of interest because of their high CE of chlorine gas evolution, due to suitable catalytic features for this kind of reaction. In fact, electrochemical corrosion of DSA anodes is limited to the electroactive surface layer. Likewise, due to the suitable shape and integrity of a DSA structure, the coating can be regenerated over and over again. Based on a study carried out by Scialdoneet et al. (2009), using DSA as an anode, the presence of chloride led to a significant increase in CE. DSA electrodes present quite a low oxygen overpotential, while they display high electrocatalytic activity for chlorine generation (Zhang et al. 2014a).

Apart from the anode materials mentioned, graphite and nonactive anodes such as BDD and PbO_2 are among favorable anode selections that provide significantly high efficiencies for the degradation of dyes in chloride-mediated systems. In EO systems applied to the treatment of chloride-containing wastewaters such as textile wastewater, potentials higher than that needed for oxygen evolution lead to the cooperation of electrochlorination and hydroxyl radical-based EOTR. In this condition, optimization of the process and prediction of how the parameters' roles will affect process efficiency and operation become complex.

In some cases, such as when the inactive BDD electrode is exerted, physisorbed hydroxyl radical species cooperate with the chlorine active species. In this way, the following reactions are expected to occur (Petrucci et al. 2015):

$$HO^{\bullet} + Cl^{-} \rightarrow ClOH^{\bullet -} \tag{3.30}$$

$$ClOH^{\bullet -} \leftrightarrow Cl^{\bullet} + OH^{-} \tag{3.31}$$

$$Cl^{\bullet} + Cl^{-} \rightarrow Cl_2^{\bullet -} \tag{3.32}$$

$$Cl_2^{\bullet -} + HO^{\bullet} \rightarrow HOCl + Cl^{-} \tag{3.33}$$

Figure 3.5 comprehensively illustrates the reactions that are hypothesized to occur during EO in the presence of chloride using BDD.

Based on Figure 3.5, it should be noted that in chloride-mediated EO, incineration reactions are considered to occur more in the volume instead of the surface of the anode (Sales Solano et al. 2013).

Based on what is stated, BDD anodes are an attractive choice, since hydroxyl radicals are formed on the surface of such anodes; nevertheless, the performance of such radicals are dramatically affected in chloride media (Palma-Goyes et al. 2015). Besides, the high cost of BDD anodes should also be considered, especially in the case of their application at the industrial scale.

Rajkumar et al. (2007) used titanium mesh coated with TiO_2–RuO_2–IrO_2 as an anode for the degradation of Reactive Blue 19 (RB19). Complete decolorization was achieved within a short electrolysis period. However, under conditions of 400 mg/L RB19 with 1.5 g/L NaCl concentration, the COD and TOC removal rates were 55.8% and 15.6%, respectively. Regarding these results, with

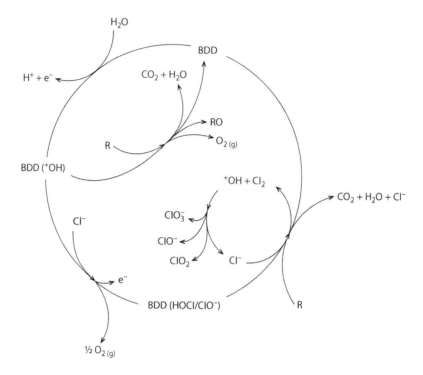

FIGURE 3.5 Diagram of HO$^\bullet$ and Cl-mediated oxidation mechanisms using BDD anode. (Reproduced from Sales Solano, A.M. et al., *Applied Catalysis B: Environmental*, 130–131, 112–20, 2013).

an increase in electrolysis time, the molecular weight of the intermediate compounds formed was lowered. Moreover, as an interesting result, electrochemical oxidation of RB19 in the presence of chloride did not generate any chlorinated organics as by-products.

Araújo et al. (2015) showed that oxidation of Rhodamine B (RhB) with a p-Si–BDD anode in NaCl electrolyte produced seven sub-product. No complete removal of all by-products formed was accomplished after 500 min of electrolysis. Figure 3.6 shows electrochemical degradation pathways for RhB in NaCl solution. Seven by-products were produced, including: (I_1) phthalic acid, (I_2) benzoic acid, (I_3) 3-dinitrobenzoic acid, (I_4) 2,5-hydroxybenzoic acid, (I_5) α-hydroxyglutaric acid, (I_7) Intermediate 7, and (I_8) chloroform. Although chloroform was formed during the electrolysis process, the authors claimed that the presence of chloroform might have been related to some gas-phase reactivity along the complex analytical path itself. Hence, the reactions that occurred at the anode surface had a partial influence on chloroform formation, since other chlorinated compounds were not formed.

Table 3.9 summarizes electrochemical oxidation with active chlorine species to decolorize or mineralize different dyes or colored wastewaters by various anodes.

3.2.2.2 Electrooxidation by Electrogenerated Hydrogen Peroxide

Hydrogen peroxide, with the chemical formula of H_2O_2, is widely used for oxidation of substances because it is a powerful oxidant, with $E^0 = 1.8$ V. Hydrogen peroxide is one of the most applicable oxidants in environmental remediation in all media (air, water, and soil). Moreover, hydrogen peroxide is well known as a green chemical, since its residuals in the environment are only oxygen and water, which do not threaten the ecosystem or public health (Neyens et al. 2003; Yu et al. 2015a). Hydrogen peroxide can be catalytically decomposed to hydroxyl radicals by transitional metals and UV light. The hydroxyl radicals decompose organic compounds as a nonselective

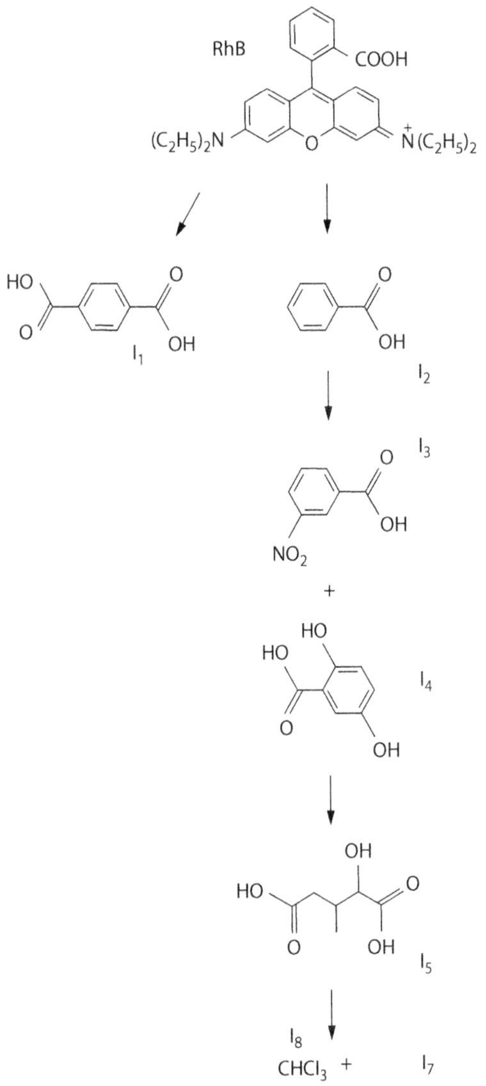

FIGURE 3.6 Electrochemical degradation pathways for RhB when NaCl used as supporting electrolyte. Intermediates detected: (I_1) phthalic acid, (I_2) benzoic acid, (I_3) 3-dinitrobenzoic acid, (I_4) 2,5-hydroxybenzoic acid, (I_5) α-hydroxyglutaric acid, (I_7) Intermediate 7, and (I_8) chloroform. (Reproduced from Araújo, D.M.D. et al., *Applied Catalysis B: Environmental*, 166–167, 454–59, 2015).

oxidant with $E^0 = 2.7$ V. The mixture of H_2O_2 and ferrous ions is well known as the Fenton reagent, which has been broadly applied for water and wastewater treatment. Ferrous ions act as a catalyst in acid media, decomposing hydrogen peroxide to hydroxyl radicals (Reaction 3.34), consequently producing ferric ions. The Fenton process is an oxidation process that has been extensively used for the degradation of dyes (Brillas et al. 2009; Nidheesh et al. 2013a).

$$Fe^{2+} + H_2O_2 + H^+ \rightarrow HO^{\bullet} + Fe^{3+} + H_2O \tag{3.34}$$

$$Fe^{3+} + H_2O_2 \rightarrow HO_2^{\bullet} + Fe^{2+} + H^+ \tag{3.35}$$

TABLE 3.9

Decolorization and Mineralization of Different Dyes or Colored Wastewater by Electrochemical Oxidation with Active Chlorine Species

Pollutant	Anode	Condition	Color Removal (%)	COD or TOC Removal (%)	References
Reactive Red 184	Pt/Ti	Dye = 10^{-2} mol/L, applied potential = 12.7 V, electrolyte: NaCl, time = 50 min	>90	N/A	Sakalis et al. 2006
Real textile wastewater	BDD	j = 40 mA/cm^2, NaCl = 5 g/L, time = 300 min	100	Almost 100 (COD), time = 600 min	Sales Solano et al. 2013
Reactive Green 19	BDD	Dye = 100 mg/L, j = 300 A/m^2, Na$_2$SO$_4$ = 0.05 M, NaCl = 0.01 M, T = 25°C, agitation rate = 250 rpm	100%	>80% (TOC)	Petrucci et al. 2015
Indigo carmine	DSA of Sb$_2$O$_5$-doped Ti/RuO$_2$-ZrO$_2$	Dye = 0.64 mM, j = 200 A/m^2, 0.05 M NaCl, flow rate: 5 L/min	N/A	90% (COD), 22% (TOC)	Palma-Goyes et al. 2015
Acid Orange 7	BDD	Dye = 209.3 mg/L, 0.30 M NaCl, pH = 3, 100 mA/cm^2 T = 35°C, time = 15 min	100%	N/A	Scialdone et al. 2014
Acid Orange 7	Ti/RuO$_2$-Pt	Dye = 50 mg/L, NaCl = 0.001 M, j = 10 mA/cm^2, pH = 6.8, Na$_2$SO$_4$ = 0.005 M, time = 4 h	100	79.48(TOC)	Zhang et al. 2014b
Acid Orange II	BDD	Dye = 50 mg/L, NaCl = 5 mM, j = 2.58/mA/cm^2, flow rate = 600 ml/L, time = 40 min	87	N/A	Zhang et al. 2014a
C.I. Reactive Orange 7	Ti/Sb–SnO$_2$	Dye = 60 mg/L, NaCl = 3.5 g/L, pH = 4, j = 19 mA/cm^2	99,70	70.3 (COD)	Basiri Parsa et al. 2013
C.I. Reactive Yellow 186	Graphite	pH = 3.9; j = 34.96 mA/cm^2, dye = 500 μM, NaCl = 0.11 M, time = 15 min	99	73 (COD)	Rajkumar et al. 2015
Reactive Black 5	Sb$_2$O$_5$-doped Ti/RuO$_2$-ZrO$_2$	Dye = 0.5 mM, NaCl = 1 M, j = 0.5 A/cm^2, T = 25°C	100	N/A	Rodríguez et al. 2014

Note: j = current density, T = temperature, I = applied current.

The high efficiency and simple operation of the Fenton process allow it to be considered as a popular process for dye removal from colored effluent. However, the process has some limitations that include the production of ferric sludge, the need for storage and shipment of concentrated H_2O_2, and the low regeneration of ferrous ions (Wang et al. 2010).

Hydrogen peroxide can be electrochemically generated in acidic media through the reduction of oxygen molecules at a carbon-based cathode (Reaction 3.36) (Nidheesh et al. 2012).

$$O_2 + 2H^+ + 2e^- \rightarrow H_2O_2 \qquad (3.36)$$

Hydrogen peroxide is continually electrogenerated in the electrochemical cell while air or oxygen is directly injected into the solution. The applied anode can be "active" or "nonactive" to generate hydroxyl radicals on its surface. Platinum and BDD are among the conventional anodes for this purpose. Indeed, the anode also participates in the degradation of pollutants through anodic oxidation. The electrogeneration of hydrogen peroxide, along with anodic oxidation, occurs in one electrochemical cell, increasing the removal efficiency of EO. Nonetheless, hydrogen peroxide alone does not fulfill the oxidation of macromolecules such as dyes, since a catalyst is required for the activation of hydrogen peroxide and, consequently, the production of hydroxyl radicals (Jiang et al. 2007; Sirés et al. 2014).

The addition of ferrous-ion salts (usually ferrous sulfate) to the electrochemical cell leads to the decomposition of electrogenerated hydrogen peroxide to hydroxyl radicals, which are responsible for the degradation of the organic compounds. The ferric ions can be reduced to ferrous ions at the cathode, based on Reaction 3.37. This reaction provides more availability of ferrous ions to react with the electrogenerated hydrogen peroxide, increasing the propagation of the reaction chain of the Fenton process (Sirés et al. 2007; Flox et al. 2006).

$$Fe^{3+} + e^- \rightarrow Fe^{2+} \quad E^0 = 0.77 \text{ V/NHE} \qquad (3.37)$$

In fact, in this process, the Fenton reagent is electrochemically produced in the electrochemical cell, called the *electro-Fenton* (EF). In the EF, H_2O_2 is electrochemically produced while ferrous ions are chemically added from outside the reactor (Rosales et al. 2012). Enric Brillas and Mehmet Oturan have been the architects of the EF process for the degradation of environmental pollutants from the 1990s to date. The mechanism of an EF is schematically presented in Figure 3.7.

During these years, several studies have been conducted based on dye degradation by the EF process. In these processes, operating parameters include applied current (current density), pH, anode and cathode materials, initial concentration of dye, air or oxygen sparging, type and dosage of catalyst, and configuration of electrochemical cell (Nidheesh et al. 2012). Recent advances in EF processes have been more focused on the synthesis of the cathode electrode and the catalysts. Graphite felt, ACF, gas diffusion cathodes, carbon cloth, polytetrafluoroethylene (PTFE), reticulated vitreous carbon (RVC), and carbon sponge have been used for hydrogen peroxide generation in the EF process, while Pt, BDD, and DSA electrodes have been applied as the anode.

Panizza and Cerisola utilized a Pt anode and gas diffusion cathodes for alizarin red degradation. They found that faster degradation occurred at pH 3.0 with COD removal efficiency of 93% in 4 h of electrolysis. This was in agreement with other studies of Fenton-based processes. An evaluation of the effect of ferrous-ion dosage indicated that an increase of Fe^{2+} from 0.25 to 1 mM developed COD removal efficiency, while a further increase of the dosage of Fe^{2+} (2 Mm) did not do so. This phenomenon is related to the scavenging effect of ferrous ions in excess amounts consuming hydroxyl radicals, based on the following reaction (Panizza et al. 2009b):

$$Fe^{2+} + HO^{\bullet} \rightarrow Fe^{3+} + HO^- \qquad (3.38)$$

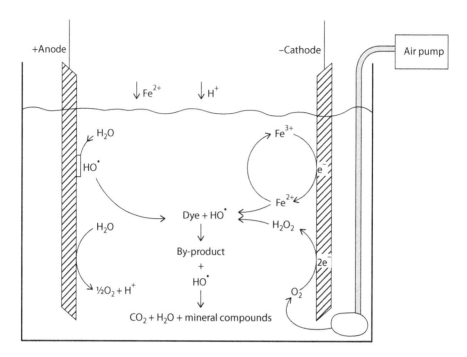

FIGURE 3.7   Scheme of EF reactor and reactions.

El-Desoky et al. (2010) studied EF for the degradation of Ponceau S azo dye with an RVC cathode and a platinum gauze anode. Under conditions of pH 2.5, cathode potential of −1.0 V (vs. SCE), 0.05 M Na_2SO_4, and 0.05 mM $FeSO_4$, complete decolorization was obtained within 30, 60, and 90 min for initial dye concentrations of 0.05, 0.1, and 0.3 mM, respectively. In this study, sodium sulfate was chosen as the best supporting electrolyte, while KCl and NaCl lowered the efficiency of the EF process due to the presence of Cl^- ions. It was advised to avoid using chloride-containing media, as they produce chlorinated organics, which are carcinogenic by-products.

Lei et al. (2010) used Cu^{2+} and Mn^{2+} as catalysts for the decoloriziation of cationic red X-GRL with an ACF cathode and an RuO_2/Ti anode. Cu^{2+} and Mn^{2+} were more effective in degrading the dye in comparison with ferrous ions. The removal efficiencies of the three catalysts followed this order: $Fe^{2+} < Cu^{2+} < Mn^{2+}$. The higher standard reduction potential of Mn^{3+} (Reaction 3.39) leads to an increase of the acceptance rate of electrons compared with Fe^{3+}. Hence, the regeneration of Mn^{2+} is faster compared with the other transitional metals.

$$Mn^{3+} + e^- \rightarrow Mn^{2+} \quad E^0 = 1.51 \text{ V/NHE} \tag{3.39}$$

Carbon black–modified graphite felt and PTFE were used in the EF process to decolorize MO. The modified graphite felt generated 472.9 mg/L H_2O_2 at pH 7.0 and current density of 50 A/m², while unmodified graphite generated only 40.3 mg/L H_2O_2. An interesting result was obtained from this study: H_2O_2 was effectively generated even without external aeration. This result indicated that the generation of enough oxygen at the anode can fulfill the generation of H_2O_2 without any external aeration. Complete decolorization of MO was achieved after 15 min by modified cathode, while mineralization reached 95.7% after 2 h of electrolysis, which was much more in comparison with the unmodified cathode (23%) (Yu et al. 2015a).

Ren et al. (2015) investigated the effect of the pore structure of mesoporous carbon–grafted activated carbon fibers (ACF@OMC) on the degradation of Rhodamine B and Orange II by the EF process. The ACF@OMC, with an average pore size of 5.4 nm, had the highest efficiency in comparison with the average pore sizes of 2.6 and 3.7 nm. The large pore size and ordered mesoporous structures improved the transformation and diffusion of O_2 on the surface of the cathode. A total of 99.7% of Rhodamine B was decolorized by ACF@OMC with an average pore size of 5.4 nm within 45 min, and 47.7% mineralization occurred in 360 min. The stability results showed that 94.2% of the dye could still be decolorized after using the cathode 10 times.

In another study (Le et al. 2015), a new cathode was introduced for the EF process, in which reduced graphene oxide (rGO) was electrochemically deposited on carbon felt. The use of a graphene-modified cathode completely decolorized AO7 after only 5 min of electrolysis. Almost 95% TOC removal was achieved in 8 h of reaction time. The stability of the cathode was examined after using 10 cycles. A fresh cathode could remove 73.9% of TOC in 2 h, while after 10 applications of the cathode, TOC removal reached 64.3%, which was a preventive for an otherwise excellent cathode.

Olvera-Vargas and coworkers (Olvera-Vargas et al. 2014) compared the efficiencies of Pt/carbon-felt and BDD/carbon-felt systems in the degradation of Azure B. Complete decolorization was achieved after 10, 15, and 45 min for BDD/carbon felt, Pt/carbon felt, and anodic oxidation (BDD) together with H_2O_2, respectively, in 0.05 M Na_2SO_4, 500 mA current, [Fe^{2+}] = 0.1 mM, pH 3.0, at room temperature. The apparent rate constants of BDD/carbon felt, Pt/carbon felt, and anodic oxidation (BDD) together with H_2O_2 were 0.37, 0.24, and 0.14 min^{-1}, respectively. Over 95% of TOC was removed in EF with BDD and anodic oxidation (BDD) together with H_2O_2, while EF with Pt could not provide the same efficiency. The formation and evolution of carboxylic acids in EF using BDD for Azure B degradation are presented in Figure 3.8. Oxamic and oxalic acids were accumulated at high concentrations, whereas formic, malic, pyruvic, and acetic acids were formed at low concentrations. Based on Figure 3.8, the carboxylic acids could be efficiently decayed in the EF process with the BDD system, which was also expected, considering the higher level of mineralization in terms of TOC removal.

Various studies applied a heterogeneous catalyst instead of ferrous ions for activation of H_2O_2. He et al. (2014) exerted Fe_3O_4 magnetic nanoparticles for the degradation of Reactive Blue 19. Optimal operating conditions were pH 3.0, current density of 3.0 mA/cm^2, 1.0 g/L Fe_3O_4, initial RB19 concentration of 100 mg/L, and reaction temperature of 35°C. Under these conditions, more than 89.7% of TOC was degraded after 180 min electrolysis. Their results demonstrated that Fe_3O_4 had a higher catalytic capacity than the ferrous ions.

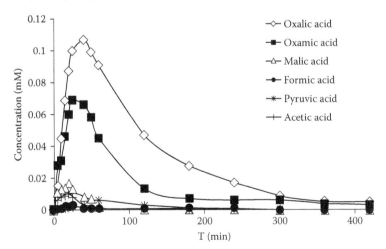

FIGURE 3.8 Time course of the concentration of the main short-chain carboxylic acids accumulated in solution during EF treatment of 200 mL 0.1 mM Azure B by BDD/carbon-felt cell using 300 mA and 0.05 M Na_2SO_4 at pH 3.0 and room temperature. (Reproduced from Olvera-Vargas, H. et al., *Environmental Science and Pollution Research*, 21, 8379–86, 2014).

Iglesias et al. (2013) used Fe alginate gel beads as a catalyst to remove Reactive Black 5 from aqueous solution. Optimum conditions were obtained by CCD, involving a voltage of 5.69 V, pH 2.24, and an iron concentration of 2.68 mM for 65% COD removal, 90% decolorization, and low energy costs of 35.61 kWh/kg_{COD} and 24.41 kWh/kg_{dye}.

Natural pyrite was used as a heterogeneous catalyst for the degradation of a new azo dye (4-amino-3-hydroxy-2-p-tolylazo-naphthalene-1-sulfonic acid) (AHPS) by EF (BDD/graphite felt). In the presence of 2 g/L pyrite, complete decolorization of 175 mg/L AHPS occurred after 20 min reaction time at 300 mA applied current, while an 8 h reaction time was required for complete mineralization. This study showed that pyrite is a low-cost and reusable solid catalyst that can perform at a favorable pH without external acidification (Labiadh et al. 2015). Several studies with their efficiencies of EF systems for dye degradation are presented in Table 3.10.

3.2.2.2.1 Photoelectro-Fenton (PEF)

To enhance the efficiency of EF, a UV-assisted electrochemical process can be employed. In this scenario, not only can UV light directly decompose electrogenerated hydrogen peroxide to $HO^\bullet$ (Reaction 3.40), but it can also regenerate ferrous ions by photolysis of ferric ions. In fact, the Fe(III) produced in Reaction 3.34 can be photo reduced by regenerating Fe^{2+} and producing more $OH^\bullet$ radicals (Reaction 3.41) (Moreira et al. 2013; Moradi et al. 2015).

$$H_2O_2 + UV \rightarrow HO^\bullet + HO^\bullet \tag{3.40}$$

$$Fe(OH)^{2+} + UV \rightarrow Fe^{2+} + HO^\bullet \tag{3.41}$$

The combination of photolysis and EF is called the *photoelectro-Fenton (PEF)* process. It should be noted that, in many PEF studies, solar energy has been utilized instead of a UV lamp, reducing the electrical energy cost of UV lamps.

Thiam et al. (2016) studied PEF (BDD anode/air diffusion cathode) for the degradation of Ponceau 4R with UVA irradiation. Three mechanisms improved the oxidation of the dye: (1) anodic oxidation by (BDD($^\bullet$OH)); (2) homogeneous hydroxyl radicals generated from the Fenton reagent; and (3) photodecomposition of the dye. Indeed, PEF has a synergistic effect on the degradation of organic pollutants. The total mineralization and color removal of 254 mg/L dye were achieved at 100 mA/cm^2 after 240 and 50 min, respectively. In this study, the oxidizing ability of PEF, EF, and EO with electrogenerated H_2O_2 (EO–H_2O_2) was assessed for decolorization of the dye in raw water. The dye solution became colorless after 180, 100, and 70 min for EO–H_2O_2, EF, and PEF, respectively.

Moreira et al. (2013) compared anodic oxidation with electrogenerated H_2O_2 (AO–H_2O_2), EF, PEF, and solar photoelectro-Fenton (SPEF) to degrade Sunset Yellow FCF (SY) azo dye. Figure 3.9 shows the removal of TOC for the ECPs. The AO–H_2O_2 could only remove 65% of TOC after 360 min, which is related to the generation of BDD($^\bullet$OH) and BDD($HO_2^\bullet$) at the anode's surface. In EF, rapid mineralization was observed during the first 120 min; after that, the TOC removal rate decreased, which could be attributed to the forming of Fe(III)–carboxylate complexes that were hardly oxidized by •OH. In addition, 82% of TOC removal was attained after 360 min. In the cases of PEF and SPEF, the removal of TOC was 95% (240 min) and 93% (150 min), respectively. This increase in efficiency was due to the photodegradation of Fe(III)–carboxylate complexes. SPEF could degrade SY in a shorter time, which was ascribed to the more potent UV intensity provided by sunlight (Moreira et al. 2013).

The comparison of EF and SPEF has been studied in other research (Ruiz et al. 2011) for the decay of Acid Yellow 36 (AY36). In both processes, an increase in applied current improved TOC removal. However, EF degraded AY36 poorly; 71% of TOC was removed in 360 min using 3 A

TABLE 3.10

Decolorization and Mineralization Degrees of Different Dyes and Colored Wastewater by Various EF Systems

EF System (Anode/Cathode)	Dye	Condition	Color Removal (%)	TOC or COD Removal (%)	References
(Pt/carbon paper)	Reactive Red 195	Dye = 50 mg/L, pH = 3.0, I = 300 mA, Fe^{3+} = 0.2 mM, oxalate$^-$ = 0.4 mM, time = 90 min	70	N/A	Djafarzadeh et al. 2013
(Pt/rotating graphite felt)	MO	Dye = 25 mg/L, pH = 3.0, j = 50 A/m², Fe^{2+} = 0.02 M, time = 15 min	100	58.7 (TOC) time = 2 h	Yu et al. 2014
(Graphite/graphite)	Rhodamine B	Dye = 10 mg/L, pH = 3.0, applied voltage = 8 V, Fe^{2+} = 10 mg/L, time = 180 min, 2.5 mg/L of $NaHCO_3$	99	N/A	Nidheesh et al. 2014
(Pt/Cu$_2$O/CNTs/PTFE)	Rhodamine B	Dye = 1.044 × 10^{-5} M, pH = 3.0, applied potential = 1.2 V, Na_2SO_4 = 0.05 M, time = 120 min	89	N/A	Ai et al. 2008
(Graphite/graphite)	Rhodamine B	Dye = 40 mg/L, pH = 3.0, I = 0.03 A, Fe^{2+} = 10 mg/L, time = 180 min, 2.5 mg/L of $NaHCO_3$	84.1	N/A	Nidheesh et al. 2013b
(RuO$_2$–IrO$_2$/carbon felt)	Orange II	Dye = 50 mg/L, pH = 3.0, j = 1.78 mA/cm², Fe^{2+} = 2 mM, time = 60 min	100	58 (TOC)	Lin et al. 2014
(Graphite/graphite)	MO	Dye = 150 mg/L, pH = 3.0, I = 2.1 A, Fe^{3+} = 2 mM, Cl$^-$ = 12 mM, time = 30 min	98.34	50 (COD)	He et al. 2013
(Pt/carbon felt)	Naphtol blue black	Dye = 0.05 mM, pH = 2.0, I = 150 mA, Fe^{3+} = 0.05 mM, Na_2SO_4 = 0.05 M, time = 40 min	97.5	N/A	Bouasla et al. 2013
(Pt/carbon felt)	Malachite green	Dye = 0.5 mM, pH = 3.0, I = 200 mA, Fe^{3+} = 0.2 mM, Na_2SO_4 = 0.05 M, time = 22 min	100	100 (COD) time = 540 min	Oturan et al. 2008a
(Pt/modified gas diffusion electrode)	Amaranth	Dye = 100 mg/L, pH = 2.0, applied potential = −0.7 V, Fe^{2+} = 0.155 mM, H_2SO_4 and K_2SO_4 = 0.1 M, time = 90 min	79.3	67.3 (TOC)	Barros et al. 2014
(Pt/Co$_3$O$_4$-graphite)	Sulforhodamine B	Dye = 0.01 mM, pH = 2–3, applied voltage = 6 V, Na_2SO_4 = 10 g/L, time = 150 min	100	55 (TOC) time = 5 h	Liu et al. 2013
(BDD/stainless steel)	Reactive Red 180	Dye = 20 ppm, pH = 3, j = 20 mA/m², Fe^{2+} = 0.02 mM, Na_2SO_4 = 10 g/L, time = 60 min	Approximately 100%	N/A	(Almomani et al. 2012)

(Continued)

TABLE 3.10 (CONTINUED)
Decolorization and Mineralization Degrees of Different Dyes and Colored Wastewater by Various EF Systems

EF System (Anode/Cathode)	Dye	Condition	Color Removal (%)	TOC or COD Removal (%)	References
(BDD/stainless steel)	Reactive Blue 19	Dye = 20 ppm, pH = 3, j = 20 mA/m², Fe²⁺ = 0.02 mM, Na₂SO₄ = 10 g/L, time = 120 min	Approximately 100%	N/A	Almomani and Baranova 2012
(Ti/PbO₂/C-PTFE-coated graphite chips)	Reactive Brilliant Red X-3B	Dye = 123 mg/L, pH = 3, I = 0.3 A, Fe²⁺ = 0.1 mM, Na₂SO₄ = 0.05 M, time = 20 min	97	87 (TOC), time = 3 h.	Lei et al. 2013
(Pt/dual gas diffusion)	Tartrazine	Dye = 1000 mg/L, pH = 3, j = 35.7 mA/cm², Fe²⁺ = 0.1 mM, Na₂SO₄ = 0.05 M, flow rate = 0.5 L/min, time = 60 min	100	66.1 (TOC), time = 2 h	Yu et al. 2015b
(BDD/granular activated carbon (GAC) packed electrode)	MO	Dye = 97 mg/L, pH = 3, I = 300 mA, Fe²⁺ = 0.5 mM. Na₂SO₄ = 0.05 M, time = 120 min	100	90 (TOC) time = 5 h	Bañuelos et al. 2014
(Pt/carbon sponge)	Basic Blue 3	Dye = 0.2 mM, pH = 3, I = 100 mA, Fe³⁺ = 0.1 mM. NaNO₃ = 0.1 M, time = 120 min, flow rate = 150 mL/min	100	91.6 (TOC) time = 8 h	Özcan et al. 2008
(BDD/Pt)	Acid Yellow 36	Dye = 60 mg/L, pH = 3, I = 100 mA, Fe²⁺ = 0.24 mM. SO₄²⁻ = 0.05 M, time = 48 min	98	N/A	Cruz-González et al. 2010
(Pt/graphite felt)	Textile wastewater	Color = 2024 ADMI, pH = 3, I = 300 mA, Fe²⁺ = 3 mM, time = 160 min	77.2	N/A	Ghanbari et al. 2015
(Pt/ACF)	Textile wastewater	COD = 1224 mg/L, pH = 3, j = 3.2 mA/cm², Fe²⁺ = 2 mM, time = 240 min, oxygen sparging = 150 cm³/min	N/A	75.2	Wang et al. 2010

Note: j = current density, T = temperature, I = applied current.

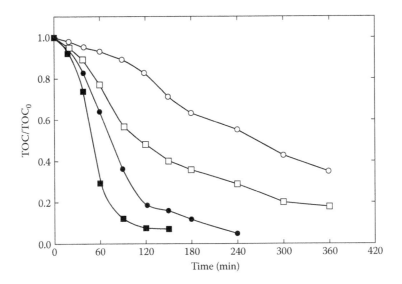

FIGURE 3.9   Descending trend of normalized TOC during electrolysis for the treatment of 100 mL SY with a concentration of 290 mg/L under the same conditions as Figure 3.1. Methods: (○) AO–H_2O_2, (□) EF, (●) PEF, and (■) SPEF. (Reproduced from Moreira, F. C. et al., *Applied Catalysis B: Environmental*, 142–143, 877–90, 2013).

applied current. SPEF, however, showed a high efficiency in TOC degradation compared with EF, since 95% of TOC decayed after only 180 min at 3 A applied current. When calculating EEC, it was revealed that SPEF was a viable method, having greater CE and lower energy cost.

Khataee and coworkers (Khataee et al. 2014) used a UVC lamp (15 W) for PEF (Pt anode/CNT-PTFE cathode) in the degradation of Acid Blue 5 (AB5). Figure 3.10 depicts the decolorization of 20 mg/L AB5 by PEF, EF, and photolysis. As can be seen, the combination of UVC with EF increased decolorization significantly with almost 100% efficiency after 20 min, while each process alone achieved less than 40% after 60 min reaction time.

PEF and SPEF processes have been considered for the degradation of Direct Yellow 4 (Garcia-Segura et al. 2014), Ponceau 4R (Thiam et al. 2015a), Reactive Red 195 (Djafarzadeh et al. 2013), Congo red (Solano et al. 2015), Allura Red AC (Thiam et al. 2015b), indigo carmine (Flox et al. 2006) and Direct Yellow 4 (Garcia-Segura et al. 2014). In all studies, it was shown that light irradiation could markedly improve the EF process.

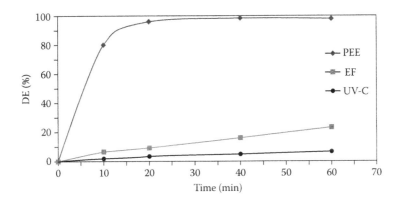

FIGURE 3.10   Decolorization efficiency (DE%) of 20 mg/L AB5 solution by EF, PEF, and photolysis using UV-C processes; pH = 3, I = 0.3 A, [Fe^{3+}] = 0.2 mM, and flow rate = 5 L/h. (Reproduced from Khataee, A. et al., *Chemical Engineering Research and Design*, 92, 362–67, 2014).

3.2.2.2.2   Sonoelectro-Fenton (SEF) Process

Recently, ultrasound irradiation has been combined with the EF process to improve system performance. In the sonoelectro-Fenton (SEF) process, regeneration of Fe^{2+} may occur through sonolysis of iron complexes, based on Reaction 3.42. Likewise, ultrasound cleans the surface of the electrode by means of dissolving the inhibiting layers, resulting in an improvement in mass transfer between solution and electrode. In sonolysis, sites of high temperature and pressure are generated with the collapse of the cavitation bubbles for short periods of time, decomposing the O–H bond, leading to the formation of $^{\bullet}$OH and H$^{\bullet}$ (Reaction 3.43). Despite the destruction of pollutants via $^{\bullet}$OH, the sonolysis decomposes the pollutants through the pyrolysis process (Babuponnusami et al. 2012; Oturan et al. 2008b; Li et al. 2010).

$$Fe - O_2H^{2+}+))) \rightarrow Fe^{2+} + HO_2^{\bullet} \tag{3.42}$$

$$H_2O+))) \rightarrow H^{\bullet} + HO^{\bullet} \tag{3.43}$$

Furthermore, sonolysis in EF can accelerate the mass transfer rate of Fe^{3+} toward the cathode for the electrogenerated Fenton reagent and consequently increase the generation of hydroxyl radicals.

Abdelsalam et al. (2002) explored the SEF process with a Pt anode, RVC cathode, and a resonance frequency of 27 kHz for the degradation of meldola blue (MDB) dye. They stated that a small amount of ferrous ions could increase the rate constant of MDB degradation. However, with an increase of ferrous ions of up to more than 1 mM, the rate constant was considerably reduced. After 100 min of sonication, 67.7% of COD was eliminated with 0.1 mM MDB, constant potential of −0.7 V versus Ag, 0.5 mM Fe^{2+}, and 124 kHz.

In another study (Martínez et al. 2012), SEF was compared with Fenton oxidation and sonolysis for the degradation of Azure B dye. The rate constant of dye degradation by SEF was 10-fold more than that of the sonolysis, while this value was double that of Fenton oxidation. Figure 3.11 shows the differences in COD removal efficiencies of the three processes. It is clear that the highest COD removal efficiency was achieved by SEF (85%). SEF provided 98% decolorization at an applied potential of −0.7 V versus SCE after 60 min of reaction time, whereas 85% decolorization was attained without sonication.

Oturan et al. (2008b) also illustrated that ultrasound irradiation improved EF efficiency significantly for 2,4-dichlorophenoxyacetic acid and 4,6-dinitro-o-cresol degradation, albeit that no improvement was observed for the degradation of the dye (azobenzene) by SEF in comparison with EF. The authors concluded that the nature of the pollutant's structure was an important factor. Moreover, the lowest ultrasound power was revealed to be more suitable for SEF.

3.2.2.2.3   Peroxi-Coagulation

In this process, both Fenton reagents can be electrochemically produced in such a way that the hydrogen peroxide and the ferrous ions are produced at the carbon-based cathode and iron anodes respectively. Indeed, an iron anode is used instead of a platinum or BDD electrode; so, this process is known as the *peroxi-coagulation (PC)* process. Despite the oxidation process by free radicals, the coagulation process also contributes to PC for dye removal via the production of $Fe(OH)_n$. The flocs produced effectively adsorb inorganic and organic pollutants such as dyes (Brillas et al. 2009; Ghanbari et al. 2015).

$$Fe \rightarrow Fe^{2+} + 2e^- \tag{3.44}$$

$$Fe^{2+} \rightarrow Fe^{3+} + e^- \tag{3.45}$$

$$Fe^{2+} \text{ or } Fe^{3+} + nOH^- \rightarrow Fe(OH)_{n(s)} \tag{3.46}$$

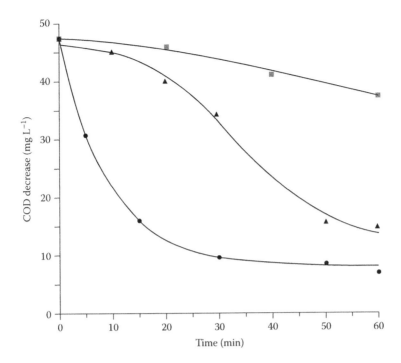

FIGURE 3.11　COD removal of 5×10^{-4} M/L Azure B in 1.5 L of solution containing 0.8×10^{-3} M/L Fe^{2+} and 50×10^{-3} M/L Na_2SO_4 as supporting electrolyte at pH 2.8, -0.7 V versus SCE at 25°C. SEF (●), Fenton (▲), and sonolysis (■). (Reproduced from Martínez, S.S. et al., *Ultrasonics Sonochemistry*, 19, 174–78, 2012).

Zarei and coworkers (Zarei et al. 2009) studied PC for the removal of Basic Yellow 2 (BY2). Carbon nanotube (CNT)-PTFE and carbon-PTFE electrodes were evaluated as cathodes. The decolorization of BY2 during the first 10 min reached 62% and 96% by the carbon-PTFE and CNT-PTFE electrodes, respectively, at 100 mA applied current. Levels of TOC removal after 6 h of electrolysis were 81% and 92% for the carbon-PTFE and CNT-PTFE electrodes, respectively. Figure 3.12 displays a probable reaction sequence for the mineralization of BY2 by PC, based on the intermediates identified by Gas chromatography-mass spectrometry (GC–MS). The intermediates produced by PC for BY2 included: (I_2) bis[4-(dimethylamino)phenyl]methanone, (I_3) 4-dimethylaminobenzoic acid, (I_4) *p*-hydroxybenzoic acid, (I_5) hydroquinone, (I_6) butenedioic acid, and (I_7) ethanedioic acid.

Zarei and coworkers (Zarei et al. 2010) also studied PC on four dyes: Basic Blue 3 (BB3), malachite green (MG), Basic Red 46 (BR46), and Basic Yellow 2 (BY2). PC was conducted by a CNT-PTFE cathode and an iron anode at pH 3. Applying 100 mA electrical current, up to 90% decolorization was obtained after only 10 min for all the dyes. Mineralization was measured in terms of TOC for mixed dyes. The results showed that 93% mineralization occurred after 6 h of electrolysis. The performance of PC was also evaluated in real water samples and the results showed that 90% of the mixed dyes was removed within 40 min reaction time. The cost of the process was US\$ 21.7/kg of removed dye.

Ghanbari and Moradi (Ghanbari et al. 2015) investigated four ECPs (electrocoagulation, electrochemical Fenton, EF, and PC) for the treatment of real textile wastewater. All the processes removed color sufficiently from the wastewater, inasmuch as between 77% and 94% decolorization efficiency was obtained. Figure 3.13 illustrates the COD removal efficiencies of the processes. As can be seen, EF and PC were able to provide 71.1% and 64.2% COD removal efficiency, respectively, at 300 mA applied current and pH 3. The BOD_5/COD ratio of textile wastewater was increased from 0.137 to 0.362 and 0.317 after treatment by EF and PC. These results showed that EF and PC could

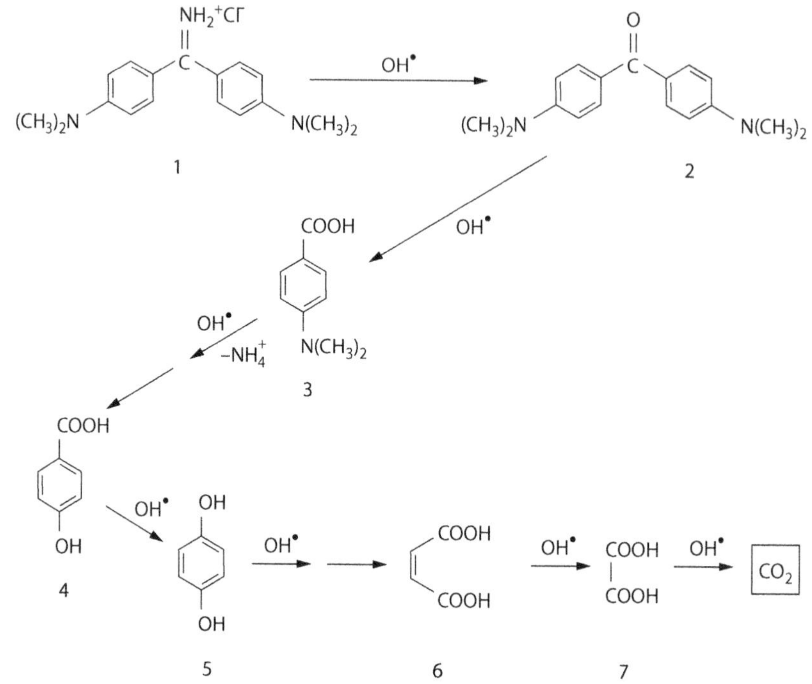

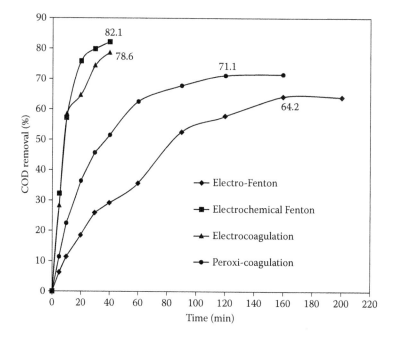

FIGURE 3.12 Proposed reaction sequence for the degradation of BY2 by PC. (Reproduced from Zarei, M. et al., *Electrochimica Acta*, 54, 6651–60, 2009.)

FIGURE 3.13 COD removal efficiency in four ECPs (EF and PC; pH 3 and 300 mA applied current). (Reproduced from Ghanbari, F. et al., *Journal of Environmental Chemical Engineering*, 3, 499–506, 2015).

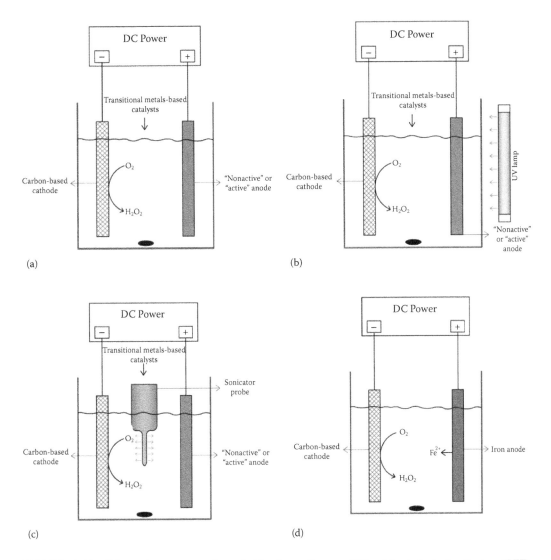

FIGURE 3.14 Schematic representation of (a) electro-Fenton (EF), (b) photoelectro-Fenton (PEF), (c) sonoelectro-Fenton (SEF), and (d) peroxi-coagulation (PC).

be used as pretreatment steps before the biological processes. However, the EECs were 63.64 and 23.19 kWh/kg COD removed for EF and PC, respectively, which were relatively high costs.

However, in the literature, other Fenton-based ECPs have also been reported that include electrochemical Fenton (ferrous ions are electrochemically produced by anode and H_2O_2 is chemically added), anodic Fenton, and Fered-Fenton (Fe^{2+} and H_2O_2 are added to the electrochemical cell with DSA electrodes). Although these ECPs are based on Fenton oxidation, they do not electrochemically generate the oxidizing agent of H_2O_2. Schematic representations of the EF, PEF, SEF, and PC processes are presented in Figure 3.14.

3.3 CHALLENGES AND FUTURE OF EO

In the case of EO, there are several factors to be considered, including process optimization, EEC and other energy resources, the need for supporting electrolytes, electrode issues (material, synthesis, fabrication, design/geometry, and operational problems such as electrode passivation and poisoning effect), the need for operators with high expertise, concerns related to intermediates, process scale-up, and so on.

As a general point of view, electrochemical advanced oxidation processes (EAOPs), including EO, are the so-called green processes that are attractive mainly due to the ability to mineralize organic compounds to harmless substances and minerals using electrons, which are clean reagents, leading to efficient detoxification without any reagent waste. Careful and precise optimization of EO and devoting full consideration to influential parameters, bring about an efficient and complete mineralization, alleviating solution toxicity. Ironically, where process optimization is ignored, such as operating in nonoptimized conditions—for example, in insufficient reaction time—this results in deficiencies in mineralization and a waste of energy. In such conditions, a dilemma arises, in the sense that the "mother" substances turn into "daughter" products, some of which have much higher toxicity. In addition, in wastewaters containing organics together with chloride and/or bromide ions, chlorinated and/or brominated organic compounds may appear within the effluent of an EO system, particularly in active chlorine-based EO systems. Indeed, halogenation of aromatic hydrocarbons during EO is of significant concern. The substitution of halides with hydrogen bonded with carbon results in the formation of more toxic and more persistent compounds that are more resistant to oxidation. Inorganic by-products also include chlorite, chlorate, perchlorate, and bromate, which are formed during the electrolysis of a solution containing chloride or bromide ions. Nonetheless, if an EO system is designed and operated properly with careful optimization and monitoring of the intermediates, concerns about the intermediates can be alleviated. One should also mention that although ease of operation is commonly indicated as one of the pros of electrochemical remediation processes, proper, safe, clean, and efficient operation of an EO system is not accomplished without the applicable contributions of scientists and experts.

Economic viability is a crucial parameter in the application of a process. In fact, when a process is optimized at laboratory scale, its feasibility and extension to the industrial scale is strongly dependent on economic issues. Within EO, the major influential economic factors involve costs related to EEC and the electrodes. As previously mentioned, EO is a green remediation process in which energy is consumed for the purpose of mineralizing organics. Although chemicals are not generally added during the operation, an electrical charge is applied, which is of paramount importance, so that researchers usually take it into account, calculating or estimating the amount of EEC and also the CE. Therefore, the need for electrical energy within EO is considered to be an important disadvantage of the process. Nevertheless, there are various strategies and solutions to decrease the amount of EEC in an EO system. For instance, attempts have been made in electrode fabrication to decrease EEC, such as the low-temperature diamond deposition method and the application of conductive diamonds. Likewise, the use of electrode materials with high oxygen overpotential prevents the drain of electricity, thereby lowering operational costs. Furthermore, the proper design of an EO cell through which mass transportation is enhanced leads to lower EEC. In addition, the application of an EO stage before biological treatment is favored, by which recalcitrant compounds are broken down into substances that are more easily biodegradable. In the case of costs related to the electrodes, numerous attempts have been made in the fabrication of electrodes and great achievements have been made in this way. BDD electrodes are supposed to be the most promising electrodes in EO, due to the features previously stated in this chapter. It is worthwhile to consider that, despite the significant cost of BDD electrodes, the advantages of applying them far outweigh the disadvantage of their high cost. Using BDD, highly efficient EO is obtained using less electrical energy, which is favored at EO scale within industries and other applications.

Electrodes and their related issues are the other challenging issues in the application of EO. Apart from their costs, as discussed, operational problems related to electrodes are also considerable. In fact, electrode passivation or fouling by oligomeric or polymeric materials is one of the most challenging issues in EO processes. Thanks to the electrogeneration of active, electrolytic hydroxyl radicals on BDD anodes, polymeric substances are oxidized on the anode surface, while anodic corrosion may be the weak point of these electrodes. In this case, many studies have also investigated and focused on solutions to overcome electrode deactivation.

Another issue in EO processes is related to the supporting electrolyte, which is required for electrical charge transfer. In fact, the addition of the supporting electrolyte is not favored, mainly because it gives rise to solution TDS and, generally, induces impurities. On the other hand, textile wastewater usually contains enough electrolytes that have been added to it previously during the dyeing function in a textile mill to achieve better fixation. In this way, sodium chloride or sodium sulfate are the most commonly used electrolytes, particularly sodium chloride, which is cheaper. Consequently, sodium chloride in the dye wastewater may inevitably act as an electrolyte during EO treatment. Likewise, the initial presence of chloride in wastewater prompts the selection and application of active chlorine-based EO. Nevertheless, there may be cases in which the addition of an electrolyte is needed. However, most often, simple supporting electrolytes are added to the solution. Research has been conducted indicating that ion-exchange resins are promising substitutions that prevent workup challenges.

In general, EO as a green, promising process is attracting much attention and is continually undergoing numerous investigations. Research is shifting toward the use of new anode materials and applying nanomaterials as heterogeneous catalysts, especially within Fenton-based electrochemical systems. The application of EO in combination with other processes such as photolysis, sonolysis, and microwave technologies is also being explored.

REFERENCES

Abdelsalam, M. E. and P. R. Birkin. 2002. A study investigating the sonoelectrochemical degradation of an organic compound employing Fenton's reagent. *Physical Chemistry Chemical Physics* 4:5340–45.

Abdessamad, N., H. Akrout, G. Hamdaoui, et al. 2013. Evaluation of the efficiency of monopolar and bipolar BDD electrodes for electrochemical oxidation of anthraquinone textile synthetic effluent for reuse. *Chemosphere* 93:1309–16.

Ai, Z., H. Xiao, T. Mei, et al. 2008. Electro-Fenton degradation of Rhodamine B based on a composite cathode of Cu_2O nanocubes and carbon nanotubes. *The Journal of Physical Chemistry C* 112:11929–35.

Akrout, H. and L. Bousselmi. 2013. Water treatment for color and COD removal by electrochemical oxidation on boron-doped diamond anode. *Arabian Journal of Geosciences* 6:5033–41.

Akrout, H., S. Jellali, and L. Bousselmi. 2015. Enhancement of methylene blue removal by anodic oxidation using BDD electrode combined with adsorption onto sawdust. *Comptes Rendus Chimie* 18:110–20.

Alfaro, M. A. Q., S. Ferro, C. A. Martínez-Huitle, et al. 2006. Boron doped diamond electrode for the wastewater treatment. *Journal of the Brazilian Chemical Society* 17:227–36.

Ali, N., A. Hameed and S. Ahmed. 2009. Physicochemical characterization and bioremediation perspective of textile effluent, dyes and metals by indigenous bacteria. *Journal of Hazardous Materials* 164:322–28.

Almomani, F. and E. A. Baranova. 2012. Kinetic study of electro-Fenton oxidation of azo dyes on boron-doped diamond electrode. *Environmental Technology* 34:1473–79.

Ammar, S., M. Asma, N. Oturan, et al. 2012. Electrochemical degradation of anthraquinone dye alizarin red: Role of the electrode material. *Current Organic Chemistry* 16:1978–85.

Anglada, Á., A. Urtiaga, and I. Ortiz. 2009. Contributions of electrochemical oxidation to waste-water treatment: Fundamentals and review of applications. *Journal of Chemical Technology & Biotechnology* 84:1747–55.

Aquino, J. M., R. C. Rocha-Filho, L. A. M. Ruotolo, et al. 2014. Electrochemical degradation of a real textile wastewater using β-PbO_2 and DSA® anodes. *Chemical Engineering Journal* 251:138–45.

Araújo, D. M. D., C. Sáez, C. A. Martínez-Huitle, et al. 2015. Influence of mediated processes on the removal of Rhodamine with conductive-diamond electrochemical oxidation. *Applied Catalysis B: Environmental* 166–167:454–59.

Asghar, A., A. A. Abdul Raman, and W. M. A. Wan Daud. 2015. Advanced oxidation processes for in-situ production of hydrogen peroxide/hydroxyl radical for textile wastewater treatment: A review. *Journal of Cleaner Production* 87:826–38.

Awad, H. S. and N. A. Galwa. 2005. Electrochemical degradation of acid blue and basic brown dyes on Pb/PbO_2 electrode in the presence of different conductive electrolyte and effect of various operating factors. *Chemosphere* 61:1327–35.

Babuponnusami, A. and K. Muthukumar. 2012. Advanced oxidation of phenol: A comparison between Fenton, electro-Fenton, sono-electro-Fenton and photo-electro-Fenton processes. *Chemical Engineering Journal* 183:1–9.

Bañuelos, J. A., A. El-Ghenymy, F. J. Rodríguez, et al. 2014. Study of an air diffusion activated carbon packed electrode for an electro-Fenton wastewater treatment. *Electrochimica Acta* 140:412–18.

Barros, W. R. P., P. C. Franco, J. R. Steter, et al. 2014. Electro-Fenton degradation of the food dye amaranth using a gas diffusion electrode modified with cobalt (II) phthalocyanine. *Journal of Electroanalytical Chemistry* 722–723:46–53.

Basiri Parsa, J., Z. Merati, and M. Abbasi. 2013. Modeling and optimizing of electrochemical oxidation of C.I. Reactive Orange 7 on the Ti/Sb–SnO_2 as anode via response surface methodology. *Journal of Industrial and Engineering Chemistry* 19:1350–55.

Bedoui, A., M. F. Ahmadi, N. Bensalah, et al. 2009. Comparative study of Eriochrome Black T treatment by BDD-anodic oxidation and Fenton process. *Chemical Engineering Journal* 146:98–104.

Bensalah, N., M. A. Q. Alfaro, and C. A. Martínez-Huitle. 2009. Electrochemical treatment of synthetic wastewaters containing Alphazurine A dye. *Chemical Engineering Journal* 149:348–52.

Bhaskar Raju, G., M. Thalamadai Karuppiah, S. S. Latha, et al. 2009. Electrochemical pretreatment of textile effluents and effect of electrode materials on the removal of organics. *Desalination* 249:167–74.

Bhatnagar, R., H. Joshi, I. D. Mall, et al. 2014. Electrochemical oxidation of textile industry wastewater by graphite electrodes. *Journal of Environmental Science and Health, Part A* 49:955–66.

Bonfatti, F., S. Ferro, F. Lavezzo, et al. 2000. Electrochemical incineration of glucose as a model organic substrate II. Role of active chlorine mediation. *Journal of the Electrochemical Society* 147:592–96.

Bouasla, C., M. E.-H. Samar, and H. Bendjama. 2013. Kinetics study and neural network modeling of degradation of naphtol blue black by electro-Fenton process: Effects of anions, metal ions, and organic compound. *Desalination and Water Treatment* 52:6733–44.

Brillas, E. and C. A. Martínez-Huitle. 2015. Decontamination of wastewaters containing synthetic organic dyes by electrochemical methods. An updated review. *Applied Catalysis B: Environmental* 166–167:603–43.

Brillas, E., I. Sirés, and M. A. Oturan. 2009. Electro-Fenton process and related electrochemical technologies based on Fenton's reaction chemistry. *Chemical Reviews* 109:6570–631.

Carneiro, P. A., C. S. Fugivara, R. F. Nogueira, et al. 2002. A comparative study on chemical and electrochemical degradation of Reactive Blue 4 dye. *Portugaliae Electrochimica Acta* 21:49–67.

Catanho, M., G. R. P. Malpass, and A. J. Motheo. 2006. Photoelectrochemical treatment of the dye Reactive Red 198 using DSA® electrodes. *Applied Catalysis B: Environmental* 62:193–200.

Chaiyont, R., C. Badoe, C. Ponce De León, et al. 2013. Decolorization of methyl orange dye at IrO_2-SnO_2-Sb_2O_5 coated titanium anodes. *Chemical Engineering & Technology* 36:123–29.

Chen, G. 2004. Electrochemical technologies in wastewater treatment. *Separation and Purification Technology* 38:11–41.

Chen, X. and G. Chen. 2006. Anodic oxidation of Orange II on Ti/BDD electrode: Variable effects. *Separation and Purification Technology* 48:45–49.

Chen, X., F. Gao and G. Chen. 2005. Comparison of Ti/BDD and Ti/SnO_2-Sb_2O_5 electrodes for pollutant oxidation. *Journal of Applied Electrochemistry* 35:185–91.

Ciríaco, L., D. Santos, M. J. Pacheco, et al. 2011. Anodic oxidation of organic pollutants on a Ti/SnO_2-Sb_2O_4 anode. *Journal of Applied Electrochemistry* 41:577–87.

Comninellis, C. 1994. Electrocatalysis in the electrochemical conversion/combustion of organic pollutants for waste water treatment. *Electrochimica Acta* 39:1857–62.

Costa, C. R., F. Montilla, E. Morallón, et al. 2009. Electrochemical oxidation of Acid Black 210 dye on the boron-doped diamond electrode in the presence of phosphate ions: Effect of current density, pH, and chloride ions. *Electrochimica Acta* 54:7048–55.

Cruz-González, K., O. Torres-López, A. García-León, et al. 2010. Determination of optimum operating parameters for Acid Yellow 36 decolorization by electro-Fenton process using BDD cathode. *Chemical Engineering Journal* 160:199–206.

Del Río, A. I., J. Fernández, J. Molina, et al. 2011. Electrochemical treatment of a synthetic wastewater containing a sulphonated azo dye. Determination of naphthalenesulphonic compounds produced as main by-products. *Desalination* 273:428–35.

Diagne, M., V. Sharma, N. Oturan, et al. 2014. Depollution of indigo dye by anodic oxidation and electro-Fenton using B-doped diamond anode. *Environmental Chemistry Letters* 12:219–24.

Djafarzadeh, N., M. Zarei, B. Behjati, et al. 2013. Optimization of the oxalate catalyzed photoelectro-Fenton process under visible light for removal of Reactive Red 195 using a carbon paper cathode. *Research on Chemical Intermediates* 39:3355–69.

Dos Santos, A. B., F. J. Cervantes, and J. B. Van Lier. 2007. Review paper on current technologies for deco-lourisation of textile wastewaters: Perspectives for anaerobic biotechnology. *Bioresource Technology* 98:2369–85.

Duan, T., Y. Chen, Q. Wen, et al. 2015. Novel composition graded Ti/Ru–Sb–SnO$_2$ electrode synthesized by selective electrodeposition and its application for electrocatalytic decolorization of dyes. *The Journal of Physical Chemistry C* 119:7780–90.

El-Desoky, H. S., M. M. Ghoneim, and N. M. Zidan. 2010. Decolorization and degradation of Ponceau S azo-dye in aqueous solutions by the electrochemical advanced Fenton oxidation. *Desalination* 264:143–50.

El-Ghenymy, A., F. Centellas, J. A. Garrido, et al. 2014. Decolorization and mineralization of Orange G azo dye solutions by anodic oxidation with a boron-doped diamond anode in divided and undivided tank reactors. *Electrochimica Acta* 130:568–76.

Eslami, A., M. Moradi, F. Ghanbari, et al. 2013. Decolorization and COD removal from real textile wastewater by chemical and electrochemical Fenton processes: A comparative study. *Journal of Environmental Health Science and Engineering* 11:1–8.

Fan, L., Y. Zhou, W. Yang, et al. 2008. Electrochemical degradation of aqueous solution of amaranth azo dye on ACF under potentiostatic model. *Dyes and Pigments* 76:440–46.

Faouzi, A. M., B. Nasr, and G. Abdellatif. 2007. Electrochemical degradation of anthraquinone dye Alizarin Red S by anodic oxidation on boron-doped diamond. *Dyes and Pigments* 73:86–89.

Feeny, D., F. Berkes, B. Mccay, et al. 1990. The tragedy of the commons: Twenty-two years later. *Human Ecology* 18:1–19.

Flox, C., S. Ammar, C. Arias, et al. 2006. Electro-Fenton and photoelectro-Fenton degradation of indigo car-mine in acidic aqueous medium. *Applied Catalysis B: Environmental* 67:93–104.

Forgacs, E., T. Cserháti, and G. Oros. 2004. Removal of synthetic dyes from wastewaters: A review. *Environment International* 30:953–71.

Fóti, G., D. Gandini, C. Comninellis, et al. 1999. Oxidation of organics by intermediates of water discharge on IrO$_2$ and synthetic diamond anodes. *Electrochemical and Solid-State Letters* 2:228–30.

Garcia-Segura, S. and E. Brillas. 2014. Advances in solar photoelectro-Fenton: Decolorization and mineral-ization of the Direct Yellow 4 diazo dye using an autonomous solar pre-pilot plant. *Electrochimica Acta* 140:384–95.

Ghalwa, N. A., M. Gaber, A. M. Khedr, et al. 2012. Comparative study of commercial oxide electrodes per-formance in electrochemical degradation of Reactive Orange 7 dye in aqueous solutions. *International Journal of Electrochemical Science* 7:6044–58.

Ghanbari, F. and M. Moradi. 2015. A comparative study of electrocoagulation, electrochemical Fenton, electro-Fenton and peroxi-coagulation for decolorization of real textile wastewater: Electrical energy consump-tion and biodegradability improvement. *Journal of Environmental Chemical Engineering* 3:499–506.

Ghanbari, F., M. Moradi, A. Eslami, et al. 2014a. Electrocoagulation/flotation of textile wastewater with simultaneous application of aluminum and iron as anode. *Environmental Processes* 1:447–57.

Ghanbari, F., M. Moradi, and M. Manshouri. 2014b. Textile wastewater decolorization by zero valent iron activated peroxymonosulfate: Compared with zero valent copper. *Journal of Environmental Chemical Engineering* 2:1846–51.

Ghanbari, F., M. Moradi, F. Mehdipour, et al. 2016. Simultaneous application of copper and PbO$_2$ anodes for electrochemical treatment of olive oil mill wastewater. *Desalination and Water Treatment* 57:5828–36.

Ghanbari, F., M. Moradi, A. Mohseni-Bandpei, et al. 2014c. Simultaneous application of iron and aluminum anodes for nitrate removal: A comprehensive parametric study. *International Journal of Environmental Science and Technology* 11:1653–60.

Grimm, J., D. Bessarabov, and R. Sanderson. 1998. Review of electro-assisted methods for water purification. *Desalination* 115:285–94.

Guenfoud, F., M. Mokhtari, and H. Akrout. 2014. Electrochemical degradation of malachite green with BDD electrodes: Effect of electrochemical parameters. *Diamond and Related Materials* 46:8–14.

Hai, F. I., K. Yamamoto, and K. Fukushi. 2007. Hybrid treatment systems for dye wastewater. *Critical Reviews in Environmental Science and Technology* 37:315–77.

Hamza, M., R. Abdelhedi, E. Brillas, et al. 2009. Comparative electrochemical degradation of the triphe-nylmethane dye methyl violet with boron-doped diamond and Pt anodes. *Journal of Electroanalytical Chemistry* 627:41–50.

Hao, O. J., H. Kim, and P.-C. Chiang. 2000. Decolorization of wastewater. *Critical Reviews in Environmental Science and Technology* 30:449–505.

Hattori, S., M. Doi, E. Takahashi, et al. 2003. Electrolytic decomposition of amaranth dyestuff using diamond electrodes. *Journal of Applied Electrochemistry* 33:85–91.

He, W., X. Yan, H. Ma, et al. 2013. Degradation of methyl orange by electro-Fenton-like process in the presence of chloride ion. *Desalination and Water Treatment* 51:6562–71.

He, Z., C. Gao, M. Qian, et al. 2014. Electro-Fenton process catalyzed by Fe_3O_4 magnetic nanoparticles for degradation of C.I. Reactive Blue 19 in aqueous solution: Operating conditions, influence, and mechanism. *Industrial & Engineering Chemistry Research* 53:3435–47.

Hmani, E., Y. Samet, and R. Abdelhédi. 2012. Electrochemical degradation of auramine-O dye at boron-doped diamond and lead dioxide electrodes. *Diamond and Related Materials* 30:1–8.

Iglesias, O., M. A. Fernández De Dios, E. Rosales, et al. 2013. Optimisation of decolourisation and degradation of Reactive Black 5 dye under electro-Fenton process using Fe alginate gel beads. *Environmental Science and Pollution Research* 20:2172–83.

Jaafarzadeh, N., F. Ghanbari, and M. Moradi. 2015. Photo-electro-oxidation assisted peroxymonosulfate for decolorization of Acid Brown 14 from aqueous solution. *Korean Journal of Chemical Engineering* 32:458–64.

Jiang, C.-C. and J.-F. Zhang. 2007. Progress and prospect in electro-Fenton process for wastewater treatment. *Journal of Zhejiang University SCIENCE A* 8:1118–25.

Jović, M., D. Stanković, D. Manojlović, et al. 2013. Study of the electrochemical oxidation of reactive textile dyes using platinum electrode. *Int. J. Electrochem. Sci.* 8:168–83.

Kapałka, A., G. Fóti, and C. Comninellis 2010. Basic principles of the electrochemical mineralization of organic pollutants for wastewater treatment. *In:* Comninellis, C. and Chen, G. (eds.) *Electrochemistry for the Environment.* 1–23. Springer: New York.

Kariyajjanavar, P., J. Narayana, and Y. A. Nayaka. 2013. Degradation of textile dye C.I. Vat Black 27 by electrochemical method by using carbon electrodes. *Journal of Environmental Chemical Engineering* 1:975–80.

Khataee, A., B. Vahid, B. Behjati, et al. 2014. Kinetic modeling of a triarylmethane dye decolorization by photoelectro-Fenton process in a recirculating system: Nonlinear regression analysis. *Chemical Engineering Research and Design* 92:362–67.

Kim, S., C. Park, T.-H. Kim, et al. 2003. COD reduction and decolorization of textile effluent using a combined process. *Journal of Bioscience and Bioengineering* 95:102–5.

Kusmierek, E., E. Chrzescijanska, M. Szadkowska-Nicze, et al. 2011. Electrochemical discolouration and degradation of reactive dichlorotriazine dyes: Reaction pathways. *Journal of Applied Electrochemistry* 41:51–62.

Labiadh, L., M. A. Oturan, M. Panizza, et al. 2015. Complete removal of AHPS synthetic dye from water using new electro-Fenton oxidation catalyzed by natural pyrite as heterogeneous catalyst. *Journal of Hazardous Materials* 297:34–41.

Le, T. X. H., M. Bechelany, S. Lacour, et al. 2015. High removal efficiency of dye pollutants by electron-Fenton process using a graphene based cathode. *Carbon* 94:1003–11.

Lei, H., H. Li, Z. Li, et al. 2010. Electro-Fenton degradation of cationic red X-GRL using an activated carbon fiber cathode. *Process Safety and Environmental Protection* 88:431–38.

Lei, Y., H. Liu, Z. Shen, et al. 2013. Development of a trickle bed reactor of electro-Fenton process for wastewater treatment. *Journal of Hazardous Materials* 261:570–76.

Li, G., M. Zhu, J. Chen, et al. 2011. Production and contribution of hydroxyl radicals between the DSA anode and water interface. *Journal of Environmental Sciences* 23:744–48.

Li, H., H. Lei, Q. Yu, et al. 2010. Effect of low frequency ultrasonic irradiation on the sonoelectro-Fenton degradation of cationic red X-GRL. *Chemical Engineering Journal* 160:417–22.

Lin, H., H. Zhang, X. Wang, et al. 2014. Electro-Fenton removal of Orange II in a divided cell: Reaction mechanism, degradation pathway and toxicity evolution. *Separation and Purification Technology* 122:533–40.

Lin, S. H. and C. F. Peng. 1994. Treatment of textile wastewater by electrochemical method. *Water Research* 28:277–82.

Liu, L., B. Li, Z. He, et al. 2015. Degradation of bromoamine acid by BDD technology: Use of Doehlert design for optimizing the reaction conditions. *Separation and Purification Technology* 146:15–23.

Liu, S., Y. Gu, S. Wang, et al. 2013. Degradation of organic pollutants by a Co_3O_4-graphite composite electrode in an electro-Fenton-like system. *Chinese Science Bulletin* 58:2340–46.

López-Grimau, V. and M. C. Gutiérrez. 2006. Decolourisation of simulated reactive dyebath effluents by electrochemical oxidation assisted by UV light. *Chemosphere* 62:106–12.

Martínez, S. S. and E. V. Uribe. 2012. Enhanced sonochemical degradation of Azure B dye by the electro-Fenton process. *Ultrasonics Sonochemistry* 19:174–78.

Martínez-Huitle, C. A. and E. Brillas. 2009. Decontamination of wastewaters containing synthetic organic dyes by electrochemical methods: A general review. *Applied Catalysis B: Environmental* 87:105–45.

Martínez-Huitle, C. A., E. V. Dos Santos, D. M. De Araújo, et al. 2012. Applicability of diamond electrode/anode to the electrochemical treatment of a real textile effluent. *Journal of Electroanalytical Chemistry* 674:103–7.

Martínez-Huitle, C. A. and S. Ferro. 2006. Electrochemical oxidation of organic pollutants for the wastewater treatment: Direct and indirect processes. *Chemical Society Reviews* 35:1324–40.

Mollah, M. Y. A., R. Schennach, J. R. Parga, et al. 2001. Electrocoagulation (EC): Science and applications. *Journal of Hazardous Materials* 84:29–41.

Moradi, M., F. Ghanbari, and E. Minaee Tabrizi. 2015. Removal of Acid Yellow 36 using Box–Behnken designed photoelectro-Fenton: A study on removal mechanisms. *Toxicological & Environmental Chemistry* 97:700–709.

Moradi, M., A. Eslami, and F. Ghanbari. 2016. Direct Blue 71 removal by electrocoagulation sludge recycling in photo-Fenton process: Response surface modeling and optimization. *Desalination and Water Treatment* 57:4659–70.

Morais, C., A. Da Silva, M. Ferreira, et al. 2013. Electrochemical degradation of methyl red using Ti/Ru0.3Ti0.7O$_2$: Fragmentation of azo group. *Electrocatalysis* 4:312–19.

Moreira, F. C., S. Garcia-Segura, V. J. P. Vilar, et al. 2013. Decolorization and mineralization of Sunset Yellow FCF azo dye by anodic oxidation, electro-Fenton, UVA photoelectro-Fenton and solar photoelectro-Fenton processes. *Applied Catalysis B: Environmental* 142–143:877–90.

Neyens, E. and J. Baeyens. 2003. A review of classic Fenton's peroxidation as an advanced oxidation technique. *Journal of Hazardous Materials* 98:33–50.

Nidheesh, P. V. and R. Gandhimathi. 2012. Trends in electro-Fenton process for water and wastewater treatment: An overview. *Desalination* 299:1–15.

Nidheesh, P. V. and R. Gandhimathi. 2013b. Removal of Rhodamine B from aqueous solution using graphite–graphite electro-Fenton system. *Desalination and Water Treatment* 52:1872–77.

Nidheesh, P., R. Gandhimathi, and S. Ramesh. 2013a. Degradation of dyes from aqueous solution by Fenton processes: A review. *Environmental Science and Pollution Research* 20:2099–132.

Nidheesh, P. V., R. Gandhimathi, and N. S. Sanjini. 2014. NaHCO3 enhanced Rhodamine B removal from aqueous solution by graphite–graphite electro Fenton system. *Separation and Purification Technology* 132:568–76.

O'Neill, C., F. R. Hawkes, D. L. Hawkes, et al. 1999. Colour in textile effluents: Sources, measurement, discharge consents and simulation: A review. *Journal of Chemical Technology & Biotechnology* 74:1009–18.

Olvera-Vargas, H., N. Oturan, C. T. Aravindakumar, et al. 2014. Electro-oxidation of the dye Azure B: Kinetics, mechanism, and by-products. *Environmental Science and Pollution Research* 21:8379–86.

Oturan, M. A., E. Guivarch, N. Oturan, et al. 2008a. Oxidation pathways of malachite green by Fe^{3+}-catalyzed electro-Fenton process. *Applied Catalysis B: Environmental* 82:244–54.

Oturan, M. A., I. Sirés, N. Oturan, et al. 2008b. Sonoelectro-Fenton process: A novel hybrid technique for the destruction of organic pollutants in water. *Journal of Electroanalytical Chemistry* 624:329–32.

Özcan, A., Y. Şahin, A. Savaş Koparal, et al. 2008. Carbon sponge as a new cathode material for the electro-Fenton process: Comparison with carbon felt cathode and application to degradation of synthetic dye Basic Blue 3 in aqueous medium. *Journal of Electroanalytical Chemistry* 616:71–78.

Palma-Goyes, R. E., J. Vazquez-Arenas, R. A. Torres-Palma, et al. 2015. The abatement of indigo carmine using active chlorine electrogenerated on ternary Sb$_2$O$_5$-doped Ti/RuO$_2$-ZrO$_2$ anodes in a filter-press FM01-LC reactor. *Electrochimica Acta* 174:735–44.

Panakoulias, T., P. Kalatzis, D. Kalderis, et al. 2010. Electrochemical degradation of Reactive Red 120 using DSA and BDD anodes. *Journal of Applied Electrochemistry* 40:1759–65.

Panizza, M. 2010. Importance of electrode material in the electrochemical treatment of wastewater containing organic pollutants. *In:* Comninellis, C. and Chen, G. (eds.) *Electrochemistry for the Environment.* 25–54. Springer: New York.

Panizza, M. and G. Cerisola. 2007. Electrocatalytic materials for the electrochemical oxidation of synthetic dyes. *Applied Catalysis B: Environmental* 75:95–101.

Panizza, M. and G. Cerisola. 2009a. Direct and mediated anodic oxidation of organic pollutants. *Chemical Reviews* 109:6541–69.

Panizza, M. and G. Cerisola. 2009b. Electro-Fenton degradation of synthetic dyes. *Water Research* 43:339–44.

Petrucci, E., L. Di Palma, R. Lavecchia, et al. 2015. Treatment of diazo dye Reactive Green 19 by anodic oxidation on a boron-doped diamond electrode. *Journal of Industrial and Engineering Chemistry* 26:116–21.

Pinheiro, H. M., E. Touraud, and O. Thomas. 2004. Aromatic amines from azo dye reduction: Status review with emphasis on direct UV spectrophotometric detection in textile industry wastewaters. *Dyes and Pigments* 61:121–39.

Rajeshwar, K., J. G. Ibanez, and G. M. Swain. 1994. Electrochemistry and the environment. *Journal of Applied Electrochemistry* 24:1077–91.

Rajkumar, D., B. J. Song, and J. G. Kim. 2007. Electrochemical degradation of Reactive Blue 19 in chloride medium for the treatment of textile dyeing wastewater with identification of intermediate compounds. *Dyes and Pigments* 72:1–7.

Rajkumar, K. and M. Muthukumar. 2015. Response surface optimization of electro-oxidation process for the treatment of C.I. Reactive Yellow 186 dye: Reaction pathways. *Applied Water Science* 1–16.

Ramírez, C., A. Saldaña, B. Hernández, et al. 2013. Electrochemical oxidation of methyl orange azo dye at pilot flow plant using BDD technology. *Journal of Industrial and Engineering Chemistry* 19:571–79.

Ren, W., Q. Peng, Z. A. Huang, et al. 2015. Effect of pore structure on the electro-Fenton activity of ACF@ OMC cathode. *Industrial & Engineering Chemistry Research* 54:8492–99.

Rocha, J. H. B., M. M. S. Gomes, E. V. D. Santos, et al. 2014. Electrochemical degradation of novacron yellow C-RG using boron-doped diamond and platinum anodes: Direct and indirect oxidation. *Electrochimica Acta* 140:419–26.

Rodríguez, F. A., E. P. Rivero, L. Lartundo-Rojas, et al. 2014. Preparation and characterization of Sb_2O_5-doped Ti/RuO_2-ZrO_2 for dye decolorization by means of active chlorine. *Journal of Solid State Electrochemistry* 18:3153–62.

Rodriguez, J., M. A. Rodrigo, M. Panizza, et al. 2009. Electrochemical oxidation of Acid Yellow 1 using diamond anode. *Journal of Applied Electrochemistry* 39:2285–89.

Rosales, E., M. Pazos, and M. A. Sanromán. 2012. Advances in the electro-Fenton process for remediation of recalcitrant organic compounds. *Chemical Engineering & Technology* 35:609–17.

Ruiz, E. J., C. Arias, E. Brillas, et al. 2011. Mineralization of Acid Yellow 36 azo dye by electro-Fenton and solar photoelectro-Fenton processes with a boron-doped diamond anode. *Chemosphere* 82:495–501.

Sakalis, A., K. Fytianos, U. Nickel, et al. 2006. A comparative study of platinised titanium and niobe/synthetic diamond as anodes in the electrochemical treatment of textile wastewater. *Chemical Engineering Journal* 119:127–33.

Sala, M., A. Del Río, J. Molina, et al. 2012. Influence of cell design and electrode materials on the decolouration of dyeing effluents. *International Journal of Electrochemical Science* 7:12470–88.

Sales Solano, A. M., C. K. Costa De Araújo, J. Vieira De Melo, et al. 2013. Decontamination of real textile industrial effluent by strong oxidant species electrogenerated on diamond electrode: Viability and disadvantages of this electrochemical technology. *Applied Catalysis B: Environmental* 130–131:112–20.

Salleh, M. A. M., D. K. Mahmoud, W. A. W. A. Karim, et al. 2011. Cationic and anionic dye adsorption by agricultural solid wastes: A comprehensive review. *Desalination* 280:1–13.

Sanromán, M. A., M. Pazos, M. T. Ricart, et al. 2004. Electrochemical decolourisation of structurally different dyes. *Chemosphere* 57:233–39.

Särkkä, H., A. Bhatnagar, and M. Sillanpää. 2015. Recent developments of electro-oxidation in water treatment: A review. *Journal of Electroanalytical Chemistry* 754:46–56.

Scialdone, O., A. Galia, and S. Sabatino. 2014. Abatement of Acid Orange 7 in macro and micro reactors. Effect of the electrocatalytic route. *Applied Catalysis B: Environmental* 148–149:473–83.

Scialdone, O., S. Randazzo, A. Galia, et al. 2009. Electrochemical oxidation of organics in water: Role of operative parameters in the absence and in the presence of NaCl. *Water Research* 43:2260–72.

Shestakova, M., M. Vinatoru, T. J. Mason, et al. 2015. Sonoelectrocatalytic decomposition of methylene blue using Ti/Ta_2O_5-SnO_2 electrodes. *Ultrasonics Sonochemistry* 23:135–41.

Simond, O., V. Schaller, and C. Comninellis. 1997. Theoretical model for the anodic oxidation of organics on metal oxide electrodes. *Electrochimica Acta* 42:2009–12.

Singh, K. and S. Arora. 2011. Removal of synthetic textile dyes from wastewaters: A critical review on present treatment technologies. *Critical Reviews in Environmental Science and Technology* 41:807–78.

Singh, M. M., Z. Szafran, and J. G. Ibanez. 1998. Laboratory experiments on the electrochemical remediation of environment. Part 4: Color removal of simulated wastewater by electrocoagulation-electroflotation. *Journal of Chemical Education* 75:1040.

Sirés, I., E. Brillas, M. Oturan, et al. 2014. Electrochemical advanced oxidation processes: Today and tomorrow. A review. *Environmental Science and Pollution Research* 21:8336–67.

Sirés, I., J. A. Garrido, R. M. Rodríguez, et al. 2007. Catalytic behavior of the Fe^{3+}/Fe^{2+} system in the electro-Fenton degradation of the antimicrobial chlorophene. *Applied Catalysis B: Environmental* 72:382–94.

Socha, A., E. Chrzescijanska, and E. Kusmierek. 2005. Electrochemical and photoelectrochemical treatment of 1-aminonaphthalene-3,6-disulphonic acid. *Dyes and Pigments* 67:71–75.

Solano, A. M. S., S. Garcia-Segura, C. A. Martínez-Huitle, et al. 2015. Degradation of acidic aqueous solutions of the diazo dye Congo red by photo-assisted electrochemical processes based on Fenton's reaction chemistry. *Applied Catalysis B: Environmental* 168–169:559–71.

Sun, J., H. Lu, L. Du, et al. 2011. Anodic oxidation of anthraquinone dye Alizarin Red S at Ti/BDD electrodes. *Applied Surface Science* 257:6667–71.

Thiam, A., E. Brillas, F. Centellas, et al. 2015a. Electrochemical reactivity of Ponceau 4R (food additive E124) in different electrolytes and batch cells. *Electrochimica Acta* 173:523–33.

Thiam, A., E. Brillas, J. A. Garrido, et al. 2016. Routes for the electrochemical degradation of the artificial food azo-colour Ponceau 4R by advanced oxidation processes. *Applied Catalysis B: Environmental* 180:227–36.

Thiam, A., I. Sirés, F. Centellas, et al. 2015b. Decolorization and mineralization of Allura Red AC azo dye by solar photoelectro-Fenton: Identification of intermediates. *Chemosphere* 136:1–8.

Trasatti, S. 1987. Progress in the understanding of the mechanism of chlorine evolution at oxide electrodes. *Electrochimica Acta* 32:369–82.

Tsantaki, E., T. Velegraki, A. Katsaounis, et al. 2012. Anodic oxidation of textile dyehouse effluents on boron-doped diamond electrode. *Journal of Hazardous Materials* 207–208:91–96.

Walsh, F. 2001. Electrochemical technology for environmental treatment and clean energy conversion. *Pure and Applied Chemistry* 73:1819–37.

Wang, C.-T., W.-L. Chou, M.-H. Chung, et al. 2010. COD removal from real dyeing wastewater by electro-Fenton technology using an activated carbon fiber cathode. *Desalination* 253:129–34.

Wang, C.-T., W.-L. Chou, Y.-M. Kuo, et al. 2009. Paired removal of color and COD from textile dyeing wastewater by simultaneous anodic and indirect cathodic oxidation. *Journal of Hazardous Materials* 169:16–22.

Wang, J. L. and L. J. Xu. 2011. Advanced oxidation processes for wastewater treatment: Formation of hydroxyl radical and application. *Critical Reviews in Environmental Science and Technology* 42:251–325.

Yahiaoui, I., F. Aissani-Benissad, F. Fourcade, et al. 2014. Combination of an electrochemical pretreatment with a biological oxidation for the mineralization of nonbiodegradable organic dyes: Basic Yellow 28 dye. *Environmental Progress & Sustainable Energy* 33:160–69.

Yavuz, Y., A. Savaş Koparal, and Ü. B. Öğütveren. 2011. Electrochemical oxidation of Basic Blue 3 dye using a diamond anode: Evaluation of colour, COD and toxicity removal. *Journal of Chemical Technology & Biotechnology* 86:261–65.

Yavuz, Y. and R. Shahbazi. 2012. Anodic oxidation of Reactive Black 5 dye using boron doped diamond anodes in a bipolar trickle tower reactor. *Separation and Purification Technology* 85:130–36.

Yi, F. and S. Chen. 2008. Electrochemical treatment of Alizarin Red S dye wastewater using an activated carbon fiber as anode material. *Journal of Porous Materials* 15:565–69.

Yu, F., M. Zhou, and X. Yu. 2015a. Cost-effective electro-Fenton using modified graphite felt that dramatically enhanced on H_2O_2 electro-generation without external aeration. *Electrochimica Acta* 163:182–89.

Yu, F., M. Zhou, L. Zhou, et al. 2014. A novel electro-Fenton process with H_2O_2 generation in a rotating disk reactor for organic pollutant degradation. *Environmental Science & Technology Letters* 1:320–24.

Yu, X., M. Zhou, G. Ren, et al. 2015b. A novel dual gas diffusion electrodes system for efficient hydrogen peroxide generation used in electro-Fenton. *Chemical Engineering Journal* 263:92–100.

Zarei, M., A. Niaei, D. Salari, et al. 2010. Removal of four dyes from aqueous medium by the peroxi-coagulation method using carbon nanotube–PTFE cathode and neural network modeling. *Journal of Electroanalytical Chemistry* 639:167–74.

Zarei, M., D. Salari, A. Niaei, et al. 2009. Peroxi-coagulation degradation of C.I. Basic Yellow 2 based on carbon-PTFE and carbon nanotube-PTFE electrodes as cathode. *Electrochimica Acta* 54:6651–60.

Zhang, C., L. Liu, W. Li, et al. 2014a. Electrochemical degradation of Acid Orange II dye with boron-doped diamond electrode: Role of operating parameters in the absence and in the presence of NaCl. *Journal of Electroanalytical Chemistry* 726:77–83.

Zhang, F., C. Feng, W. Li, et al. 2014b. Indirect electrochemical oxidation of dye wastewater containing Acid Orange 7 using Ti/RuO$_2$-Pt Electrode. *International Journal of Electrochemical Science* 9:943–54.

Zhao, H.-Z., Y. Sun, L.-N. Xu, et al. 2010. Removal of Acid Orange 7 in simulated wastewater using a three-dimensional electrode reactor: Removal mechanisms and dye degradation pathway. *Chemosphere* 78:46–51.

Zhou, M. and J. He. 2007. Degradation of azo dye by three clean advanced oxidation processes: Wet oxidation, electrochemical oxidation and wet electrochemical oxidation: A comparative study. *Electrochimica Acta* 53:1902–10.

Zhou, M., H. Särkkä, and M. Sillanpää. 2011. A comparative experimental study on methyl orange degradation by electrochemical oxidation on BDD and MMO electrodes. *Separation and Purification Technology* 78:290–97.

Zongo, I., A. H. Maiga, J. Wéthé, et al. 2009. Electrocoagulation for the treatment of textile wastewaters with Al or Fe electrodes: Compared variations of COD levels, turbidity and absorbance. *Journal of Hazardous Materials* 169:70–76.

4 Advanced Oxidation Processes Using Nanomaterials

Emilio Rosales, Marta Pazos, and M. Ángeles Sanromán

CONTENTS

ABSTRACT

Advanced oxidation processes (AOPs) are currently being proposed as an alternative to the treatment of emerging pollutants in wastewater. These processes are known for their ability to mineralize a wide range of organic compounds. AOPs involve the generation of highly reactive radical species, predominantly the hydroxyl radical, which is strong enough to nonselectively oxidize most organic compounds through chain reactions. There are different processes in which this radical can be generated; notwithstanding, nowadays, the use of the Fenton reagent is attracting the attention of the scientific community. The Fenton reaction is a catalytic AOP that combines hydrogen peroxide (H_2O_2) and ferrous iron, as the catalyst, to obtain the hydroxyl radical that can oxidize specific contaminants or wastewaters. Furthermore, the Fenton process can be coupled with electrochemical treatment—the electro-Fenton process—which has a synergetic effect in the efficiency of the degradation.

In the last decades, the manipulation of matter on an atomic, molecular, and supramolecular scale has provided useful applications for different sectors. Therefore, nanotechnology can enhance diverse processes as a result of the variety of potential applications of the new designed materials. As expected, nanotechnology has been introduced into AOPs; specifically, in the Fenton treatment. In this chapter, the use of iron nanoparticles in the Fenton and electro-Fenton processes has been reviewed for their potential application in the treatment of emerging pollutants in wastewater by AOPs. As a general overview, it was found that these new materials can increase the efficacy of the process by acting as catalyst or even in their incorporation as electrode material. Thus, the use of nanomaterials opens up new methods for the treatment of recalcitrant pollutants.

4.1   INTRODUCTION

Wastewater usually contains various pollutants, depending on what it was used for. Thus, the composition of wastewater from industrial sources varies from one industry to another, and therefore requires treatment based on the specific industry that produces a particular effluent. Moreover, the specifications for discharging industrial effluents into the public sewer network or surface watercourses depend on several factors, but the most important is related to the toxicity and the content of organic and inorganic elements. In many parts of the world, health problems and diseases have often been caused by discharging untreated or inadequately treated wastewater. Nowadays, there are several conventional physical, biological, and chemical processes used for its treatment. However, some contaminants found in wastewater are recalcitrant to commonly applied processes. In this context, the application of new and combined technologies in which the synergic effect of the different strategies of decontamination can provide an alternative that increases the efficiencies of individual treatments is being pursued.

In the last years, techniques such as AOPs have offered effective and rapid alternative methodology for the treatment of emerging pollutants in wastewater. AOPs are based on the generation of highly reactive radical species, predominantly the hydroxyl radical ($^{\bullet}OH$), which is strong enough to nonselectively oxidize most organic compounds through chain reactions. The *in situ* generation of highly reactive species, such as the hydroxyl radical, allows the degradation of organic matter to take place (Table 4.1: Equations 4.1 through 4.19).

A wide variety of AOPs are available:

- Chemical oxidation processes using hydrogen peroxide, ozone, ozone and peroxide combined, hypochlorite, Fenton's reagent, and so on
- Ultraviolet (UV) radiation-enhanced oxidation such as UV/ozone, UV/hydrogen peroxide, UV/air
- Wet air oxidation and catalytic wet air oxidation

AOPs are particularly appropriate for effluents containing refractory, toxic, or nonbiodegradable materials. The processes offer several advantages over biological or physical processes, including

- Process operability
- Absence of secondary wastes
- Ability to handle fluctuating flow rates and compositions

Among the AOPs, the processes based on Fenton's reagent have attracted particular attention in the research community. The Fenton process consists of the reaction of Fe^{2+} species with H_2O_2 under strong acid conditions to generate highly reactive hydroxyl radicals (Table 4.1: Equation 4.1).

It is important to note that, similarly to other AOPs, the processes based on homogeneous Fenton reactions present some disadvantages, among which are: (1) risks and high costs associated with storage and transport of H_2O_2; (2) management of large amounts of chemicals used to acidify the medium at pH values between 2 and 4; (3) accumulations of iron sludge that need to be removed after the treatment—the separation is not simple, due to the colloidal characteristics of the resulting dispersion; and that (4) complete mineralization is not always achieved, due to the formation of iron complexes. Recent studies have attempted to overcome these drawbacks by using heterogeneous catalysts (Table 4.1: Equation 4.23), coupled with UV light (Table 4.1: Equation 4.25) or electric fields (Table 4.1: Equation 4.20). These processes are known as heterogeneous Fenton, photo-Fenton, and electro-Fenton, respectively, and by applying of these combined processes, it is possible to abate most of the drawbacks that have been detected.

Our group has been studying the degradation of chloramphenicol by different AOPs: anodic oxidation, Fenton, Fenton with successive H_2O_2 additions, and electro-Fenton. As shown in Figure 4.1, the best results were obtained by a combination of the electrochemical process and the Fenton

TABLE 4.1
Reactions Involved in Some AOPs (Fenton, Electro-Fenton, and Photo-Fenton)

Reaction	Equation	Description	References
$Fe^{2+} + H_2O_2 \rightarrow Fe^{3+} + {}^{\bullet}OH + OH^-$	4.1	CFR: initial	Garrido-Ramírez et al. 2010; Ng et al. 2014
$Fe^{3+} + H_2O_2 \rightarrow Fe^{2+} + {}^{\bullet}OOH + H^+$	4.2	CFR: propagation	Garrido-Ramírez et al. 2010; Ng et al. 2014
$Fe^{2+} + H_2O_2 \rightarrow Fe(OOH)^{2+} + H^+$	4.3	CFR: propagation	De Laat and Gallard 1999
$Fe^{2+} + {}^{\bullet}OOH \rightarrow Fe^{3+} + HOO^-$	4.4	CFR: propagation	Garrido-Ramírez et al. 2010
$Fe^{3+} + OOH^- \rightarrow Fe^{2+} + {}^{\bullet}OOH$	4.5	CFR: propagation	De Laat and Gallard 1999
$Fe^{2+} + {}^{\bullet}OOH \rightarrow Fe^{3+} + H_2O_2$	4.6	CFR: propagation	Duesterberg et al. 2005; Duesterberg and Waite 2007
$H_2O_2 + {}^{\bullet}OH \rightarrow H_2O$	4.7	CFR: propagation	Garrido-Ramírez et al. 2010
$H_2O_2 + {}^{\bullet}OH \rightarrow H_2O + {}^{\bullet}OOH$	4.8	CFR: propagation	Duesterberg et al. 2005; Duesterberg and Waite 2007
${}^{\bullet}OOH + H_2O_2 \rightarrow H_2O + O_2 + {}^{\bullet}OH$	4.9	CFR: propagation	De Laat and Gallard 1999
$RH + {}^{\bullet}OH \rightarrow H_2O + CO_2$	4.10	CFR: propagation	Duesterberg et al. 2005; Duesterberg and Waite 2007
${}^{\bullet}R + Fe^{2+} \rightarrow RH + Fe^{3+}$	4.11	CFR: propagation	Duesterberg et al. 2005; Duesterberg and Waite 2007
${}^{\bullet}R + Fe^{3+} \rightarrow R^+ + Fe^{2+}$	4.12	CFR: propagation	Duesterberg et al. 2005; Rosales et al. 2012a
${}^{\bullet}R + {}^{\bullet}R \rightarrow R\text{-}R$	4.13	CFR: termination	Duesterberg et al. 2005; Duesterberg and Waite 2007
$Fe^{3+} + {}^{\bullet}OOH \rightarrow Fe^{2+} + O_2 + H^+$	4.14	CFR: termination	Duesterberg et al. 2005; Duesterberg and Waite 2007
$Fe^{2+} + {}^{\bullet}OH \rightarrow Fe^{3+} + OH^-$	4.15	CFR: termination	Mousset et al. 2016
${}^{\bullet}OOH + {}^{\bullet}OOH \rightarrow H_2O_2 + O_2$	4.16	CFR: termination	Duesterberg et al. 2005; Duesterberg and Waite 2007
${}^{\bullet}OH + {}^{\bullet}OH \rightarrow H_2O + O_2$	4.17	CFR: termination	Duesterberg et al. 2005; Duesterberg and Waite 2007
${}^{\bullet}OH + {}^{\bullet}OH \rightarrow H_2O_2$	4.18	CFR: termination	Buxton et al. 1988
${}^{\bullet}OOH + {}^{\bullet}OH \rightarrow H_2O + O_2$	4.19	CFR: termination	Buxton et al. 1988
$O_2 + 2H^+ + 2e^- \rightarrow H_2O_2$	4.20	Electrochemical generation of H_2O_2 (cathode)	Mousset et al. 2016
$H_2O \rightarrow H^+ + {}^{\bullet}OH + e^-$	4.21	Electrochemical generation of ${}^{\bullet}OH$ (anode)	Qiu et al. 2015
$Fe^{3+} + e^- \rightarrow Fe^{2+}$	4.22	Electrochemical regeneration of Fe^{2+} (cathode)	Li et al. 2011; Zhang et al. 2012
$S_c - Fe^{2+} + H_2O_2 \rightarrow S_c - Fe^{3+} + OH^- + {}^{\bullet}OH$	4.23	Heterogeneous Fenton reactions in catalyst surface	Hammouda et al. 2016
$S_c - Fe^{3+} + H_2O_2 \rightarrow S_c - Fe^{2+} + {}^{\bullet}OOH + H^+$	4.24	Heterogeneous Fenton reactions in catalyst surface	Hammouda et al. 2016
$Fe^{3+} + H_2O + h\nu \rightarrow Fe^{2+} + {}^{\bullet}OH + H^+$	4.25	Photo-Fenton reaction	Machulek et al. 2012

CFR, classical Fenton reaction.

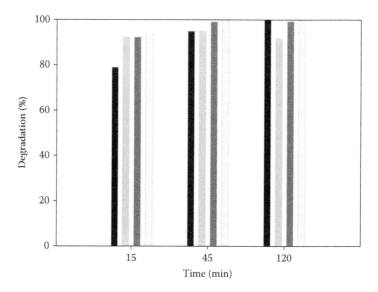

FIGURE 4.1 Comparison among anodic oxidation (black), Fenton (gray), Fenton with successive additions of H_2O_2 (dark gray), and electro-Fenton (light gray) treatments in the degradation of chloramphenicol (90 mg/L).

reaction, reaching near complete degradation after 45 min of treatment. It is clear that the worst degradation levels were obtained by the Fenton treatment alone, due to the limitations of the hydroxyl radical in the bulk solution. For this reason, the efficiency of the process was increased when H_2O_2 was added into the solution every 15 min.

4.2 HETEROGENEOUS FENTON PROCESS

In the homogeneous Fenton process, the regeneration rate of ferrous ions is slow, and this fact provokes an increase of concentration of ferric ions as insoluble complexes with reaction time. In addition, an increase in the pH of the solution is detected, and this fact results in further production of iron sludge. Therefore, the rate of sludge production can be decreased by increasing the regeneration rate of ferrous ions and stabilizing the pH level at 3. Another solution is the use of insoluble iron oxides or iron immobilized in different supports that facilitate iron separation. This process has been referred to as the heterogeneous Fenton process or the Fenton-like catalyst, and it seems an attractive process from the point of view of an industrial application (Figure 4.2).

In the heterogeneous Fenton process, iron salts are supported in a solid matrix, and in a suitable aqueous medium the reduction–oxidation reactions between Fe^{2+} and Fe^{3+} take place in the presence of hydrogen peroxide which promote the formation of reactive components such as •OH and hydroperoxyl (•OOH) radicals (Table 4.1: Equations 4.1 and 4.2). These radicals can oxidize organic compounds adsorbed over the catalyst or near to active iron ions present both at the catalyst surface and in the bulk liquid phase.

A wide range of materials have been reported to carry out heterogeneous catalysis, such as zeolites, sepiolite, alginate, hydrogels, clays, and activated carbon (Bocos et al. 2014; Bounab et al. 2015; Fernández de Dios et al. 2014; Iglesias et al. 2013; Rosales et al. 2012b). In addition, these catalysts must fulfill several requirements such as low cost, high activity, conversion of hydrogen peroxide with minimum decomposition, marginal leaching of active cations, and high stability over a wide range of conditions (Garrido-Ramírez et al. 2010).

Recently, the heterogeneous Fenton process has been improved by applying iron nanomaterials (Hansson et al. 2012; Muñoz et al. 2015). At nanosize (particles with a diameter <100 nm), physical and chemical properties become very different from those of the same material in larger bulk form.

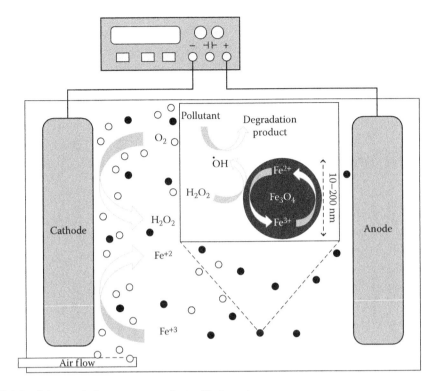

FIGURE 4.2   Scheme of a heterogeneous electro-Fenton setup.

The activity and selectivity of catalyst nanoparticles are strongly dependent on their size, shape, and surface structure, as well as their bulk and surface composition. Advances in physical methods for characterizing the structure and composition of such nanoparticles contribute to a molecular-level understanding of the structure–performance relationships. Advances have also occurred in the synthesis of nanoparticles and nanoporous supports with controlled size and shape (Bell 2003). The small particle size increases the proportion of atoms located at the surface, increasing the possibility for the atoms to adsorb, interact, and react with other atoms and molecules (Crane and Scott 2012).

In the heterogeneous Fenton process, the use of ferromagnetic nanoparticles, mainly zerovalent iron (nZVI), magnetite (Fe_3O_4), and maghemite (γ-Fe_2O_3), have attracted much interest for the degradation of a wide range of organic compounds (Table 4.2). As pointed out by Muñoz et al. (2015), the use of magnetic materials in this field only started around 2008, and continues to grow exponentially year by year. Zhang et al. (2008) studied the degradation of phenol by Fenton oxidation using ferromagnetic nanoparticles (diameter 13 nm), prepared according to the method of Molday (1984). In this case, a solution of 10 mM $FeCl_3$ and 6 mM $FeSO_4$ was mixed at pH 1.7 under N_2 protection. Then, an ammonia aqueous solution (1.5 M) was dropped into the mixture, with violent stirring, until the pH of the solution increased to 9 (Equation 4.27). The ferromagnetic nanoparticles obtained were washed immediately with water and ethanol and dried at room temperature under vacuum. After the optimization of operational conditions, the researchers determined that about 30% of the phenol was mineralized, while the rest became small molecular organic acids. In addition, they pointed out three significant advantages of using ferromagnetic nanoparticles in wastewater treatment: (1) ferromagnetic nanoparticles are cheap and easy to prepare increasing the cost efficiency of the Fenton process, (2) these nanoparticles are robust and thermostable and, for this reason, may be recycled and regenerated, and (3) they are nontoxic and environmentally friendly.

$$Fe^{2+} + Fe^{3+} + 8\ OH^- \rightarrow Fe_3O_4 + 4H_2O \qquad (4.26)$$

TABLE 4.2

Degradation of Different Organic Compounds Using Heterogeneous Fenton Technique

Pollutant	Catalyst	Size (nm)	Working Parameters		Treatment Time (min)	Removal Efficiency (%)	References
			H_2O_2/Catalyst	Operational Conditions			
Phenol (200 mg/L)	Nanoscale zerovalent iron (nZVI)	60–120	500 mg/L H_2O_2/0.5 g/L	1 L pH 6.2	60	65.7	Babuponnusami and Muthukumar 2012
Chlorpheniramine (15 mg/L)	Nanoscale zerovalent iron (nZVI)	40–70	0.01 mM H_2O_2/22.4 mg/L	300 mL pH 3	60	100	Wang et al. 2016
Nitrobenzene (40 mg/L)	nZVI/granular activated carbon	40–100	5 mM H_2O_2/0.4 g/L nZVI/ GAC	200 mL pH 4	100	87.5	Hu et al. 2015
Methyl Violet 10B (100 mg/L)	Maghemite	<50	2–4 mM/4 g/L	430 mL	60	55.6	Tiya-Djowe et al. 2015
2,4,6-Trichlorophenol (100 mg/L)	Fe_3O_4/CeO_2 composite (ratio 1:1)	5–10	30 mM H_2O_2/2.5 g/L	10 mL free	120	99.05	Xu and Wang 2015
Nalidixic acid (2.9 mg/L)	nanomagnetite	12	Generated H_2O_2/15.6 g/L	32 mL pH 6.5	30	60	Ardo et al. 2015
Phenol (80 mg/L)	γ-Fe_2O_3@Cu/Al-MCM-41	—	48.9 mM H_2O_2/1 g/L	60 mL pH 4	120	60[a]	Ling et al. 2014
Paracetamol (100 mg/L)	Fe_3O_4 < 50 nm	29	28 mM H_2O_2/6 g/L	500 mL pH 2.6 60°C	300	100; 43[a]	Velichkova et al. 2013
Paracetamol (100 mg/L)	Fe_3O_4 < 5 μm	67	28 mM H_2O_2/6 g/L	500 mL pH 2.6 60°C	300	100; 34[a]	Velichkova et al. 2013
Paracetamol (100 mg/L)	Fe_2O_3	48	28 mM H_2O_2/6 g/L	500 mL pH 2.6 60°C	300	100; 39[a]	Velichkova et al. 2013
Imidacloprid (10 mg/L)	Meso-$CuFe_2O_4$	50–150	40 mM H_2O_2/0.3 g/L	50 mL pH 3 30°C	300	98	Wang et al. 2014b
Clofibric acid (100 mg/L)	FeOOH	—	0.5 g/L H_2O_2/2 g/L	100 mL pH free	360	95; 39[a]	Sable et al. 2015
Clofibric acid (100 mg/L)	0.5%Pd/FeOOH	—	0.5 g/L H_2O_2/2 g/L	100 mL pH free	360	100; 65[a]	Sable et al. 2015
Clofibric acid (100 mg/L)	25% FeOOH/γ-Al_2O_3	—	0.5 g/L H_2O_2/2 g/L	100 mL pH free	360	90; 60[a]	Sable et al. 2015
Clofibric acid (100 mg/L)	25% FeOOH/ZrO_2	—	0.5 g/L H_2O_2/2 g/L	100 mL pH free	360	92; 44[a]	Sable et al. 2015
Clofibric acid (100 mg/L)	Lepidocrocite	—	0.5 g/L H_2O_2/2 g/L	100 mL pH free	360	86; 32[a]	Sable et al. 2015
2,4-Dichlorophenol (100 mg/L)	Nanoscale zerovalent iron (nZVI)	0–50	10 mM H_2O_2/1.5 g/L	15 mL pH 3	180	57.87	Li et al. 2015
2,4-Dichlorophenol (100 mg/L)	n-Ni/Fe	0–60	10 mM H_2O_2/1.5 g/L	15 mL pH 3	180	34.23	Li et al. 2015
2,4-Dichlorophenol (100 mg/L)	n-Pd/Fe	0–60	10 mM H_2O_2/1.5 g/L	15 mL pH 3	180	27.94	Li et al. 2015

[a] TOC reduction.

4.2.1 ZEROVALENT IRON NANOPARTICLES (nZVI)

nZVI have been used for the treatment of different polluted wastewaters due to their great ability to degrade different types of pollutants at much lower operational cost (Babuponnusami and Muthukumar 2012, 2014; Pradhan and Gogate 2010). They produce hydroxyl radicals by corroding their surfaces and also have high reactivity due to larger specific surface areas (Kallel et al. 2009). Thus, an integrated nZVI reduction process coupled with an oxidation process, nZVI-H_2O_2, was employed in an attempt to achieve both complete degradation and mineralization of organic contaminants (Moon et al. 2011). In addition, nZVI have the capacity to remain in suspension and, hence, aqueous slurries containing nZVI can be injected directly into contaminated zones (Xu and Zhao 2007). To date, there have been nearly 60 pilot or full-scale groundwater and soil remediation projects conducted worldwide, and more than 34 projects have been carried out in the United States (Yan et al. 2013; Yang et al. 2014).

Recently, Li et al. (2015) reported several iron-based nanoparticles (nZVI, n-Ni/Fe, n-Pd/Fe) used as heterogeneous catalysts in the Fenton oxidation of 2,4-dichlorophenol. The Fenton oxidation efficiencies using nZVI, n-Ni/Fe, n-Fe/Pd, and Fe^{2+} were 57.87%, 34.23%, 27.94%, and 19.61%, respectively, after 180 min. These results demonstrated the high potential of nZVI as catalysts.

The main advantage of the use of these catalysts is their high removal efficiency at near-neutral pH conditions that are comparable to high acidic conditions, which may alleviate the problem of neutralizing the treated solution at the end of the treatment process (Babuponnusami and Muthukumar 2012). In the literature, there are several examples of this fact. Thus, the removal of several dyes (methylene blue, methyl orange, and Acid Red 73) using nZVI as a catalyst has been reported (Fu et al. 2012; Moon et al. 2011). Furthermore, other complex organic pollutants were also treated by this system; thus, the complete degradation of biocide 4-chloro-3-methyl phenol in aqueous solution was also achieved by nZVI operating at pH levels greater than 6 (Xu and Wang 2011).

Xia et al. (2014) prepared nZVI particles via a liquid-phase reduction method described by Wang and Zhang (1997). In this method, nZVI were produced by adding 1.6 M $NaBH_4$ aqueous solution dropwise to a 1.0 M $FeCl_3$ aqueous solution at ambient temperature with vigorous stirring (Equation 4.28). The Fenton process was employed with these nZVI to investigate the removal of chloramphenicol from aqueous solution, and the effects of key parameters such as nZVI dosage and initial pH were evaluated. It was found that removal efficiency could be enhanced with an increase in nZVI dosage and a decrease in the initial pH. Under optimal conditions, the researchers determined that the chloramphenicol was completely removed. In addition, Raman analyses of nZVI particles before and after the process revealed that the pollutant was adsorbed and reduced on the surface of the nZVI.

$$Fe^{3+} + 3BH_4^- + 9H_2O \rightarrow Fe^0 \downarrow + 3B(OH)_3 + 10.5H_2 \uparrow \tag{4.27}$$

One of the disadvantages of these nZVI is in relation to their possible inactivation in water, because they can be oxidized to form a core-shell structure (Equation 4.29) where the core is zerovalent iron and the shell consists of iron oxides or hydroxides (Martin et al. 2008; Yang et al. 2015). To avoid this negative effect, Yan et al. (2015) immobilized nZVI on magnetic Fe_3O_4-reduced graphene oxide (Fe^0-Fe_3O_4-RGO). This catalyst showed superparamagnetic properties and had a high Brunauer-Emmett-Teller (BET) surface area. It was found that a possible reaction process occurred on the surface of the Fe_3O_4, while the Fe^0 was a source of electrons that could enhance reactions, and the graphene sheets acted as a good supporter and electron conductor. The results obtained showed that Fe^0-Fe_3O_4-RGO had good potential as a Fenton catalyst, due to its high reactivity, reusability, and easy magnetic separation.

$$Fe^0_{(s)} + 2H_2O_{(aq)} \rightarrow Fe^{2+}_{(aq)} + H_{2(g)} + 2OH^-_{(aq)} \tag{4.28}$$

Green methods can be used for the production of nZVI, from extracts of grape marc, black tea, and vine leaves. These nZVI have adequate nanometric size and have been used to degrade organic compounds in aqueous solutions and sandy soils, and showed encouraging results even in comparison with other common chemical oxidants. Machado et al. (2013) produced nZVI that were able to degrade ibuprofen via the nZVI Fenton process, achieving degradation efficiencies above 95%, which shows the great potential of this remediation strategy as an alternative to traditional and more aggressive technologies. Similarly, iron nanoparticles were synthesized using tea extracts (green tea, oolong tea, and black tea extracts) as catalysts for the Fenton-like oxidation of monochlorobenzene (Kuang et al. 2013). The findings indicated that the best results were obtained when green tea extracts, which contain a high concentration of caffeine/polyphenols, were used as both reducing and capping agents in the synthesis of nZVI. These facts are in accordance with those reported by Wu et al. (2015), who studied the morphology, size, and changes in the iron nanoparticles' surface using scanning electron microscopy (SEM), x-ray diffraction (XRD), and Fourier transform infrared spectroscopy (FTIR) techniques and determined that green tea extracts generate the formation of nanoparticles of ZVI, Fe_2O_3, and Fe_3O_4. In all these studies, it was demonstrated that the use of iron nanoparticles, synthetized by green methods, has great potential as a nanoremediation strategy.

4.2.2 Magnetite (Fe_3O_4)

Iron oxide minerals such as magnetite (Fe_3O_4) have attracted attention concerning the replacement of iron salts in Fenton reactions (Huang et al. 2012). The potential of magnetite derives from its high ability for degradation of recalcitrant pollutants compared with conventional iron-supported catalysts. Due to the octahedral site in the magnetite structure, it can easily accommodate both Fe^{2+} and Fe^{3+}, which means that Fe^{2+} can be reversibly oxidized and reduced back within the same structure (Xu and Zhao 2007). Thus, magnetite exhibits unique electric and magnetic properties, good structural stability, and excellent reusability. Recently, nanomagnetite has been seen as an attractive alternative to micrometer-size magnetite, as it can provide a large exposed surface area in the reaction medium and thus improve the decomposition rate. A wide variety of methods has been reported to synthesize nanomagnetite with different morphologies (tubes, wires, films, spheres, rods, and other novel structures) as a result of different manufacturing processes: sol-gel, wet-chemical, hydrothermal, coprecipitation, and microemulsion (Ardo et al. 2015; Du et al. 2010; Peng et al. 2006; Prakash et al. 2004). Recently, Hou et al. (2014) reported an environmentally benign "green" route to the synthesis of Fe_3O_4 nanoparticles with various morphologies, achieved by the thermal reduction of α-Fe_2O_3 materials synthesized via ionic liquids assisted by the hydrothermal process. They prepared several magnetites with different porous structures (microcubes, nanospheres, and nanorods) and concluded that there was a relationship between the catalytic activity and the porous structures exhibiting the synthesized nanorods' greatest activity.

Among the different methods, coprecipitation is one most used. Wei et al. (2011) synthesized magnetic Fe_3O_4 nanoparticles by coprecipitation of Fe^{2+} and Fe^{3+} in basic solution. They demonstrated the high catalytic efficiency of these nanoparticles in the degradation of phenol under a much wider pH range (pH 3–7) compared with the classical Fenton's reagent. Recently, Ardo et al. (2015) prepared nanomagnetites with a particle size around 12 nm by aqueous coprecipitation in O_2-free deionized water at pH 12 by NaOH. They demonstrated the nanomagnetites' ability to catalyze the degradation of a recalcitrant quinolone antibacterial agent, such as nalidixic acid, at pH 6.5. However, Zhang et al. (2011) reported that the degradation of norfloxacin by system nanomagnetite/H_2O_2 was strongly pH dependent and favored when the operation took place in acidic conditions. These results indicate that it is necessary to study each process, because the effect of the process variables depends on the type of nanomagnetite used and its properties.

Xu et al. (2012b) also synthesized Fe_3O_4 magnetic nanoparticles by coprecipitation in argon based on Massart's method. To do so, a mixture solution of 100 mL 0.01 M $FeSO_4 \cdot 7H_2O$ and 0.02 M $Fe_2(SO_4)_3$ with 0.2 mL H_2SO_4 was added dropwise to 100 mL 0.2 M NaOH solution. After 1.5 h of vigorous stirring under an argon stream at 80°C, the precipitate was deposited and then washed two times with deionized water and ethanol, and nanoparticles were dried at room temperature under vacuum. These nanoparticles were used as a catalyst for the Fenton degradation of 2,4-dichlorophenol in aqueous solution (Xu and Wang 2012a,b). The researchers concluded that the process fitted well to a two-stage first-order kinetic, consisting of an induction period (first stage) followed by a rapid degradation stage (second stage). Both stages depended on pH level, nanoparticle dosage, and temperature used, but the effect of the concentrations of H_2O_2 was irrelevant in the range studied.

To increase the catalytic activity of these nanoparticles, other authors have proposed the use of nanomagnetite composites. Thus, Xu and Wang (2015) studied the oxidation of 4-chlorophenol via the heterogeneous Fenton system using a superparamagnetic nanoscaled Fe_3O_4/CeO_2 composite. In this case, the component CeO_2 facilitated the dissolution of Fe_3O_4, and $\bullet OH$ was generated by the reaction of Fe^{2+} and Ce^{3+} with H_2O_2, leading to the enhanced Fenton chemistry.

Several authors (Guo and Al-Dahhan 2006; Xu and Wang 2012a; Zhang et al. 2009a) have reported that catalyst reusability is limited and initial activity decreased gradually over consecutive runs. They attributed catalyst deactivation to a number of diverse factors, including reduction of the catalyst specific area, poisoning of the active catalytic sites by adsorbed organic species, conglomeration of nanoparticles, and the discarding of supernatants during the rinsing of nanoparticles.

To reduce this negative effect, several authors (Guo et al. 2013; Zubir et al. 2015) have reported the deposition of nanoparticles onto the surface of unique two-dimensional graphene or graphene oxide to increase the surface area and deal with the lack of ink-bottle pores (Guo et al. 2013). Therefore, by immobilization of iron oxide on graphene oxide sheets, a $GO-Fe_3O_4$ nanocomposite is obtained, which exhibits a highly heterogeneous electronic structure determined by the interplay of the oxygenated functional groups. The application of this new catalyst has recently been tested, and several organic pollutants such as dyes (Acid Orange 7 and Rhodamine B) and phenolic compounds (bisphenol and 4-nitrophenol) were efficiently degraded (Guo et al. 2013; Hua et al. 2014; Zubir et al. 2014). As described by Zubir et al. (2015), the active sites of Fe^{2+} in Fe_3O_4 are not oxidized to Fe^{3+} in the Fenton reaction. Graphene oxide plays a sacrificial role via the oxidation of C=C bonds and transfers the electrons to Fe_3O_4 to maintain the active Fe^{2+} sites. For this reason, changes are not detected in the ratio of Fe^{3+}/Fe^{2+} of $GO-Fe_3O_4$. Therefore, $GO-Fe_3O_4$ confers superior catalytic efficiency, recyclability, and longevity, which are otherwise not available in Fe_3O_4. Another alternative is its use as a support of multiwalled carbon nanotubes. These have attracted special attention because of their unique properties, such as large specific surface area, chemical stability, and supermechanical strength. Hu et al. (2011) prepared a novel catalyst—inverse-spinel ferroferric oxide nanoparticles decorated with multiwalled carbon nanotubes ($Fe_3O_4/MWCNTs$)—and applied it to the removal of traces of androgen 17α-methyltestosterone in a Fenton system. The TEM images showed the octahedron Fe_3O_4 nanoparticles growing regularly on the multiwalled carbon nanotube surfaces, with diameters ranging from 40 to 100 nm, and most of the Fe_3O_4 nanoparticles were strung along the multiwalled carbon nanotubes. They demonstrated that the degradation process mainly occurred on the surface of the catalyst with insignificant iron leaching in a two-stage process: adsorption and degradation. All results indicated that the novel catalyst has good structural stability and shows little iron leaching.

Another new system was proposed to increase Fenton reaction efficiency by use of poly(acrylic acid)-co-sodium carboxymethyl cellulose/magnetite nanocomposites (Wang et al. 2014a). This nanocomposite is prepared by the *in situ* encapsulation method, in which sodium carboxymethyl cellulose was used to bond on to the surfaces of generated γ-Fe_2O3 nanoparticles; then, the acrylic acid monomers were grafted onto the chains of sodium carboxymethyl cellulose and copolymerized to form a pH-responsive polymer. Results showed that the nanocomposite exhibited excellent catalytic properties, stability, and reusability, operating within a wide pH range (3–7). This behavior could be associated with the pH-sensitive characteristic of the polymer coating in aqueous solutions. In the correct pH range, the hydrophilic polymer chain is highly swollen, which promotes the dispersion of Fe_3O_4 nanoparticles and a solid–liquid interface mass transfer that accelerates the catalytic reaction. Similarly, Hammouda et al. (2015) synthesized magnetic Fe_3O_4 nanoparticles embedded with iron in alginate beads. The nanoparticles showed a good response to the magnetic field, which made their separation from the liquid phase easier. Concerning reusability, the results revealed that the catalyst possessed good reusability properties and could be used three times without affecting its efficiency and without significant change of its magnetic properties. Moreover, the amount of the active species leached from the catalyst into the solution during the Fenton process was lower after three reuses, which is minimal compared with some industrial heterogeneous catalysts.

4.2.3 Maghemite (γ-Fe$_2$O$_3$)

Maghemite (γ-Fe$_2$O$_3$) is considered as the second most stable polymorph of iron oxide and exhibits a spinel crystal structure, in which all the iron cations are in a trivalent state. The neutrality of the cell is guaranteed by the presence of cation vacancies (Dronskowski 2001). For this reason, excellent results in the decolorization of methyl violet were obtained when γ-Fe$_2$O$_3$ nanoparticles (particle size <50 nm) acted as a heterogeneous catalyst for the Fenton decomposition of H_2O_2 (Tiya-Djowe et al. 2015). In addition, the characterization by FTIR and XRD of the catalyst after three cycles showed that the catalyst retains its properties and can be reused for many treatments.

A comparative study between two magnetic catalysts, γ-Fe$_2$O$_3$ nanoparticles and γ-Fe$_2$O$_3$/SiO$_2$ nanocomposite microspheres, was carried out by Ferroudj et al. (2013). Their objective was to study how the encapsulation of the magnetic nanoparticles within the silica beads influences their catalytic activity, stability, and separation properties. The γ-Fe$_2$O$_3$ nanoparticles were synthesized by oxidation, adding HNO_3 and $Fe(NO_3)_3$ successively to magnetite nanoparticles prepared by Massart's method. The γ-Fe$_2$O$_3$/SiO$_2$ nanocomposite microspheres were prepared via a modified emulsion solvent evaporation method, obtaining a polydispersed spherical shape with a mean diameter of about 2 μm. Small dispersed nanoparticles should be highly active, but their colloidal stability causes problems in their separation from the effluent. Although this problem can be overcome by encapsulation, at the same time, diffusion of the pollutants to the catalytic sites may be slowed down. The results revealed that both catalysts exhibit a superparamagnetic behavior with a strong magnetic susceptibility, although the strong colloidal stability of the free γ-Fe$_2$O$_3$ nanoparticles restricts their recovery by magnetic settlement. In addition, a negligible iron leaching was detected, which is indicative that the degradation process is mainly heterogeneous. The mineralization and decolorization of aqueous solutions containing model pollutants indicated that the degradation rates seemed to be correlated to their adsorption on to the catalyst, followed by catalytic activity, which is influenced by various parameters such as the solution pH, the weight fraction of γ-Fe$_2$O$_3$, the amount of catalyst, and the initial concentration of H_2O_2. It is pointed out that both catalysts showed high efficiency and stability during repeated experiments, despite a relative decrease in reaction rates.

The use of a nanocomposite with γ-Fe$_2$O$_3$ as the magnetic core requires that the whole synthesis process should be operated under anoxic conditions that increase complexity in terms of scalability. For this reason, Xia et al. (2011) synthesized a novel magnetic nanocomposite catalyst with a superparamagnetic maghemite core and an ordered mesoporous silica shell operating in ambient conditions. It was found that, by application of an external magnetic field, the catalyst used could be collected and recycled efficiently, and that this magnetic nanocomposite had enough active sites to support Fenton reactions and exhibited remarkable catalytic performance.

Recently, Lin et al. (2014) carried out synthesis by means of a nanoassembling procedure, with γ-Fe$_2$O$_3$ nanoparticles as the core layer and the incorporation of copper and aluminum as the shell layer. In this case, the aluminum contained in the core-shell catalyst retained a high BET surface area; however, an increase in the copper incorporated would lead to a significantly decreased BET surface area and a less ordered mesophase morphology, although the catalytic activity could be improved. Therefore, it was demonstrated that it is possible to prepare magnetically separable core-shell composites with multiple metal ions, with the optimization of their incorporation being necessary to increase their stability, reusability, and catalytic activity.

4.3 ELECTRO-FENTON PROCESS

The electro-Fenton process has drawn considerable attention as a promising alternative technology, by overcoming the drawbacks of the classical Fenton process mentioned in the introduction. By the combination of Fenton's reagent and the electrochemical system, continuous electrogeneration of H_2O_2 (Table 4.1: Equation 4.20) and regeneration of Fe^{3+} is possible, generated from the Fenton reaction (Table 4.1: Equation 4.22), by direct reduction on the cathode, reduction processes involving H_2O_2, or intermediate organic radicals. This allows the catalytic propagation of the Fenton reaction. Therefore, it is of great advantage in relation to the conventional Fenton process, since the addition of H_2O_2 is not necessary. Moreover, the generation of hydroxyl radicals does not involve the use of harmful chemicals which can be hazardous for the environment; for this reason, this process is environmentally friendly for wastewater treatment. Furthermore, the rate of organic pollutant removal and the mineralization yield are usually significantly higher than for the classical chemical Fenton treatment (Brillas et al. 2009; Hou et al. 2015).

In the literature, it is possible to find electro-Fenton processes that have been effectively applied to the removal of several pollutants, including dyes, amines, herbicides, fertilizers, insecticides (Alfaya et al. 2015), aromatic acids, phenolic compounds, organic acids, petrochemical wastes, persistent organic pollutants (POPs) (Zhang et al. 2015a), pharmaceuticals (Bocos et al. 2015a; Isarain-Chávez et al. 2010), personal care products (PPCPs) (Su et al. 2012), and endocrine disrupters (EDCs) (Wang et al. 2015). Several key operative parameters play an important role in assuring the efficiency of this process, such as the electrode material, catalyst dosage, and pH level (Rosales et al. 2012a). In this chapter, attention is focused on the use of nanomaterials to increase the efficiency of the electro-Fenton process. Recently, these materials have been included in the electro-Fenton treatment in different parts of the degradation system, such as the cathode electrode and catalyst (Table 4.3). For this reason, the effect of nanomaterials in these will be evaluated.

4.3.1 Cathode Material

As shown in Table 4.3, electrochemical reactions depend strongly on the electrode material used in the electrochemical cell. Traditionally, carbon-based and gas diffusion electrodes, as cathodes, have been reported to be active in the production of hydrogen peroxide. Petrucci et al. (2016) have compared three different carbon-based materials—graphite, carbon felt, and reticulated vitreous carbon—as cathodes in the electro-Fenton process. They demonstrated that all the cathodes presented good ability to electrogenerate hydrogen peroxide. However, their behavior differed significantly in the electroregeneration of ferrous ions. In this case, reticulated vitreous carbon and carbon felt presented efficiencies and reduction rates higher than graphite, and results were almost unaffected by the operative conditions.

Graphite felt activated by KOH at high temperature (AGF-900) was adopted as the electro-Fenton cathode to degrade organic pollutants contained in wastewater. Wang et al. (2015) demonstrated that, through KOH activation, the raw graphite felt could be provided with: (1) a higher surface area, (2) more oxygen-containing functional groups, and (3) more hydrophilicity. The comparison with other commonly used materials confirms the effectiveness of AGF-900 as an electro-Fenton cathode. This fact can be explained by the treatment provided to graphite felt adding more functional groups, making it more applicable as a cathode for the removal of organic pollutants in wastewater.

A polyacrylonitrile-based carbon-fiber brush (PAN-CFB) cathode was manufactured by Xia et al. (2015) using an organic fiber precursor with a higher carbonization temperature ($1200°C–1600°C$), which resulted in the generation of a turbostratic graphite structure with lower electrical resistivity compared with activated carbon fiber (ACF) (Xia et al. 2015). In addition, cyano groups with strong polarity existed in the PAN precursor; some residue nitrogen content still remained in the carbon felt after carbonization at high temperature. The doped-nitrogen carbon materials favoring oxygen

TABLE 4.3

Use of Nanomaterials as Catalysts/Electrodes in Electro-Fenton Process

Pollutant	Catalyst	Size (nm)	Fe Concentration	Working Parameters		Treatment Time (min)	Removal Efficiency (%)	References
				Cathode/Anode	Operational Conditions			
Rhodamine B (5 mg/L)	FeOOH/graphite felt	50–150	18.5 wt%	Carbon paper coated with Pt/graphite felt	50 mM Na$_2$SO$_4$; 175 mL; pH 6.8; air 1.5 L/min; 2 V	120	50.4	Sun et al. 2015
Rhodamine B (5 mg/L)	Fe$_2$O$_3$/graphite felt	30	18.9 wt%	Carbon paper coated with Pt/graphite felt	50 mM Na$_2$SO$_4$; 175 mL; pH 6.8; air 1.5 L/min; 2 V	120	69.5	Sun et al. 2015
Rhodamine B (5 mg/L)	Fe$_3$O$_4$/graphite felt	10–100	17.7 wt%	Carbon paper coated with Pt/graphite felt	50 mM Na$_2$SO$_4$; 175 mL; pH 6.8; air 1.5 L/min; 2 V	120	77.4	Sun et al. 2015
Catechol (110 mg/L)	Nano-Fe$_3$O$_4$	40	1.0 g/L	Activated carbon fiber/Pt	None electrolyte; 200 mL; pH 3; air saturated; 10 mA/cm^2; 25°C	120	100	Hou et al. 2015
Methylene blue (2.4·10^{-4} M)	Carbon nanotubes/Fe$_3$O$_4$	5–10	10 g/L composite	Graphite electrode/Ti/SnO$_2$	0.1 M Na$_2$SO$_4$; 150 mL; pH 5; air saturated; 5 V; 20°C	150	97.8	Shen et al. 2014
Metalaxyl (500 mg/L)	Ferrite-carbon aerogel	80–120	0.05–6%	Ferrite-carbon aerogel/BDD	0.1 M Na$_2$SO$_4$; 100 mL; O$_2$ 0.02 m^3/h; 10 mA/cm^2; 25°C	60	73	Wang et al. 2013
Phenol (200 mg/L)	nZVI	60–120	0.5 g/L	Stainless steel	1 g/L Na$_2$SO$_4$; 1 L; pH 6.2; 12 mA/cm^2	60	87.38	Babuponnusami and Muthukumar 2012
Amaranth (80 mg/L)	FeOOH/activated carbon	—	3.12 wt%	Graphite felt/Pt sheet	0.2 M Na$_2$SO$_4$; 400 mL; pH 4.0; saturated O$_2$; −0.64 V	180	47.98	Zhang et al. 2012
Orange II (10 mg/L)	Nano-Fe^0/activated carbon fibre	100–600	—	Nano-Fe^0/ACF/RuO$_2$/Ti mesh	0.05 M Na$_2$SO$_4$; 500 mL; pH 7; air: 300 mL/min; 0.2 A	120	93	Li et al. 2011

reduction reaction have been proved to be efficient nonplatinum oxygen reduction reaction catalysts in neutral and alkaline systems (Liu et al. 2010).

The applicability of nickel foam as a cathode in electro-Fenton treatment has been demonstrated (Bocos et al. 2015b; Liu et al. 2012). Initial linear sweep voltammetry studies showed the ability of nickel foam to increase H_2O_2 generation during electro-Fenton treatment, enhancing the removal yields achieved in comparison with other carbonaceous materials (Bocos et al. 2015b). However, in this electrode, nickel leaching must be controlled and, for this reason, several proposals have been developed. Carbon-coated nickel foam electrodes were prepared via a simple and effective method, the hydrothermal-carbonization cycle coating process (Song et al. 2015). The results of the experiments indicated that nickel leaching can be effectively controlled at a carbon-coated nickel foam cathode (the coating process cycle is carried out four times) and it still presents an excellent and effective performance in the degradation process. Furthermore, the current efficiency for H_2O_2 generation at the carbon-coated nickel foam-4 coating was enhanced 12-fold compared with untreated nickel foam and, consequently, the degradation level was higher. Bocos et al. (2015c) used a new cathode in which the iron was fixed on nickel foam and coated with chitosan. Different chitosan coatings were tested to optimize the manufacturing process of this new cathode. It was concluded that the best electrode was obtained using a half-coated electrode covered with iron-chitosan of medium molecular weight. The physicochemical characteristics of this electrode and its reusability permitted the successful performance of continuous treatment of organic compounds.

Recently, the use of mesoporous carbon has attracted much attention due to its superior properties as an electrode material, because of its tunable pore structure (Hosseini et al. 2011), high BET surface area (Zhang et al. 2009b), large pore volume, and better electrical conductivity (Park et al. 2013). Ren et al. (2015) prepared different mesoporous carbon-grafted ACFs with average pore sizes of 2.6, 3.7, and 5.4 nm, respectively. In this study, the effects of the pore structure and pore size of the mesoporous carbon cathode materials on electro-Fenton efficiency have been evaluated by the generation rates of hydrogen peroxide and hydroxyl radicals. It was found that the process efficiency of the cathode materials followed the increasing order of 2.6 < 3.7 < 5.4. Even after 10 consecutive experiments, the reactivity of mesoporous carbon-grafted ACF cathode materials (5.4 nm) remained almost unchanged, indicating their promising application in practical wastewater treatment.

A new cathode for the electro-Fenton process was set up by the electrochemical deposition of reduced graphene oxide on the surface of carbon felt (Le et al. 2015). The structure and properties of the modified electrode were investigated. The reduced graphene oxide modified cathode presented enhanced electrochemical properties, such as an increase of the redox current and a decrease of the charge transfer resistance in the presence of the redox probe $[Fe(CN)_6]^{3-}/[Fe(CN)_6]^{4-}$, showing better kinetic properties compared with raw carbon felt. This improvement enhanced the production of hydrogen peroxide significantly, which was confirmed by linear scanning voltammetry analysis. Moreover, the new cathode exhibited good stability as mineralization occurred after 10 cycles of degradation.

The development of new nanomaterials permits their use in the fabrication of different composite electrodes: a $Fe@Fe_2O_3$/CNTs composite electrode was prepared by a combination of core-shell $Fe@Fe_2O_3$ nanowires and multiwalled carbon nanotubes (CNTs) by using polytetrafluoroethylene (PTFE) (Ai et al. 2007). A $Fe@Fe_2O_3$/ACF composite electrode was fabricated using loaded nanostructured $Fe@Fe_2O_3$ on ACF by using PTFE (Li et al. 2009). These electrodes showed excellent mechanical integrity, making them attractive as stable electrodes to electrogenerate hydrogen peroxide and control the release of iron, showing much higher activity than other electro-Fenton systems under a neutral pH.

In other studies, Fe nanoparticles were included in the cathode by electrochemical deposition. Fu et al. (2007) prepared an Fe-modified multiwalled carbon nanotube (Fe-MWNT) electrode, Chen et al. (2008) fabricated Fe cuboid nanoparticles supported on glassy carbon, and Li et al. (2011) obtained an Fe^0/ACF cathode by electrochemical deposition of Fe nanoparticles onto the

ACF substrate using a potentiostat in the chronoamperometry mode. In all cases, these cathodes exhibited much higher catalytic activity than conventional electrodes.

Nanotechnology has been used to fabricate other types of cathode, such as ferrite-carbon aerogel (FCA) (Wang et al. 2015). This was synthesized directly from metal-resin precursors, accompanied by phase transformation. Self-doped ferrite nanocrystals and carbon matrices were formed synchronously via moderate condensation and sol-gel processes, leading to a homogeneous texture. An FCA of optimal 5% ferric content was composed of coin-like carbon nanoplates with a continuous porous structure, and the ferric particles with diameters of dozens of nanometers were uniformly embedded into the carbon framework. The FCA exhibited good conductivity, high catalytic efficiency, and distinguished stability after prolonged recycling.

Graphite felt (GF) was selected by Sun et al. (2015) as the supporting material of iron oxides (FeOOH/GF, Fe_2O_3/GF, and Fe_3O_4/GF), due to its high surface area, good electrochemical stability, and high electroconductivity. Among the three types of composites, Fe_3O_4/GF exhibited the highest electro-Fenton catalytic activity, whereas the lowest activity was observed with FeOOH/GF. These results are in agreement with the redox activities of the corresponding composites observed in the cyclic voltammetry analysis. The decomposition of H_2O_2 on the iron oxides followed a completely surface-catalyzed reaction, in which the iron oxides maintained their structures without leaching of iron species.

Due to the influence of oxygen concentration on the cathode surface, several electrode configurations have been developed to increase hydrogen peroxide generation. Thus, Zhang et al. (2015b) developed a rotating graphite disk electrode (RDE) as a cathode. Experimental results showed that the rotation of the RDE resulted in the efficient production of H_2O_2 without oxygen aeration, and excellent ability to degrade organic pollutants compared with the electro-Fenton system with a fixed cathode. Chronoamperometry analysis demonstrated that the diffusion coefficient of dissolved oxygen was 19.45×10^{-5} cm^2/s, which is greater than that reported in the literature.

4.3.2 Heterogeneous Electro-Fenton

As mentioned in the section on the heterogeneous Fenton process, during the last few years there has been a widespread development of catalysts for industrial wastewater treatment. In this section, the attention is focused on the studies in which nanoparticles have been used as catalysts in the heterogeneous electro-Fenton system.

The degradation of phenol was carried out by the Fenton and electro-Fenton processes using nZVI as the catalyst, which was prepared by the sodium borohydride ($NaBH_4$) reduction method (Babuponnusami and Muthukumar 2012). In this study, experiments were carried out at an initial pH level of 6.2 (natural pH of solution), and the results showed that after 60 min of reaction time, 65.7% of phenol removal was observed in the Fenton process, which improved up to 87.38% when the electro-Fenton process was used. These results showed the superiority of the electro-Fenton process over the Fenton process.

Expósito et al. (2007) evaluated the use of magnetite (Fe_3O_4) and wüstite (FeO) with a size of <5 μm as catalysts in the treatment of aniline solution by the heterogeneous electro-Fenton process. It was demonstrated that these two processes followed a purely homogeneous catalytic pathway, where only the iron ions released into the solution took part in the production of hydroxyl radicals via the Fenton reaction. Thus, the catalyst is an iron source in the electro-Fenton process and its ability to self-regulate the supply of a constant amount of iron ions over the reaction time increases the efficiency of the electro-Fenton process. The main advantages of the use of these catalysts are their magnetic properties, which permit the remaining catalyst to be easily collected after decontamination treatment by applying a magnetic field.

Zeng et al. (2011) reported the successful removal of 4,6-dinitro-o-cresol using nanomagnetite as the catalyst and a cathodic Fenton process to generate hydrogen peroxide. The large surface area of nanomagnetite enhances the dissolution of iron at lower pH levels, and thus increases the rate of degradation. For this reason, it was identified that the target compound was effectively degraded under acidic pH conditions (2.8) mainly due to iron leaching from the oxide surface, which favored the homogeneous Fenton reaction. However, this system degraded only half the target compound in 1 h at a near-neutral pH (5.9), primarily due to the dominance of direct electrolysis under these conditions. To completely degrade the target compound, either an acidic condition or further time is required.

To increase the efficiency of the electro-Fenton process, Zhang et al. (2012) prepared, by air oxidation of the ferrous hydroxide suspension method, an activated carbon-supported nano-FeOOH catalyst. This was because FeOOH combines some attractive properties such as its wide range of operating pH levels, the controllable leaching of iron into the solution, and its excellent adsorption performance for large-scale applications. The heterogeneous catalyst possesses an admirable absorbability, and it was demonstrated that the degradation of azo dye amaranth took place through the heterogeneous Fenton reaction on the catalyst/solution interface, but also through the homogeneous Fenton reaction in the bulk solution. In this case, the introduction of iron ions into the solution occurred due to the reaction of ferric ions on the nano-FeOOH surface under acidic conditions, and the electrochemical reductive liberation of iron species on the surface and in the liquid under external electric field conditions, which facilitated the regeneration of ferrous ions and produced a beneficial enhancement to the homogeneous and heterogeneous Fenton reaction rates.

Shi et al. (2014) demonstrated that $Fe@Fe_2O_3$ core-shell nanowires can improve the efficiency of Fenton oxidation twofold, with Rhodamine B as a model pollutant operating at pH > 4. Active species-trapping experiments revealed that the enhancement to Rhodamine B oxidation was attributed to molecular oxygen activation induced by $Fe@Fe_2O_3$ core-shell nanowires. The molecular oxygen activation process could generate superoxide radicals to assist the iron core in the reduction of ferric ions to accelerate the Fe^{3+}/Fe^{2+} cycles, which favored H_2O_2 decomposition to produce more hydroxyl radicals for the oxidation of Rhodamine B. The combination of $Fe@Fe_2O_3$ core-shell nanowires and ferrous ions ($Fe@Fe_2O_3/Fe^{2+}$) offered a superior Fenton catalyst to decompose H_2O_2, producing •OH. By employing benzoic acid as a probe reagent to check the generation of hydroxyl radicals, the researchers found that the generation rate in the presence of $Fe@Fe_2O_3/Fe^{2+}$ was two to four orders of magnitude greater than those of commonly used iron-based Fenton catalysts, and 38 times greater than those of Fe^{2+}. In addition, the high reusability and stability of $Fe@Fe_2O_3$ core-shell nanowires in the electro-Fenton process were also proven.

He et al. (2014) applied an Fe_3O_4 magnetic nanoparticle electro-Fenton process, using ACF felt as the cathode and platinum as the anode, for the degradation of Reactive Blue 19 dye in aqueous solution. It was found that the electro-Fenton process, catalyzed by Fe_3O_4 magnetic nanoparticles, had a higher degradation capacity than commercial Fe_3O_4 and Fe^{2+}. In addition, it was demonstrated that nanomagnetite-electro-Fenton degradation followed a two-stage first-order kinetic, with a slow induction period and a subsequent stage of rapid degradation. Moreover, removal efficiency depended on the initial pH of the solution, current density, catalyst loading, initial concentration of target pollutant, and electrolysis temperature. It was found that nearly 90% of mineralization was obtained when operating at the optimal conditions: an initial pH of 3.0, current density of $3.0\ mA/cm^2$, 1.0 g/L concentration of added nanomagnetite, and a reaction temperature of 35°C.

Recently, Hou et al. (2015) studied the degradation of catechol in a heterogeneous electro-Fenton system, catalyzed by nanomagnetite prepared via the chemical coprecipitation method, with precursors of ammonia, ferric sulfate, and ferrous sulfate (Hong et al. 2009). It was demonstrated that current density was essential and the obtained degradation levels were satisfactory at a pH of 6.0, showing high effectiveness of the catalyst over a wide range of pH values, although the degradation of catechol decreased with an increase in the pH. Furthermore, scavenging effects on the radical hydroxyls were detected at a high catalyst load. The system presented high stability and reusability, providing evidence for its potential application. The kinetic study revealed that the interface reaction mechanism was dominant in the heterogeneous electro-Fenton system using nanomagnetite

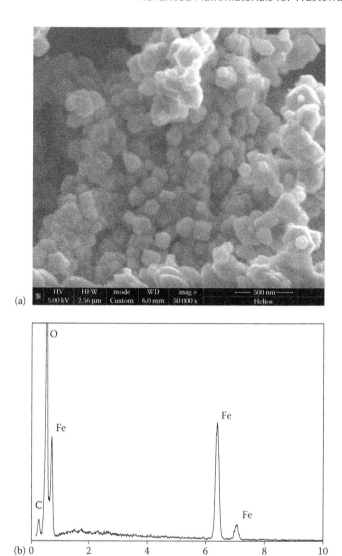

FIGURE 4.3 SEM picture of (a) iron oxide nanoparticles and (b) energy dispersive spectroscopy (EDS) analysis.

as a catalyst, and a semiempirical kinetic model based on the Fermi equation was suitable for the description of this heterogeneous electro-Fenton reaction.

To compare the effects of catalysts on the electro-Fenton system, commercial nanomagnetite was used both free and embedded into alginate beads, and the degradation of chloramphenicol was evaluated. The nanomagnetite alginate beads were prepared by a mixture of 0.1 M nanoparticles of size <5 μm (supplied by Sigma Aldrich) (Figure 4.3) and a suspension of Na-alginate (2%). This solution was dropped on to a solution of 0.15 M $BaCl_2$ to create spherical alginate beads with nano-magnetite trapped within their structure. As shown in Figure 4.4, the wet beads were found to be spherical and black due to the presence of magnetite nanoparticles. The diameters of 100 wet beads were measured using an optical microscope. Different sizes of these beads can be achieved, depending on the synthesis conditions.

Several batch experiments were carried out in a glass cylindrical reactor with a working volume of 0.15 L (Rosales et al. 2012b). The reactor was filled with the catalyst and an aqueous solution of chloramphenicol, at an initial concentration of 90 mg/L. These solutions were agitated with a

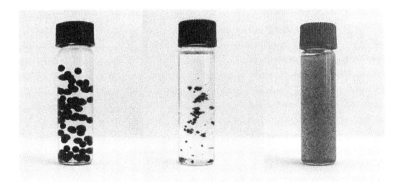

FIGURE 4.4 Different sizes of iron oxide alginate nanoparticles obtained using different diameters of needle.

mechanical system and a glass mixer to avoid concentration gradients. For all experiments, the electrolyte utilized was 0.01 M Na_2SO_4. A constant current intensity was applied by two electrodes connected to a direct power supply (HP model 3662). Carbon-felt and boron-doped diamond (BDD) were selected as cathode and anode, respectively. The carbon felt was placed on the inner wall of the cell (6 × 12 cm), covering the total internal perimeter, while the BDD (3 × 6 cm) was centered in the reactor. A continuous bubbling (1 L/min) of air at atmospheric pressure was established in the reactor surrounding the cathode; this flow started 10 min before the electro-Fenton process, to reach solution saturation of air at atmospheric pressure. This was carried out for the *in situ* production of H_2O_2 by the electrochemical reduction of oxygen. Figure 4.5 shows the profile of H_2O_2 obtained in an electro-Fenton system. It is clear that the H_2O_2 generation is constant over treatment time. This fact permits a continuous Fenton reaction in the medium to take place under catalyst action, and explains the high efficiency of this system in comparison with the conventional Fenton process as shown in Figure 4.1.

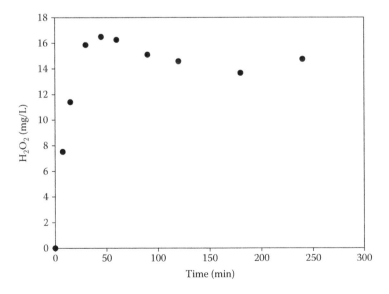

FIGURE 4.5 Electrochemical generation of H_2O_2. Operational conditions: volume 150 mL, intensity 150 mA, pH 3, electrolyte Na_2SO_4 0.01 M, air 1 L/min, and cathode/anode carbon felt/BDD.

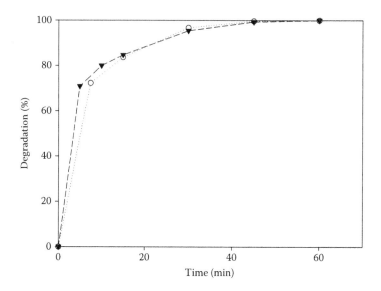

FIGURE 4.6 Chloramphenicol degradation profiles with time obtained during an electro-Fenton treatment using magnetic nanoparticles (○) and alginate immobilized nanoparticles (▼). Operational conditions: volume 150 mL, intensity 150 mA, pH 3, Fe 50 mg/L, electrolyte Na_2SO_4 0.01 M, air 1 L/min, chloramphenicol 90 mg/L, and cathode/anode carbon felt/graphite.

After the electro-Fenton treatment of chloramphenicol, using both catalysts, a negligible difference between free and nanomagnetite alginate beads was detected (Figure 4.6). In both cases, near complete degradation was obtained after 45 min with a mineralization level around 80%. However, when the homogeneous electro-Fenton process was carried out using free iron, the maximum mineralization level was about 58% after 1 h of treatment. These results demonstrated the high efficiency of the nanomagnetite electro-Fenton system. To confirm the mineralization of the chloramphenicol, the evolution of acid carboxylics was performed. As can be seen in Figure 4.7,

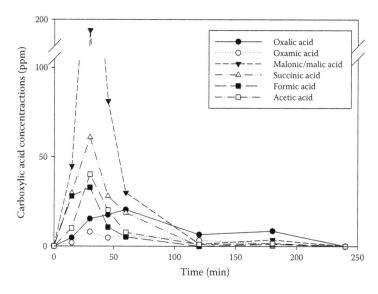

FIGURE 4.7 Generation of carboxylic acids during the degradation of chloramphenicol (90 mg/L) by electro-Fenton. Operational conditions: volume 150 mL, intensity 150 mA, pH 3, Fe nanoparticles 50 mg/L, electrolyte Na_2SO_4 0.01 M, air 1 L/min, and cathode/anode carbon felt/BDD.

the carboxylic acid concentration increased at the initial time until reaching a maximum after 30 min and, after that, decreased as a result of oxidation to CO_2 and water.

4.4　CONCLUSIONS

Over the last decades, nanotechnology has contributed to widespread technological innovations in different processes. In this chapter, the use of nanomaterials in the treatment of recalcitrant pollutants by different AOPs such as the Fenton and electro-Fenton processes was reviewed and discussed. The application of iron nanoparticles (ZVI, Fe_3O_4, and γ-Fe_2O_3) as catalyst has improved the effectiveness of the processes, as a result of the change in physical properties when compared with macroscopic systems. Hence, it was found that the development of iron catalysts based on nanomaterials provided an increase in the surface area-to-volume ratio, increasing their reactive and catalytic activities. The reported studies demonstrated a high ability for the developed materials to improve the production of hydroxyl radicals, which favors the treatment of pollutants. Furthermore, some nanomaterials can be manufactured to provide magnetic properties to the catalyst, which favors their recovery after the process. On the other hand, recent research has demonstrated that the inclusion of nanomaterials in electrodes can increase pollutant degradation and/or generation of reactive species. Based on the studies disclosed in this chapter, it can be established that the use of nanomaterials is a promising approach in the Fenton and electro-Fenton processes, opening new alternatives in the treatment of recalcitrant pollutants.

REFERENCES

Ai, Zhihui, Mei, Tao, Liu, Juan et al. Fe@Fe$_2$O$_3$ core-shell nanowires as an iron reagent. 3. Their combination with CNTs as an effective oxygen-fed gas diffusion electrode in a neutral electro-Fenton system. *Journal of Physical Chemistry C* 111(40) (2007): 14799–803.

Alfaya, Elena, Iglesias, Olalla, Pazos, Marta, and Sanromán, M. Ángeles. Environmental application of an industrial waste as catalyst for the electro-Fenton-like treatment of organic pollutants. *RSC Advances* 5(19) (2015): 14416–24.

Ardo, Sady G., Nélieu, Sulvie, Ona-Nguema, Georges et al. Oxidative degradation of nalidixic acid by nano-magnetite via Fe^{2+}/O$_2$ mediated reactions. *Environmental Science and Technology* 49(7) (2015): 4506–14.

Babuponnusami, Arjunan and Muthukumar, Karuppan. Removal of phenol by heterogenous photo electro Fenton-like process using nano-zero valent iron. *Separation and Purification Technology* 98 (2012): 130–135.

Babuponnusami, Arjunan and Muthukumar, Karuppan. A review on Fenton and improvements to the Fenton process for wastewater treatment. *Journal of Environmental Chemical Engineering* 2(1) (2014): 557–72.

Bell, Alexis T. The impact of nanoscience on heterogeneous catalysis. *Science* 299(5613) (2003): 1688–91.

Bocos, Elvira, Alfaya, Elena, Iglesias, Olalla, Pazos, Marta, and Sanromán, M. Ángeles. Application of a new sandwich of granular activated and fiber carbon as cathode in the electrochemical advanced oxidation treatment of pharmaceutical effluents. *Separation and Purification Technology* 151 (2015a): 243–50.

Bocos, Elvira, Iglesias, Olalla, Pazos, Marta, and Sanromán, M. Ángeles. Nickel foam a suitable alternative to increase the generation of Fenton's reagents. *Process Safety and Environmental Protection*. In press, doi:10.1016/j.psep.2015.04.011. (2015b).

Bocos, Elvira, Pazos, Marta, and Sanromán, M. Ángeles. Electro-Fenton decolourization of dyes in batch mode by the use of catalytic activity of iron loaded hydrogels. *Journal of Chemical Technology and Biotechnology* 89(8) (2014): 1235–42.

Bocos, Elvira, Pérez-Álvarez, David, Pazos, Marta, Rodríguez-Argüelles, M. Carmen, and Sanromán, M. Ángeles. Coated nickel foam electrode for the implementation of continuous electro-Fenton treatment. *Journal of Chemical Technology and Biotechnology* 91 (2015c): 685–92.

Bounab, Loubna, Iglesias, Olalla, González-Romero, Elisa, Pazos, Marta, and Sanromán, M. Ángeles. Effective heterogeneous electro-Fenton process of m-cresol with iron loaded actived carbon. *RSC Advances* 5(39) (2015): 31049–56.

Brillas, Enric, Sirés, Ignasi, and Oturan, Mehmet A. Electro-Fenton process and related electrochemical technologies based on Fenton's reaction chemistry. *Chemical Reviews* 109(12) (2009): 6570–631.

Buxton, George V., Greenstock, Clive, Helman, W. Philips, and Ross, Alberta B. Critical review of rate constants for reactions of hydrated electrons, hydrogen atoms and hydroxyl radicals (OH/O⁻) in aqueous solution. *Journal of Physical and Chemical Reference Data* 17(2) (1988): 513–886.

Chen, Yan-Xin, Chen, Sheng-Pei, Chen, Qing-Song, Zhou, Zhi-You, and Sun, Shi-Gang. Electrochemical preparation of iron cuboid nanoparticles and their catalytic properties for nitrite reduction. *Electrochimica Acta* 53(23) (2008): 6938–43.

Crane, Richard A. and Scott, Thomas B. Nanoscale zero-valent iron: Future prospects for an emerging water treatment technology. *Journal of Hazardous Materials* 211–212 (2012): 112–25.

De Laat, Joseph and Gallard, Hervé. Catalytic decomposition of hydrogen peroxide by Fe(III) in homogeneous aqueous solution: Mechanism and kinetic modeling. *Environmental Science and Technology* 33(16) (1999): 2726–32.

Dronskowski, Richard. The little maghemite story: A classic functional material. *Advanced Funtional Materials* 11(1) (2001): 27–29.

Du, Ning, Xu, Yanfang, Zhang, Hui, Zhai, Chuanxin, and Yang, Deren. Selective synthesis of Fe_2O_3 and Fe_3O_4 nanowires via a single precursor: A general method for metal oxide nanowires. *Nanoscale Research Letters* 5(8) (2010): 1295–300.

Duesterberg, Christopher K., Cooper, William J., and Waite, T. David. Fenton-mediated oxidation in the presence and absence of oxygen. *Environmental Science and Technology* 39(13) (2005): 5052–58.

Duesterberg, Christopher K. and Waite, T. David. Kinetic modeling of the oxidation of p-hydroxybenzoic acid by Fenton's reagent: Implications of the role of quinones in the redox cycling of iron. *Environmental Science and Technology* 41(11) (2007): 4103–10.

Expósito, Eduardo, Sánchez-Sánchez, Carlos M. and Montiel, Vicente. Mineral iron oxides as iron source in electro-Fenton and photoelectro-Fenton mineralization processes. *Journal of the Electrochemical Society* 154(8) (2007): 116–22.

Fernández de Dios, M. Ángeles, Iglesias, Olalla, Pazos, Marta, and Sanromán, M. Ángeles. Application of electro-Fenton technology to remediation of polluted effluents by self-sustaining process. *The Scientific World Journal* 2014 (2014): 1–8.

Ferroudj, Nassira, Nzimoto, Jimmy, Davidson, Anne et al. Maghemite nanoparticles and maghemite/silica nanocomposite microspheres as magnetic Fenton catalysts for the removal of water pollutants. *Applied Catalysis B: Environmental* 136–137 (2013): 9–18.

Fu, Fenglian, Chen, Zhihong, and Wang, Qi. Removal of a di-azo dye acid red 73 synthetic wastewater by advanced Fenton process based on nanoscale zero-valent iron. *Advanced Materials Research* 518–523 (2012): 2718–21.

Fu, Jian-Liang, Zhang, Xing-Wang, and Lei, Le-Cheng. Fe-modified multi-walled carbon nanotube electrode for production of hydrogen peroxide. *Acta Physico-Chimica Sinica* 23(8) (2007): 1157–62.

Garrido-Ramírez, Elizabeth G., Theng, Benny K. G., and Mora, María L. Clays and oxide minerals as catalysts and nanocatalysts in Fenton-like reactions: A review. *Applied Clay Science* 47(3–4) (2010): 182–92.

Guo, Jing and Al-Dahhan, Muthanna. Activity and stability of iron-containing pillared clay catalysts for wet air oxidation of phenol. *Applied Catalysis A: General* 299 (2006): 175–84.

Guo, Sheng, Zhang, Gaoke, Guo, Yadan, and Yu, Jimmy C. Graphene oxide–Fe_2O_3 hybrid material as highly efficient heterogeneous catalyst for degradation of organic contaminants. *Carbon* 60 (2013): 437–44.

Hammouda, Samia B., Adhoum, Nafaâ, and Monser, Lotfi. Chemical oxidation of a malodorous compound, indole, using iron entrapped in calcium alginate beads. *Journal of Hazardous Materials* 301 (2016): 350–61.

Hammouda, Samia Ben, Adhoum, Nafaâ, and Monser, Lotfi. Synthesis of magnetic alginate beads based on Fe_3O_4 nanoparticles for the removal of 3-methylindole from aqueous solution using Fenton process. *Journal of Hazardous Materials* 294 (2015): 128–36.

Hansson, Henrik, Kaczala, Fabio, Marques, Marcia, and Hogland, William. Photo-Fenton and Fenton oxidation of recalcitrant industrial wastewater using nanoscale zero-valent iron. *International Journal of Photoenergy* 2012(531076) (2012): 1–11.

He, Zhiqiao, Gao, Chao, Qian, Mengqian, Shi, Yuanqiao, Chen, Jianmeng, and Song, Shuang. Electro-Fenton process catalyzed by Fe_3O_4 magnetic nanoparticles for degradation of C.I. Reactive Blue 19 in aqueous solution: Operating conditions, influence, and mechanism. *Industrial and Engineering Chemistry Research* 53(9) (2014): 3435–47.

Hong, Ruo-Yu, Li, Jian-Hua, Zhang, Shi-Zhong et al. Preparation and characterization of silica-coated Fe_3O_4 nanoparticles used as precursor of ferrofluids. *Applied Surface Science* 255(6) (2009): 3485–92.

Hosseini, Soraya, Khan, Moonis A., Malekbala, Mohamad R., Cheah, Willie, and Choong, Thomas S. Y. Carbon coated monolith, a mesoporous material for the removal of Methyl Orange from aqueous phase: Adsorption and desorption studies. *Chemical Engineering Journal* 171(3) (2011): 1124–31.

Hou, Baolin, Han, Hongun, Jia, Shenyong, Zhuang, Haifeng, Xu, Peng, and Wang, Dexin. Heterogeneous electro-Fenton oxidation of catechol catalyzed by nano-Fe_3O_4: Kinetics with the Fermi's equation. *Journal of the Taiwan Institute of Chemical Engineers* 56 (2015): 138–147.

Hou, Liwei, Zhang, Qinghua, Jérôme, François, Duprez, Daniel, Zhang, Hui, and Royer, Sébastien. Shape-controlled nanostructured magnetite-type materials as highly efficient Fenton catalysts. *Applied Catalysis B: Environmental* 144 (2014): 739–49.

Hu, Sihai, Yao, Hairui, Wang, Kaifeng, Lu, Cong, and Wu, Yaoguo. Intensify removal of nitrobenzene from aqueous solution using nano-zero valent iron/granular activated carbon composite as Fenton-like catalyst. *Water, Air, and Soil Pollution* 226(155) (2015): 1–13.

Hu, Xiaobin, Liu, Benzhi, Deng, Yuehua, Chen, Hongzhe, Luo, Si, Sun, Cheng, Yang, Po, Yang, Shaogui. Adsorption and heterogeneous Fenton degradation of 17α-methyltestosterone on nano Fe_3O_4/MWCNTs in aqueous solution. *Applied Catalysis B: Environmental* 107(3–4) (2011): 274–83.

Hua, Zulin, Ma, Wenqiang, Bai, Xue et al. Heterogeneous Fenton degradation of bisphenol a catalyzed by efficient adsorptive Fe_3O_4/GO nanocomposites. *Environmental Science and Pollution Research* 21(12) (2014): 7737–45.

Huang, Ruixiong, Fang, Zhanqiang, Yan, Xiaomin, and Cheng, Wen. Heterogeneous sono-Fenton catalytic degradation of bisphenol a by Fe_3O_4 magnetic nanoparticles under neutral condition. *Chemical Engineering Journal* 197 (2012): 242–49.

Iglesias, Olalla, Fernández de Dios, M. Ángeles, Pazos, Marta, and Sanromán, M. Ángeles. Using iron-loaded sepiolite obtained by adsorption as a catalyst in the electro-Fenton oxidation of Reactive Black 5. *Environmental Science and Pollution Research* 20(9) (2013): 5983–93.

Isarain-Chávez, Eloy, Arias, Conchita, Cabot, Pere L. et al. Mineralization of the drug β-blocker atenolol by electro-Fenton and photoelectro-Fenton using an air-diffusion cathode for H_2O_2 electrogeneration combined with a carbon-felt cathode for Fe^{2+} regeneration. *Applied Catalysis B: Environmental* 96(3–4) (2010): 361–69.

Kallel, Monem, Belaid, Chokri, Boussahel, Rachdi, Ksibi, Mohamed, Montiel, Antoine, and Elleuch, Boubaker. Olive mill wastewater degradation by Fenton oxidation with zero-valent iron and hydrogen peroxide. *Journal of Hazardous Materials* 163(2–3) (2009): 550–54.

Kuang, Ye, Wang, Qinping, Chen, Zuliang, Megharaj, Mallavarapu, and Naidu, Ravendra. Heterogeneous Fenton-like oxidation of monochlorobenzene using green synthesis of iron nanoparticles. *Journal of Colloid and Interface Science* 410 (2013): 67–73.

Le, Thi X. H., Bechelany, Mikhael, Lacour, Stella, Oturan, Nihal, Oturan, Mehmet A., and Cretin, Marc. High removal efficiency of dye pollutants by electron-Fenton process using a graphene based cathode. *Carbon* 94 (2015): 1003–11.

Li, Hualiang, Lei, Hhengyi, Chen, Kai et al. A Nano-Fe^0/ACF cathode applied to neutral electro-Fenton degradation of orange II. *Journal of Chemical Technology and Biotechnology* 86(3) (2011): 398–405.

Li, Jinpo, Ai, Zhihui, and Zhang, Lizhi. Design of a neutral electro-Fenton system with Fe@Fe_2O_3/ACF composite cathode for wastewater treatment. *Journal of Hazardous Materials* 164(1) (2009): 18–25.

Li, Renchao, Gao, Ying, Jin, Xiaoying, Chen, Zuliang, Megharaj, Mallavarapu, and Naidu, Ravendra. Fenton-like oxidation of 2,4-DCP in aqueous solution using iron-based nanoparticles as the heterogeneous catalyst. *Journal of Colloid and Interface Science* 438 (2015): 87–93.

Ling, Yuhan, Long, Mingce, Hu, Peidong, Chen, Ya, and Huang, Juwei. Magnetically separable core–shell structural γ-Fe_2O_3@Cu/Al-MCM-41 nanocomposite and its performance in heterogeneous Fenton catalysis. *Journal of Hazardous Materials* 264 (2014): 195–202.

Liu, Ruili, Wu, Dongqing, Feng, Xinliang, and Müllen, Klaus. Nitrogen-doped ordered mesoporous graphitic arrays with high electrocatalytic activity for oxygen reduction. *Angewandte Chemie: International Edition* 49(14) (2010): 2565–69.

Liu, Wei, Ai, Zihui, and Zhang, Lizhi. Design of a neutral three-dimensional electro-Fenton system with foam nickel as particle electrodes for wastewater treatment. *Journal of Hazardous Materials* 243 (2012): 257–64.

Machado, Susana, Stawinski, W., Slonina, P. et al. Application of green zero-valent iron nanoparticles to the remediation of soils contaminated with ibuprofen. *Science of the Total Environment* 461–462 (2013): 323–29.

Machulek, Amilcar, Quina, Frank H., Gozzi, Fabio et al. Fundamental mechanistic studies of the photo-Fenton reaction for the degradation of organic pollutants. In ed. Tomasz Puzyn, *Organic Pollutants Ten Years After the Stockholm Convention: Environmental and Analytical Update*, pp. 271–92, InTech, 2012. Rijeka, Croatia.

Martin, John E., Herzing, Andrew A., Yan, Weile et al. Determination of the oxide layer thickness in core-shell zerovalent iron nanoparticles. *Langmuir* 24(8) (2008): 4329–34.

Molday, Robert S. *Magnetic Iron-Dextran Microspheres US4452773 A.* Canadian Patents and Development Limited (CA), assignee (1984).

Moon, Byung-Hyun, Park, Young-Bae, and Park, Kyung-Hun. Fenton oxidation of Orange II by pre-reduction using nanoscale zero-valent iron. *Desalination* 268(1–3) (2011): 249–52.

Mousset, Emmanuel, Frunzo, Luigi, Esposito, Giovanni, Hullebusch, Eric D. V., Oturan, Nihal, and Oturan, Mehmet A. A complete phenol oxidation pathway obtained during electro-Fenton treatment and validated by a kinetic model study. *Applied Catalysis B: Environmental* 180 (2016): 189–98.

Muñoz, Macarena, de Pedro, Zahara M., Casas, Jose A., and Rodriguez, Juan J. Preparation of Magnetite-based catalysts and their application in heterogeneous Fenton oxidation: A review. *Applied Catalysis B: Environmental* 176–177 (2015): 249–65.

Ng, Yee-Sern, Sen Gupta, Bhaskar, and Hashim, Mohd A. Stability and performance enhancements of electrokinetic-Fenton soil remediation. *Reviews in Environmental Science and Biotechnology* 13(3) (2014): 251–63.

Park, Jin-Bum, Lee, Jinwoo, Yoon, Chong Seung, and Sun, Yang-Kook. Ordered mesoporous carbon electrodes for Li-O_2 batteries. *ACS Applied Materials and Interfaces* 5(24) (2013): 13426–31.

Peng, Sheng, Wang, Chao, Xie, Jin, and Sun, Shouheng. Synthesis and stabilization of monodisperse Fe nanoparticles. *Journal of the American Chemical Society* 128(33) (2006): 10676–77.

Petrucci, Elisabetta, Da Pozzo, Anna, and Di Palma, Luca. On the ability to electrogenerate hydrogen peroxide and to regenerate ferrous ions of three selected carbon-based cathodes for electro-Fenton processes. *Chemical Engineering Journal* 283 (2016): 750–58.

Pradhan, Amey A. and Gogate, Parag R. Degradation of p-nitrophenol using acoustic cavitation and Fenton chemistry. *Journal of Hazardous Materials* 173(1–3) (2010): 517–22.

Prakash, Anand, McCormick, Alon V., and Zachariah, Michael R. Aero-sol-gel synthesis of nanoporous iron-oxide particles: A potential oxidizer for nanoenergetic materials. *Chemistry of Materials* 16(8) (2004): 1466–71.

Qiu, Shan, He, Di, Ma, Jinxing, Liu, Tongxu, and Waite, T. David. Kinetic modeling of the electro-Fenton process: Quantification of reactive oxygen species generation. *Electrochimica Acta* 176 (2015): 51–58.

Ren, Wei, Peng, Qiaoli, Huang, Ze'ai et al. Effect of pore structure on the electro-Fenton activity of ACF@OMC cathode. *Industrial and Engineering Chemistry Research* 54(34) (2015): 8492–99.

Rosales, Emilio, Iglesias, Olalla, Pazos, Marta, and Sanromán, M. Ángeles. Decolourisation of dyes under electro-Fenton process using Fe alginate gel beads. *Journal of Hazardous Materials* 213–214 (2012b): 369–77.

Rosales, Emilio, Pazos, Marta, and Sanromán, M. Ángeles. Advances in the electro-Fenton process for remediation of recalcitrant organic compounds. *Chemical Engineering and Technology* 35(4) (2012a): 609–17.

Sable, Shailesh S., Ghute, Pallavi P., Álvarez, Pedro, Beltrán, Fernando J., Medina, Francesc, and Contreras, Sandra. FeOOH and derived phases: Efficient heterogeneous catalysts for clofibric acid degradation by advanced oxidation processes (AOPs). *Catalysis Today* 240 (2015): 46–54.

Shen, Lihua, Yan, Pei, Guo, Xiaobin, Wei, Haixia, and Zheng, Xiaofeng. Three-dimensional electro-Fenton degradation of methyleneblue based on the composite particle electrodes of carbon nanotubes and nano-Fe_3O_4. *Arabian Journal for Science and Engineering* 39(9) (2014): 6659–64.

Shi, Jingu, Ai, Zhihui, and Zhang, Lizhi. Fe@Fe_2O_3 core-shell nanowires enhanced Fenton oxidation by accelerating the Fe(III)/Fe(II) cycles. *Water Research* 59 (2014): 145–53.

Song, Shuqin, Wu, Mingmei, Liu, Yuhui, Zhu, Qiping, Tsiakaras, Panagiotis, and Wang, Yi. Efficient and stable carbon-coated nickel foam cathodes for the electro-Fenton process. *Electrochimica Acta* 176 (2015): 811–18.

Su, Chia-Chi, Chang, An-Tzu, Bellotindos, Luzvisminda M., and Lu, Ming-Chun. Degradation of acetaminophen by Fenton and electro-Fenton processes in aerator reactor. *Separation and Purification Technology* 99 (2012): 8–13.

Sun, Min, Ru, Xiao-Ru, and Zhai, Lin-Feng. *In situ* fabrication of supported iron oxides from synthetic acid mine drainage: High catalytic activities and good stabilities towards electro-Fenton reaction. *Applied Catalysis B: Environmental* 165 (2015): 103–10.

Tiya-Djowe, Antoine, Acayanka, Elie, Lontio-Nkouongfo, G., Laminsi, Samuel, and Gaigneaux, Eric M. Enhanced discolouration of methyl violet 10B in a gliding arc plasma reactor by the maghemite nanoparticles used as heterogeneous catalyst. *Journal of Environmental Chemical Engineering* 3(2) (2015): 953–60.

Velichkova, F., Julcour-Lebigue, Carine, Koumanova, Bogdana, and Delmas, Henri. Heterogeneous Fenton oxidation of paracetamol using iron oxide (nano)particles. *Journal of Environmental Chemical Engineering* 1(4) (2013): 1214–22.

Wang, Chuan-Bao and Zhang, Wei-Xian. Synthesizing nanoscale iron particles for rapid and complete dechlorination of TCE and PCBs. *Environmental Science and Technology* 31(7) (1997): 2154–56.

Wang, Lin, Yang, Juan, Li, Yongmei, Lv, Juan, and Zou, Jinte. Removal of chlorpheniramine in a nanoscale zero-valent iron induced heterogeneous Fenton system: Influencing factors and degradation intermediates. *Chemical Engineering Journal* 284 (2016): 1058–67.

Wang, Wei, Liu, Ying, and Li, Tielong. Catalytic property of smart polymer coated nano-Fe_3O_4. *Journal of Tianjin University Science and Technology* 47(5) (2014a): 459–63.

Wang, Yanbin, Zhao, Hongying, Li, Mingfang, Fan, Jiaqi, and Zhao, Guoha. Magnetic ordered mesoporous copper ferrite as a heterogeneous Fenton catalyst for the degradation of imidacloprid. *Applied Catalysis B: Environmental* 147 (2014b): 534–45.

Wang, Yi, Liu, Yuhui, Wang, Kun, Song, Shuqin, Tsiakaras, Panagiotis, and Liu, Hong. Preparation and characterization of a novel KOH activated graphite felt cathode for the electro-Fenton process. *Applied Catalysis B: Environmental* 165 (2015): 360–68.

Wang, Yujing, Zhao, Guohua, Chai, Shouning, Zhao, Hongying, and Wang, Yanbin. Three-dimensional homogeneous ferrite-carbon aerogel: One pot fabrication and enhanced electro-Fenton reactivity. *ACS Applied Materials and Interfaces* 5(3) (2013): 842–52.

Wei, Wang, Li, Tie Long, Liu, Ying, and Zhou, Ming Hua. Highly active heterogeneous Fenton-like systems based on Fe_3O_4 nanoparticles. *Advanced Materials Research* 233–235 (2011): 487–90.

Wu, Yan, Zeng, Shenliang, Wang, Feifeng, Megharaj, Mallavarapu, Naidu, Ravendra, and Chen, Zuliang. Heterogeneous Fenton-like oxidation of Malachite Green by iron-based nanoparticles synthesized by tea extract as a catalyst. *Separation and Purification Technology* 154 (2015): 161–67.

Xia, Guangsen, Lu, Yonghong, Gao, Xueli, Gao, Congjie, and Xu, Haibo. Electro-Fenton degradation of methylene blue using polyacrylonitrile-based carbon fiber brush cathode. *Clean—Soil, Air, Water* 43(2) (2015): 229–36.

Xia, Min, Chen, Chen, Long, Mingce, Chen, Chao, Cai, Weimin, and Zhou, Baoxue. Magnetically separable mesoporous silica nanocomposite and its application in Fenton catalysis. *Microporous and Mesoporous Materials* 145(1–3) (2011): 217–23.

Xia, Siqing, Gu, Zaoli, Zhang, Zhiqiang, Zhang, Jiao, and Hermanowicz, Slawomir W. Removal of chloramphenicol from aqueous solution by nanoscale zero-valent iron particles. *Chemical Engineering Journal* 257 (2014): 98–104.

Xu, Lejin and Wang, Jianlong. A heterogeneous Fenton-like system with nanoparticulate zero-valent iron for removal of 4-chloro-3-methyl phenol. *Journal of Hazardous Materials* 186(1) (2011): 256–64.

Xu, Lejin and Wang, Jianlong. Degradation of 2,4,6-trichlorophenol using magnetic nanoscaled Fe_3O_4/CeO_2 composite as a heterogeneous Fenton-like catalyst. *Separation and Purification Technology* 149 (2015): 255–64.

Xu, Lejin and Wang, Jianlong. Fenton-like degradation of 2,4-dichlorophenol using Fe_3O_4 magnetic nanoparticles. *Applied Catalysis B: Environmental* 123–124 (2012a): 117–26.

Xu, Lejin and Wang, Jianlong. Magnetic nanoscaled Fe_3O_4/CeO_2 composite as an efficient Fenton-like heterogeneous catalyst for degradation of 4-chlorophenol. *Environmental Science and Technology* 46(18) (2012b): 10145–53.

Xu, Yinhui and Zhao, Dongye. Reductive immobilization of chromate in water and soil using stabilized iron nanoparticles. *Water Research* 41(10) (2007): 2101–8.

Xue, Xiaofei, Hanna, Khalil, Abdelmoula, Mustapha, and Deng, Nansheng. Adsorption and oxidation of PCP on the surface of magnetite: Kinetic experiments and spectroscopic investigations. *Applied Catalysis B: Environmental* 89(3–4) (2009): 432–40.

Yan, Hao, Zhang, Jiancheng, You, Chenxia, Song, Zhenwei, Yu, Benwei, and Shen, Yue. Influences of different synthesis conditions on properties of Fe_3O_4 nanoparticles. *Materials Chemistry and Physics* 113(1) (2009): 46–52.

Yan, Weile, Lien, Hsing-Lung, Koel, Bruce E., and Zhang, Wei-Xian. Iron nanoparticles for environmental clean-up: Recent developments and future outlook. *Environmental Sciences: Processes and Impacts* 15(1) (2013): 63–77.

Yang, Bo, Tian, Zhang, Zhang, Li, Guo, Yaopeng, and Yan, Shiqiang. Enhanced heterogeneous Fenton degradation of methylene blue by nanoscale zero valent iron (nZVI) assembled on magnetic Fe_3O_4/reduced graphene oxide. *Journal of Water Process Engineering* 5 (2015): 101–11.

Yang, Jiacheng, Wang, Xiangyu, Zhu, Minping, Liu, Huiling, and Ma, Jun. Investigation of PAA/PVDF–NZVI Hybrids for metronidazole removal: Synthesis, characterization, and reactivity characteristics. *Journal of Hazardous Materials* 264 (2014): 269–77.

Zeng, Xia, Hanna, Khalil, and Lemley, Ann T. Cathodic Fenton degradation of 4,6-dinitro-o-cresol with nano-magnetite. *Journal of Molecular Catalysis A: Chemical* 339(1–2) (2011): 1–7.

Zhang, Chao, Zhou, Minghua, Ren, Gengbo et al. Heterogeneous electro-Fenton using modified iron-carbon as catalyst for 2,4-dichlorophenol degradation: Influence factors, mechanism and degradation pathway. *Water Research* 70 (2015a): 414–24.

Zhang, Di, Wang, Yi-Xuan, Niu, Hong-Yun, and Meng, Zhao-Fu. Degradation of norfloxacin by nano-Fe_3O_4/H_2O_2. *Huanjing Kexue/Environmental Science* 32(10) (2011): 2943–48.

Zhang, Guoquan, Wang, Shuai, and Yang, Fenglin. Efficient adsorption and combined heterogeneous/homogeneous Fenton oxidation of amaranth using supported nano-FeOOH as cathodic catalysts. *Journal of Physical Chemistry C* 116(5) (2012): 3623–34.

Zhang, Jinbin, Zhuang, Jie, Gao, Lizeng et al. Decomposing phenol by the hidden talent of ferromagnetic nanoparticles. *Chemosphere* 73(9) (2008): 1524–28.

Zhang, Junyong, Deng, Yonghui, Wei, Jing et al. Design of amphiphilic ABC triblock copolymer for templating synthesis of large-pore ordered mesoporous carbons with tunable pore wall thickness. *Chemistry of Materials* 21(17) (2009b): 3996–4005.

Zhang, Shengxiao, Zhao, Xiaoli, Niu, Hongyun, Shi, Yali, Cai, Yaqi, and Jiang, Guibin. Superparamagnetic Fe_3O_4 nanoparticles as catalysts for the catalytic oxidation of phenolic and aniline compounds. *Journal of Hazardous Materials* 167(1–3) (2009a): 560–6.

Zhang, Yan, Gao, Ming-Ming, Wang, Xi-Hua, Wang, Shu-Guang, and Liu, Rui-Tig. Enhancement of oxygen diffusion process on a rotating disk electrode for the electro-Fenton degradation of tetracycline. *Electrochimica Acta* 182 (2015b): 73–80.

Zubir, Nor A., Yacou, Christelle, Motuzas, Julius, Zhang, Xiwang, and Diniz Da Costa, Joao C. Structural and functional investigation of graphene oxide-Fe_3O_4 nanocomposites for the heterogeneous Fenton-like reaction. *Scientific Reports* 4(4594) (2014): 1–8.

Zubir, Nor A., Yacou, Christelle, Motuzas, Julius, Zhang, Xiwang, Zhao, Xiu S., and Diniz Da Costa, Joao C. The sacrificial role of graphene oxide in stabilising a Fenton-like catalyst GO-Fe_3O_4. *Chemical Communications* 51(45) (2015): 9291–93.

5 Applications of Synthetic Nanocomposite Ion-Exchange Materials as Chemical and Vapor Sensors

Asif Ali Khan, Shakeeba Shaheen, and Nida Alam

CONTENTS

ABSTRACT

This chapter reports the detrimental effects of various heavy metal ions and a few chemical species present in water. Their removal from wastewaters can be accomplished through various treatment options such as chemical precipitation, complexation, coagulation, ion exchange, adsorption, solvent extraction, electrodeposition, and membrane operations. The chapter describes the various methodologies employed for the synthesis of nanocomposite materials. The formation of the nanocomposites is further characterized by Fourier transform infrared (FTIR) spectroscopy, x-ray diffraction (XRD), scanning electron microscopy (SEM), transmission electron microscopy (TEM), energy-dispersive x-ray (EDX), and thermogravimetric analysis (TGA). Further, the application of the nanocomposites as ion exchangers, ion-selective electrodes, and vapor sensors are explored and illustrated.

5.1 INTRODUCTION

Pollution by different chemical species remains an important environmental issue. The increasing demand for chemical surveillance in environmental protection and in many industrial processes has created the need for remediation, with features such as high selectivity, sensitivity, reliability, and sturdiness. The presence of Cd(II), Hg(II), Pb(II), Ni(II), Cr(II), As(IV), some pesticides and surfactants such as zinc(II) dimethyldithiocarbamate, and cetyltrimethylammonium bromide, respectively, in potable water above the permissible limit is harmful to human beings. The various chemical species with their harmful effects are discussed as follows:

5.1.1 CADMIUM

Cadmium is a toxic heavy metal that mainly appears in the environment due to industrial processes. Cadmium strongly adsorbs to organic matter in soils. The National Institute of Occupational Safety and Health (NIOSH) has set the danger to life and health level (IDLH) as 9 mg/m³. The U.S. Environmental Protection Agency (EPA) has found cadmium to potentially cause the following health effects when people are exposed to it at levels above the minimum concentration level (MCL) for relatively short periods of time: nausea, vomiting, diarrhea, muscle cramps, salivation, sensory disturbances, liver injury, convulsions, shock, and renal failure. In the long term, cadmium has the potential to cause the following effects from a lifetime of exposure at levels above the MCL: kidney, liver, bone, and blood damage (Rezaei et al. 2008). The major sources of cadmium exposure are the smelting process, electroplating, manufacture of cadmium alloys, and production of nickel-cadmium batteries and welding.

5.1.2 MERCURY

Mercury is highly toxic in nature when inhaled or ingested into the body. An increased level of mercury in the body can lead to mercury poisoning and also cause permanent damage to the brain and kidneys. Inorganic mercuric compounds mainly attack liver and kidneys. Mercuric chloride is corrosive when ingested; it precipitates proteins of the mucous membrane, causing an ashen appearance of the mouth, pharynx, and gastric mucus. Organic mercurials are also toxic substances; Hg(II) can pass through the placental barrier and enter the fetal tissues (Khan et al. 2012). Mercury also shows strong affinity for ligands containing sulfur atoms, and thus causes the blocking of sulfhydryl groups (–SH) of proteins, enzymes, and membranes. Hg(II) ions are also responsible for injuries of the kidney and gastrointestinal tract (Khan et al. 2008). Therefore, considering all the health and environmental hazards associated with mercury compounds, their use has been brought under the control of various regulations in many countries.

5.1.3 ARSENIC

A few years ago, the presence of arsenic in groundwater in the Indian state of West Bengal created an alarming situation, as many people residing there were detected with serious health problems. The EPA set the standard for arsenic in drinking water as 10 ppb, to protect consumers served by public water systems from the effects of long-term, chronic exposure to arsenic. The noncancer effects of arsenic are thickening and discoloration of skin, stomach pain, vomiting, diarrhea, numbness in hands and feet, partial paralysis, blindness, and birth defects. Arsenic has also been linked to cancer of the bladder, lungs, skin, kidney, nasal passages, liver, and prostate (Khan and Shaheen 2014a). Anthropogenic activities, including mining, smelting, and the widespread use of pesticides in the farming industries, have contributed to arsenic pollution in the environment (MacDonald 2001). Arsenic pollution in drinking water is a serious environmental and health concern because of the toxicity of arsenic to humans and other living organisms. Elevated levels of arsenic in the environment can enter into the food chain and cause deleterious effects on humans and other organisms.

5.1.4 CHROMIUM

As per the standards, the maximum permissible limits of Cr(VI) for discharge to inland surface water and potable water are 0.1 and 0.05 mg/L, respectively. Chromium and its compounds have been widely used in different commercial ventures, such as mining, tanning, concrete, the generation of steel or other metal compounds, electroplating operations, photographic materials, and destructive painting and metal businesses. Studies have indicated that workers in the chromate generation industry have a high risk of respiratory illness, fibrosis, puncturing of the nasal septum, advancement of nasal polyps, and lung diseases (Amdur et al. 1991).

5.1.5 LEAD

Man was familiar with lead as early as 4000 BC. Both the Egyptians and Hebrews mined lead ore in around 2000 BC. The earliest written accounts of lead toxicity have been found on Egyptian parchments. Perhaps, even more importantly, it is now suggested that there may be no level of Pb that does not produce a toxic effect, particularly in the developing central nervous system. The definition of an action or intervention level is 10 µg/dL of lead. This concern applies to the fetus *in utero* and women of childbearing age (Goyer 1993). The major sources of lead emissions have historically been from fuels in on-road motor vehicles and industrial sources. Humans are exposed to lead through contaminated drinking water and food. Lead, in general, is a metabolic poison and enzyme inhibitor, causes damage to the nervous system and kidneys, and is a suspected carcinogen (Khan and Baig 2012).

5.1.6 Copper

Copper is a vital component because of its utilization as a key supplement to amphibian organic entities and is dangerous at a high fixation of >15 mg/day; the maximum permissible concentration limit is 2 mg/day. Copper insufficiency brings about weakness, while its collection brings about dyslexia, hypoglycemia, liver and kidney harm, severe mucosal irritation, central nervous system irritation, gastrointestinal disorders, and Wilson's disease (Khan and Baig 2015).

5.1.7 Nickel

Nickel is a toxic heavy metal found in the environment due to various materials and industrial activities. The acceptable limit of nickel in drinking water is 0.01 mg/L and the industrial discharge limit in wastewater is 2 mg/L. Higher concentrations of nickel cause poisonous effects that can lead to headache, dizziness, nausea, tightness of the chest, dry cough, vomiting, chest pain, shortness of breath, rapid respiration, cyanosis, and extreme weakness. Therefore, it is essential to remove nickel from the environment beyond its permissible limit (Revathi et al. 2005).

5.1.8 Pesticide

Zinc(II) dimethyldithiocarbamate (Ziram) belongs to the carbamate family and has been gradually replacing more persistent species (mainly organophosphates) due to their low persistence in the environment and biological activity (Khan and Akhtar 2011a). Ziram is an agricultural fungicide. It is usually applied to the foliage of plants, but is also used for soil and seed treatment. Ziram is used primarily on almonds and stone fruits. It is also used as an accelerator in the manufacturing of rubber, packaging materials, adhesives, and textiles. Another use of the compound is as a bird and rodent repellent. Ziram can cause skin and mucous membrane irritation. Humans with prolonged inhalation exposure to Ziram have reported developing nerve and visual disturbances (Meister 1992). Ziram is harmful to the eyes and may cause irreversible eye damage (MSDS 1991). Metal salts of dithiocarbamates are more effective as fungicides, because the toxicity of the dithiocarbamate is high, due to the presence of metal in it.

5.1.9 Surfactant

Surfactants as environmental pollutants are found in various sources, including natural and wastewaters, process solutions of industrial enterprises, and household consumer products. Surfactants are widely used in industrial processes for their physicochemical characteristics such as detergency, emulsification, foaming, dispersion, and solubilization effects (Lawrence and Ress 2000; Czapla and Bart 2000; Lin et al. 1999; Sabah et al. 2002). The experimental data has revealed that surfactants can kill microorganisms at quite low concentrations of 1–5 mg/L and harm them at even lower concentrations of 0.5 mg/L (Rozzi et al. 2002).

Various conventional technologies used to separate and remove the various chemical species include solvent extraction, chemical precipitation, ion exchange, adsorption, flocculation, membrane separation, and electrochemical methods (Peters et al. 1985). Among these, ion exchange is a promising technique, due to its simple operating method and potential recovery. One effective approach is to develop ion-exchange materials of high selectivity toward different chemical species present in water.

5.1.10 Ion Exchangers

Ion exchange is a process that allows the separation of ions and polar molecules based on their charge. It can be used for various kinds of charged molecules, including proteins. This separation process occurs between the mobile phase and ion-exchange groups bonded to the support material. Additional, nonionic adsorption processes contribute to the separation mechanism in highly polar ions. Different types of ion-exchange materials may be discussed as follows:

5.1.10.1   Inorganic Ion-Exchange Materials

The last 40 years or more have seen a great upsurge in the researches on synthetic inorganic ion exchangers. The main emphasis has been given to the development of new materials possessing chemical stability, reproducibility in ion-exchange behavior, and selectivity for certain metal ions important from the analytical and environmental point of view. Synthetic inorganic ion exchangers are generally produced as gelatinous precipitates by rapidly mixing elements of groups III, IV, V, and VI of the periodic table, usually at room temperature.

Inorganic ion exchangers are generally the oxides, hydroxides, and insoluble acid salts of polyvalent metals, heteropolyacid salts, and insoluble metal ferrocyanides.

5.1.10.2   Organic Ion-Exchange Materials

Organic ion-exchange materials have, as a rule, a three-dimensional polymeric structure. The network is called the *framework* or *matrix*, which is a flexible random network of hydrocarbon chains. This matrix carries the ionic groups such as $-SO_3^-$, $-COO^-$, $-PO_3^{2-}$, $-AsO_3^{2-}$, and so on in cation exchangers, and $-NH_3^+$, $-NH_2^+$, $-N^+$, $-S^+$, and so on in anion exchangers. Ion-exchange resins are thus cross-linked polyelectrolytes (Andrei 2007). The resins are made insoluble by crosslinking the various hydrocarbon chains. Besides the cross-linked three-dimensional networks, noncross-linked polymers must also be named. Insolubility and stability are provided to these materials by physical tangling and knotting of the polymeric chains. Hydrogen bonds and van der Waals forces also contribute to insolubility (Hellferich 1962).

The degree of cross-linking determines the mesh width of the matrix, swelling ability, movement of mobile ions, hardness, and mechanical durability. Highly cross-linked resins are harder, more resistant to mechanical degradation, less porous, and swell less in solvents. When an organic ion exchanger is placed in a solvent or solution, it will expand or swell. The main advantages of synthetic organic ion-exchange resins are their high capacity, wide applicability, wide versatility, and low cost relative to some synthetic inorganic media. The main limitations are their limited radiation and thermal stabilities.

5.1.10.3   Organic–Inorganic Composite Ion Exchangers

Organic–inorganic ion-exchange materials formed by the combination of organic polymers and inorganic materials are attractive for the purpose of creating high-performance or high-functional polymeric materials, termed as *organic–inorganic ion-exchange materials*. The conversion of organic ion-exchange materials into hybrid ion exchangers is the latest development in this discipline. The preparation of hybrid ion exchangers is carried out by the binding of organic polymers; for example, polyaniline, polyacrylonitrile, and polystyrene. These polymer-based hybrid ion exchangers show an improvement in a number of properties: chemical, mechanical, radiation stability, improvement in ion-exchange properties, and also a selective nature for heavy toxic metal ions. One of their important properties is their granulometric nature, which makes them more suitable for application in column operations.

5.1.10.4   Fibrous Ion Exchangers

Recently, fibrous ion-exchange materials have drawn the attention of researchers and scientists, as they exhibit a high efficiency in the process of sorption from gaseous and liquid media. Fibrous ion exchangers can be used in the form of various textile goods such as cloth, conveyer belts, nonwoven materials, staples, nets, and so on, thus opening up many possibilities for new technological processes. They consist of monofilaments of uniform size ranging from 5 to 50 µm. This predetermines the short diffusion path of the sorbent and the high rate of sorption that can be about 100 times higher than that of the granular resins that are normally used with a particle diameter of 0.25–1 μm. Hence, they are more useful in large-scale processes where a high resistance of filtering layers is needed.

5.1.10.5 Composite Ion Exchangers

The concept of composites dates back to the times of the Israelites (800 BC) and Egyptians (in the third millennium BC), who used straw in brick manufacturing as a reinforcement material. Naturally occurring composites are bone, feathers, natural fibers, bamboo, and wood. Bone is an organic–inorganic composite of protein (collagen) and minerals (calcium apatite), and bamboo is cellulose reinforced by silica. Composite materials are engineering materials made from two or more constituent materials with significantly different physical or chemical properties, which remain separate and distinct on a macroscopic level within the finished structure (Richardson 1987). The properties of the new material are dependent on the properties of the constituent materials as well as the properties of the interfaces.

Nanocomposites are composites in which at least one of the phases shows dimensions in the nanometer range (Roy et al. 1986). It has been reported that changes in particle properties can be observed when the particle size is less than a particular level, called the *critical size* (Kamigaito 1991). Additionally, as dimensions reach the nanometer level, interactions at phase interfaces become largely improved, and this is important for the enhancement of materials' properties. Nanocomposite materials can be classified according to their matrix material and also the nature of association between the inorganic and organic components.

5.1.10.6 Applications of Nanocomposite Ion-Exchange Materials

A nanocomposite material prepared by the incorporation of organic polymers into the inorganic microporous precipitate is an emerging field of research. The chemical, thermal, and mechanical stabilities of such materials encourage the researcher to explore new possibilities for their applications. A recent challenge is to analyze the threat to human life and other living beings due to the hazardous effects of environmental pollutants. The importance of monitoring environmental pollutants has recently led to increasing interest in the development of novel sensors for the detection of toxic metals.

Ion-exchanger-incorporated potentiometric membrane sensors are well-established analytical tools, routinely used for the selective and direct measurement of a wide variety of different ions in environmental samples. The key ingredient of any potentiometric ion-selective sensor is its ion-selective membrane. A number of organic–inorganic composite ion-exchange materials have been successfully prepared and used in making ion-selective membrane electrodes (Khan and Shaheen 2014b,c). Ion-selective electrodes are mainly membrane-based devices; they consist of permselective ion-conducting materials, which separate the sample from the inside of the electrode. On the inside is a filling solution containing the ion of interest at constant activity. The membrane is usually nonporous, water insoluble, and mechanically stable.

The ability of nanomaterials to conduct electricity varies widely, allowing their classification into good conductors, semiconductors, and nonconductors or insulators. Conducting polymers are a unique class of materials exhibiting the electrical and optical properties of metals or semiconductors. These materials have great potential for use in various devices as a result of their architectural diversity and flexibility, low cost, and ease of synthesis (Hundt et al. 2010). Conducting polymers have extended π-systems and are highly susceptible to chemical or electrochemical oxidation or reduction. Thus, the electrical and optical properties of such polymers may be precisely altered by controlling the process of oxidation and reduction. Conducting polymers have emerged as remarkable gas-sensing materials (Khan et al. 2015). A number of conducting nanocomposite ion-exchange materials have been synthesized and show an appreciable sensing ability for various gas sensors (Khan and Shaheen 2013a; Shakir et al. 2014).

5.2 EXPERIMENTAL

5.2.1 Synthesis of Nanocomposite Material

The nanocomposites investigated by Khan et al. (Khan and Baig 2013; Khan and Shaheen 2013a; Shakir et al. 2014) are mainly synthesized by two processes:

5.2.1.1 Sol-Gel Method

The sol-gel method is a process of synthesizing nanomaterial from a chemical solution or sol, which acts as the precursor for an integrated network (gel) of either discrete particles or network polymers. Typical precursors are metal alkoxides and metal chlorides, which undergo hydrolysis and poly-condensation reactions to form either a network "elastic solid" or a colloidal suspension—a system composed of discrete, often amorphous, submicrometer particles dispersed to various degrees in a host fluid. The formation of a metal oxide involves connecting the metal centers with oxo (M–O–M) or hydroxo (M–OH–M) bridges, therefore generating metal-oxo or metal-hydroxo polymers in solution. Thus, the sol evolves toward the formation of a gel-like diphasic system, containing both a solid phase and a liquid phase, whose morphologies range from discrete particles to continuous polymer networks. In the case of the colloid, particle density may be so low that a significant amount of fluid may initially need to be removed for the gel-like properties to be recognized. This can be accomplished in a number of ways. The most widely used method is to allow time for sedimentation to occur, and then to pour off the remaining liquid. Centrifugation can also be used to accelerate the process of phase separation.

5.2.1.2 *In Situ* Polymerization Method

The *in situ* polymerization method is generally employed in the preparation of composites where uniform dispersion of nanoparticles occurs in the polymer matrix; the reaction is performed either in the presence of an inorganic acid such as HCl, HNO_3, H_2SO_4, and so on or in the presence of surfactants such as dodecylbenzene sulfonic acid or *p*-toluenesulfonic acid. These high molecular surfactants not only affect the morphology of the composites, but they also improve the processability of the nanocomposites and enhance various applicable properties such as increased conductivity to several orders of magnitude, which is generally unachievable in the case of solely inorganic acids.

Table 5.1 lists the various nanocomposites prepared by the above-mentioned processes.

5.2.2 Distribution Studies

The distribution behavior of metal ions plays an important role in the determination of the selectivity of the nanocomposite. In certain practical applications, equilibrium is most conveniently expressed in terms of the distribution coefficients of the counterions (Khan et al. 2010).

The distribution coefficient (K_d values) of various metal ions on the nanocomposites are determined by the batch method in various solvent systems. An amount of 200 mg of the nanocomposite cation exchanger beads are kept in the H^+-form in Erlenmeyer flasks with 20 mL of different metal nitrate solutions in the required medium, and left for 24 h with continuous shaking in a temperature-controlled incubator shaker at $25 \pm 2°C$ to attain equilibrium. The metal-ion concentrations in the solution are determined before and after equilibrium by titrating against a standard 0.005 M solution of ethylenediaminetetraacetic acid (Reilley et al. 1959). The distribution coefficient is the measure of the fractional uptake of metal ions competing for H^+ ions by an ion-exchange material and, hence, can be calculated mathematically using the formula given as:

$$K_d = \frac{(I - F)}{F} * \frac{V}{M} (mL/g)$$ (5.1)

TABLE 5.1

Synthesis, IEC, and Selectivity of Nanocomposites

S. No.	Nanocomposite Ion-Exchanger Material	Inorganic Material	Organic Material	Mixing Volume Ratio of Organic–Inorganic Material	Synthesis Method	IEC (meq/gm)	Selectivity	References
1	Poly-o-anisidineSn(IV) tungstate	0.1 M stannic chloride, 0.1 M sodium tungstate	2.5% o-anisidine, 0.05 M ammonium persulfate	1:1:8:2	Sol-gel	2.25	Hg(II)	Khan et al. 2012
2	Poly-o-toluidine Sn(IV) tungstate	0.1 M stannic chloride, 0.1 M sodium tungstate	5% o-Toluidine, 0.4 M ammonium persulfate	8:2:1:1	Sol-gel	2.50	Cd(II)	Khan and Shaheen 2013b
3	Polyaniline Zr(IV) molybdophosphate	0.1 M zirconium oxychloride, 0.1 M sodium molybdate, 0.1 M ortho-phosphoric acid	5% aniline, 0.1 M ammonium persulfate	8:4:8:1:1	Sol-gel	2.50	Ni(II)	Khan and Shaheen 2014b
4	Poly-o-toluidine/multiwalled carbon nanotubes/Sn(IV) tungstate	Stannic chloride, Sodium tungstate, MWCNTs	o-Toluidine, Ammonium persulfate		In situ	1.25	Pb(II)	Khan and Shaheen, 2014c
5	Poly-o-anisidineSn(IV) arsenophosphate	0.1 M stannic chloride, 0.1 M sodium arsenate, 0.1 M o-phoshphoric acid	2.5% o-anisidine, 0.05 M ammonium persulfate	1:1:1:1:1	Sol-gel	1.82	Pb(II)	Khan et al. 2009
6	Polypyrrole-zirconium(IV) selenoiodate	0.1 M zirconium oxychloride, 0.2 M potassium iodate, 0.2 M sodium selenite	33% pyrrole, 0.1 M ferric chloride	1:2:1:1:40	Sol-gel	2.49	Pb(II)	Khan et al. 2015

No.	Material	Reagent 1	Reagent 2	Reagent 3	Monomer	Oxidant/reagent	Ratio	Method	Value	Metal	Reference
7	Poly-o-toluidine Th(IV) phosphate	0.1 M Th(NO$_3$)$_3$	1 M H$_3$PO$_4$		10% o-toluidine	0.1 M ammonium persulfate	1:2:1:1	Sol-gel	1.90	Hg(II)	Khan and Khan 2009
8	Poly-o-anisidine Sn(IV) phosphate	SnCl$_4$·5H$_2$O	Na$_2$HPO$_4$		2% o-anisidine	0.1 M (NH$_4$)$_2$S$_2$O$_8$	1:1:1:1	Sol-gel	2.18	Cd(II)	Khan and Khan 2010
9	Polyaniline–zirconium titanium phosphate	0.2 M TiCl$_4$	0.1 M Na$_2$HPO$_4$	0.2 M ZrOCl$_2$	5% aniline	0.1 M (NH$_4$)$_2$S$_2$O$_8$	1:1:2:05:1	Sol-gel	4.52	Pb(II) and Hg(II)	Khan and Paquiza 2011
10	Poly-o-toluidine Ce(IV) phosphate	0.1 M Ce(SO4)$_2$·4H$_2$O	2 M H$_3$PO$_4$		20% o-toluidine	0.4 M (NH$_4$)$_2$S$_2$O$_8$	1:1::1	Sol-gel	1.04	Cd(II)	Khan and Akhtar 2011b
11	Polypyrrole–zirconium titanium phosphate	0.1 M TiCl$_4$	0.1 M ZrOCl$_2$	0.2 M Na$_2$HPO$_4$	33% pyrrole	0.1 M FeCl$_3$	1:1:2:0.05:2	Sol-gel	3.86	Th(IV)	Khan et al. 2011
12	Poly-o-toluidine Zr(IV) Phosphate	0.1 M ZrOCl$_2$·8H$_2$O	3 M H$_3$PO$_4$		20% o-toluidine	0.4 M K$_2$S$_2$O$_8$	1:2::1	Sol-gel	1.71	Hg(II)	Khan et al. 2008
13	Poly(methyl methacrylate)–cerium molybdate	0.1 M ceric ammonium nitrate	0.1 M ammonium molybdate		1 M MMA in	0.5 M Na$_2$HPO$_4$	1:3:1:0.5	In situ	1.60	Pb(II)	Khan and Paquiza 2013
14	Poly(3-methylthiophene)-titanium(IV) molybdophosphate	Titanium sulfate	0.5 M molybdo-phosphoric acid		Iron(III) chloride	3-methyl thiophene monomer (7.5%)	2:1:2	In situ	2.39		Khan and Shaheen 2013a
15	Poly:carbazole-titanium dioxide	TiO$_2$ nanoparticles			Carbazole	APS	2:2:7	In situ			Shakir et al. 2014

where:
   I   is the initial amount of metal ion in the aqueous phase
   F   is the final amount of metal ion in the aqueous phase
   V   is the volume of the solution (mL)
   M   is the amount of cation exchanger (g)

On the basis of higher K_d values for the nanocomposites in the various solvent systems, they are then selected for the specific heavy metal ion. Various metal ions are shown to be selective for the synthesized nanocomposites in Table 5.1.

5.2.3 ION-EXCHANGE PROPERTIES OF SYNTHESIZED NANOCOMPOSITES

Ion-exchange capacity (IEC) is taken as a measure of the hydrogen ion liberated by a neutral salt to flow through the nanocomposite exchanger and is determined by a standard column process. An amount of 1 g of the dry cation exchanger in H$^+$-form is placed into a glass column having an internal diameter of 1 cm and fitted with glass wool support at the bottom. An amount of 1 M of alkali metal nitrates are used to elute the H$^+$ ions completely from the cation-exchange column, maintaining a very slow flow rate (~0.5 mL/min). The effluent is titrated against a standard (0.1 M) NaOH solution using a phenolphthalein indicator (Khan et al. 2013b). Figure 5.1 depicts the column method employed to elucidate the IEC.

5.2.4 CHARACTERIZATION OF THE NANOCOMPOSITE

X-ray diffraction (XRD) data of nanocomposites were recorded by a Bruker D8 diffractometer with Cu K α radiation at 1.540°A in the range of $20° \leq 2\theta \leq 80°$ at 40 keV. For Fourier transform infrared (FTIR) spectroscopic studies of composites, a Perkin Elmer Spectrum-BX spectrophotometer of

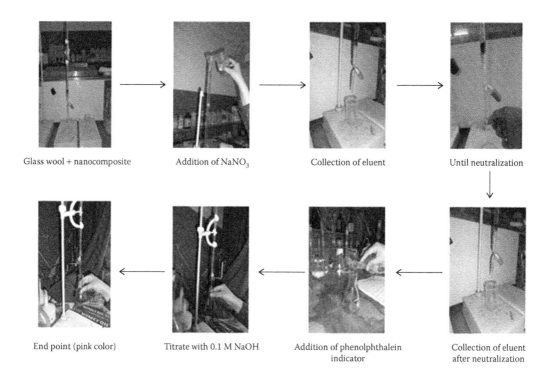

Glass wool + nanocomposite   Addition of NaNO₃   Collection of eluent   Until neutralization

End point (pink color)   Titrate with 0.1 M NaOH   Addition of phenolphthalein   Collection of eluent
                                                              indicator   after neutralization

FIGURE 5.1 **(See color insert)** Experimental setup for the determination of ion-exchange capacity.

range 4000–400 nm was used. An LEO 435-VF scanning electron microscopy (SEM) micrograph and energy-dispersive x-ray (EDX) studies were also used to obtain atomic composition of composites. Thermogravimetric analysis (TGA) was performed by a Perkin Elmer instrument heating from ~10°C to 1000°C at a rate of 10°C/min in a nitrogen atmosphere with a flow rate of 30 mL/min.

5.2.5 PREPARATION OF NANOCOMPOSITE MEMBRANE

Nanocomposite membranes were prepared in various weight ratios of the constituent components by a simple solution casting method. Different weight ratios were dispersed in different solvents at room temperature. Mechanical stirring was applied to obtain a well-dispersed composite solution, and this solution was further cast into clean glass plates and kept at room temperature to achieve complete evaporation of the solvent. The resultant composite membranes were cautiously peeled out of the glass plate, rinsed with doubly distilled water on both sides, dried at room temperature, and later stored in a desiccator for further experiments.

The *IEC of the membrane* can be determined by a similar procedure to that mentioned (Khan and Paquiza 2011).

Characterization of the nanocomposite membrane can also be performed by the techniques mentioned such as SEM, transmission electron microscopy (TEM), XRD, FTIR, and TGA.

5.2.6 PHYSICOCHEMICAL CHARACTERIZATION

- Water content (% total wet weight)

  Membranes were first conditioned and then soaked in water to elute diffusible salts, blotted with Whatman filter paper to remove surface moisture, and immediately weighed. These were further dried and weighed after 24 h. The water content (% total wet weight) was calculated as:

$$\% \text{Total wet weight} = \frac{W_w - W_d}{W_w} * 100 \tag{5.2}$$

  where:
  W_d is the weight of the dry membrane
  W_w is the weight of the wet membrane

- Porosity

  Porosity (ε) was determined as the volume of water incorporated in the cavities per unit of membrane volume from the water content data:

$$\varepsilon = \frac{W_w - W_d}{AL\rho_w} \tag{5.3}$$

  where:
  W_d is the weight of the wet membrane
  W_w is the weight of the dry membrane
  A is the area of the membrane
  L is the thickness of the membrane
  ρ_w is the density of water

- Thickness and swelling

  The thickness of the membrane was measured by taking the average thickness of the membrane by using a screw gage. Swelling is measured as the difference between the average thickness of the membrane equilibrated with 1 M NaCl for 24 h and that of the dry membrane.

• Fabrication of ion-selective membrane electrode

The ion-selective membranes obtained by this procedure were cut into the shape of disks and mounted at the lower end of a Pyrex glass tube (outer diameter 0.8 cm, inner diameter 0.6 cm) with Araldite. Finally, the assembly was allowed to dry in air for 24 h. The glass tube was filled with a solution of the ion (as reference) toward which the ion exchanger was selective and kept in an identical solution of the same ion at room temperature. A saturated calomel electrode (SCE) was inserted into the tube to give electrical contact and another SCE was used as an external reference electrode. The whole arrangement can be shown as:

Internal reference electrode (SCE)	Internal electrolyte 1×10^{-1} M metal ion	Ion-exchange membrane	Sample solution 1×10^{-10} to 1×10^{-1} M metal ion	External reference electrode (SCE)

The following parameters were evaluated to study the characteristics of the electrode such as lower detection limit, electrode response curve, response time, and working pH range.

5.2.7 Electrode Response

The response of the electrode in terms of the electrode potential (at $25 \pm 2°C$), corresponding to the concentration of a series of standard solutions of 0.1 M selective metal nitrate (10^{-10} to 10^{-1} M), was determined at a constant ionic strength as described by the International Union of Pure and Applied Chemistry (IUPAC) Commission for Analytical Nomenclature. Potential measurements of the ion-exchange electrode were plotted against the selected concentrations of the respective ions in an aqueous medium using the electrode assembly. The calibration graphs were plotted three times to check the reproducibility of the system. Figure 5.2 shows the experimental setup for the potentiometric response used in the membrane studies.

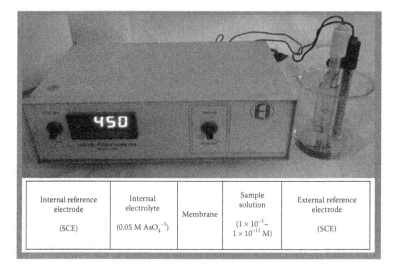

FIGURE 5.2 **(See color insert)** A simple set up of an ion-selective electrode.

5.2.8 RESPONSE TIME

The response time was measured by recording the electromotive force (EMF) of both the electrodes as a function of time when they were immersed in the solution to be studied. The method of determining the response time is outlined as follows: The electrode is first dipped in a 1×10^{-1} M solution of the ion concerned and immediately shifted to another solution (pH ~ 4) of 1×10^{-2} M ion concentration of the same ion (a 10-fold higher concentration). The potential of the solution is read at 0 s, that is, just after dipping the electrode into the second solution, and is subsequently recorded at intervals of 5 s. The potentials were then plotted versus time. The time at which the potentials attain a constant value represents the response time of the electrode.

5.2.9 SELECTIVITY COEFFICIENT

The response of the primary ion in the presence of other foreign ions is measured in terms of the potentiometric selectivity coefficient by mixed solution method (MSM). The selectivity coefficient was calculated using the equation:

$$K_{AB}^{POT} = \frac{a_A}{(a_B)^{z_A/z_B}} \tag{5.4}$$

where:

a_A and a_B are the activities of primary and interfering ions of varying concentration of primary ions and fixed concentration of interfering ions

z_A and z_B are the charges on the ions

5.2.10 ELECTRICAL CONDUCTIVITY

The direct current (dc) electrical conductivity of the conducting composite was measured by using a four-in-line probe. The sample of nanocomposites was dried at 40°C–50°C in an oven for 24 h. An amount of 200 mg of the composite material was pelletized at room temperature with the help of a hydraulic press at 25 kN for 10 min (Khan et al. 2015). The conductivity (σ) was calculated using the following equations:

$$\rho = \rho^\circ / G_7(W/S) \tag{5.5}$$

$$G_7(W/S) = (2S/W)\ln 2 \tag{5.6}$$

$$\rho^\circ = (V/I)2\pi S \tag{5.7}$$

$$\sigma = 1/\rho \tag{5.8}$$

where:

$G_7(W/S)$ is a correction divisor, which is a function of the thickness of the sample, as well as probe spacing

I is the current (A)
V is the voltage (V)
W is the thickness of the film (cm)
S is the probe spacing (cm)

FIGURE 5.3   Set up of four-in-line probe resistivity measurement instruments.

5.2.11   Sensing Measurements

The vapor-sensitive characteristics of the composite were investigated by recording their electrical responses when exposed alternately to different concentrations of gas vapors at room temperature (25°C).

Different concentrations of vapors were taken in liquid form. The sensing material was placed into the glass chamber and gently pressed by a four-in-line probe to record the current–voltage characteristics using a digital microvoltmeter (DMV 001) and low current sources (LCS 02). The distance between the sensing material and the solvent was kept at 3–4 cm at the time of exposure of different concentrations of vapors on the sensing material at room temperature (25°C). The required concentrations of the liquid were poured into the chamber through a funnel. The sensing material was exposed to the vapors in the glass chamber for an appropriate time and was then exposed to air before the next concentration of the liquid was poured into the glass chamber. Figure 5.3 shows the instrumental setup of the four-in-line resistivity probe employed for sensing studies.

5.3   RESULTS AND DISCUSSION

Organic–inorganic nanocomposite cation exchangers were introduced into the field of hybrid composites by the incorporation of organic polymers into the matrix of inorganic precipitates, using sol-gel mixing and the *in situ* oxidative polymerization method.

The binding of polymers into the matrix of inorganic precipitates is possible, as reported earlier (Shakir et al. 2014), due to ionic interactions between the radical cations of the polymers and anionic groups of inorganic material, as given in Schemes 5.1a through f:

The composite ion-exchange material possessed varying IEC for Na^+ measured under similar conditions in Table 5.1. However, in the case of polyaniline-zirconium titanium phosphate (Khan and Paquiza 2011) and polypyrrole-zirconium titanium phosphate (Khan et al. 2010) IECs of 4.52 and 3.86 meq/g were observed, respectively.

To explore the potentiality of the material in the separation of metal ions, distribution studies for metal ions were performed in different solvent systems. It is apparent that the K_d values may

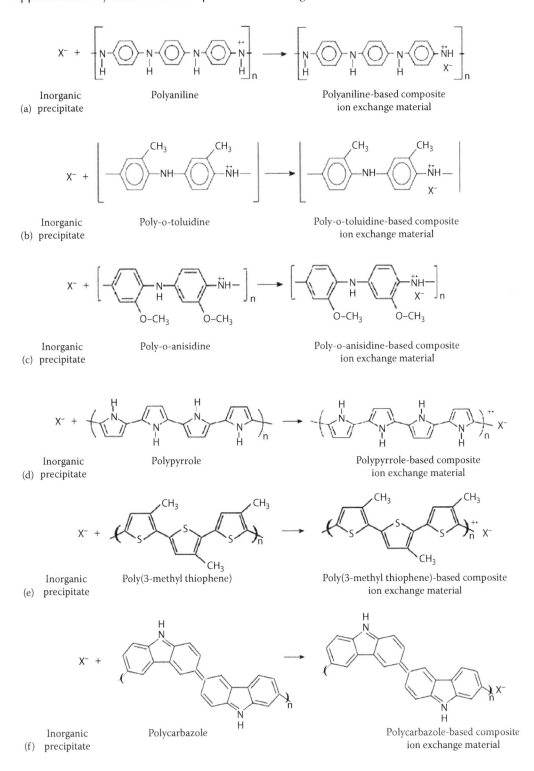

SCHEME 5.1   (a–f) Structure of polymers with their inorganic binding entities (X^-).

vary with the composition and nature of the working solvents (Table 5.1). In the case of each nanocomposite, the K_d value of the selective metal ion is the highest among the other metal ions. The high uptake of certain metal ions demonstrates not only the ion-exchange properties but also the adsorption and ion-sieve characteristics of the cation exchanger.

5.3.1 CHARACTERIZATIONS

5.3.1.1 Scanning Electron Microscopy

Figures 5.4a through f show the SEM micrographs of the prepared composite cation exchangers where there is a change of morphology from the parent components, which indicates the formation of organic–inorganic nanocomposite cation exchangers (a: polyaniline-titanium(IV) molybdophosphate [PANI-TMP], b: poly-o-toluidine Sn(IV) tungstate [POTST], c: poly-o-anisidine Sn(IV) tungstate [POAST], d: polypyrrole-zirconium(IV) selenoiodate [Ppy/ZSI], e: poly(3-methylthiophene)-titanium(IV) molybdophosphate [P3MTh-TMP], f: polycarbazole-titanium dioxide [PCz/TiO$_2$]).

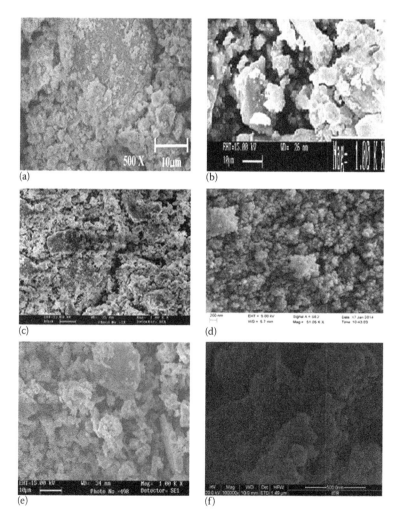

FIGURE 5.4 Scanning electron microphotograph of (a) PANI-TMP, (b) POTST, (c) POAST, (d) Ppy/ZSI, (e) P3MTh-TMP, and (f) PCz/TiO$_2$ nanocomposite cation exchangers.

TABLE 5.2
Percentage Composition of the Various Elements in Nanocomposites

S. No.	Name of Composite Material	Elements	Percentage Composition
1	Polyaniline-titanium(IV) molybdophosphate	Ti	19.92
		Mo	18.50
		P	15.17
		C	18.51
		H	3.165
		N	3.80
		O	20.935
2	Poly-o-toluidine Sn(IV)tungstate	Sn	2.876
		W	19.72
		C	12.20
		N	5.146
		H	3.722
		O	56.336
3	Poly-o-anisidine Sn(IV)tungstate	Sn	2.79
		W	24.39
		C	12.20
		N	3.72
		H	2.14
		O	42.45
4	Polypyrrole-zirconium(IV) selenoiodate	C	24.97
		N	10.25
		O	15.33
		Cl	3.14
		Fe	1.32
		Zr	20.76
		Se	14.63
		I	9.60
5	Poly(3-methythiophene)-titanium(IV) molybdophosphate	Ti	20.12
		Mo	17.20
		P	13.18
		C	15.31
		H	4.12
		S	3.56
		N	0.00
		O	26.51

5.3.1.2   Energy-Dispersive X-Ray Studies

Table 5.2 lists the percentage composition of the various elements found to be present in the nano-composites as prepared in the previous subsection.

5.3.1.3   Transmission Electron Microscopy

From Figures 5.5a through f showing TEM, the composite cation exchangers show particle sizes ranging between 1 and 100 nm. This completes the preparation of the composites as nanocomposites.

5.3.1.4   X-Ray Diffraction

Figures 5.6a through f shows the XRD pattern of the various nanocomposites. In the cases of the PANI-TMP, POTST, and POAST nanocomposites, the semicrystalline nature of the nanocomposites were observed due to the dominant amorphous nature of the organic polymer.

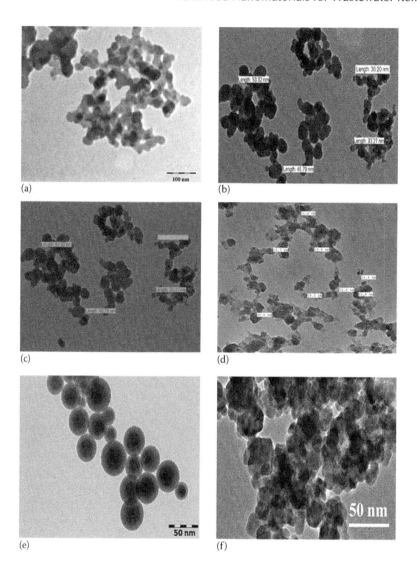

FIGURE 5.5 Transmission electron microphotographs of (a) PANI-TMP, (b) POTST, (c) POAST, (d) Ppy/ZSI, (e) P3MTh-TMP, and (f) PCz/TiO$_2$ nanocomposite cation exchangers showing different particle sizes.

With Ppy/ZSI, the characteristic peak at the $2\theta = 20°–30°$ value indicated the amorphous nature of Ppy; this also showed the crystalline nature of the inorganic precipitate, with sharp peaks at $2\theta = 35°–40°$ and $40°–50°$; however, the peaks at $2\theta = 26°$ and $56°$, as found in case of the Ppy/ZSI nanocomposite cation exchanger, indicated its amorphous structure, which may be due to some interaction of the inorganic precipitate with the dominant amorphous nature of the polymer (Khan et al. 2015).

With P3MTh-TMP, the characteristic peak at a 2θ value of ~24 demonstrated the semicrystalline nature of TMP. The XRD pattern of pure organic components showed a broad diffraction peak, attributed to its amorphous nature, at $2\theta \sim 20°–40°$. The diffraction peaks of TMP in the composite showed a decrease in crystallinity, which may be attributed to the dominant amorphous nature of P3MTh (Khan and Shaheen 2013a).

The analysis of the XRD spectra of PCz/TiO$_2$ nanocomposites clearly exhibited the major peaks of PCz and TiO$_2$, indicating the successful incorporation of TiO$_2$ in the PCz/TiO$_2$ nanocomposite. The characteristic peaks of PCz at the 2θ values of 17.66°, 19.51°, 21.75°, 23.12°, and 29.44° are in good agreement with previously reported XRD spectra of PCz (Shakir et al. 2014).

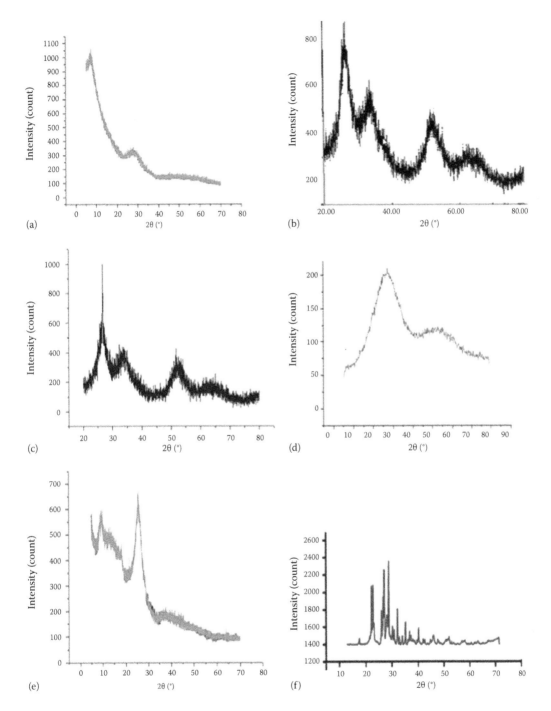

FIGURE 5.6 Powder x-ray diffraction pattern of (a) PANI-TMP, (b) POTST, (c) POAST, (d) Ppy/ZSI, (e) P3MTh-TMP, and (f) PCz/TiO$_2$ nanocomposite cation exchangers.

5.3.1.5 Thermogravimetric Analysis

Figures 5.7a through f suggest that the PANI-TMP nanocomposite is more thermally stable than polyaniline (Khan et al. 2013). At ~1000°C, the percentage of residual weights of the PANI and PANI-TMP nanocomposites are 38.4% and 54.7%, respectively. Thus, the addition of the TMP

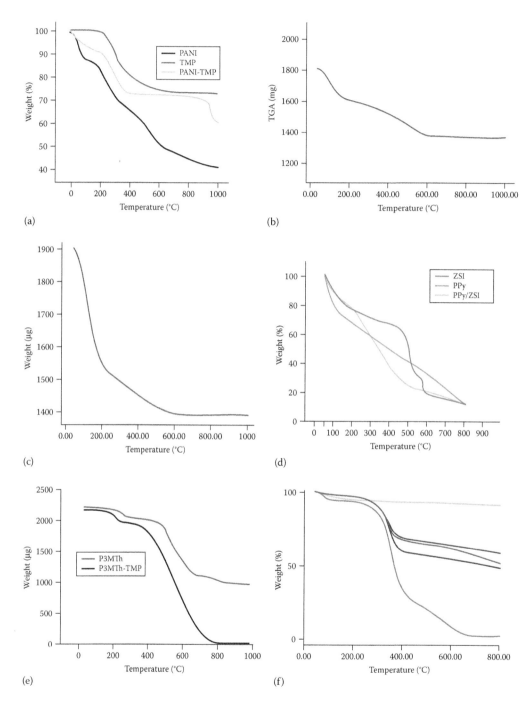

FIGURE 5.7 **(See color insert)** TGA curves of (a) PANI-TMP, (b) POTST, (c) POAST, (d) Ppy/ZSI, (e) P3MTh-TMP, and (f) PCz/TiO$_2$ nanocomposite cation exchangers.

cation exchanger improves the thermal stability of the PANI-TMP nanocomposite. It is clear from the TGA curve (Figure 5.7a) of the nanocomposite that there is weight loss of mass (of about 5.9%) up to 100°C, which may be due to the removal of external water molecules. Further weight loss of mass (of about 11.3%) from 100°C to 300°C may be due to the loss of interstitial water molecules by the condensation of −OH groups. Still further loss of weight (about 27.5%) at 984°C onward predicts the complete decomposition of the organic part and the formation of metal oxides of the exchange material.

The thermogravimetric analysis curve of the POAST nanocomposite material (Khan and Shaheen 2013b) showed rapid weight loss (9.05%) up to 100°C, due to the removal of external water molecules. Gradual weight loss from 150°C to about 400°C may be due to the formation of pyrophosphate groups by the condensation of phosphate. Further, the inclination point was observed at about 550°C, which indicates the complete decomposition of the material and the formation of metal oxides. From about 600°C to 1000°C, a sharp weight loss indicated by the curve may be due to the decomposition of the metal oxides.

The inorganic ZSI shows a 20% weight loss up to a temperature of 150°C, which may be attributed to the loss of gel water or water of crystallization. The weight loss occurring in the temperature range 150°C–450°C is 12%, which can be ascribed to the loss of the hydroxyl group. Furthermore, the weight loss observed from 450°C to 600°C is ~40%, corresponding to the vaporization of the IO_3^- group. Henceforth, stability is achieved until 800°C is reached.

In the case of the Ppy/ZSI nanocomposite, initially, a 20% weight loss is seen up to 200°C; thereafter, it decomposes gradually up to 510°C (60% weight loss) because of degradation of Ppy and subsequently remains stable up to 800°C. The residual weights of Ppy/ZSI and Ppy at 800°C are 83% and 90%, respectively. This suggests that the addition of inorganic precipitate improves the thermal stability of the nanocomposite Ppy/ZSI as compared with pure Ppy.

The TGA curves of P3MTh and the P3MTh-TMP nanocomposite are shown in Figure 5.7f. In the case of P3MTh, there are two stages of weight loss: The first, up to 300°C, can be attributed to the loss of physisorbed water molecules and volatile impurities. The second weight loss, up to 800°C, can be ascribed to the degradation of the polymer's unsaturated groups; whereas, from 800°C to 980°C, the polymer shows some stability and hardly any weight loss is seen. The P3MTh-TMP nanocomposite was initially stable up to 400°C (2.5% weight loss); thereafter, it decomposed gradually at 850°C because of the degradation of P3MTh and subsequently remained stable up to 980°C. The total weight loss up to 980°C has been estimated to be about 100% and 51.80% for P3MTh and the P3MTh-TMP nanocomposite, respectively. These results confirm that the presence of TMP in the P3MTh-TMP nanocomposite is responsible for the higher thermal stability of the composite material in comparison with pristine P3MTh.

The TGA graphs of the PCz/TiO_2 nanocomposites exhibit similar patterns, where weight loss takes place between 300°C and 400°C, which may be due to the degradation of PCz; they subsequently remain stable up to 800°C. The total weight loss has been estimated to be about 30%−40% for the PCz/TiO_2 nanocomposite. The comparison of the TGA graph of pure TiO_2 nanoparticles (which shows no degradation up to 800°C) with that of the PCz/TiO_2 nanocomposites indicates that the presence of TiO_2 nanoparticles in PCz/TiO_2 nanocomposites is responsible for higher thermal stability with respect to pristine PCz (Shakir et al. 2014).

5.3.1.6 Fourier Transform Infrared Spectroscopy

Table 5.3 lists the various synthesized nanocomposites with the wavenumber value of the respective bond vibrations and Figures 5.8a through f represent the graphs associated with it.

5.3.2 Nanocomposite Membrane

Different samples of membranes were synthesized with the nanomaterials listed in Table 5.3 and their varying mixing ratios with the binder are mentioned in Table 5.4. The electroactive cation exchanger was ground to a fine powder and was mixed thoroughly with the binder, dissolved in 10 mL of tetrahydrofuran (THF), and finally mixed with two to three drops of dioctylphthalate, which act as a plasticizer. The mixing ratio of the ion exchanger was varied with a fixed content of the binder to obtain a composition with the highest-performing membrane, and the resulting solutions were carefully poured into a glass-casting ring (of diameter 5 mm) resting on a glass plate.

TABLE 5.3
Peak Values of FTIR Spectra in Nanocomposites

S. No.	Name of Composite Material	Functional Group	Peak Value (cm⁻¹)
1	Polyaniline-titanium(IV) molybdophosphate	C–N aromatic stretching vibrations	1249
		Quinoid (Q) and benzenoid (B) structures of the polyaniline	1579, 1487
		–OH stretching	3323
		Deformation vibration of H–O–H	1633
		M–O	734
		Symmetric and antisymmetric stretching of the P–O bond in PO₃ groups	1200, 900
2	Poly-o-toluidine Sn(IV)tungstate	C≡N stretching	2361.7
		H–O–H bending bands	1609.1
		C–H	1491
		C–N	1400
		PO₄⁻³	1029
		Sn–O	803–554
3	Poly-o-anisidine Sn(IV)tungstate	N–H stretching	3628
		C≡C stretching	13,427.1
		H–O–H	1634–1055
		Sn–O	799–529
4	Polypyrrole-zirconium(IV) selenoiodate	C–N stretching vibration	1200
		C–C out of plane ring deformation	784
		Heterocyclic aromatic ring C–C conjugation	1650–1400
		Superposition of metal oxygen stretching vibrations	980
		H–O–H	880–610
5	Poly(3-methythiophene)-titanium(IV) molybdophosphate	C–H out of plane bending	701
		C=C	1652
		O–H stretching	3363
		Symmetric and antisymmetric stretching of the P–O bond in PO₃ groups	1016
6	Polycarbazole-titanium dioxide	C=C out of plane bending	590
		Ring deformation vibrations of aromatic structure of PCz	744
		C–H out of plane bending vibration	1199
		Ring stretching vibration	1406, 1456
		N–H stretching	3149
		C=C stretching modes of benzenoid and quinoid rings	1600, 1495

These rings were left untouched, so that the THF could evaporate slowly and leave thin films. For each nanocomposite membrane, the ratio that exhibited maximum IEC for the membrane when using this procedure was observed (Table 5.5). The conditions of preparation of the nanocomposite ion-exchanger membrane are listed in Table 5.4.

5.3.3 Physicochemical Characterization of the Membrane

The performance of an ion-exchange membrane depends on its complete physicochemical properties, which involves the determination of all those parameters that affect its electrochemical properties. These parameters are the thickness of the membrane, porosity, water content, and swelling, which were determined after conditioning of the membrane (Table 5.5) (Khan and Paquiza 2011).

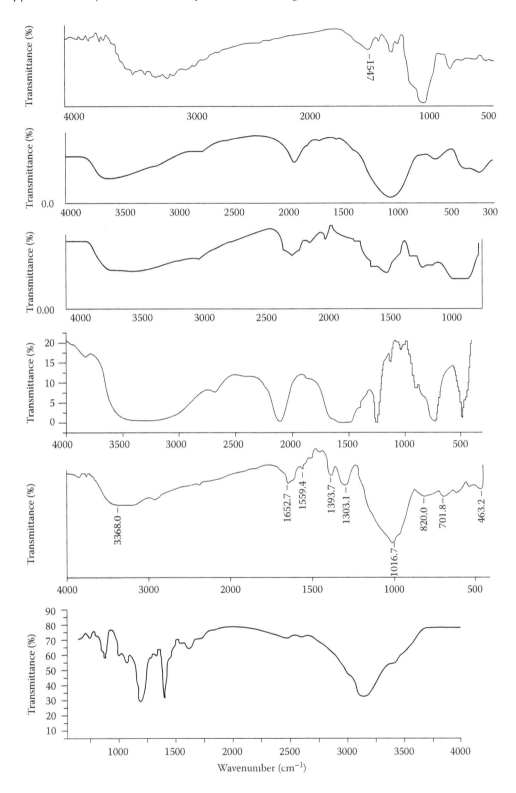

FIGURE 5.8 FTIR spectra of (a) PANI-TMP, (b) POTST, (c) POAST, (d) Ppy/ZSI, (e) P3MTh-TMP, and (f) PCz/TiO$_2$ nanocomposite cation exchangers.

TABLE 5.4

Varying Mixing Ratios of Prepared Membranes

S. No.	Name of Composite Membrane	Composite	Binder	Composite: Binder Ratio	References
1	Polyaniline-zirconium(IV) molybdophosphate	Polyaniline-zirconium(IV) molybdophosphate	Araldite	1:1	Khan and Shaheen 2014b
2	Poly-o-toluidine Sn(IV) tungstate	Poly-o-toluidine Sn(IV) tungstate	Araldite	1:1	Khan and Shaheen 2013b
3	Poly-o-anisidine Sn(IV) tungstate	Poly-o-anisidine Sn(IV) tungstate	Araldite	1:1	Khan et al. 2012
4	Polypyrrole Th(IV) phosphate	Polypyrrole Th(IV) phosphate	Araldite	1:1	Khan et al. 2005
5	Poly-o-toluidine/multiwalled carbon nanotubes/Sn(IV) tungstate (POT/MWCNT/ST)	POT/MWCNT/ST	Polystyrene	0.2:0.5	Khan and Shaheen 2014c

The nanocomposite membrane samples were selected to make the ion-selective electrode. The low order of water content, the swelling, porosity, and lesser thickness of these membranes suggest that interstices are negligible and diffusion across the membranes occurs mainly through the exchanger sites. The sensitivity and selectivity of the ion-selective electrode depend on the nature of the electroactive material. When the membrane of such materials is placed between two electrolyte solutions of the same nature, at the same pressure and temperature, but at different concentrations, some ions to which the membrane is selective pass from the solution of higher concentration through the membrane to that of the lower concentration, thus producing an electrical potential difference (Khan et al. 2013).

5.3.4 Electrode Response

The potentiometric response of the membrane electrode over a wide concentration range of 1×10^{-1} to 1×10^{-10} M was observed. The electrode showed a linear response at a varying range of concentrations, with an average Nernstian slope per decade change in concentrations of the selective metal ions, as mentioned in Table 5.6 for different nanomembranes. The limit of detection was determined from the intersection of the two extrapolated segments of the calibration graph (Khan et al. 2013).

The pH effect on the potential response of the electrode was measured for a fixed concentration of the metal ions at different pH values. The electrode potential remains unchanged within a specified pH range (Table 5.6) known as the working pH range for the electrode. Another important factor is the response time of the ion-selective electrode. The average response time was defined as the time required for the electrode to reach a stable potential after successive immersions of the electrode in different ion solutions, each having a 10-fold difference in concentration.

All the membranes that were synthesized could be used successfully for up to 9–12 months without any notable drift in potential, during which the potential slope was reproducible within ±1 mV per concentration decade. Whenever a drift in the potential was observed, the membrane was reequilibrated with 0.1 M metal nitrate solutions for 3–4 days.

The selectivity behavior is one of the important characteristics of the ion-selective electrodes, determining whether reliable measurement in the target sample was possible or not. It was determined by the MSM. It was observed that most of the interfering ions showed low values of the selectivity coefficient, indicating no interference in the performance of the membrane electrode assembly. Any noted remarkable selectivity for a specific metal ion over other ions reflects the high affinity of the membrane toward the metal ion.

TABLE 5.5
Various Physiochemical Characterizations of Ion-Exchanger Nanocomposite Membrane

S. No.	Name of Nanocomposite Membrane	Thickness of Membrane (mm)	Water Content as Percentage Weight of Wet Membrane	Porosity	Swelling of Percentage Weight of Wet Membrane	Na + IEC of Membrane (meq/g)	References
1	Polyaniline-zirconium(IV) molybdophosphate	0.12	9.7×10^{-2}	0.087	0.065	0.65	Khan and Shaheen 2014b
2	Poly-o-toluidine Sn(IV) tungstate	0.65	2.15	3.38×10^{-2}	0.02	—	Khan and Shaheen 2013b
3	Poly-o-anisidine Sn(IV) tungstate	0.60	1.50	8.8×10^{-4}	0.005	—	Khan et al. 2012
4	Polypyrrole Th(IV) phosphate	0.50	5.61	0.094	0.2	—	Khan et al. 2005
5	Poly-o-toluidine/multiwalled carbon nanotubes/Sn(IV) tungstate (POT/MWCNT/ST)	0.014	1.3×10^{-1}	1.16×10^{-1}	No swelling	1.05	Khan and Shaheen 2014c

TABLE 5.6
Various Electrode Characterizations of Ion-Exchanger Nanocomposite Membranes

S. No	Name of Nanocomposite Membrane	Nernstian Response	Slope (mV per decade change of activity)	Response Time (s)	pH Range
1	Polyaniline-zirconium(IV) molybdophosphate	1×10^{-1} to 1×10^{-7} M	26.25	50	4.0–7.0
2	Poly-o-toluidine Sn(IV) tungstate	1×10^{-1} to 1×10^{-8} M	27.42	20	4.0–8.0
3	Poly-o-anisidine Sn(IV) tungstate	1×10^{-1} to 1×10^{-7} M	21	30	4.0–8.0
4	Polypyrrole Th(IV) phosphate	1×10^{-1} to 5×10^{-6} M	29.17	35	3.0–8.5
5	Poly-o-toluidine/multiwalled carbon nanotubes/Sn(IV) tungstate (POT/MWCNT/ST)	1×10^{-1} to 1×10^{-9} M	—	50	—

5.3.5 ELECTRICAL CONDUCTIVITY

The percolation threshold was observed to occur at around 10% loading of aniline monomer in the case of polyaniline-titanium(IV) phosphate (Khan et al. 2011). The electrical conductivity measurements for various nanocomposites are tabulated in Table 5.7. There was high improvement in conductivity at this concentration. Conductivity did not change significantly with further increases of the concentration of aniline monomer until there was 20% of the monomer content in the nanocomposite. The increase in conductivity can be well understood from the percolative path, in which the concentration of conducting particles increases; thus, the conductivity depends significantly on the carrier transport through the conducting fillers.

The electrical conductivity of the poly-o-toluidine zirconium(IV) phosphate (Khan et al. 2008) nanocomposite was observed with increasing temperatures from 30°C to 130°C, as shown in Figure 5.9. On examination, it was observed that the electrical conductivity of the nanocomposite increases with the increase in temperature in all nanocomposite materials discussed. This increase in conductivity with an increase in temperature is the characteristic of "thermal activated behavior" (Parvatikar et al. 2006). To explain the conduction mechanism in the conducting polymers, the concepts of polaron and bipolaron were introduced. A low level of oxidation of the polymer produces a polaron and a higher level of oxidation produces a bipolaron. Both polarons and bipolarons are mobile, and could move along the polymer chain by the rearrangement of double and single bonds in the conjugated system. Conduction by polarons and bipolarons was supposed to be the dominant factor that determines the mechanism of charge transport in the polymer with nondegenerate ground states. The magnitude of the conductivity was determined by the number of charge carriers available for conduction and the rate at which they move, that is, mobility. In conducting polymers that could be considered as semiconductors, the charge carrier concentration increased with increasing temperature. Since the charge carrier concentration was much more temperature dependent than mobility, it was therefore the dominant factor, and conductivity increased with the increase in temperature (Khan and Baig 2013).

5.3.6 SENSING MEASUREMENTS

In the cases of polyaniline-titanium(IV) phosphate (PANI-TiP) (Khan et al. 2011), poly(3-methylthiophene)-titanium(IV) molybdophosphate (P3MTh-TMP) (Khan and Shaheen 2013a), and polycarbazole-titanium dioxide (PCz/TiO$_2$) (Shakir et al. 2014) (Figures 5.10a through c), there were remarkable changes in the resistivity on exposure of different concentrations of aqueous ammonia

TABLE 5.7

Electrical Conductivity Measurements for Nanocomposites and Their Application as Specific Vapor Sensors

S. No.	Name of Nanocomposite	Amount of Polymer (%)	Electrical Conductivity at 25°C (S/cm)	Vapor Sensor	Reference
1	Polyaniline-titanium(IV) phosphate	2.5	1.23	Ammonia	Khan et al. 2011
		5.0	1.86		
		7.5	2.17		
		10	4.68		
		15	4.74		
		20	4.79		
2	Poly-o-toluidine Zr(IV) phosphate	2.5	0.54	Humidity	Khan et al. 2008
		5.0	0.86		
		7.5	0.98		
		10	1.23		
		15	1.27		
		20	1.29		
3	Poly-o-anisidine Sn(IV) phosphate	0.5	5.40×10^{-3}	—	Khan and Khan 2010
		1.0	4.65×10^{-3}		
		1.5	3.46×10^{-3}		
		2.0	5.40×10^{-2}		
		8.0	9.65×10^{-3}		
		10	7.65×10^{-3}		
		20	1.65×10^{-3}		
4	Polypyrrole-zirconium(IV) selenoiodate	8	0.42	Formaldehyde	Khan et al. 2015
5	Poly(3-methylthiophene)-titanium(IV) molybdophosphate	1.0	2.78	Ammonia	Khan and Shaheen 2013a
		2.5	5.79		
		5.0	15.02		
		7.5	26.55		
		10.0	27.04		
		12.0	27.10		
6	Polycarbazole-titanium dioxide	7	0.13	Ammonia	Shakir et al. 2014
7	Poly-o-toluidine/multiwalled carbon nanotubes/Sn(IV) tungstate (POT/MWCNT/ST)	—	0.2	Amine	Khan and Shaheen 2015

at room temperature as a function of time. The resistivity increases with increasing ammonia concentrations. This can be recovered with a flush of air. The response time of the sensor decreases with increasing ammonia concentrations; however, on repeating the experiment, the recovery time increases when the sensor is exposed to air for a few minutes. For PANI-TiP, the response time of the sensor is around 10 s for 3%–6% aqueous ammonia, while for 12% it is less than 10 s. The reversibility of the composite was also investigated, where the response of the composite to 3%–6% aqueous ammonia is highly reversible during the cyclic measurements test, as depicted in Figures 5.10a(i) and a(ii). Similarly, the response behavior studies were carried out in the presence of 12% aqueous ammonia (Khan et al. 2011). It was observed that the response cycle showed poor performance and the time taken to regain a sensitivity value close to the original one was quite large. This poor response may be attributed to the complete consumption of the reaction sites of the polymer, or because of the insufficient numbers of sites available for the ammonia moiety to

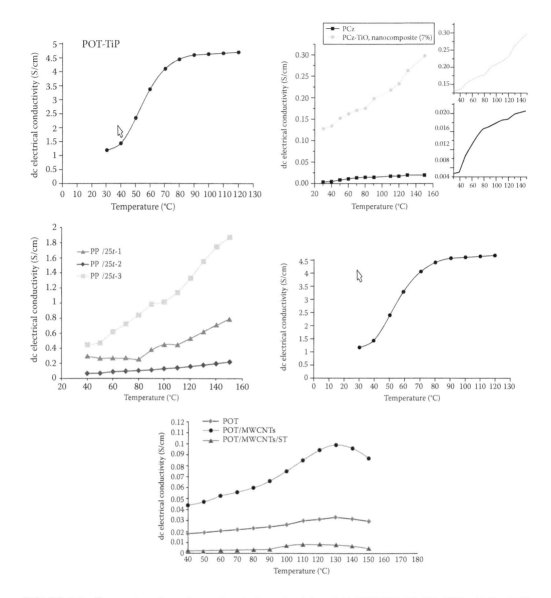

FIGURE 5.9 Temperature dependence electrical conductivity of (a) POT-TiP, (b) PCz/TiO$_2$, (c) Ppy/ZSI, (d) P3MTh–TMP, and (e) POT/MWCNT/ST cation exchange nanocomposites.

form the complex structure necessary to obtain the response behavior. The range between 3% and 6% showed a better reversible response for the results obtained by successive cyclic measurements.

It can be inferred from Figures 5.10b(i) and b(ii) that, for P3MTh-TMP, the sensor works best at concentrations of 0.2 to 0.6 M and, at higher concentrations, slight irreversibility takes place, which may be due to the electrical compensation of the polymer backbone by ammonia.

The formaldehyde-sensing performance of polypyrrole-zirconium(IV) selenoiodate (Ppy/ZSI) (Figures 5.10c[i] and c[ii]) cation-exchange nanocomposite was monitored by measuring resistivity changes on exposure to formaldehyde vapors of different concentrations (Khan et al. 2015). The response of the gas sensor rapidly improved under the exposure of aqueous formaldehyde, and obviously reduced on the change of the environment to air. With the increase in the concentration of formaldehyde, the response of resistivity also improved. The response of the sensor was around 20–10 s for 5%–10% aqueous formaldehyde, respectively. In Figures 5.10c(i) and c(ii), the response

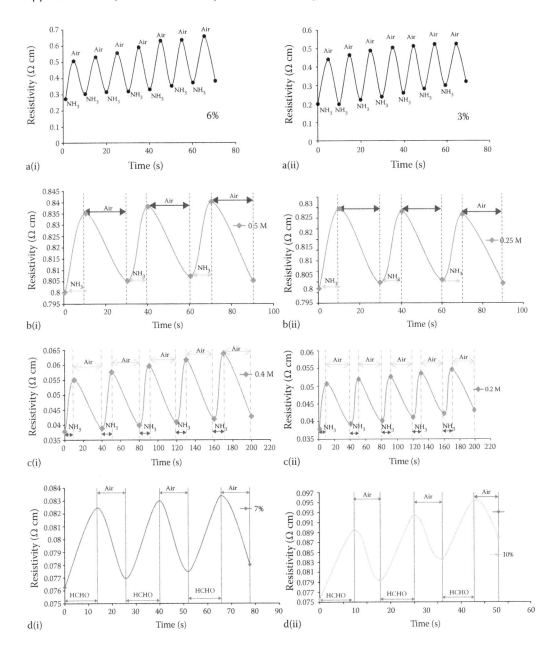

FIGURE 5.10 Reversible resistivity response curves of (a) PANI-TiP, (b) P3MTh–TMP, (c) PCz/TiO$_2$, and (d) Ppy/ZSI cation-exchange nanocomposites toward different concentrations of aqueous ammonia and formaldehyde.

time showed a decreasing trend with an increase in formaldehyde concentration. At lower concentrations, the number of formaldehyde molecules striking per second on the surface were low; as a result, longer response times were observed. Also, the recovery time was 12 s for 7% aqueous formaldehyde. The decreasing trend of the recovery time may be due to the fast dissolution of the complex structure formed at the surface of the nanocomposite with an increase in formaldehyde concentration.

The characteristic response of poly-o-toluidine zirconium(IV) phosphate nanocomposite as a function of the percentage level of relative humidity was observed (Khan et al. 2008). It was seen that the resistivity of the nanocomposite decreased as the percentage level of relative humidity increased. The decrease in the resistivity or increase in the conductivity with increasing humidity can be attributed to the mobility of the dopant ions in the polymer, which are loosely attached to the polymer chain by weak van der Waals forces of attraction. At low humidity, the mobility of the dopant ion is restricted because, under dry conditions, the polymer chain would tend to curl up into a compact coil form. Also, it has been reported that the conductivity of the conducting polymer increases when the sample absorbs moisture. A decrease in resistivity with an increase in the humidity proves the adsorption of the water molecules, which make the polymer more p-type in nature.

There were remarkable changes in the resistivity of the composite ion exchangers (Figure 5.10d) on exposure to 0.5 M aqueous $-NH_3$ and $-CH_2NH_2$ solutions with vapor concentrations of 3.53% and 3.24%, respectively, of their original solutions at room temperature as a function of time for poly-o-toluidine/multiwalled carbon nanotubes/Sn(IV) tungstate (POT/MWCNT/ST) composite ion exchangers. The response and recovery time of POT/MWCNT/ST sensors were around 5 and 20 s for 0.5 M aqueous $-CH_2NH_2$ vapors, respectively. The reversible sensing responses of the composites to the 0.5 M CH_2NH_2 vapors was also investigated and were found to be highly reversible during cyclic measurements (Khan and Shaheen 2015).

5.4 CONCLUSION

In this chapter, the authors have mentioned nanocomposite ion exchangers based on different polymeric materials such as polyaniline, polyanisidine, polytoluidine, polypyrrole, poly(3-methythiophene), and polycarbazole. The preparation conditions of the nanocomposites and their functioning as better ion exchangers are described, proposed by their IEC. The characterization of these nanocomposites was performed by various techniques such as FTIR, XRD, SEM, TEM, EDX, and TGA. The application of the synthesized nanocomposites was applied for the formation of, firstly, an ion-selective membrane electrode and, secondly, for vapor sensing.

The nanocomposite membranes were fabricated into ion-selective electrodes having appreciable operating characteristics, including Nernstian response, reasonable detection limit, relatively high selectivity, wide dynamic range, and fast response. Electrical conductivity and sensing performance of the nanocomposites was also investigated and showed an increasing electrical response with increasing temperature. The nanocomposite-based sensor also showed an increase in resistivity for chemical vapors and produced reversible behavior.

REFERENCES

Amdur, M. O., Doull, J., and Klaassen, C. D., 1991. *Toxicology: The Basic Science of Poisons*, 4th ed., Pergamon Press, New York.

Andrei, A. Z., 2007. *Ion Exchange Materials Properties and Applications*, Elsevier, Amsterdam, The Netherlands.

Czapla, C., and Bart, H. J., 2000. Characterization and modeling of the extraction kinetics of organic acids considering boundary layer charge effects. *Chem. Eng. Technol.* 23:1058–1062.

Goyer, R. A., 1993. Lead toxicity: Current concerns. *Environ. Health Persp.* 100:177–187.

Hellferich, F., 1962. *Ion Exchange*, McGraw-Hill, New York.

Hundt, N., Palaniappan, K., and Sista, P. et al., 2010. Synthesis and characterization of polythiophenes with alkenyl substituents. *Polym. Chem.* 1:1624.

Kamigaito, O., 1991. What can be improved by nanometer composites? *J. Jpn. Soc. Powder Metall.* 38:315.

Khan, A. A., and Akhtar, T., 2011a. Adsorption and electroanalytical studies of a poly-o-toluidine Zr(IV) phosphate nanocomposite for zinc(II) dimethyldithiocarbamate. *J. Appl. Polym. Sci.* 119:1393–1397.

Khan, A. A., and Akhtar, T., 2011b. Synthesis, characterization and analytical application of nano-composite cation-exchange material, poly-o-toluidine Ce(IV) phosphate: Its application in making Cd(II) ion selective membrane electrode. *Solid State Sci.* 13:559–568.

Khan, A. A., and Baig, U., 2012. Electrically conductive membrane of polyaniline–titanium(IV)phosphate cation exchange nanocomposite: Applicable for detection of Pb(II) using its ion-selective electrode. *J. Ind. Eng. Chem.* 18:1937–1944.

Khan, A. A., and Baig, U., 2013. Electrical conductivity and humidity sensing studies on synthetic organic-inorganic poly-o-toluidine-titanium(IV)phosphate cation exchange nanocomposite. *Solid State Sci.* 15:47–52.

Khan, A. A., and Baig, U., 2015. Polyurethane-based cation exchange composite membranes: Preparation, characterization and its application in development of ion-selective electrode for detection of copper (II). *J. Ind. Eng. Chem.* 29:392–399.

Khan, A. A., and Khan, A., 2009. Adsorption thermodynamics and kinetics of Mancozeb onto poly-o-toluidine Th(IV) phosphate: A nano-composite cation-exchanger—and its application as a Mancozeb-sensitive membrane electrode. *Adsorpt. Sci. Technol.* 27:567–578.

Khan, A. A., and Khan, A., 2010. Ion-exchange studies on poly-o-anisidine Sn(IV) phosphate nanocomposite and its application as Cd(II) ion-selective membrane electrode. *Cent. Eur. J. Chem.* 8:396–408.

Khan, A. A., and Paquiza, L., 2011. Analysis of mercury ions in effluents using potentiometric sensor based on nanocomposite cation exchanger polyaniline–zirconium titanium phosphate, *Desalination* 272:278–285.

Khan, A. A., and Paquiza, L., 2013. Synthesis and characterization of in situ polymerized poly (methyl methacrylate)–cerium molybdate nanocomposite for electroanalytical application, *J. Appl. Polym. Sci.* 127:3737–3748.

Khan, A. A., and Shaheen, S., 2013a. Electrical conductivity and ammonia sensing studies on in situ polymerized poly(3-methythiophene)–titanium(IV)molybdophosphate cation exchange nanocomposite, *Sensors Actuat B-Chem.* 177:1089–1097.

Khan, A. A., and Shaheen, S., 2013b. Synthesis and characterization of a novel hybrid nanocomposite cation exchanger poly-o-toluidine Sn(IV) tungstate: Its analytical applications as ion-selective electrode. *Solid State Sci.* 16:158–167.

Khan, A. A., and Shaheen, S., 2014a. Determination of arsenate in water by anion selective membrane electrode using polyurethane–silica gel fibrous anion exchanger composite. *J. Hazard. Mater.* 264:84–90.

Khan, A. A., and Shaheen, S., 2014b. Chronopotentiometric and electroanalytical studies of Ni(II) selective polyaniline Zr(IV)molybdophosphate ion exchange membrane electrode, *J. Electroanal. Chem.* 714:38–44.

Khan, A. A., and Shaheen, S., 2014c. Preparations and characterizations of poly-o-toluidine/multiwalled carbon nanotubes/Sn(IV)tungstate composite ion exchange thin films and their application as a Pb(II) selective electrode. *RSC Adv.* 4:23456–23463.

Khan, A. A., and Shaheen, S., 2015. Electrical conductivity, isothermal stability and amine sensing studies of a synthetic poly-o-toluidine/multiwalled carbon nanotube/Sn(IV)tungstate composite ion exchanger doped with p-toluene sulfonic acid. *Anal. Methods.* 7:2077.

Khan, A. A., Baig, U., and Khalid, M., 2011. Ammonia vapor sensing properties of polyaniline–titanium(IV) phosphate cation exchange nanocomposite. *J. Hazard. Mater.* 186:2037–2042.

Khan, A. A., Baig, U., and Khalid, M., 2013. Electrically conductive polyaniline-titanium(IV)molybdophosphate cation exchange nanocomposite: Synthesis, characterization and alcohol vapour sensing properties. *J. Ind. Eng. Chem.* 19:1226–1233.

Khan, A. A., Inamuddin, and Akhtar, T., 2008. Organic–inorganic composite cation-exchanger: Poly-o-toluidine Zr(IV) phosphate-based ion-selective membrane electrode for the potentiometric determination of mercury. *Anal. Sci. Int. J. Jpn. Anal. Chem.* 24:881–887.

Khan, A. A., Inamuddin, and Alam, M. M., 2005. Determination and separation of Pb^{2+} from aqueous solutions using a fibrous type organic–inorganic hybrid cation-exchange material: Polypyrrolethorium(IV) phosphate. *React. Funct. Polym.* 63:119–133.

Khan, A. A., Khan, A., and Habiba, U., 2009. Synthesis and characterization of organic-inorganic nanocomposite poly-o-anisidine Sn(IV) arsenophosphate: Its analytical applications as Pb(II) ion-selective membrane electrode. *Int. J. Anal. Chem.* 2009:659215.

Khan, A. A., Paquiza, L., and Khan, A. 2010. An advanced nano-composite cation-exchanger polypyrrole zirconium titanium phosphate as a Th(IV)-selective potentiometric sensor: Preparation, characterization and its analytical application. *J. Mater. Sci.* 45:3610–3625.

Khan, A. A., Rao, R. A. K., Alam, N., and Shaheen, S., 2015. Formaldehyde sensing properties and electrical conductivity of newly synthesized polypyrrole-zirconium(IV)selenoiodate cation exchange nanocomposite, *Sensors Actuat. B-Chem.* 211:419–427.

Khan, A. A., Shaheen, S., and Habiba, U., 2012. Synthesis and characterization of poly-o-anisidineSn(IV) tungstate: A new and novel "organic–inorganic" nano-composite material and its electro-analytical applications as Hg (II) ion-selective membrane electrode. *J. Adv. Res.* 3:269–278.

Lawrence, M. J., and Ress, G. D., 2000. Microemulsion-based media as novel drug delivery systems. *Adv. Drug Delivery Rev.* 45:89–121.

Lin, S. H., Lin, C. M., and Leu, H. G., 1999. Operating characteristics and kinetics studies of surfactant wastewater treatment by Fenton oxidation. *Water Res.* 33:1735–1741.

MacDonald, R., 2001. Providing clean water: Lessons from Bangladesh: Large parts of the world face an unwelcome choice between arsenic and micro-organisms. *Br. Med. J.* 322:626–627.

Meister, R.T., 1992. *Farm Chemicals Handbook*, Meister, Willoughby, OH.

MSDS for Ziram., 1991. FMC Corporation. Philadelphia, PA.

Parvatikar, N., Jain, S., Bhoraskar, S. V., and Prasad, M. V. N. A., 2006. Spectroscopic and electrical properties of polyaniline/CeO$_2$ composites and their application as humidity sensor. *J. Appl. Polym. Sci.* 102:5533–5537.

Peters, R. W., Ku, Y., and Bhattacharyya, D., 1985. Evaluation of recent treatment techniques for removal of heavy metals from industrial wastewaters. AICHE Symposium Series.

Reilley, C. N., Schmidt, R. W., and Sadek, F. S., 1959. Chelon approach to analysis: I. Survey of theory and application. *J. Chem. Educ.* 36:555.

Revathi, M., Kavitha, B., and Vasudevan, T., 2005. Removal of nickel ions from industrial plating effluents using activated alumina as adsorbent. *J. Environ. Eng.* 47:1–6.

Rezaei, B., Meghdadi, S., and Zarandi, R. F., 2008. A fast response cadmium-selective polymeric membrane electrode based on N, N′-(4-methyl-1, 2-phenylene)diquinoline-2-carboxamide as a new neutral carrier. *J. Hazard. Mater.* 153:179–186.

Richardson, T., 1987. *Composites: A Design Guide*, Industrial Press, New York.

Roy, R., Roy, R. A., and Roy, D. M., 1986. Alternative perspectives on "quasi-crystallinity": Non-uniformity and nanocomposites. *Mater. Lett.* 4:323.

Rozzi, A., Antonelli, M., and Angeretti, C., 2002. Removal of non ionic surfactants used in the tannery by an adsorbent resin. In: *Proceedings of the CIWEM Conference Wastewater Treatment: Standards and Technologies to Meet the Challenge of the 21st Century*, Leeds, UK: 4–6.

Sabah, E., Turan, M., and Celik, M. S., 2002. Adsorption mechanism of cationic surfactants onto acid- and heat-activated sepiolites. *Water Res.* 36:3957–3964.

Shakir, M., Noor-e-Iram, and Khan, M. S., 2014. Electrical conductivity, isothermal stability, and ammonia-sensing performance of newly synthesized and characterized organic–inorganic polycarbazole–titanium dioxide nanocomposite, *Ind. Eng. Chem. Res.* 53:8035–8044.

6 Nanomaterial-Supported Biopolymers for Water Purification

Nalini Sankararamakrishnan

CONTENTS

ABSTRACT

In developing countries, due to rapid industrial growth and increases in human population, concerns over safe drinking water have become a critical issue. To provide safe drinking water, it is imperative to tackle both chemical and bacteriological contaminants. In most drinking water treatment plants, a combination of chemical coagulation and disinfection processes are adopted. The main disinfection procedures include ultraviolet (UV) irradiation, ozonolysis, chlorination, and electromagnetic radiation. It is well known that disinfection using chlorine produces carcinogenic by-products, while disinfection using ozone and UV irradiation is expensive. Thus, the important challenge faced by the water purification sector is a cost-effective and efficient method for providing safe water without endangering human health.

To address this challenge, water treatment using biopolymeric materials is expected to play an increasingly important role in drinking water treatment. Chitosan is one of the abundantly available biopolymers after cellulose. It is biocompatible, renewable, biodegradable, nontoxic, possessing antibacterial properties, making it environmentally friendly. The main challenges in the use of chitosan include stability and selectivity toward various pollutants. A new class of nanocomposites are fabricated by combining biopolymeric materials such as chitosan with nanomaterials. To circumvent the problems associated with the selectivity and stability of chitosan, these nanocomposites can be tuned for their physicochemical and structural properties such as charge density, porosity, hydrophilicity, and thermal and mechanical stability, and introduce attractive functionalities such as antibacterial, photocatalytic, or adsorptive capabilities. These advanced nanocomposite biopolymers could be fabricated to meet specific water treatment applications. Thus, this chapter will focus on the recent scientific and technological advances in the development of chitosan-based nanocomposites for water treatment.

Keywords: Chitosan, Nanocomposites, Contaminants, Water treatment

6.1 INTRODUCTION

Water is an essential and important component of our earth and plays a pivotal role in the functioning of our ecosystems. Though several water bodies are present, the availability of potable drinking water for millions of people around the world is still lacking. The main cause of this scenario may be associated with urbanization and enormous population growth. More than 700 inorganic and organic micropollutants have been reported in water. While some are highly carcinogenic and toxic, others have very long residence times in the environment and are neither biotransformable nor biodegradable. Several techniques are available to remove these contaminants from potable water and industrial wastewater, which include conventional coagulation, chemical precipitation, ion exchange, reverse osmosis, electrolysis, electrodialysis, and adsorption (Ho and McKay 1999). Among these, the two electrochemical techniques, namely electrodialysis and electrolysis, are not cost-effective and, thus, their use in developing countries is limited. Around 90% of water is wasted in reverse osmosis and, thus, this technique is not suitable for places with limited water availability. Ion exchange targets only particular oxidation states of the pollutants, and membrane fouling is the commonest problem associated with it. Conventional coagulation techniques and chemical precipitation leave behind huge amounts of sludge, which is a secondary pollutant that needs posttreatment(s), thereby increasing the final water treatment cost. Adsorption, however, is a cost-effective technique that removes both inorganic and organic contaminants from water. The use of biopolymers as adsorption materials is most cost-effective.

Around the world, numerous seafood processing industries generate chitosan as a waste product. It is obtained from the crustacean shells of shellfish, crabs, and prawns. It is the most abundant natural biopolymer available after cellulose. Chitosan is obtained by the de-N-acetylation of chitin. It is a transformed polysaccharide, consisting of cationic aminopolysaccharide copolymers of glucosamine and N-acetylglucosamine. It is biocompatible, renewable, biodegradable, and nontoxic, possessing antibacterial properties, making it environmentally friendly. It contains both amino and hydroxyl groups, making it an efficient heavy metal scavenger (Guibal 2004). The structure of chitosan is shown in Figure 6.1. Although chitosan has been extensively used in the removal of various heavy metals and dyes, research is still focused on its successful application as an efficient adsorbent for water treatment. Among nanomaterials, surface morphology, pore diameter, specific surface area, and surface functionalization with specific groups are significantly relevant. Thus, to improve the sorption capacity, various nanomaterials including silver nanoparticles (to enhance antibacterial properties) (Zhu et al. 2010), zerovalent iron (ZVI) (heavy metal scavenging) (Chauhan et al. 2014), TiO_2 (photocatalytic reduction of dyes) (Lučić et al. 2014), Fe_3O_4 (to improve magnetic properties) (Rahimi et al. 2015), and carbon nanotubes (CNTs) (to enhance mechanical strength) (Chaterjee et al. 2010) are doped to the chitosan matrix to form bionanocomposites. This chapter pertains to the recent developments of the nanomaterial-supported biopolymer—namely chitosan—toward water purification.

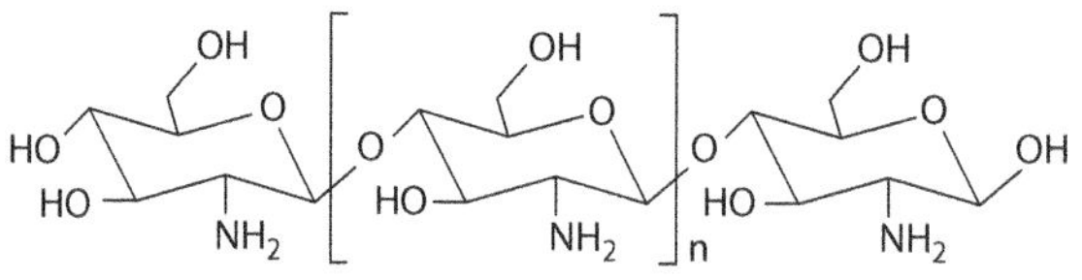

FIGURE 6.1 Structure of chitosan.

6.2 REMOVAL OF CONTAMINANTS

6.2.1 REMOVAL OF HEAVY METALS

Industrial wastes from various industries such as pigments and paints, extraction and mining, electroplating, glass production, and battery manufacturing plants release various heavy metals into water bodies. These heavy metals are not biodegradable and their presence in water causes bioaccumulation in living organisms, leading to chronic diseases and creating health problems. Heavy metal ions such as Cd^{2+}, Cr^{6+}, Cu^{2+}, Pb^{2+}, Zn^{2+}, As^{3+}, As^{5+}, or Hg^{2+}, originating from natural geological or industrial sources, cause hazardous water pollution. As discussed in the introduction section, the most viable option for the removal of these contaminants is the adsorption technique using novel sorbents.

6.2.1.1 Membranes

A copper removal membrane, incorporating chitosan beads inside an ethylene vinyl alcohol matrix, was developed by Tetala and Stamatialis (2013). The resulting membrane exhibited fast sorption kinetics, and the maximum adsorption capacity was reported to be 225.7 mg/g.

Electrospinning is a proven technique for the fabrication of nonwoven nanofibers of diameters less than 100 nm (Li and Xia, 2004). These fibrous mats are used as membranes and find a wide range of applications in environmental areas, owing to their high porosity, good permeability, and small interfibrous pore sizes. Chauhan et al. (2014) evaluated a novel electrospun chitosan/polyvinyl alcohol (PVA)/ZVI (CPZ) nanofibrous mat for the removal of both inorganic forms of arsenic at neutral pH. A very high removal capacity of 200.0 and 142.9 mg/g for As(V) and As(III), respectively, has been reported. The schematic representation of the formation of fibers and their application is shown in Figure 6.2. The nanofibrous mat was characterized by various spectral techniques, and the adsorption mechanism has been explained in terms of electrostatic attraction between the protonated amino groups of chitosan/arsenate ions and oxidation of arsenite to arsenate by Fentons generated from ZVI and subsequent complexation of the arsenate with the oxidized iron.

Min et al. (2015) prepared chitosan-based electrospun nanofiber membrane (CS-ENM) and used it as an adsorbent for arsenate removal from water. The adsorption isotherm data correlated well with the Langmuir model, and the maximum adsorption capacity was found to be 30.8 mg/g. x-ray photoelectron spectroscopy (XPS) analysis suggested amine groups on the chitosan were involved in the complexation of As(V).

6.2.1.2 Magnetic Nanocomposites

The removal of metal contaminants using magnetic adsorbents has recently attracted the attention of the water treatment industry. Figure 6.3 represents how magnetic separation facilitates the removal of heavy metals using magnetic nanocomposites.

Fan and coworkers (2013) synthesized magnetic nanocomposites of graphene oxide and chitosan by an easy, rapid process. The green composite was used to adsorb Pb^{2+} ions with an adsorption capacity of 76.94 mg/g. In a similar study, chitosan-coated magnetite nanoparticles were successfully used to remove Pb^{2+} ions (Gregorio-Jauregui et al. 2012). A removal efficiency of 53.6% was reported. The magnetization value of 70.1 m^3/kg was achieved for Fe_3O_4 nanoparticles in the absence of chitosan, while composites with increased chitosan content were characterized by decreased magnetization (66.4 and 45.1 m^3/kg). Recently, Rahimi et al. (2015) reported goethite/chitosan nanocomposites for the removal of Pb(II) ions from aqueous solutions. The maximum removal efficiency of 98.26% was observed at pH 6, using 0.05 g of the adsorbent at an initial concentration of 74.4 mg/L. Magnetic hydrogel beads, consisting of amine-functionalized magnetite nanoparticles, carboxylated cellulose nanofibrils, and poly(vinyl alcohol)-blended chitosan, were prepared by an instantaneous gelation method and used as adsorbents for Pb(II) removal. The complexing ability of

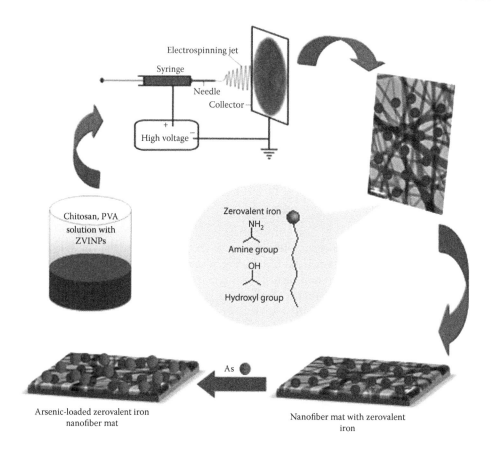

FIGURE 6.2 **(See color insert)** Schematic representation of the formation of a nonfibrous mat and its application toward arsenic removal.

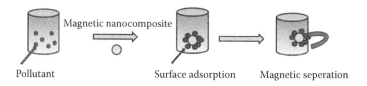

FIGURE 6.3 Schematic representation of the magnetic nanocomposite toward pollutant removal.

carboxylate groups of carboxylated cellulose nanofibrils plays an important role in Pb(II) sorption, and the prepared magnetic hydrogels exhibit higher adsorption capacity with a value of 171.0 mg/g (Zhou et al. 2014).

An amino-functionalized magnetic cellulose composite was synthesized by coating cellulose over magnetic silica nanoparticles, followed by the grafting of glycidyl methacrylate and ring-opening reaction with ethylenediamine to yield amino groups. A prepared nanocomposite was evaluated for Cr(VI) removal from an aqueous solution. The adsorption isotherms fitted the Langmuir model with a maximum adsorption capacity of 171.5 mg/g at 25°C. The nanocomposite material was reported to be a promising adsorbent for Cr(VI) removal, with the advantages of a rapid adsorption rate, high adsorption capacity, and convenient recovery under magnetic field (Gregorio-Jauregui et al. 2012).

6.2.1.3 Others

A nanochitosan (nano-Ch) was prepared by ionic gelation of chitosan and tripolyphosphate, and nanochitosan-actetophenone (nano-Ch-Act) was prepared by Schiff-base condensation of nano-Ch and acetophenone (Mahmoud et al. 2015). The prepared nano-Ch and nano-Ch-Act exhibited a very high capacity, ranging from 1298 to 1608 and 810 to 1236 μmol/g at pH 7.0, respectively. The ions tested included Co^{2+}, Cd^{2+}, Cu^{2+}, Pb^{2+}, and Hg^{2+}. Further, the adsorbents were applied to the removal of the aforesaid ions from wastewater, seawater, and tap water. Alsabagh et al. (2015) prepared a multifunctional nanocomposite comprising silver nanoparticles, copper nanoparticles, and carbon nanotubes. The adsorption efficiency of the nanocomposite was found to be superior for Cu(II), Cd(II), and Pb(II) compared with plain chitosan. Further, the authors claim that within 10 min, complete removal of metal ions took place. These composites were regenerated by ethylenediiaminetetraacetic acid (EDTA) and were used for five cycles.

A nanocomposite comprising chitosan-poly(vinyl alcohol)/bentonite (CTS-PVA/BT) with high selectivity for Hg(II) was synthesized by doping BT into the CTS-PVA polymer matrix. The nanocomposites possessed a mesoporous structure with a narrow-sized distribution, and it was found that BT effectively enhanced thermal stability. The adsorption capacities for Hg(II) ions with a BT content of 0%, 10%, 30%, and 50% were found to be 460.18, 455.12, 392.19, and 360.73 mg/g, respectively, which are much higher than those for Pb(II), Cd(II), and Cu(II) ions. It was reported that the presence of BT improved its selectivity toward Hg(II) (Wang et al. 2014a,b).

Recently, Zavareh and coworkers (2015) reported As(III) adsorption combined with enhanced antimicrobial properties using a Cu-chitosan/alumina nanocomposite. It was observed that the antimicrobial activity of Cu-chitosan/nano-Al_2O_3 was found to be higher than that of chitosan/nano-Al_2O_3 and pure chitosan.

6.2.2 REMOVAL OF FLUORIDE, NITRATE, AND PHOSPHATE

Fluoride is one of the important micronutrients required by the human body for strong teeth and bones. However, excessive amounts of fluoride may lead to dental and skeletal fluorosis. Because of its deleterious effects on humans, the Bureau of Indian Standards (BIS) and the World Health Organization (WHO) have set the permissible limit in drinking water as 1.5 mg/L. Various techniques have been proposed for the removal of fluoride. Novel chitosan/montmorillonite/zirconium oxide (CTS/MMT/ZrO_2) nanocomposites were prepared and evaluated for fluoride removal (Teimouri et al. 2015). The fluoride adsorption capacity of CTS/MMT/ZrO_2 was experimentally found to be 23 mg/g at pH 4. The composite was characterized by various spectral techniques. The adsorption capacity of the CTS/MMT/ZrO_2 nanocomposite was higher than those of CTS (52 mg/kg for fluoride removal), ZrO_2, CTS/ZrO_2, MMT, and CTS/MMT. Pandi and Viswanathan (2015) reported an ecomagnetic biosorbent prepared by uniform deposition of magnetic Fe_3O_4 particles on the surface of a nanohydroxyapatite (n-HAp)/chitosan (CS) nanocomposite and demonstrated its applicability in fluoride sorption.

A novel nanocomposite with a surface area of 212.9 m^2/g was prepared from chitosan and Fe_3O_4/ZrO_2 under mild conditions. The nanocomposite had the ability to adsorb both nitrate and phosphate. The maximum adsorption amount of phosphate and nitrate was 26.5 mg P/g and 89.3 mg/g, respectively. The adsorption process fitted well to the pseudo-first-order kinetic rate model, and the mechanism involved simultaneous adsorption and intraparticle diffusion. The experimental results suggested that the composite is a promising adsorbent for treating nutrient-contaminated aqueous streams (Jiang et al. 2013).

6.2.3 REMOVAL OF ORGANIC POLLUTANTS AND DYES

Rapid industrialization has resulted in the discharge of various dyes and other colored toxic effluents into aqueous streams. The presence of these dyes in water bodies is highly visible and undesirable.

They are produced to be resistant to light, water, weather, and detergents. Further, the dyes are not biodegradable under aerobic conditions or degrade very slowly, producing vividly colored treated effluents. These organic dyes are harmful to aquatic plants and animals, and have a carcinogenic and mutagenic action on human beings (Aksu 2005). The most easy and cost-effective method to remove dyes from aqueous solutions is adsorption (Barquist and Larsen 2010; Ratnamala et al. 2012; Liu et al. 2008a).

The development of new composite adsorbents with excellent adsorption capacities are gaining increased momentum in recent research. Titanium dioxide-incorporated nanocomposites have been extensively used in the removal of organic pollutants, owing to their photocatalytic degradation of organic pollutants. Lučić and coworkers (2014) studied TiO_2/hydrogel nanocomposites for the photocatalytic degradation of three different groups of anionic azo dyes, namely, C.I. acid red 18, C.I. acid blue 113, C.I. reactive black 5, C.I. direct blue 78, and C.I. reactive yellow 17. Under sunlike illumination, nanocomposites with immobilized colloidal TiO_2 nanoparticles completely removed C.I. acid blue 113, C.I. acid red 18, C.I. direct blue 78, and C.I. reactive black 5, while the removal degree of C.I. reactive yellow 17 was 55%. Another study (Xiao et al. 2015) developed nano-TiO_2 doped chitosan by the molecular imprinting technique and demonstrated the removal of methyl orange dye. The sorbent showed enhanced photocatalytic selectivity for methyl orange. The removal of the dye was reported to be through photocatalytic degradation rather than adsorption, and it was reported to retain 60% of the activity even after 10 sorption–desorption cycles.

Similarly to TiO_2, ZnO has also been found useful for degradation of dyes. Pandiselvi et al. (2015) synthesized polyaniline-ZnO/chitosan (PZO/chitosan) nanocomposite using chitosan $ZnCl_2$ and aniline, through the precipitation–oxidation method. The photocatalytic and adsorption properties of the sorbent were evaluated for reactive orange and methylene blue dyes. Results indicated that the maximum dye removal rates were 96% and 88.5% for reactive orange and methylene blue, respectively, with initial dye concentration of 50 mg/L, dose rate of 0.3 g/L, and neutral pH, in 120 min of sunlight exposure. Thus, it is concluded that the prepared nanocomposite possessed both the advantages of both photodegradation and the adsorption process and could find use in the treatment of various pollutants.

Novel monodispersed pompon-like magnetite/chitosan (Fe_3O_4/CS) nanocomposites were prepared by a solvothermal method and evaluated for the removal of toxic sodium pentachlorophenate (PCP-Na) from aqueous media (Liu et al. 2013a,b). The results of XPS studies suggested that hydrogen bonding, electrostatic attraction, and π–π interactions were all believed to play a role in PCP-Na adsorption on Fe_3O_4/CS. The magnetization capacity of the sorbent was found to be 22.2 emu/g, and thus the composite could be easily separated from water with magnets within 2 min. The equilibrium was achieved within 30 min and the maximum removal of PCP-Na (91.5%) was obtained at pH 6.5. A polyaniline-coated chitosan-functionalized magnetic nanocomposite was prepared and applied for the removal of endocrine-disrupting phenols from water and fruit juices (Jiang et al. 2015). Initially, chitosan magnetic spheres (Fe_3O_4@CHI) were prepared by coprecipitation followed by polymerization on aniline to the magnetic core (Fe_3O_4@CHI@PANI). The prepared microspheres were of uniform size of about 100 nm, with core diameter ranging from 20 to 30 nm. The magnetic microspheres exhibited a saturation magnetization of 32 emu/g. Endocrine-disrupting phenols such as Bisphenol A, 2,4-dichlorophenol, and trichlosan were used as analytes. The extraction of these phenols occurred via π–π interaction between a polyaniline shell and aromatic compounds. Jaiswal and coworkers (2012) demonstrated the removal of organophosphorous pesticide malathion from agricultural runoff using a copper-coated chitosan nanocomposite (CuCH). The adsorption capacity of CuCH toward malathion was found to be 322.6 mg/g at pH 2. Using Fourier transform infrared (FTIR) spectra, GC-MS, and energy dispersive x-ray spectroscopy (EDAX) spectra as tools, it was postulated that malathion undergoes hydrolysis at pH 2 to malathion mono- and diacid, and then to dithiophosphonate. The dithiophosphonate formed complexes with the copper present on the adsorbent to form a copper-dithiophosphonate complex. The schematic representation of this mechanism is shown in Figure 6.4. Further, the removal of malathion from surface waters using CuCH was also demonstrated. The surface waters were spiked with 0.5 and 1.0 µg/mL and equilibrated with 0.02 g of the adsorbent for 16 h, and the efficiency of removal of the pesticide was tested using a GC-MS instrument after extraction with dichloromethane. The results obtained are shown in Table 6.1. It is evident from the data that complete removal was achieved.

Biopolymer-based magnetic nanocomposites have been successfully utilized for the decontamination of organic dyes in wastewater (Reddy et al. 2013; Jiang et al. 2013; Chaterjee et al.

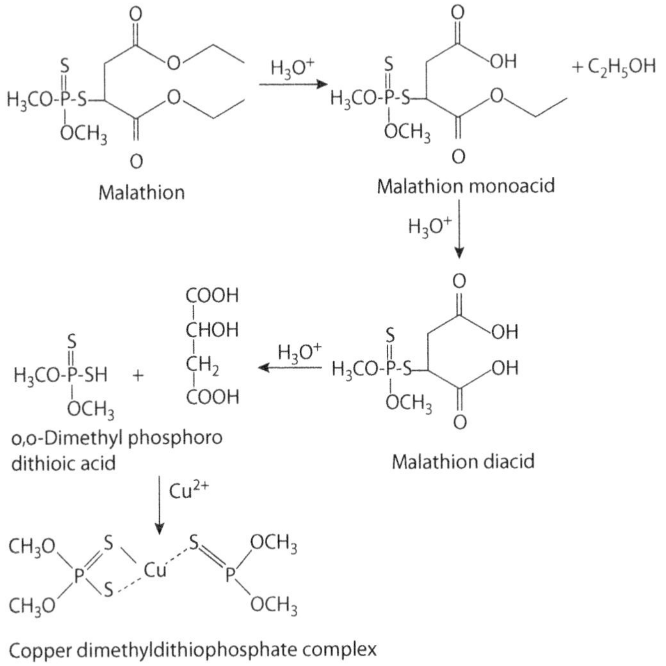

FIGURE 6.4 Proposal mechanism of absorption for malathion on CuCH. (From Jaiswal, M., et al., *Environ. Sci. Pollut. Res.* 19, 2055–2062, 2012.)

TABLE 6.1

Mean Recovery (%) and Relative Standard Deviation (in parentheses) of Pesticide Residue at Two Concentration Levels

S. No.	Sample	Concentration Levels (µg/L)
1	Surface water	N.D.
2	Surface water + 0.5 µg/mL malathion	92.5 (3.2)
3	Surface water + 1.0 µg/mL malathion	95.50 (2.21)
4	Surface water + 1.0 µg/mL malathion treated with CuCH	N.D.
5	Surface water + 2.0 µg/mL malathion treated with CuCH	N.D

Source: Jaiswal, M., et al., *Env. Sci. Pollut. Res.* 19, 2055–2062, 2012. With permission of Springer.

Note: N.D.: not detected.

2010; Pourjavadi et al. 2013). Using a magnetic chitosan/graphene oxide nanocomposite, methylene blue was removed by adsorption (Fan et al. 2012). The sorption fitted the Langmuir adsorption isotherm, and a maximum adsorption capacity of 180.83 mg/g was reported. Similarly, Zhu et al. (2012) reported magnetic chitosan/poly(vinyl alcohol) hydrogel beads with a saturation magnetization of 21.96 emu/g for the removal of Congo red. The maximum adsorption capacity was found to be 470.1 mg/g, which was higher than for the previously reported chitosan/carbon nanotube nanocomposite (Chaterjee et al. 2010). In another study, magnetic chitosan nanocomposites (MCNCs) with a magnetic saturation capacity of 17.5 emu/g were synthesized by an inexpensive reduction–precipitation technique and applied to the removal of acid red 2 (Kadam et al. 2015). Chitosan-embedded organoclay (Cloisite 15A and 30B) was applied to dye removal (Daraei et al. 2013). Recently, a novel nanocomposite comprising a multiwalled carbon nanotube (MWCNT) functionalized (f) with chitosan (CS) and poly-2-hydroxyethyl methacrylate (pHEMA) was prepared and evaluated for the removal of methyl orange (Mahmoodian et al. 2015). The adsorption kinetic was reported to follow pseudo-second-order kinetics with a correlation coefficient >0.9986.

6.2.4 Antimicrobial Property

Biofouling of membranes due to microbial growth and biofilm formation is one of the most challenging issues in water and wastewater treatment (Zhu et al. 2010). Biofouling reduces permeate quality, decreases membrane permeability, and increases the energy costs of the purification process. The development of antimicrobial membranes will increase membrane efficiency significantly. Another added advantage is the provision of pathogen-free water by the use of antimicrobial membranes.

The antimicrobial property of chitosan and chitosan-based nanocomposites is well established (Rabea et al. 2003; Liu et al. 2008b; Honarkar and Barikani 2009; Ignatova et al. 2013; Chi et al. 2007; Deng et al. 2012). A schematic representation of the antimicrobial activity of chitosan is shown in Figure 6.5. The antibacterial properties of chitosan may be attributed to the electrostatic interaction between the negatively charged components in the microbial cell membranes and the

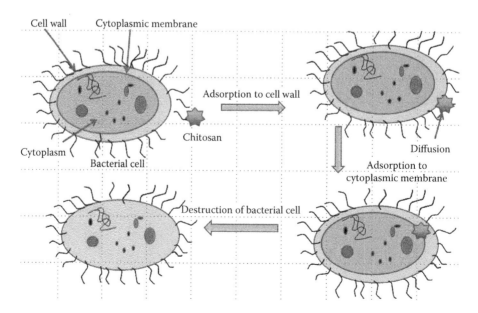

FIGURE 6.5 **(See color insert)** Schematic representation of antibacterial activity chitosan.

positively charged amine groups on the chitosan backbone. The barrier property is altered due to this binding between cell wall components and chitosan and, thus, leads to cell death (Angelova et al. 1995). Chitosan has several advantages over other types of disinfectants because it possesses a broader spectrum of activity, a higher antibacterial activity, and a lower toxicity toward mammalian cells. Rhim et al. (2006) showed that the montmorillonite–chitosan (Mt–CS) nanocomposite exhibited higher antimicrobial activity against *Staphylococcus aureus* (Gram+/KCCM11764) and *Escherichia coli* (Gram−/KCCM 11600) than Na-montmorillonite and pure chitosan (Han et al. 2010). It was suggested that chitosan-based materials are more effective toward gram-negative bacteria compared with gram-positive species, owing to their thinner murine wall, which may allow them to be more rapidly absorbed (Rhim et al. 2006). Shameli et al. (2011) reported a nanocomposite consisting of silver-doped Mt–CS to exhibit significant antibacterial activity toward gram-negative bacteria, that is, *E. coli* O157:H7 and *Pseudomonas aeruginosa*, and gram-positive bacteria, that is, *S. aureus* and methicillin-resistant *S. aureus*.

Silver (Ag) possesses excellent biocidal properties, and has been the most widely used antimicrobial agent in nanocomposite membranes (Liu et al. 2008). For example, Chou et al. (2005) doped AgNPs into a cellulose acetate (CA) matrix for antibacterial applications. A novel clay-polydimethyloxane-chitosan-silver (clay-PDMS-Ct-Ag) nanocomposite completely destroyed various infectious bacteria (*P. aeruginosa*, ATCC27853; *Candida albicans*, ATCC14053; *E. coli*, ATCC25922; *S. aureus*, ATCC25923) (Zhou et al. 2007). Recently, Kumar-Krishnan and coworkers (2015) demonstrated that chitosan-silver (CS/Ag) nanocomposites, either in the form of nanoparticles (AgNP) or as ionic dendritic structures (Ag+), exhibited excellent antibacterial potency toward gram-positive *S. aureus* and gram-negative *E. coli*. The results demonstrated that, in contrast to CS/Ag+ ion films, the CS/AgNP composite film, with a particle size of <10 nm, showed a higher antibacterial property. A similar study was reported using chitosan and silver nanoparticles (An et al. 2014). Gluteraldehyde was used as a cross-linking agent. Antimicrobial assays were performed using fungi and typical Gram bacteria. The inhibitory effect indicated that the nanocomposites with higher concentrations of AgNPs and of smaller size exerted a stronger antibacterial activity. The mechanism of antimicrobial action is also detailed using XPS studies.

A rapid and green method has been reported, using quaternized chitosan (QCS), montmorillonite (MMT), and gemini surfactants, to synthesize silver nanoparticles (AgNPs) and simultaneously achieve an exfoliated chitosan/clay nanocomposite under microwave irradiation (Liu et al. 2013). The Ag NP-loaded QCS/clay nanocomposites exhibited excellent antimicrobial activity. The lowest minimum inhibition concentration against various microorganisms was 0.00001% wt and the antimicrobial mechanism was evaluated by transmission electron microscopy (TEM) and scanning electron microscopy (SEM) studies. In a similar study, Na+ ions of montmorillonite were replaced by silver ions and chitosan was intercalated into this matrix (Lavorgna et al. 2014). This multifunctional bionanocomposite exhibited a significant reduction in the growth of *Pseudomonas* spp.

Similarly to silver, ZnO nanoparticles have also been reported to exhibit antibacterial properties. In a recent study, ZnO-decorated chitosan-grapheneoxide nanocomposites were prepared and characterized by various spectral techniques. The material showed significantly enhanced anitmicrobial properties toward both gram-negative bacteria *E. coli* and gram-positive bacteria *S. aureus* (Chowdhuri et al. 2015).

6.2.5 Concurrent Removal of Heavy Metal and Disinfection Control

To provide safe drinking water, it is imperative to remove water pollutants and eliminate bacterial infection. Thus, a composite adsorbent to tackle both chemical and biological contaminants is a challenging task. Recently, Chauhan et al. (2014) developed novel biopolymeric nanofibers consisting of chitosan/Fe(III)/PVA to eliminate both heavy metal contamination and bacterial growth. Initially, the prepared nanofibers were cross-linked with gluteraldehyde to improve mechanical strength. The mechanism involved in cross-linking is shown in Scheme 6.1. The prepared nanofibrous mat

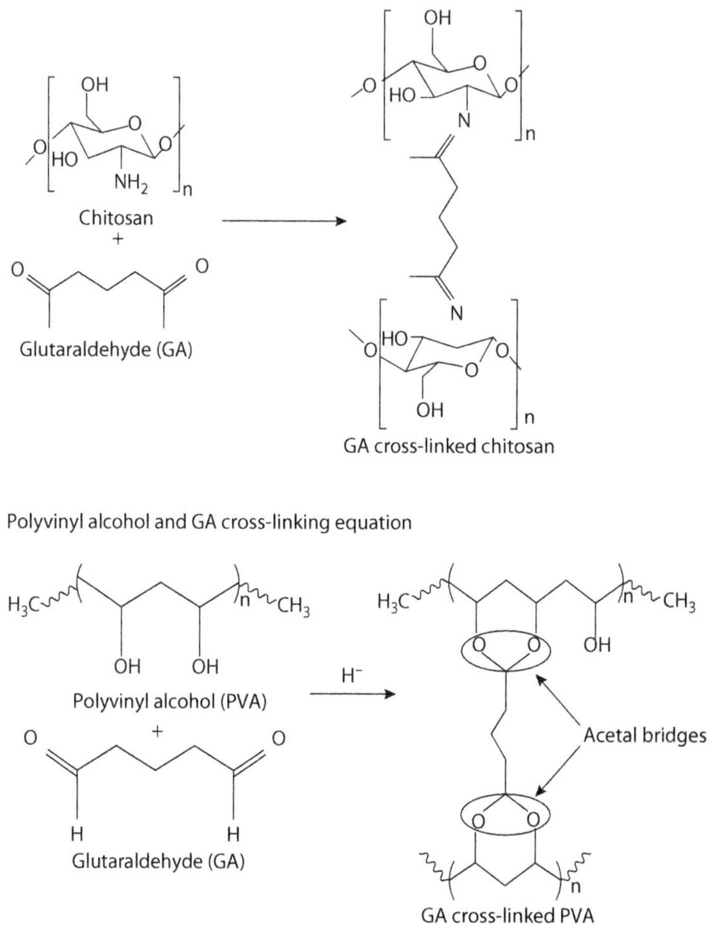

SCHEME 6.1   Cross-linking of chitosan and PVA with glutaraldehyde. (Source: Reproduced from Chauhan, D., et al., *RSC Adv.* 4, 54694–54702, 2014.)

removed As(III), As(V), Cr(VI), and F⁻ ions. The mechanism of interaction of the nanofiber with various ions tested is shown in Scheme 6.2. The selection order of the investigated anions toward the nanofibrous mats followed the order $Cr(VI) > As(V) > As(III) > F$. Further, the prepared nanocomposite mat also exhibited 100% disinfection toward *E. coli* at an initial concentration of 10^{4}–10^{5} CFU/mL, as shown in Figure 6.6.

6.3   CONCLUSIONS AND PERSPECTIVES

In recent years, there has been tremendous progress in the development of chitosan-based nanocomposite materials. The incorporation of nanomaterials on to chitosan provides unique properties and attractive functionalities such as antibacterial, photocatalytic, or adsorptive capabilities. These nanocomposite materials provide a new dimension to fabricate the next generation of materials targeting various pollutants and enhanced antibacterial properties.

To scale up these bionanocomposites for large-scale practical applications, several factors still need to be addressed. First, the stability of chitosan under various conditions. Though cross-linking

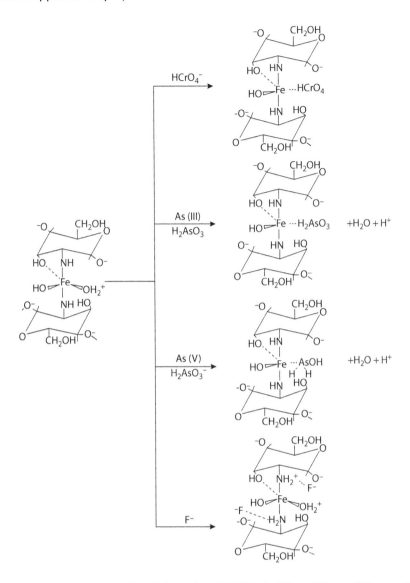

SCHEME 6.2 Schematic representation of absorption of Cr(VI), As(III), and F onto CPF mats. (Reproduced from Chauhan, D., et al., *RSC Adv.* 4, 54694–54702, 2014.)

and the addition of nanomaterials such as CNTs enhance stability, systematic studies regarding performance changes are still lacking. Secondly, the most common problem with doped nanomaterials is aggregation. Hence, a proper strategy needs to be employed for the appropriate dispersion of these nanomaterials. Another problem that needs to be addressed is the leaching of doped nanomaterials, as this can induce secondary pollution. Thus, the potential effect of nanomaterials leached to the environment should be systematically evaluated. Finally, the practical application of nanomaterial-supported biopolymers for water treatment is still in its infancy. Though there are many laboratory-based studies regarding the application of these nanocomposites, reports on large-scale production and industrial application is still lacking. Research needs to be focused on the design of nanocomposites that can tackle both chemical and biological contamination. Further, the sustainability and cost-effectiveness of nanocomposites should also be taken into account.

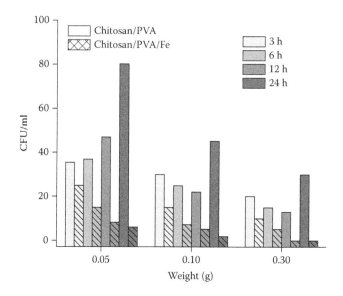

FIGURE 6.6 **(See color insert)** Antibacterial activity of CPF mats. (Reproduced from Chauhan, D., et al., *RSC Adv.* 4, 54694–54702, 2014.)

REFERENCES

Aksu, Z. (2005). Application of biosorption for the removal of organic pollutants: A review, *Process Biochem.* 40:997–1026.

Alsabagh, A.M., Fathy, M., and R.E. Morsi (2015). Preparation and characterization of chitosan/silver nanoparticle/copper nanoparticle/carbon nanotube multifunctional nano-composite for water treatment: Heavy metals removal; kinetics, isotherms and competitive studies, *RSC Adv.* 5:55774–55783.

An, J., Ji, Z., Wang, D., Luo, Q., and X. Li (2014). Preparation and characterization of uniform-sized chitosan/ silver microspheres with antibacterial activities, *Mater. Sci. Eng. C.* 36:33–41.

Angelova, N.M., Rashkov, I., Maximova, V., Bogdanova, S., and A. Domard (1995). Preparation and properties of modified chitosan films for drug release, *J. Bioact. Compat. Pol.* 10:285–298.

Barquist, K., and S.C. Larsen (2010). Chromate adsorption on bifunctional, magnetic zeolite composites. *Microporous Mesopor. Mater.* 130:197–202.

Chaterjee, S., Lee, M.W., and S.H. Woo (2010). Adsorption of Congo red by chitosan hydogel beads impregnated with carbon nanotubes, *Bioresour. Technol.* 101:1800–1806.

Chauhan, D., Dwivedi, J., and N. Sankararamakrishnan (2014). Facile synthesis of smart biopolymeric nano fibers towards heavy metal removal and disinfection control, *RSC Adv.* 4:54694–54702.

Chi, W.L., Qin, C.Q., Zeng, L.T., Li, W., and W.J. Wang (2007). Microbiocidal activity of chitosan- N-2-hydroxypropyl trimethyl ammonium chloride. *Appl. Polym. Sci.* 103:3851–3856.

Chou, W.L., Yu, D.G., and M.C. Yang (2005). The preparation and characterization of silver- loading cellulose acetate hollow fiber membrane for water treatment, *Polym. Adv. Technol.* 16:600–607.

Chowdhuri, A.R., Tripathy, S., Chandra, S., Roy, S., and S.K. Sahu (2015). A ZnO decorated chitosan-graphene oxide nanocomposite shows significantly enhanced antimicrobial activity with ROS generation, *RSC Adv.* 5:49420–49428.

Daraei, P., Madaeni, S.S., Salehi, E., Ghaemi, N., Ghari, H.S., Khadivi, M.A., and E. Rostami (2013). Novel thin film composite membrane fabricated by mixed matrix nanoclay/chitosan on PVDF microfiltration support: Preparation, characterization and performance in dye removal, *J. Membr. Sci.* 436:97–108.

Deng, H., Lin, P., Xin, S., Huang, R., Li, W., Du, Y., Zhou, X., and J. Yang (2012). Quaternized chitosan-layered silicate intercalated composites based nanofibrous mats and their antibacterial activity, *Carbohyd. Polym.* 89:307–313.

Fan, L., Luo, C., Sun, M., Li, X. and Lu, F., and H. Qiu (2012). Preparation of novel magnetic chitosan/graphene oxide composite as effective adsorbents toward methylene blue, *Bioresour. Technol.* 114:703–706.

Fan, L., Luo, C., Sun, M., Li, X., and H. Qiu (2013). Highly selective adsorption of lead ions by water-dispersible magnetic chitosan/graphene oxide composites, *Colloids Surf. B* 103:523–529.

Gregorio-Jauregui, K.M., Pineda, M.G., Rivera-Salinas, J.E., Hurtado, G., Saade, H., Martinez, J.L., Ilyina, A., and R.G. Lopez (2012). One-step method for preparation of magnetic nanoparticles coated with chitosan, *J. Nanomater.* Article ID 813958.

Guibal, E. (2004). Interactions of metal ions with chitosan based adsorbents, *Sep. Purif. Technol.* 38:43–74.

Han, Y.-S., Lee, S.-H., Choi, K.H., and I. Park (2010). Preparation and characterization of chitosan–clay nanocomposites with antimicrobial activity, *J. Phys. Chem. Solids* 71:464–467.

Ho, Y.S., and G. Mckay (1999). Pseudo second order model for sorption processes, *Process Biochem.* 34:451–465.

Honarkar, H., and M. Barikani (2009). Applications of biopolymers I: Chitosan. *Monatsh. Chem.* 140:1403–1420.

Ignatova, M., Manolova, N., and I. Rashkov (2013). Electrospun antibacterial chitosan-based fibers. *Macromol. Biosci.* 13:860–872.

Jaiswal, M., Chauhan, D., and N. Sankararamakrishnan (2012). Copper chitosan nanocomposite: synthesis, characterization, and application in removal of organophosphorous pesticide from agricultural runoff, *Environ. Sci. Pollut. Res.* 19:2055–2062.

Jiang, H., Chen, P., Luo, S., Tuo, X., Cao, Q., and M. Shu (2013). Synthesis of novel nanocomposite Fe_3O_4/ZrO_2/chitosan and its application for removal of nitrate and phosphate, *App. Surf. Sci.* 284:942–949.

Jiang, X., Cheng, J., Zhou, H., Li, F., Wu, W., and K. Ding (2015). Polyaniline-coated chitosan-functionalized magnetic nanoparticles: Preparation for the extraction and analysis of endocrine-disrupting phenols in environmental water and juice sample, *Talanta* 141:239–246.

Kadam, A.A., and D.S. Lee (2015). Glutaraldehyde cross-linked magnetic chitosan nanocomposites: Reduction precipitation synthesis, characterization, and application for removal of hazardous textile dyes, *Biores. Technol.* 193:563–567.

Kumar-Krishnan, S., Prokhorov, E., Hernández-Iturriaga, M., Mota-Morales, J.D., Kovalenco, Y., Vázquez-Lepe, M., Sánchez, I.C., and G. Luna-Bárcenas (2015). Chitosan/silver nanocomposites: Synergistic antibacterial action of silver nanoparticles and silver ions, *Eur. Polym. J.* 67:242–251.

Lavorgna, M., Attianese, I., Buonocore, G.G., Conte, A., Del Nobile, M.A., Tescione, F., and E. Amendola (2014). MMT-supported Ag nanoparticles for chitosan nanocomposites: Structural properties and antibacterial activity, *Carbohyd. Polym* 102:385–392.

Li, D., and Y.N. Xia (2004). Electrospinning of nanofibres: Reinventing the wheel? *Adv. Mater.* 16:1151–1170.

Liu, B., Shen, S., Luo, J., Wang, X., and R. Sun (2013a). One-pot green synthesis and antimicrobial activity of exfoliated Ag NP-loaded quaternized chitosan/clay nanocomposites, *RSC Adv.* 3:9714–9722.

Liu, H., You, L., Ye, X., Li, W., and Z. Wu (2008a). Adsorption kinetics of an organic dye by wet hybrid gel monoliths, *J. Sol–Gel Sci. Technol.* 45:279–290.

Liu, Y., Wang, X., Yang, F., and X. Yang (2008b). Excellent antimicrobial properties of mesoporous anatase TiO_2 and Ag/TiO_2 composite films, *Microporous Mesoporous Mater.* 114:431–439.

Liu, Y., Yu, H., Zhan, S., Li, S., Yang H., and B. Liu (2013b). Removal of PCP-Na from aqueous systems using monodispersed pompon-like magnetic nanoparticles as adsorbents, *Water Sci. Technol.* 68:2704–2711.

Lučić, M., Milosavljević, N., Radetić, M., Radoičić, M., and M.K. Krušić (2014). The potential application of TiO_2/hydrogel nanocomposite for removal of various textile azo dyes, *Sep. Purif. Technol.* 122:206–216.

Mahmoodian, H., Moradi, O., Shariatzadeha, B., Asif, M., and V.K. Gupta (2015). Enhanced removal of methyl orange from aqueous solutions by poly HEMA-chitosan-MWCNT nano-composite, *J. Mol. Liq.* 202:189–198.

Mahmoud, M.E., Abou Kana, M.T.H. and A.A. Hendy (2015). Synthesis and implementation of nano-chitosan and its acetophenone derivative for enhanced removal of metals, *Int. J. Biol. Macromol.* 24:725–734.

Manjarrez Nevárez, L., Ballinas Casarrubias, L., Canto, O.S., Celzard, A., Fierro, V., Ibarra Gómez, R., and G. González Sánchez (2011). Biopolymers-based nanocomposites: Membranes from propionated lignin and cellulose for water purification, *Carbohyd. Polym.* 86:732–741.

Min, L.-L., Yuan, Z.-H., Zhong, L.-B., Liu, Q., Wu, R.-X., and Y.-M. Zheng (2015). Preparation of chitosan based electrospun nanofiber membrane and its adsorptive removal of arsenate from aqueous solution, *Chem. Eng. J.* 267:132–141.

Pandi, K., and N. Viswanathan (2015). Synthesis and applications of eco-magnetic nano-hydroxyapatite chitosan composite for enhanced fluoride sorption, *Carbohyd. Polym.* 134:732–739.

Pandiselvi, K., and S. Thambidurai (2015). Synthesis of adsorption cum photocatalytic nature of polyaniline-ZnO/chitosan composite for removal of textile dyes, *Desal Water Treat.* 58:8343–8357.

Pourjavadi, A., Hosseini, S.H., Seidi, F., and R. Soleyman (2013). Magnetic removal of crystal violet from aqueous solutions using polysaccharide-based magnetic nanocomposite hydrogels. *Polym. Int.* 62:1038–1044.

Rabea, E.I., Badawy, M.E.-T., Stevens, C.V., Smagghe, G., and W. Steurbaut (2003). Chitosan as antimicrobial agent: Applications and mode of action. *Biomacromolecules* 4:1457–1465.

Rahimi, S., Moattari, R.M., Rajabi, L., and A.A. Derakhshan (2015). Optimization of lead removal from aqueous solution using goethite/chitosan nanocomposite by response surface methodology, *Colloid Surf. A* 484:216–225.

Ratnamala, G.M., Vidya, K., and G. Srinikethan (2012). Removal of remazal brilliant blue dye from dye-contaminated water by adsorption using red mud: Equilibrium, kinetic and thermodynamic studies, *Soil Pollut.* 223:6187–6199.

Reddy, D.H.K., and S.M. Lee (2013). Application of magnetic chitosan composites for the removal of toxic metal and dyes from aqueous solutions, *Adv. Colloid Interf. Sci.* 201–202:68–93.

Rhim, J.W., Hong, S.I., Park, H.M., and P.K.W. Ng (2006). Preparation and characterization of chitosan-based nanocomposite films with antimicrobial activity, *J. Agric. Food Chem.* 54:5814–5822.

Shameli, K., Ahmad, M.B., Zargar, M., Yanus, W.M., Ibrahim, N.A., Shabanzadeh, P., and M.G. Moghaddam (2011). Synthesis and characterization of silver/montmorillonite/chitosan bionanocomposites by chemical reduction method and their antibacterial activity, *Int. J. Nanomed.* 6:271–284.

Teimouri, A., Ghanavati Nasab, S., Habibollahi, S., Fazel-Najafabadi, M., and A.N. Chermahini (2015). Synthesis and characterization of a chitosan/montmorillonite/ZrO$_2$ nanocomposite and its application as an adsorbent for removal of fluoride, *RSC Adv.* 5:6771–6781.

Tetala, K.K.R., and D.F. Stamatialis (2013). Mixed matrix membranes for efficient adsorption of copper ions from aqueous solutions, *Sep. Purif. Technol.* 104:214–220.

Wang, X., Yang, L., Zhang, J., Wang, C., and Q. Li (2014a). Preparation and characterization of chitosan-poly(vinyl alcohol)/bentonite nanocomposites for adsorption of Hg(II) ions, *Chem. Eng. J.* 251:404–412.

Wang, Z., Ma, H., Hsiao, B.S., and B. Chu (2014b). Nanofibrous ultrafiltration membranes containing cross-linked poly(ethyleneglycol) and cellulose nanofiber composite barrier layer, *Polymer* 55:366–372.

Xiao, G., Su, H., and T. Tan (2015). Synthesis of core-shell bioaffinity chitosan-TiO$_2$ composite and its environmental applications, *J. Hazard. Mater.* 283:888–896.

Zavareh, S., Zarei, M., Darvishi, F., and Azizi, H (2015). As(III) adsorption and antimicrobial properties of Cu-chitosan/alumina nanocomposite, *Chem. Eng. J.* 273:610–621.

Zhou, N.-L., Liu, Y., Li, L., Meng, N., Huang, Y.-X., Zhang, J., Wei, S.-H., and J. Shen, (2007). A new nano-composite biomedical material of polymer/Clay–Cts–Ag nanocomposites. *Curr. Appl. Phys.* 7S1:e58.

Zhou, Y., Fu, S., Zhang, L., Zhan, H., and M.V. Levit (2014). Use of carboxylated cellulose nanofibrils-filled magnetic chitosan hydrogel beads as adsorbents for Pb(II). *Carbohyd. Polym.* 101:75–82.

Zhu, H.Y., Fu, Y.Q., Jiang, R., Yao, J., Xiao, L., and G.M. Zeng (2012). Novel magnetic chitosan/poly(vinyl alcohol) hydrogel beads: Preparation, characterization and application for adsorption of dye from aqueous solution. *Bioresour. Technol.* 105:24–30.

Zhu, X., Bai, R., Wee, K.H., Liu, C., and S.L. Tang (2010). Membrane surfaces immobilized with ionic or reduced silver and their anti-biofouling performances, *J. Membr. Sci.* 363:278–286.

7 Nanomaterial-Based Sorbents for the Removal of Heavy Metal Ions from Water

Vinod Kumar Garg and Navish Kataria

CONTENTS

ABSTRACT

Nanomaterials are often used in different applications such as drug delivery, as ceramic materials, semiconductors, and electronics; in medicine, cosmetics, water purification, and so on. Nanotechnology has also modernized water and wastewater treatment in recent years, as nanomaterials are gaining momentum globally in the field of water and wastewater purification. Previously, nanoparticles, nanomembranes, and nanopowder were used for the removal of heavy metals such as chromium, copper, nickel, cadmium, lead, mercury, and zinc; nutrients such as phosphate, ammonia, nitrate and nitrite, cyanide, and organics; and microbes including algae, bacteria, viruses, and so on. Nanomaterials owe their importance to their extremely small size, high reactivity, high surface-to-volume ratio, and their modification with various functional groups to increase their affinity toward a particular compound. These properties of nanomaterials aid in degrading and scavenging pollutants from water. Nanomaterials of different shapes, sizes, and morphologies can be customized in the laboratory. The method of preparation plays an important role in obtaining shape-/size-controlled, highly stable, and reactive nanomaterials. They can be synthesized using different methods such as coprecipitation, thermal decomposition, and the hydrothermal route, which are widely used and easily scalable with high yields. They have high adsorption capacities and can be designed to target specific contaminants, thus providing local and practical solutions to tackle global water pollution problems. This chapter focuses on different methods of preparation of nanomaterials, their application in the removal of heavy metals, and the mechanisms responsible for their removal.

7.1   INTRODUCTION

Nanotechnology is a multidisciplinary scientific field that is undergoing unprecedented growth at the global level. It involves the application of nanosized materials to produce products and services for the welfare of humankind (Sattler 2010). Nanomaterials usually have a size range of 1–100 nm. Nanometer-sized particles offer novel structural, optical, and electronic properties that are otherwise unattainable with individual molecules or bulk solids (Huang et al. 2011a). Nanomaterials are often used in different applications: in the pharmaceutical industry, as ceramic materials, as semiconductors, in electronics, cosmetics, water purification, and so on.

The research work conducted previously has shown that nanotechnology also has the potential to play a key role in the modernization of water and wastewater treatment; hence, nanomaterials are gaining momentum globally in these fields. A bibliographic survey has shown that nanoparticles, nanomembranes, and nanopowders have been used for the removal of heavy metals such as chromium, copper, nickel, cadmium, lead, mercury, zinc; nutrients like phosphate, ammonia, nitrate and nitrite, cyanide, organics; and microbes including algae, bacteria, viruses, and so on. Nanomaterials owe their importance to their extremely small size, high reactivity, high surface-to-volume ratio, and their modification with various functional groups to increase their affinity toward a particular compound. These properties of nanomaterials help in degrading and removing pollutants from water and wastewater. This chapter focuses on the application of nanomaterials in the removal of heavy metals from water.

7.2   SYNTHESIS OF NANOMATERIALS

Several methods to synthesize nanomaterials have been reported in the literature. These methods are based on two basic approaches: one is "top-down" and the other is "bottom-up." In the "top-down" approach, the bulk material is cut into smaller and smaller sizes and finally attains nanosized particles; whereas the "bottom-up" approach of nanoparticle synthesis involves the condensation and cluster formation of atoms or molecular entities in a gas phase or liquid phase to form the material in nanosize. The latter approach is far more popular and significant in the synthesis of nanoparticles, due to several advantages associated with it. These are higher purity, better particle size and surface chemistry. The general overview of these approaches is shown in Figure 7.1.

Several other physical, chemical, and biological methods can be used for the synthesis of nanoparticles (Figure 7.2). With physical methods, the "top-down" approach is used for nanoparticle preparation; whereas with chemical methods, nanomaterials are synthesized with strict control over their crystalline structure, particle size, and morphology.

Biological methods of nanoparticle synthesis include the use of biological agents such as bacteria, fungi, yeast, actinomycetes, algae, and plants (Quester et al. 2013). Biological synthesis is an economically and environmentally sustainable method than physical and chemical methods.

7.3   CHARACTERIZATION OF NANOMATERIALS

Nanomaterials are characterized based on their size and method of preparation, employing a number of techniques. They are mainly characterized by size, size distribution, shape, structure, surface area, surface charge, porosity, composition, concentration, surface functionalities, surface speciation, stability, crystallinity, magnetic properties, surface texture, and so on. Cong et al. (2011)

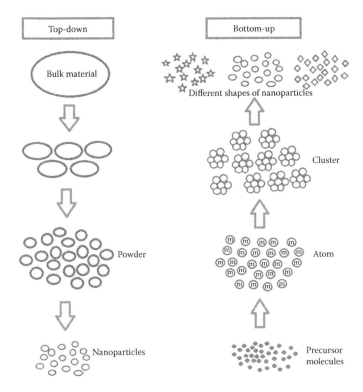

FIGURE 7.1   (See color insert) Schematic presentation of "top-down" and "bottom-up" approaches used for nanomaterial synthesis.

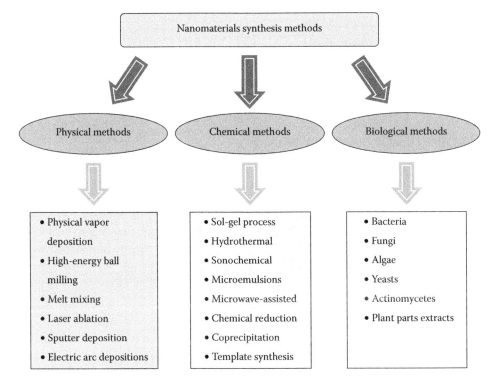

FIGURE 7.2   (See color insert) Flow chart of nanoparticle synthesis methods.

reported the necessity of proper and accurate characterization of nanomaterials to ensure the reliability and reproducibility of the results. If nanomaterials are used without characterization, this may lead to variability and errors in the results. Without characterization, applications of nanomaterials in various fields have limited value, due to the unknown variability of the nanoparticles in experimental conditions (Warheit 2008). Many authors have reported the synthesis and use of nanomaterials for different applications with little or no characterization. In such a situation, the reported results are of little importance and may be misleading. Warheit (2008) has reported that there is a long list of physicochemical characteristics, but it is recommended to prioritize them based on the potential application of nanomaterials. Various techniques used for nanomaterial characterization are as follows:

1. X-ray photoelectron spectroscopy (XPS)
2. High resolution transmission electron microscope (HR-TEM)
3. Scanning transmission electron microscope (STEM)
4. Energy-dispersive x-ray spectroscopy (EDX)
5. Field emission scanning electron microscopy (FESEM)
6. Raman spectroscopy (RS)
7. Attenuated total reflectance infrared (ATR-IR)
8. Brunauer–Emmett–Teller (BET)
9. Thermogravimetric analysis (TGA)
10. Selected area electron diffraction (SAED)
11. Thermogravimetry differential scanning calorimetry (TG-DSC)
12. Electrophoretic mobilities (EPMs)
13. X-ray absorption fine structure (XAFS)
14. Differential gravimetric analysis (DTG)
15. Small angle x-ray diffraction (SAXS)
16. Barrett–Joyner–Halanda (BJH)
17. Atomic force microscopy (AFM)
18. Vibrating sample magnetization (VSM)
19. Pore size distribution (PSD)
20. Physical property measurement system (PPMS)
21. Nuclear magnetic resonance (NMR)
22. Superconducting quantum interference device (SQUID)

7.4 PROPERTIES OF NANOMATERIALS

The fundamental properties of matter change at the nanoscale; hence, the properties of nanoparticles can be significantly different from those of atoms and bulk materials of the same substance. Rao et al. (2005) have mentioned that nanomaterials are the boundary between atoms and molecules and the macroworld, where, ultimately, the properties are dictated by the fundamental behavior of atoms. Wilson et al. (2008) have reported that unique properties of nanomaterials include (a) a large fraction of surface atoms; (b) high surface energy; (c) spatial confinement; (d) reduced imperfections; and (e) a great level of porosity, which make them special candidates for many novel applications in almost every field of technology. A brief overview of different properties of nanomaterials follows:

7.4.1 OPTICAL PROPERTIES

The optical properties of nanomaterials depend on size, shape, surface characteristics, and other variables, including doping and interaction with the surrounding environment or other nanostructures. Based on their optical properties, nanomaterials are employed in optical detectors, lasers, sensors, imaging, phosphors, display, solar cells, photocatalysis, photoelectrochemistry, and bio-medicine (Lue 2007).

7.4.2 ELECTRICAL PROPERTIES

The electrical properties of nanomaterials depend on their diameter and vary between metallic and semiconducting materials, mainly due to the electrical conductivity/photoconductivity in nanotubes, nanorods, and nanocomposites. Based on their electrical properties, nanomaterials have increased the capabilities of electronic devices and reduced power consumption.

7.4.3 MECHANICAL PROPERTIES

Mechanically, nanomaterials are very strong and can withstand acute strain, as they have minimum defects in their structure. Guo et al. (2014a) have reviewed the special mechanical properties of nanoparticles that allow their novel applications in surface engineering, tribology, and nanomanu-facturing. Some of the main applications of nanoparticles as a result of their mechanical properties include lubricant additives, nanoparticle-reinforced composite coatings, and so on.

7.4.4 MAGNETIC PROPERTIES

Very small particles have special atomic structures with discrete electronic states, which give rise to special properties in addition to superparamagnetism behavior. The magnetic properties of nano-materials are useful in imaging, bioprocessing, and refrigeration, as well as high storage-density magnetic memory media. Magnetic nanocomposites have been used for mechanical force transfer (ferrofluids), high-density information storage, and magnetic refrigeration. For example, bulk gold and platinum are nonmagnetic, but at nanosize they are magnetic.

7.5 ENVIRONMENTAL APPLICATIONS OF NANOMATERIALS

Nanomaterials have significant applications in engineering, pharmaceutics, cosmetics, food and agriculture, electronics, precision mechanics, optics, energy production, and environmental sci-ences (Gross 2001; Kim et al. 2005; Moore 2006). Nanomaterials also have wide scope in remediat-ing environmental problems. Remediation is the process of transforming pollutants from toxic to less toxic in air, water, and soil. A wide variety of nanomaterials have been developed for the moni-toring and remediation of pollution, including silver nanoparticles, carbon nanotubes with enzymes, as sensors, as building blocks, and magnetic nanoparticles coated with antibodies (Pandey and Fuelekar 2012). The application of nanomaterials to heavy metal removal from water and wastewa-ter is given in the following sections.

7.5.1 HEAVY METALS IN WATER AND WASTEWATER

Haphazard industrialization, mining, urbanization, and so on lead to the release of a wide range of contaminants in surface water and groundwater, which may create severe health problems and are considered to be a serious environmental concern in the scientific community in particular and in society in general. The predominant contaminants in water and wastewater mainly include heavy metals, inorganic compounds, organic pollutants, and other complex compounds (O'Connor 1996; Fatta et al. 2011; Li et al. 2011; Cundy et al. 2008; Chong et al. 2010; Zeng et al. 2011). Among these, heavy metals released into water bodies are a serious health issue, because high concentrations of heavy metals in water can damage human health and the environment.

Metals having a density ≥ 5.0 g cm^{-3} are called *heavy metals* (Lozet and Mathieu 1991; Morris 1992). Unlike some organic pollutants, toxic heavy metals such as Cu, Co, Ni, Pb, Cd, Hg, As(V),

Cr(VI) Cu, Pb, and Cd are nonbiodegradable, and can accumulate in living organisms. The maximum permissible limits of heavy metals in drinking water, prescribed by different agencies, are given in Table 7.1. Exposure to these heavy metals is toxic, hazardous to the health of fauna, flora, the aquatic ecosystem, and human beings. These heavy metals are easily soluble, and persist in the environment for long periods. They are carcinogenic, mutagenic, and teratogenic to living systems even at low concentrations. These cause serious effects on health, leading to many fatal diseases. The effects of heavy metals on health are given in Table 7.2.

Disposal of heavy metal–laden wastewater also affects the environmental matrices (water, soil, and air). Because of this, heavy metals should be removed from wastewater before their discharge into the environment (Cundy et al. 2008; Chong et al. 2010; Zeng et al. 2011). Several technologies, such as membrane filtration, reverse osmosis, advance oxidation, photocatalysis, reduction, ion exchange, adsorption, flocculation, and precipitation have been investigated for the removal of

TABLE 7.1
Guidelines for Maximum Permissible Limits of Heavy Metals in Drinking Water

Heavy Metal	Maximum Permissible Limits (mg L^{-1})			
	BIS	ICMR	WHO	USEPA
Arsenic (As)	0.05	0.05	0.01	0.05
Cadmium (Cd)	0.003	0.003	0.003	0.005
Chromium (Cr)	0.05	—	0.05	0.1
Copper (Cu)	1.5	1.5	2.0	1.3
Iron (Fe)	0.3	1.0	0.3	0.3
Lead (Pb)	0.01	0.05	0.01	0.15
Manganese (Mn)	0.3	0.5	0.5	.05
Mercury (Hg)	0.001	0.001	0.001	0.002
Nickel (Ni)	0.02	—	0.07	—
Selenium (Se)	0.01	—	0.01	0.05
Zinc (Zn)	15.0	5.0	5.0	5.0

Note: BIS: Bureau of Indian Standards; ICMR: Indian Council of Medical Research; WHO: World Health Organization; USEPA: US Environmental Protection Agency.

TABLE 7.2
Health Effect and Sources of Heavy Metals

Heavy Metals	Diseases/Health Effects	Source(s)
Mercury	Minimata, circulatory system, nervous system	Industrial waste, mining, burning of coal, paper industry
Cadmium	Itai-itai, kidney damage, hypertension	Metal plating, mining waste, welding, pesticides, fertilizers, Cd and Ni battery, nuclear fission plant
Arsenic	Blackfoot disease, cancer, abdominal pain	Chemical waste, pesticides, fungicides, metal smelters
Lead	Plumbusis, joint disease, ailments of the kidney	Plumbing, gasoline, water pipes, paint, pesticides, smoking, burning of coal, paper industry
Chromium	Lung damage and cancer, lung and respiratory tract cancer, as well as kidney diseases	Metal plating, cooling tower water additive (chromate) normally found as Cr (VI) in polluted water, mining activity

heavy metals from water and wastewater. But, these technologies have some disadvantages, such as high maintenance costs, incomplete metal removal, costly equipment, high energy and chemical requirements, and the generation of toxic sludge. Disposal and discharge of toxic sludge is another major problem for the scientific community and users of these technologies. In recent years, various advanced technologies and methods have been explored to provide a better aquatic environment for future generations in a sustainable manner. The use of nanomaterials for water treatment is among these advanced technologies. Their use offers vast application for water and wastewater treatment and overcomes major challenges faced by existing treatment technologies. It also provides new treatment capabilities that could allow economic utilization of unconventional water sources to expand the water supply (Qu et al. 2013).

7.5.2   WHY NANOMATERIALS AS SORBENTS FOR HEAVY METAL REMOVAL?

With the development of nanotechnology, nanomaterials have been extensively used as adsorbents in wastewater treatment. Adsorption is one of the widely used processes that play an important role in wastewater treatment, based on the physical interaction between metal ions and adsorbents. It can be inferred from the available literature that nanomaterials are excellent adsorbents for the removal of heavy metal ions from wastewater, due to their unique structural features. The important characteristics of nanomaterials that make them ideal adsorbents are

- High catalytic potential and high porosity
- Large surface area and small size
- High reactivity and large number of active sites for interaction with different heavy metals
- Sequester heavy metals with varying molecular size, hydrophobicity, and speciation behavior
- Work rapidly
- Good metal binding capacity
- Ease of separation
- Can be regenerated for reuse

These features are responsible for the better adsorption capacity of nanomaterials and may be the reason behind growing scientific interest in nanomaterials for heavy metal removal from aqueous systems (Sharma et al. 2011). The diverse range of nanoparticles commonly used for the treatment of wastewater are made of alumina, anatase, akaganeite, cadmium sulfide, cobalt ferrite, copper oxide, gold, maghemite, iron, iron oxide, iron hydroxide, nickel oxide, silica, stannous oxide, titanium oxide, titanium oxide, zinc sulfide, zinc oxide, zirconia, and some alloys.

Scientists are also in the process of developing nanostructured filters that can remove virus cells from water, and the deionization method using nanosized fiber electrodes. Gold and silver nanoparticles utilized in water purification are used to remove low concentrations of halogenated compounds such as pesticides and heavy metals, with several products being commercially available (Pradeep and Anshup 2009). Zodrow et al. (2009) have reported that nanofiltration is more efficient than traditional micron-scale filtration, and greatly improved the removal of foreign substances from water. Aal et al. (2009) have reported the use of ZnO nanofibers on thin films for photocatalytic degradation of organic chemicals such as trichlorophenol.

7.5.3   NANOADSORBENTS FOR HEAVY METAL REMOVAL

A variety of nanomaterials, such as carbon nanotubes, carbon-based material composites, graphene, nanometal or metal oxides, and polymeric sorbents, have been previously studied for the removal of heavy metal ions from aqueous solutions, and the results indicate that these nanomaterials show high adsorption capacity. A brief review of these studies is given in this section.

7.5.3.1 Carbon-Based Nanosorbents

Carbon-based nanosorbents have been widely used in wastewater treatment processes for heavy metal adsorption due to their high sorption capacities, small size, and large surface area (Li et al. 2013a). Although activated carbon is a commercially used adsorbent for the removal of heavy metals from wastewater, it is difficult to remove heavy metals at ppb levels. In such a scenario, nanomaterials can play an effective role. A large number of carbon-based nanomaterials, including carbon nanotubes, fullerenes, graphene, nanodiamonds, graphene oxides, carbon nanocones/disks, nanofibers, and nanohorns, as well as their functionalized forms, have been reported as effective nanoadsorbents for the removal of heavy metals from wastewater (Perez-Aguilar et al. 2011).

7.5.3.1.1 Carbon Nanotubes

Carbon nanotubes (CNTs) have been used by many authors as adsorbents to remove various environmental pollutants, because of their unique one-dimensional structure, large surface area, and amount of micropores. However, the adsorption capacity of unmodified CNTs toward heavy metal ions is still limited, due to their poor dispersibility and lack of functional groups. So, authors are modifying CNTs to improve their metal removal efficiency (Xie et al. 2015). Anitha et al. (2015) reported the removal of heavy metal ions, namely, Cd^{2+}, Cu^{2+}, Pb^{2+}, and Hg^{2+}, from aqueous media using functionalized single-walled carbon nanotubes (SWCNT) with carboxyl, hydroxyl, and amide functional groups. The results showed that the adsorption capacity of CNTs was improved significantly after surface modification with functional groups. The adsorption capacity was 150%–230% more with $CNT\text{-}COO^-$ than with pure CNTs. Li et al. (2015) reported the synthesis of a new carbon nanotube composite—dithiocarbamate groups functionalized multiwalled CNT (DTC-MWCNT)—by the reaction of oxidized MWCNT with ethylene diamine and carbon disulfide, and used it for the removal of heavy metal ions from aqueous solutions. DTC-MWCNT was found to have appreciable adsorption capacities for Cd(II), Cu(II), and Zn(II), of 167.2, 98.1, and 11.2 mg g^{-1}, respectively. Xie et al. (2015) successfully prepared polyethylenimine-functionalized CNTs via the combination of mussel-inspired chemistry and the Michael addition reaction. The Langmuir adsorption capacity of this CNT was found to be 70.9 mg g^{-1} for Cu(II). The results showed that the adsorption capacity of amine-functionalized CNTs was significantly enhanced toward Cu(II). Yaghmaeian et al. (2015) reported the removal of inorganic mercury from aquatic environments by MWCNTs. The Langmuir adsorption capacity of MWCNTs was found to be 25.6 mg g^{-1} for mercury. The original CNTs and oxygen containing surface functionalized CNT-O have been tested for the adsorption of anionic chromate. The maximum adsorption capacity of CNT-O for CrO_4^{2-} was found to be 249 μmol g^{-1}, whereas the original CNTs had a lesser adsorption capacity (16 μmol g^{-1}) as compared with CNT-O. The results showed that the adsorption capacity of CNT-O for anionic chromate is 15 times higher than that of CNTs. Almost similar results can be predicted on the basis of surface area of the adsorbents. The estimated maximum adsorption capacity per unit surface area on CNT-O is 0.83 μmol m^{-2}, which is higher than the 0.06 μmol m^{-2} predicted for CNTs (Xu et al. 2011). Nanoporous carbon (NPC), with a unique surface morphology, exhibited superior sorption of heavy metals. It demonstrated a high sorption capacity for all the heavy metals including zinc, lead (2000 μeq g^{-1}), nickel and cadmium (1428.6 μeq g^{-1}). (Ruparelia et al. 2008). Studies of various authors are summarized in Table 7.3.

7.5.3.1.2 Graphenes

Graphene is a single atomic layer of sp^2–hybridized bonded carbon atoms, arranged in two-dimensional honeycomb networks (Ivanovskii 2012). A number of graphene-based nanomaterials, including graphene oxide (GO), reduced GO, multilayer GO, and their modified/functionalized GOs have been developed since their discovery in 2004. Graphenes have shown great potential in various technologies, namely, semiconductors, energy storage materials, quantum dots, electronics, environmental remediation, solar cells, and sensors. The large theoretical specific surface area of graphene (2630 m^2 g^{-1}) makes them excellent nanosorbents for metal ions, and a number of studies

TABLE 7.3

Carbon-Based Nanomaterials for Heavy Metals Removal from Water

Nanosorbents	Methods	Characterization	Heavy Metals	Removal Process	Efficiency (%)	Other Condition	Adsorption Capacity ($mg\ g^{-1}$)	Surface Area ($m^2\ g^{-1}$)	References
Carbon nanotubes/$CoFe_2O_4$ magnetic hybrid material	Polyol method	TEM, FTIR, XPS VSM, BET, BJH	Pb (II)	Adsorption	91%	pH-6.0, Dose-0.5 g L^{-1}, Temp. 30°C	140	157.5	Zhou et al. 2014
Graphene-carbon nanotube-iron oxide 3-D nanostructures	Microwaves	XPS, HR-TEM, SEM, EDX, Raman spectra	As(V), As(III)	Adsorption	—	—	—	63	Vadahanambi et al. 2013
SWCNT–graphene hybrid	Vacuum-assisted filtration	FE-SEM, FTIR, UV-VIS, Raman spectra	Cu(II)	Adsorption	—	pH 6.8, Temp. 20°C, Time 3 h	250	—	Dichiara et al. 2015
Reduced graphite oxide–Fe_3O_4 NP composites	Hummer's method	XRD, FTIR, Raman spectra, XPS	Hg(II), CH_3Hg^+, $C_2H_5Hg^+$	Adsorption	>99%	Time 30 min, Temp. 20°C, Conc. 10 µM	47.6	—	Shih et al. 2014
Magnetic graphene nanocomposites (MGNCs)	Thermodecomposition method	HR-TEM, XRD, TGA, SAED, BET	Cr(VI)	Sonication	100%	pH 7, Dose 3 g L^{-1}, Time 5 min Conc. 1000 µg L^{-1}	—	42.1	Zhu et al. 2012
Graphene magnetic material (Fe_3O_4-GS)	Hydrothermal method	SEM, TEM, FTIR, XRD, XPS, BET	Cr(VI), Pb(II), Hg(II), Cd(II), Ni(II)	Adsorption	—	—	17.29, 27.95, 23.03, 27.83, 22.07	62.43	Guo et al. 2014b
Cysteine-functionalized graphene oxide (GO)	Modified Hummer's method	UV–VIS, FTIR, TEM, XPS	Pb(II)	Electrochemical deposition	99.2%	pH 5.0, Conc. 14 ppm, Time 10 min	—	—	Seenivasan et al. 2015
Sulfur/reduced graphene oxide nanohybrid (SRGO)	Ultrasonication	XRD, UV–VIS, SEM, TEM, TGA	Hg(II)	Adsorption	90%	pH 6–8, Time 30 min	—	—	Thakur et al. 2013
HA-coated MWCNTs (25% HAeCNTs and 50% HAeCNTs)	Precipitates	SEM-EDS, BET, XAFS, FTIR, TG-DSC, EPMs	Pb(II)	Adsorption	—	Temp. 25°C, Time 24 h	318, 333	199, 189	Lin et al. 2012
Surface-activated carbon nanospheres	Hydrothermal	HR-TEM, TEM, XRD, FTIR, TGA, SAED, BET	Ag(I)	Adsorption	>99.9	Time 6 min Conc. 7 ppm, Temp. 25°C	152	<15	Song et al. 2011
Thiol-derivatized single-walled carbon nanotube (SWCNT-SH)	Ultrasonication	FTIR, TGA, Raman spectra	Hg(II)	Adsorption	99.8%	Dose 0.25 g L^{-1}, Conc. 30 mg L^{-1}, Time 1 h, Temp. 25°C	131	—	Bandaru et al. 2013

(Continued)

TABLE 7.3 (CONTINUED)
Carbon-Based Nanomaterials for Heavy Metals Removal from Water

Nanosorbents	Methods	Characterization	Heavy Metals	Removal Process	Efficiency (%)	Other Condition	Adsorption Capacity (mg g⁻¹)	Surface Area (m² g⁻¹)	References
NN-mSiO2@ MWCNTs	Hydrothermal	TEM, XRD, FTIR	Cu(II)	Adsorption	—	Time 30 min, Temp. 25°C	66.577	—	Yang et al. 2013
Thiol-functionalized CNTs/Fe₃O₄	Hydrothermal	TEM, XRD, FTIR, TGA, EDX. BET	Hg(II), Pb(II)	Adsorption	—	pH 6.5, Temp. 25°C, Conc. 50 mg L⁻¹	65.52, 65.40	97.163	Zhang et al. 2012
Carbon nanotubes (CNTs)	UV-light oxidation and ultrasonication method	SEM, XRD, FTIR, DTG	Pb(II)	Adsorption	—	pH 5.0	UV-light 511.99, ultrasonication 342.36	—	Bayazit and Inci 2013
Single-walled carbon nanotube-doped walnut shell composite (SWCNTs/WSh)	CVD method	TEM, FTIR, SEM, BET	Pb(II)	Adsorption	99%	pH 5.0, Time 30 min, Conc. 50 mg L⁻¹	294.1	738	Saadat et al. 2014
Montmorillonite-supported carbon nanosphere adsorbent (Mt-spC)	Hydrothermal	SEM, TEM, XRD, FTIR, TGA, SAXS. BET, BJH	Cr(VI)	Adsorption	80%	pH 2.0, Time 60 min, Conc. 80 mg L⁻¹	156.25	34.41	Li et al. 2014
Graphite oxide nanofilled with chitosan (GO/CS) and magnetic chitosan (GO/mCS)	Ultrasonication	SEM/EDAX, FTIR, XRD, DTG	Hg(II)	Adsorption	96%	pH 6, Time 150 min, Temp. 25°C, Conc. 100 mg L⁻¹	GO/CS-381, GO/mCS-397	—	Kyzas et al. 2014
Graphene oxide/Fe₃O₄ (GO/Fe₃O₄) and PAA/GO/Fe₃O₄	Ultrasonication	SEM, TEM, XRD FTIR, TGA, AFM, VSM, Raman spectra	Cu(II), Cd(II), Pb(II)	Sonication adsorption	55%, 70%, 45%	pH 3, Time 30 min, Conc. 10 mg L⁻¹		—	Zhang et al. 2013
Fe-grown carbon nanofibers containing porous carbon microbeads	Chemical vapor deposition	SEM, EDX, BET, PSD	Cr(VI)	Adsorption	Batch 86%, Column 90%	Conc. 10–150 mg L⁻¹, Temp. 30°C	41	425	Talreja et al. 2014
Magnetic carbonaceous nanoparticles (MNPs@C)	Hydrothermal	XRD, TEM, BET, zeta potential analysis	Pb(II)	Adsorption	—	pH 6.0, Time 3 days	123	60	Nata et al. 2010
N-doped porous carbon with magnetic particles (RHC-mag-CN)	Milling and ultrasonication	SEM, TEM, XRD, FTIR, XPS, EDS, TGA, SAED, zeta potential, Raman spectra	Cr(VI)	Chemical adsorption	92%	Time 10 min, Dose 2.0 g L⁻¹	16	1136	Li et al. 2013b
Carbon-encapsulated magnetic nanoparticles (CEMNPs)	Carbon arc plasma route	XRD, TEM, Raman spectra	Cd(II), Co(II), Cu(II)	Adsorption	95%, 80%, 95%	pH 9.0, Conc. 1–20 mg L⁻¹, Time 4 h	1.77, 1.23, 3.21	—	Bystrzejewski et al. 2009

are available that have inferred their potential application in heavy metal removal from aqueous systems (Zhu et al. 2010). Graphene and its modified products also meet the conditions required to be a good adsorbent, hence gaining much attention in heavy metal removal. Even highly toxic ions, such as arsenate and arsenite, can also be removed from seawater by graphenes with good adsorption capacities of 142 and 122 mg g^{-1}, respectively (Mishra and Ramaprabhu 2011). The adsorption capacities of Pb(II) ions onto few-layered GO nanosheets (GONSs) were approximately 842, 1150, and 1850 mg g^{-1} at 20°C, 40°C, and 60°C, respectively (Zhao et al. 2011). Gopalakrishnan et al. (2015) reported the removal of heavy metal ions from the use of GO nanosorbents. The results showed that GO nanosorbents are highly effective in removing heavy metals from effluent at low electrical conductivity. Wang et al. (2015) reported that GOs may have strong applications in heavy metal removal from aqueous media, due to the availability of large amounts of oxygen-containing functional groups on their surfaces to form strong surface complexes with metal ions. Based on this hypothesis, they studied the sorption of Pb(II) on GOs under different experimental conditions. The results showed that the maximum adsorption capacity of GOs for Pb(II) was 2.27 mmol g^{-1}. The binding of Pb(II) on GOs has been attributed mainly to the COOH groups, which was further established by far-infrared spectroscopy analysis and density functional theory calculations.

Cui et al. (2015) reported the synthesis of ethylene diamine tetraacetic acid (EDTA) functionalized magnetic graphene oxide (EDTA-mGO) and its use in heavy metals Pb(II), Hg(II), and Cu(II) removal from aqueous media. The results showed that the adsorption performance of EDTA-mGO was better than that of EDTA, Fe_3O_4, GO, EDTA-Fe_3O_4, mGO, and EDTA-GO. The adsorption capacities of EDTA-mGO for Pb(II), Hg(II), and Cu(II) were found to be 508.4, 268.4, and 301.2 mg g^{-1}, respectively. Authors compared their results with available data and found that adsorption of heavy metals onto EDTA-mGO is several times higher than other adsorbents.

GO sheets tend to aggregate after adsorption, due to the decreased electrostatic repulsion between oxygen-containing groups, and their separation after adsorption is very difficult. To overcome these issues, Tan et al. (2015) reported the preparation of a GO membrane and its application in Cu(II), Cd(II), and Ni(II) removal from aqueous media. The GO membrane had strong mechanical properties and production was easily achieved. The membrane was prepared from purified natural graphite using the modified Hummer's method. The maximum adsorption capacities for Cu(II), Cd(II), and Ni(II) were found to be 72.6, 83.8, and 62.3 mg g^{-1}, respectively. A comparison of the data with the available literature showed that the adsorption capacity of the GO membrane was significantly higher than other recently studied adsorbents. Ren et al. (2012) studied the adsorption of Cu(II) and Pb(II) on a graphite nanosheet (GNS)–MnO_2. The maximum adsorption capacities for Cu(II) and Pb(II) were 1637.9 and 793.65 mmol g^{-1}, respectively.

Some other carbon-based nanosorbents such as nanomembranes, nanoparticles, and nanosheets were synthesized by various chemical and physical methods, and were characterized by different techniques, and their role in toxic metal removal from water was reported in the literature, as shown in Table 7.3.

7.5.3.2 Polymer-Based Nanosorbents

In the past few years, polymer-based adsorbents have emerged as an alternative to traditional sorbents/biosorbents for removal of metals from water. The development of polymer-based hybrid nanosorbents has opened up their new prospective role in the removal of metal ions from aqueous systems (Deosarkar and Pangarkar 2004; Yang et al. 2003). These nanosorbents exhibit large surface areas and perfect backbone strength, and their basic physicochemical properties such as internal surface area, high dispersion, attached functional groups, pore size, and volume distribution are modifiable by altering the polymerization conditions and processes (Kunin 1977; Pan et al. 2009).

Fe_3O_4 nanoparticles coated with the polymer polyethylenimine (PEI) were intercalated between sodium-rich montmorillonite (MMT) layers under an optimized pH of 2.0 as an adsorbent for the removal of Cr(VI) from an aqueous solution (Larraza et al. 2012). Chitosan-g-poly (acrylic acid)/attapulgite (CTS-g-PAA/APT) composites were prepared and used for the removal of Cu(II) and Hg(II) metal ions from an aqueous system. Removal experiments were carried out under optimization conditions (pH = 5.5, contact time = 15 min). The composites removed more than 90% of Cu(II) from the water. FTIR spectra of the CTS–g–PAA/APT compound before and after the adsorption of Cu(II) showed that the $-NH_2$ and $-OH$ groups of chitosan and the $-COOH$ groups of PAA in the composites played a significant role in the removal of Cu(II) ions (Wang et al. 2009). The maximum adsorption capacity for Hg(II) was found to be 785.2 mg g^{-1} (Wang and Wang 2010). Zhang et al. (2011) synthesized the hybrid natural polymer containing compounds for the adsorption of Pb(II) from solution. The adsorption capacity of the hybrid polymer for Pb(II) was in the range of 702.35–843.86 mg g^{-1}.

Some researchers synthesized polymer-based nanocomposites using natural (e.g., chitosan) as well as synthetic polymers (e.g., PEG, PEE, PVP, PVA). Various studies related to the preparation of these adsorbents and their applications in heavy metal removal from aqueous systems are given in Table 7.4.

7.5.3.3 Metal-Based Nanosorbents

The preparation and use of metal-based nanosorbents such as metal or metal oxides, bimetallics, ferrites, doped-metal oxides, functionalized metal oxides, and other inorganic nanomaterials for the removal of heavy metal ions from water/wastewater has been widely reported. Nanosized metals or metal oxides include nanosized ferric oxides (Feng et al. 2012), zinc oxide (Kikuchi et al. 2006), silver nanoparticles (Fabrega et al. 2011), manganese oxides (Gupta et al. 2011), titanium oxides (Luo et al. 2010), copper oxides (Goswami et al. 2012), and so on; all these materials provide a large surface area, small size, high regeneration efficiency, and a specific affinity for heavy metal adsorption from water/wastewater.

Hu et al. (2007) studied the removal of Cr(VI) by different magnetic nanoparticles prepared using the coprecipitation method. It has been reported that the equilibrium time for Cr(VI) removal by different magnetic nanoparticles ranged from 5 to 60 min. The size of these nanoparticles was ≈20 nm. It has also been reported that among different magnetic nanoparticles—$CoFe_2O_4$, $ZnFe_2O_4$, $MnFe_2O_4$, $CuFe_2O_4$, $NiFe_2O_4$, and $MgFe_2O_4$—the magnetic nanoparticles of $MnFe_2O_4$ had the highest adsorption efficiency (99.5%), and maximum adsorption of Cr(VI) was obtained at pH 2.0. Lazaridis et al. (2005) reported that the adsorption capacity of nanocrystalline akaganeite for Cr(VI) removal from aqueous systems is 80 mg g^{-1} at pH 5.5.

Mahdavi et al. (2015a) reported the preparation of ZnO nanoparticles (NPs) modified with humic acid (Zn-H), extractant of walnut shell (Zn-W) and 1,5-diphenyl-carbazon (Zn-C). The mean diameter of these NPs was in the range 52–76 nm. The metal removal efficiency of the NPs in solutions was in the order: $Cu^{2+} > Cd^{2+} > Ni^{2+}$. The adsorption efficiencies of Zn-H, Zn-W, and Zn-C for Cd^{2+} were found to be 50.0, 27.8, and 28.5 mg g^{-1}, respectively. Mahdavi et al. (2015b) also reported the removal of Cd^{2+}, Cu^{2+}, Ni^{2+}, and Pb^{2+} from aqueous solutions using Fe_3O_4, ZnO, and CuO nanoparticles. These nanoparticles had mean diameters of about 50 nm (spheroid), 25 nm (rod shape), and 75 nm (spheroid), respectively. The maximum uptake values (the sum of four metals) in multimetal solutions were 360.6, 114.5, and 73.0 mg g^{-1} for ZnO, CuO, and Fe_3O_4, respectively. Based on the average metal removal by the three nanoparticles, the following order was determined for single component solutions: $Cd^{2+} > Pb^{2+} > Cu^{2+} > Ni^{2+}$; while the following order was determined for multiple component solutions: $Pb^{2+} > Cu^{2+} > Cd^{2+} > Ni^{2+}$. Finally, ZnO nanoparticles were identified as the most promising sorbent due to their high metal uptake efficiency. Joodaki et al. (2015) investigated the removal of Cu(II) from aqueous solution by activated carbon cloth and carbon cloth loaded with nanostructured zinc oxide. The microwave-assisted chemical bath deposition method was used to deposit ZnO nanoparticles onto activated carbon cloth. SEM images showed that the structure of the deposited ZnO was flowerlike with a size of 2–5 μm. The maximum adsorption capacity of the ZnO-loaded carbon cloth was 769 mg g^{-1}, which was much higher than activated carbon cloth. Tu et al. (2013) reported a novel technique for manufacturing Mn-Zn ferrite nanoparticles by acid dissolution and ferrite processes. The powders of waste dry batteries (PWDBs) were used as starting raw materials, because Mn and Zn content inside the PWDBs is potentially high. At 0.005 g

TABLE 7.4

Polymer-Based Nanomaterials for Heavy-Metal Removal from Water

Nanosorbents	Methods	Characterization	Heavy Metals	Removal Process	Efficiency	Other Conditions	Adsorption Capacity (mg g⁻¹)	Surface Area (m² g⁻¹)	References
Polymer-supported nanoiron oxides	Precipitated	XRD, TEM, XPS	Cu(II)	Adsorption	80%	pH 5.5	3.4	194.2	Qiu et al. 2012
Poly (oligo (ethylene glycol) methacrylate) (POEGMA) brushes	Chemical	AFM, FESEM, XPS, DLS, UV-Vis TEM	Pb(II)	Sensing	—	—	—	—	Ferhan et al. 2013
Poly (ethylene-covinyl alcohol) nanofibers	Melt mixing	SEM, XPS, FTIR, XPS, UV-Vis	Cr(VI)	Adsorption	86.6%	pH 2, Time 120 min. Temp. 25°C. Conc. 100 mg L⁻¹	90.74	—	Xu et al. 2015
Polyaniline-polystyrene (PANI–PS) composite	Chemical	SEM, UV-Vis, FTIR	Pb(II), Cu(II), Hg(II), Cd(II), Cr(VI)	Adsorption	—	pH 4.0, Time12 h	312, 171, 148, 124, 58	—	Alcaraz-Espinoza et al. 2015
Plasma polymer-functionalized silica particles	Chemical	XPS, FESEM, zeta potential	Cu(II)	Adsorption	96.7%	pH 5.5, Time 60 min	25.0	—	Akhavan et al. 2015
Polyaniline-coated ethyl cellulose	Sonication	SEM, XPS, FTIR, TGA	Cr(VI)	Adsorption	100%	pH 1.0, Time 30 min, Dose 5 g L⁻¹	38.76	—	Qiu et al. 2014
PEI-coated Fe₃O₄ NPs	Precipitation	XPS, DLS, TEM	Cu(II)	Sonication	100%	pH 5.5, Time 60 min	—	—	Goon et al. 2010
Poly-sulfoamino anthraquinone nanosorbents	Chemical oxidative polymerization	IR, UV-Vis, XRD, FESEM, AFM, DSC, TG, DTG	Pb(II), Hg(II)	Adsorption	99.6% 99.8%	Temp. 30°C, Time 60 min, Conc. 200 mg L⁻¹, Dose 2.0 g L⁻¹	10,350, 14,140	115.15	Huang et al. 2011b
Poly(vinyl alcohol)/poly(4-vinylpyridine) (PVA/P4VP)	Chemical	FTIR	Hg(II)	Membrane filtration	100%	pH 2.5	450	—	Bessbousse et al. 2009
Poly(MMA-co MA)/APTMS-Fe3O4	Chemical	FTIR, XRD, FESEM, TEM, AFM	Co(II), Cr(III), Zn(II), Cd(II)	Adsorption	47.35%, 67.45%, 76.65%, 81.75%	pH 6.0, Time 60 min, Conc. 20 mg L⁻¹, Temp. 25°C	90.09, 90.91, 109.89, 111.11	—	Masoumi et al. 2014
Bromomethylated poly(2,6-dimethyl-1,4-phenylene oxide)	Chemical	FESEM, SEM, EDX	Cu(II)	Membrane filtration	—	pH 4.75, Conc. 10 ppm	69.12	—	Cheng et al. 2010a

(Continued)

TABLE 7.4 (CONTINUED)
Polymer-Based Nanomaterials for Heavy-Metal Removal from Water

Nanosorbents	Methods	Characterization	Heavy Metals	Removal Process	Efficiency	Other Conditions	Adsorption Capacity (mg g⁻¹)	Surface Area (m² g⁻¹)	References
Thiol-functionalized nanosilica PES (MNS-PES-1000) membranes	Ultrasonication	FESEM, Raman spectra, EDS, SEM, EDS, TGA, FTIR	Hg(II), Pb(II)	Membrane separation	73.2%, 71.2%	Time 24 h, Area 2 cm², Conc. 500 ppm	1725 mg m⁻², 1328 mg m⁻²	—	Rezvani-Boroujeni et al. 2015
Resin-supported polyethyleneimine nanoclusters	Chemical	FTIR, TEM, BJH, SEM	Cu(II)	Adsorption	—	Conc. 100 mg L⁻¹, Temp. 25°C, Dose 1 g L⁻¹	99	17.08	Chen et al. 2010
MNPs-polyAEMA •DTC	Chemical	FTIR, XPS, TGA, TEM, PPMS	Hg(II)	Adsorption	100%	Conc. 10 ppm, Temp. 40°C, Time 12 h	59.45	—	Farrukh et al. 2013
CM-β-CD polymer-coated magnetic nanoparticles (CDpoly-MNPs)	Coprecipitation	FTIR, XPS, FETEM, Zeta potential	Pb(II), Cd(II), Ni(II)	Adsorption	99.5%, 55.9%, 24.3%	pH 5.5–6, Time 45 min, Temp. 25°C, Conc. 300 ppm	64.5, 27.7, 13.2	—	Badruddoza et al. 2013
Dithiocarbamate-incorporated PS microspheres	Chemical	SEM	Cd(II), Cu(II), Pb(II), Hg(II)	Adsorption	—	pH 7.0, Conc. 30 ppm, Temp. 25°C	1.92, 3.76, 6.17, 22.92	—	Denizli et al. 2000
Fe₃O₄@PANI-AmAzoTCA[4]	Coprecipitation	NMR, FTIR, XRD, SEM, TEM, AFM VSM	Cu(II), Cd(II), Co(II), Cr(III)	Adsorption	99.5%, 99%, 98%, 97%	pH 8. Time 60 min, Conc. 100 ppm, Dose 1 g L⁻¹, Temp. 30°C	312.5, 285.7, 312.5, 303.0	—	Norouzian et al. 2015
Polydopamine polymer decorated with magnetic nanoparticles (Fe₃O₄/PDA)	Hydrothermal	HR-TEM, XPS, FTIR, TGA	Cu(I), Ag(I), Hg(II)	Adsorption	—	Dose 0.1 g L⁻¹, Temp. 30°C, Time 4 h	112.9, 259.1, 467.3	—	Zhang et al. 2014b
Polydopamine nanospheres (PDA)	Template synthesis	XPS and TEM	Hg(II)	Adsorption	100% at pH 1.0	pH 6, Time 4 h	1861 at 25°C, 2037 at 40°C, 2076 at 55°C	—	Zhang et al. 2014a
Poly(ethyleneimine) in poly(vinyl alcohol) matrix	Solvent evaporation	—	Pb(II), Cu(II), Cd(II)	Membrane separation	96–80%, 99%, 99.5%	pH 5, Conc. 100 mg L⁻¹, Temp. 21°C	123, 30, 37	—	Bessbousse et al. 2008
Polyaniline/humic acid nanocomposite	Chemical oxidation	BET, FTIR, TEM, XPS	Hg(II)	Adsorption	95%	Dose 0.5 g L⁻¹, pH 5.0, Conc.50 mg L⁻¹, Temp. 24±1°C	671	35.4	Zhang et al. 2010
Bioinspired polydopamine coated natural zeolites	Chemical	FTIR, TGA, XPS, BET, TEM, SEM, Zeta potential	Cu(II)	Adsorption	91.4%	pH 5.5, Time 24 h	28.58	4.57	Yu et al. 2014
Polymer-based zirconium phosphate	Chemical	XPS, BET	Pb(II)	Adsorption	—	—	0.67 meq g⁻¹	24.0	Pan et al. 2007

TABLE 7.5

Metal-based Nanosorbents for Heavy-Metal Removal from Water

Nanosorbents	Methods	Characterization	Heavy Metals	Removal Process	Efficiency (%)	Conditions	Adsorption Capacity (mg g^{-1})	Surface Area (m^2 g^{-1})	References
Chitosan-bound Fe$_3$O$_4$ magnetic NPs	Coprecipitation	XRD, TEM, zeta potential	Cu(II)	Adsorption	—	pH 5.0, Temp. 27°C, Conc. 1100 mg L^{-1}	21.5	—	Chang and Chen 2005
CuO nanoparticles	—	SEM, TEM	As(V)	Sand filtration	99%	—	—	86.51	Reddy et al. 2013
ZnO and SnO$_2$ nanoparticles	Precipitation	XRD, SEM, TEM, SAED, FTIR, BET	Cr(VI)	Adsorption	90%, 87%	pH 2.0, Temp. 30°C, Dose 2.0 g L^{-1}, Conc. 3 mg L^{-1}	9.38, 3.09	15.75, 24.48	Kumar et al. 2013b
CeO$_2$ NPs	—	DLS, XRD, SEM, TEM, zeta potential	Cu(VI)	Adsorption	96.5%	pH 7, Dose 0.32 g L^{-1}, Time 3 h	121.95	65	Recillas et al. 2010
CdS/ZnS core–shell nanoparticles	Sonochemical reaction	XRD, PL, TEM, HR-TEM, EDX, SEM	Hg(II), Pb(II)	Ultrasonic power adsorption	—	Time 40 min, Conc. 20 mg L^{-1}, Temp. 25°C	—	—	Amiri et al. 2014
ZnO nanorods	Hydrothermal	XRD, TEM, EDX, SEM	Pb(II), Cd(II)	Adsorption	—	—	160.7, 147.25	15.7	Kumar et al. 2013a
Chitosan-stabilized bimetallic Fe/Ni nanoparticles	Liquid-phase reduction method	XRD, TEM, EDX, SEM	Cd(II)	Adsorption	93%	pH 5.0, Dose 2.0 g L^{-1}, Conc. 60 mg L^{-1}, Temp. 25°C	29.63	—	Weng et al. 2013
Chitosan-coated MnFe$_2$O$_4$ nanoparticles	Chemical precipitation	XRD, TEM, FTIR, SQUID	Cu(II), Cr(VI)	Adsorption	99%	pH 6.0, Conc. 1.0 mg L^{-1}, Temp. 25°C	22.6, 15.4	—	Xiao et al. 2013
Maghemite-magnetite nanoparticles	—	XPS, BET	Cd(II)	Adsorption	74.8%	Temp. 22°C, Dose 0.8 g Conc. 1.5 mg L^{-1}, Time 120 min	2.7	49.5	Chowdhury and Yanful 2013
Fe$_3$O$_4$ loaded tea waste (Fe$_3$O$_4$_TW)	Chemical precipitation	XRD, TEM, EDX, SEM	Ni(II)	Adsorption	99%	Dose 0.25 g L^{-1}, Temp. 25°C, Time 120 min	38.3	27.5	Panneerselvam et al. 2011
Fe$_3$O$_4$ (100) bound cubic Fe$_3$O$_4$ (111) bound octahedral	Precipitation, Hydrothermal	XRD, TEM, EDX, SEM, HR-TEM	Zn(II), Cd(II), Pb(II), Cu(II), Hg(II)	Electrochemical sensing		pH 7.0, Dose 1.0 g L^{-1},Time 24 h, Temp. 25°C	—		Yao et al. 2014

(Continued)

TABLE 7.5 (CONTINUED)
Metal-based Nanosorbents for Heavy-Metal Removal from Water

Nanosorbents	Methods	Characterization	Heavy Metals	Removal Process	Efficiency (%)	Conditions	Adsorption Capacity (mg g^{-1})	Surface Area (m^2 g^{-1})	References
Humic acid (HA)-coated Fe$_3$O$_4$ nanoparticles (Fe$_3$O$_4$/HA)	Precipitation	XRD, TEM, EDX, SEM	Hg(II), Pb(II), Cd(II), Cu(II)	Adsorption	99%, 95%	Conc. 0.1 mg L^{-1}, pH 6.0, Dose 0.1 g L^{-1}, Temp. 20°C	46.3–97.7	64	Liu et al. 2008
Flowerlike α-Fe$_2$O$_3$ nanostructures	Microwave-assisted solvothermal	XRD, TEM, EDX, SEM, HR-TEM	As(V), Cr(VI)	Adsorption	—	Conc. 25 mg L^{-1}, pH 3, Temp. 20°C	51, 30	130	Cao et al. 2012
Carboxyl functional magnetite nanoparticles (CMNPs)	Solvothermal	XRD, TEM, SEM, FTIR	Pb(II), Cd(II), Cu(II)	Adsorption	100%	pH 6.0, Temp. 25°C, Time 120 min	74.63, 45.66, 44.84	—	Shi et al. 2015
Fe$_3$O$_4$ magnetic nanoparticles	Coprecipitation	XRD, TEM, SEM, FTIR	Ni(II), Cr(VI)	Adsorption	90.5%, 99.1%	pH 2–8, Conc. 5 mg L^{-1}, Temp. 25°C, Dose 4 g L^{-1}	3.31, 3.30	86.5	Sharma and Srivastava 2011
Hydrous Zr(IV) oxide-based nanocomposite NZP	Liquid precipitation and fabrication	XRD, TEM, SEM, FTIR, XPS	Pb(II), Cd(II)	Adsorption	—	pH 5.3, Temp. 30°C, Conc. 0.5 mM, Dose 0.5 g L^{-1}	319.4, 214.7	20.4	Hua et al. 2013
Ferricyanide embedded conductive polypyrrole (FCN/PPy)	Unipolar pulse electropolymerization	EDS, SEM, FTIR, XPS	Ni(II)	Ion exchange	98.5%	Time 50 s, Temp. 10°C	—	—	Du et al. 2014

Mn-Zn ferrite dose, 10 mL volume at 27°C, and contact time 60 min, As, Cd, and Pb removal was 99.9%, 99.7%, and 99.8%, respectively. Moussa et al. (2013) reported the synthesis of ferrite-coated apatite magnetic nanomaterial by a coprecipitation method and applied it in the removal of Eu(III) ions from aqueous solution. The synthesized magnetic nanoadsorbent had a crystalline structure and possessed a surface area amounting to 85.11 $m^2 g^{-1}$. The maximum adsorbed amount of Eu(III) was attained at pH 2.5 with a value reaching 157.14 $mg\ g^{-1}$. Various other such metal-based nano-sorbents used for heavy metal removal are given in Table 7.5.

7.6 CONCLUSION

Removal of heavy metals from water and wastewater is a challenge for scientists and policymakers at the global level. A number of techniques/methods have been suggested for the removal of heavy metals from aqueous media, including ion exchange, biosorption, membrane filtration, photocatalysis, reverse osmosis, advance oxidation, precipitation, and so on. But, each method has its own advantages and limitations. So, the scientific community is constantly in search of new methods or materials which can be sustainably employed for water treatment. In the preceding two decades, scientists have been working on the removal of heavy metals using different nanomaterials, including carbon nanotube-based nanomaterials, graphene-based nanocomposites, metal oxide nanoparticles, bimetallics, metal-organic nanocomposites, polymer-based nanocomposites/nanomembranes, and so on. Most of the studies have shown that that nanomaterials are effective for removing target metal ions from multicomponent systems. Nanomaterials have a high regeneration efficiency and reusability as an adsorbent that make them an economically feasible and effective material for the removal of heavy metals in future. The results of these studies are promising and may lead to commercial application of nanomaterials for heavy metal removal from aqueous media. But, this bibliographic survey has shown that most of the studies conducted on this subject are at laboratory scale. In future, scientists should undertake studies at pilot scale to prove the worth of nanomaterials for removal of heavy metals from water and wastewater.

REFERENCES

Aal, A.A., Mahmoud, S.A., and Boul-Gheit, A.K. 2009. Nanocrystalline ZnO thin film for photocatalytic purification of water, *Materials Science and Engineering: C Materials for Biological Applications* 29: 831–835.

Akhavan, B., Jarvis, K., and Majewski, P. 2015. Plasma polymer-functionalized silica particles for heavy metals removal, *ACS Applied Materials and Interfaces* 7: 4265–4274.

Alcaraz-Espinoza, J.J., Chaávez-Guajardo, A.E., Medina-Llamas, J.C., Andrade, C.A.S., and Melo, C.P. 2015. Hierarchical composite polyaniline–(electrospun polystyrene) fibers applied to heavy metal remediation, *ACS Applied Materials and Interfaces* 7: 7231–7240.

Amiri, O., Hosseinpour-Mashkani, S.M., Rad, M.M., and Abdvali, F. 2014. Sonochemical synthesis and characterization of CdS/ZnS core–shell nanoparticles and application in removal of heavy metals from aqueous solution, *Superlattices and Microstructures* 66: 67–75.

Anitha, K., Namsani, S., and Singh, J.K. 2015. Removal of heavy metal ions using a functionalized single-walled carbon nanotube: A molecular dynamics study, *Journal of Physical Chemistry A* 119: 8349–8358.

Badruddoza, A.Z.M., Shawon, Z.B.Z., Daniel, T.W.J., Hidajat, K., and Uddin, M.S. 2013. Fe_3O_4/cyclodextrin polymer nanocomposites for selective heavy metals removal from industrial wastewater, *Carbohydrate Polymers* 91: 322–332.

Bandaru, N.M., Reta, N., Dalal, H., Ellis, A.V., Shapter, J., and Voelcker, N.H. 2013. Enhanced adsorption of mercury ions on thiol derivatized single wall carbon nanotubes, *Journal of Hazardous Materials* 261: 534–541.

Bayazit, S.S. and Inci, I. 2013. Adsorption of Pb(II) ions from aqueous solutions by carbon nanotubes oxidized different methods, *Journal of Industrial and Engineering Chemistry* 19: 2064–2071.

Bessbousse, H., Rhlalou, T., Verchère, J.-F., and Lebrun, L. 2008. Removal of heavy metal ions from aqueous solutions by filtration with a novel complexing membrane containing poly(ethyleneimine) in a poly(vinyl alcohol) matrix, *Journal of Membrane Science* 307: 249–259.

Bessbousse, H., Rhlalou, T., Verchère, J.-F., and Lebrun, L. 2009. Novel metal-complexing membrane containing poly(4-vinylpyridine) for removal of Hg(II) from aqueous solution, *Journal of Physical Chemistry B* 113: 8588–8598.

Bystrzejewski, M., Pyrzynska, K., Huczko, A., and Lange, H. 2009. Carbon-encapsulated magnetic nanoparticles as separable and mobile sorbents of heavy metal ions from aqueous solutions, *Carbon* 47: 1189–1206.

Cao, C.-Y., Qu, J., Yan, W.-S., Zhu, J.-F., Wu, Z.-Y., and Song, W.-G. 2012. Low-cost synthesis of flower like α-Fe$_2$O$_3$ nanostructures for heavy metal ion removal: Adsorption property and mechanism, *Langmuir* 28: 4573–4579.

Chang, Y.-C. and Chen, D.-H. 2005. Preparation and adsorption properties of monodisperse chitosan-bound Fe3O4 magnetic nanoparticles for removal of Cu(II) ions, *Journal of Colloid and Interface Science* 283: 446–451.

Chen, Y., Pan, B., Li, H., Zhang, W., Lv, L., and Wu, J. 2010. Selective removal of Cu(II) ions by using cation-exchange resin-supported polyethyleneimine (PEI) nanoclusters, *Environmental Science and Technology* 44: 3508–3513.

Cheng, Z., Wu, Y., Wang, N., Yang, W., and Xu, T. 2010. Development of a novel hollow fiber cation-exchange membrane from bromomethylated poly (2,6-dimethyl-1,4-phenylene oxide) for removal of heavy-metal ions, *Industrial Engineering and Chemistry Research* 49: 3079–3087.

Chong, M.N., Jin, B., Chow, C.W.K., and Saint, C. 2010. Recent developments in photocatalytic water treatment technology: A review, *Water Research* 44: 2997–3027.

Chowdhury, S.R. and Yanful, E.K. 2013. Kinetics of cadmium (II) uptake by mixed maghemite-magnetite nanoparticles, *Journal of Environmental Management* 129: 642–651.

Cong, Y., Banta. G., and Selek. H. 2011. Toxic effects and bioaccumulation of nano-micron and ionic Ag in the polychaete, *Nereis diversicolor, Aquatic Toxicology* 105: 403–411.

Cui, L., Wang, Y., Gao, L., Hu, L., Yan, L., Wei, Q., and Du, B. 2015. EDTA functionalized magnetic graphene oxide for removal of Pb(II), Hg(II) and Cu(II) in water treatment: Adsorption mechanism and separation property, *Chemical Engineering Journal* 281: 1–10.

Cundy, A.B., Hopkinson, L., and Whitby, R.L.D. 2008. Use of iron-based technologies in contaminated land and groundwater remediation: A review, *Science of Total Environment* 400: 42–51.

Denizli, A., Kesenci, K., Arica, Y., and Piskin, E. 2000. Dithiocarbamate-incorporated monosize polystyrene microspheres for selective removal of mercury ions, *Reactive & Functional Polymers* 44: 235–243.

Deosarkar, S.P. and Pangarkar, V.G. 2004. Adsorptive separation and recovery of organics from PHBA and SA plant effluents, *Separation and Purification Technology* 38: 241–254.

Dichiara, A.B., Webber, M.R., Gorman, W.R., and Rogers, R.E. 2015. Removal of copper ions from aqueous solutions via adsorption on carbon nanocomposites, *ACS Applied Materials and Interfaces* 7: 15674–15680.

Du, X., Zhang, H., Hao, X., Guan, G., and Abudula, A. 2014. Facile preparation of ion-imprinted composite film for selective electrochemical removal of nickel (II) ions, *ACS Applied Materials and Interfaces* 6: 9543–9549.

Fabrega, J., Luoma, S.N., Tyler C.R., Galloway, T.S., and Lead, J.R. 2011. Silver nanoparticles: Behaviour and effects in the aquatic environment, *Environment International* 37: 517–531.

Farrukh, A., Akram, A., Ghaffar, A., Hanif, S., Hamid, A., Duran, H., and Yameen, B. 2013. Design of polymer-brush-grafted magnetic nanoparticles for highly efficient water remediation, *ACS Applied Materials and Interfaces* 5: 3784–3793.

Fatta, K.D., Kalavrouziotis, I.K., Koukoulakis, P.H., and Vasquez, M.I. 2011. The risks associated with wastewater reuse and xenobiotics in the agroecological environment, *Science of Total Environment* 409: 3555–63.

Feng, L., Cao, M., Ma, X., Zhu, Y., and Hu, C. 2012. Super paramagnetic high-surface area Fe$_3$O$_4$ nanoparticles as adsorbents for arsenic removal, *Journal of Hazardous Materials* 217–218: 439–446.

Ferhan, A.R., Guo, L., Zhou, X., Chen, P., Hong, S., and Kim, D.-H. 2013. Solid-phase colorimetric sensor based on gold nanoparticle-loaded polymer brushes: Lead detection as a case study, *Analytical Chemistry* 85: 4094–4099.

Goon, I.Y., Zhang, C., Lim, M., Gooding, J.J., and Amal, R. 2010. Controlled fabrication of polyethylenimine-functionalized magnetic nanoparticles for the sequestration and quantification of free Cu, *Langmuir* 26: 12247–12252.

Gopalakrishnan, A., Krishnan, R., Thangavel, S., Venugopal, G., and Kim, S. 2015. Removal of heavy metal ions from pharma-effluents using graphene-oxide nanosorbents and study of their adsorption kinetics, *Journal of Industrial and Engineering Chemistry* 30: 14–19.

Goswami, A., Raul, P.K., and Purkait, M.K. 2012. Arsenic adsorption using copper (II) oxide nanoparticles, *Chemical Engineering Research and Design* 90: 1387–1396.

Gross, M. 2001. *Travels to the Nanoworld: Miniature Machinery in Nature and Technology*, New York: Plenum Trade.

Guo, D., Xie, G., and Luo, J. 2014. Mechanical properties of nanoparticles: Basics and applications, *Journal of Physics D: Applied Physics* 47: 013001.

Guo, X., Du, B., Wei, Q., Yang, J., Hu, L., Yan, L., and Xu, W. 2014b. Synthesis of amino functionalized magnetic graphenes composite material and its application to remove Cr(VI), Pb(II), Hg(II), Cd(II) and Ni(II) from contaminated water, *Journal of Hazardous Materials* 278: 211–220.

Gupta, K., Bhattacharya, S., Chattopadhyay, D., Mukhopadhyay, A., and Biswas, H. 2011. Ceria associated manganese oxide nanoparticles: Synthesis, characterization and arsenic (V) sorption behaviour, *Chemical Engineering Journal* 172: 219–229.

Hu, J., Lo, I.M.C., and Chen, G. 2007. Comparative study of various magnetic nanoparticles for Cr(VI) removal, *Separation and Purification Technology* 56: 249–256.

Hua, M., Jiang, Y., Wu, B., Pan, B., Zhao, X., and Zhang, Q. 2013. Fabrication of a new hydrous Zr(IV) oxide-based nanocomposite for enhanced Pb(II) and Cd(II) removal from waters, *ACS Applied Materials and Interfaces* 5: 12135–12142.

Huang, C., Notten, A., and Rasters, N. 2011a. Nanoscience and technology publications and patents: A review of social science studies and search strategies, *Journal of Technology Transfer* 36: 145–172.

Huang, M.-R., Huang, S.-J., and Li, X.-G. 2011b. Facile synthesis of polysulfoaminoanthraquinone nanosorbents for rapid removal and ultrasensitive fluorescent detection of heavy metal ions, *Journal of Physical Chemistry C* 115: 5301–5315.

Ivanovskii, A.L. 2012. Graphene and graphene-like materials, *Russian Chemical Reviews* 81: 571–605.

Joodaki, F., Azizian, S., and Sobhanardakani, S. 2015. Synthesis of nanostructured ZnO loaded on carbon cloth as high potential adsorbent for copper ion, *Desalination and Water Treatment* 55: 596–603.

Kikuchi, Y., Qian, Q., Machida, M., and Tatsumoto, H. 2006. Effect of ZnO loading to activated carbon on Pb(II) adsorption from aqueous solution, *Carbon* 44: 195–202.

Kim, D., El-Shall, H., Dennis, D., and Morey, T. 2005. Interaction of PLGA nanoparticles with human blood constituents, *Colloids and Surfaces B: Interfaces* 40: 83–91.

Kumar, K.Y., Muralidhara, H.B., Nayaka, Y.A., Balasubramanyam, J., and Hanumanthappa, H. 2013a. Hierarchically assembled mesoporous ZnO nanorods for the removal of lead and cadmium by using differential pulse anodic stripping voltammetric method, *Powder Technology* 239: 208–216.

Kumar, K.Y., Muralidhara, H.B., Nayaka, Y.A., Balasubramanyam, J., and Hanumanthappa, H. 2013b. Low-cost synthesis of metal oxide nanoparticles and their application in adsorption of commercial dye and heavy metal ion in aqueous solution, *Powder Technology* 246: 125–136.

Kunin, R. 1977. Polymeric adsorbents for treatment of waste effluents, *Polymer Engineering Science* 17: 58–62.

Kyzas, G.Z., Travlou, N.A., and Deliyanni, E.A. 2014. The role of chitosan as nanofiller of graphite oxide for the removal of toxic mercury ions, *Colloids and Surfaces B* 113: 467–476.

Larraza, I., López-Gónzalez, M., Corrales, T., and Marcelo, G. 2012. Hybrid materials: Magnetite-polyethylenimine-montmorillonite, as magnetic adsorbents for Cr(VI) water treatment, *Journal of Colloid and Interface Science* 385: 24–33.

Lazaridis, N.K., Bakoyannakis, D.N., and Deliyanni, E.A. 2005. Chromium (VI) sorptive removal from aqueous solutions by nanocrystalline akaganeite, *Chemosphere* 58: 65–73.

Li, Q., Yu, J., Zhou, F., and Jiang, X. 2015. Synthesis and characterization of dithiocarbamate carbon nanotubes for the removal of heavy metal ions from aqueous solutions, *Colloids and Surfaces A: Physicochemical and Engineering Aspects* 482: 306–314.

Li, S., Anderson, T.A., Maul, J.D., Shrestha, B., Green, M.J., and Cañas-Carrell, J.E. 2013a. Comparative studies of multi-walled carbon nanotubes (MWNTs) and octadecyl (C18) as sorbents in passive sampling devices for biomimetic uptake of polycyclic aromatic hydrocarbons (PAHs) from soils, *Science of Total Environment* 560:461–462.

Li, T., Shen, J., Huang, S., Li, N., and Ye, M. 2014. Hydrothermal carbonization synthesis of a novel montmorillonite supported carbon nanosphere adsorbent for removal of Cr (VI) from waste water, *Applied Clay Science* 93/94: 48–55.

Li, X., Zeng, G.M., Huang, J.H., Zhang, D.M., Shi, L.J., and He, S.B. 2011. Simultaneous removal of cadmium ions and phenol with MEUF using SDS and mixed surfactants, *Desalination* 276: 136–41.

Li, Y., Zhu, S., Liu, Q., Chen, Z., Gu, J., Zhu, C., Lu, T., Zhang, D., and Ma, J. 2013b. N-doped porous carbon with magnetic particles formed in situ for enhanced Cr(VI) removal, *Water Research* 47: 4188–4197.

Lin, D., Tian, X., Li, T., Zhang, Z., He, X., and Xing, B. 2012. Surface-bound humic acid increased Pb^{2+} sorption on carbon nanotubes, *Environmental Pollution* 167: 138–147.

Liu, J.-F., Zhao, Z.-S., and Jiang, G.-B. 2008. Coating Fe_3O_4 magnetic nanoparticles with humic acid for high efficient removal of heavy metals in water, *Environmental Science & Technology* 42: 6949–6954.

Lozet, J. and Mathieu, C. 1991. *Dictionary of Soil Science*, 2nd ed., Rotterdam, The Netherlands: A.A. Balkema.

Lue, J.T. 2007. Physical properties of nanomaterials, in: *Encyclopedia of Nanoscience and Nanotechnology*, edited by H.S. Nalwa, New York: American Scientific, 1–46.

Luo, T., Cui, J., Hu, S., Huang, Y. and Jing, C. 2010. Arsenic removal and recovery from copper smelting wastewater using TiO_2, *Environmental Science & Technology* 44: 9094–9098.

Mahdavi, S., Afkhami, A., and Merrikhpour, H. 2015a. Modified ZnO nanoparticles with new modifiers for the removal of heavy metals in water, *Clean Technologies and Environmental Policy* 17: 1645–1661.

Mahdavi, S., Jalali, M., and Afkhami, A. 2015b. Removal of heavy metals from aqueous solutions using Fe_3O_4, ZnO, and CuO nanoparticles, *Journal of Nanoparticle Research* 14: 846.

Masoumi, A., Ghaemy, M., and Bakht, A.N. 2014. Removal of metal ions from water using poly(MMA-co-MA)/modified-Fe_3O_4 magnetic nanocomposite: Isotherm and kinetic study, *Industrial Engineering Chemistry Research* 53: 8188–8197.

Mishra, A.K. and Ramaprabhu, S. 2011. Functionalized graphene sheets for arsenic removal and desalination of sea water, *Desalination* 282: 39–45.

Moore, M.N. 2006. Do nanoparticles present ecotoxicological risks for the health of the aquatic environment? *Environment International* 32: 967–976.

Morris, C. 1992. *Academic Press Dictionary of Science and Technology*, San Diego, CA: Academic Press.

Moussa, S.I., Sheha, R.R., Saad, E.A., and Tadros, N.A. 2013. Synthesis and characterization of magnetic nano-material for removal of Eu^{3+} ions from aqueous solutions, *Journal of Radioanalytical and Nuclear Chemistry* 295: 929–935.

Nata, I.F., Salim, G.W., and Lee, C.-K. 2010. Facile preparation of magnetic carbonaceous nanoparticles for Pb^{2+} ions removal, *Journal of Hazardous Materials* 183: 853–858.

Norouzian, R.-S. and Lakouraj, M.M. 2015. Preparation and heavy metal ion adsorption behavior of novel supermagnetic nanocomposite based on thiacalix[4]arene and polyaniline: Conductivity, isotherm and kinetic study, *Synthetic Metals* 203: 135–148.

O'Connor, G.A. 1996. Organic compounds in sludge-amended soils and their potential for uptake by crop plants, *Science of Total Environment* 185: 71–81.

Pan, B., Pan, B., Zhang, W., Lv, L., Zhang, Q., and Zheng, S. 2009. Development of polymeric and polymer-based hybrid adsorbents for pollutants removal from waters, *Chemical Engineering Journal* 151: 19–29.

Pan, B.C., Zhang, Q.R., Zhang, W.M., Pan, B.J., Du, W., Lv, L., Zhang, Q.J., Xua, Z.W., and Zhang, Q.X. 2007. Highly effective removal of heavy metals by polymer-based zirconium phosphate: A case study of lead ion, *Journal of Colloid and Interface Science* 310: 99–105.

Pandey, B. and Fulekar, M.H. 2012. Nanotechnology: Remediation technologies to clean up the environmental pollutants, *Research Journal of Chemical Sciences* 2(2): 90–96.

Panneerselvam, P., Morad, N., and Tan, K.A. 2011. Magnetic nanoparticle (Fe_3O_4) impregnated onto tea waste for the removal of nickel (II) from aqueous solution, *Journal of Hazardous Materials* 186: 160–168.

Perez-Aguilar, N.V., Diaz-Flores, P.E., and Rangel-Mendez, J.R. 2011. The adsorption kinetics of cadmium by three different types of carbon nanotubes, *Journal of Colloid and Interface Science* 364: 279–287.

Pradeep, T. and Anshup. 2009. Noble metal nanoparticles for water purification: A critical review, *Thin Solid Films* 517: 6441–6478.

Qiu, B., Xu, C., Sun, D., Yi, H., Guo, J., Zhang, X., Qu, H., Guerrero, M., Wang, X., Noel, N., Luo, Z., Guo, Z., and Wei, S. 2014. Polyaniline coated ethyl cellulose with improved hexavalent chromium removal, *ACS Sustainable Chemical Engineering* 2: 2070–2080.

Qiu, H., Zhang, S., Pan, B., Zhang, W., and Lv, L. 2012. Effect of sulfate on Cu (II) sorption to polymer-supported nano-iron oxides: Behavior and XPS study, *Journal of Colloid and Interface Science* 366: 37–43.

Qu, X., Alvarez, P., and Li, Q. 2013. Applications of nanotechnology in water and wastewater treatment, *Water Research* 47: 931–946.

Quester, K., Avalos-Bojra, M., and Castro-Longoria. 2013. Biosynthesis and microscopic study of metallic nanoparticles, *Micron* 54/55: 1–27.

Rao, K.J., Mahesh, K., and Kumar, S. 2005. A strategic approach for preparation of oxide nanomaterials, *Bulletin of Materials Science* 28: 19–24.

Recillas, S., Colón, J., Casals, E., González, E., Puntesb, V., Sáncheza, A., and Fonta, X. 2010. Chromium VI adsorption on cerium oxide nanoparticles and morphology changes during the process, *Journal of Hazardous Materials* 184: 425–431.

Reddy, K.J., McDonald, K.J., and King, H. 2013. A novel arsenic removal process for water using cupric oxide nanoparticles, *Journal of Colloid and Interface Science* 397: 96–102.

Ren, Y., Yan, N., Feng, J., Ma, J., Wen, Q., Nan, L., and Qing, D. 2012. Adsorption mechanism of copper and lead ions onto graphene nanosheet/ɗMnO$_2$, *Materials Chemistry and Physics* 136: 538–544.

Rezvani-Boroujeni, A., Javanbakht, M., Karimi, M., Shahrjerdi, C., and Akbari-adergani, B. 2015. Immobilization of thiol-functionalized nanosilica on the surface of poly(ether sulfone) membranes for the removal of heavy-metal ions from industrial wastewater samples, *Industrial Engineering Chemistry Research* 54: 502–513.

Ruparelia, J.P., Duttagupta, S.P., Chatterjee, A.K., and Mukherji, S. 2008. Potential of carbon nanomaterials for removal of heavy metals from water, *Desalination* 232: 145–156.

Saadat, S., Karimi-Jashni, A., and Doroodmand, M.M. 2014. Synthesis and characterization of novel single-walled carbon nanotubes-doped walnut shell composite and its adsorption performance for lead in aqueous solutions, *Journal of Environmental Chemical Engineering* 2: 2059–2067.

Sattler, K.D. 2010. *Handbook of Nanophysics, Principles and Methods*, New York: CRC Press.

Seenivasan, R., Chang, W.-J., and Gunasekaran, S. 2015. Highly sensitive detection and removal of lead ions in water using cysteine-functionalized graphene oxide/polypyrrole nanocomposite film electrode, *ACS Applied Materials & Interfaces* 7: 15935–15943.

Sharma, Y.C. and Srivastava, V. 2011. Comparative studies of removal of Cr(VI) and Ni(II) from aqueous solutions by magnetic nanoparticles, *Journal of Chemical Engineering Data* 56: 819–825.

Shi, J., Li, H., Lu, H., and Zhao, X. 2015. Use of carboxyl functional magnetite nanoparticles as potential sorbents for the removal of heavy metal ions from aqueous solution, *Journal of Chemical Engineering Data* 60: 2035–2041.

Shih, Y.-C., Ke, C.-Y., Yu, C.-J., Lu, C.-Y., and Tseng, W.-L. 2014. Combined tween 20-stabilized gold nanoparticles and reduced graphite oxide–Fe$_3$O$_4$ nanoparticle composites for rapid and efficient removal of mercury species from a complex matrix, *ACS Applied Materials & Interfaces* 6: 17437–17445.

Song, X., Gunawan, P., Jiang, R., Leong, S.S.J., Wang, K., and Xu, R. 2011. Surface activated carbon nanospheres for fast adsorption of silver ions from aqueous solutions, *Journal of Hazardous Materials* 194: 162–168.

Talreja, N., Kumar, D., and Verma, N. 2014. Removal of hexavalent chromium from water using Fe-grown carbon nanofibers containing porous carbon microbeads, *Journal of Water Process Engineering* 3: 34–45.

Tan, P., Sun, J., Hu, Y., Fang, Z., Bi, Q., Chen, Y., and Cheng, J. 2015. Adsorption of Cu^{2+}, Cd^{2+} and Ni^{2+} from aqueous single metal solutions on graphene oxide membranes, *Journal of Hazardous Materials* 297: 251–260.

Thakur, S., Das, G., Raul, P.K., and Karak, N. 2013. Green one-step approach to prepare sulfur/reduced graphene oxide nanohybrid for effective mercury ions removal, *Journal of Physical Chemistry* 117: 7636–7642.

Tu, Y.J., You, C.F., and Chang, C.K. 2013. Conversion of waste Mn-Zn dry battery as efficient nano-adsorbents for hazardous metals removal, *Journal of Hazardous Materials* 258–259: 102–108.

Vadahanambi, S., Lee, S.H., Kim, W.J., and Oh, I.K. 2013. Arsenic removal from contaminated water using three- dimensional graphene-carbon nanotube-iron oxide nanostructures, *Environmental Science & Technology* 47: 10510–10517.

Wang, X., Chen, Z., and Yang, S. 2015. Application of graphene oxides for the removal of Pb(II) ions fromaqueous solutions: Experimental and DFT calculation, *Journal of Molecular Liquids* 211: 957–964.

Wang, X. and Wang, A. 2010. Adsorption characteristics of chitosan-g-poly(acrylic acid)/attapulgite hydrogel composite for Hg(II) ions from aqueous solution, *Separation Science and Technology* 45: 2086–2094.

Wang, X., Zheng, Y., and Wang, A. 2009. Fast removal of copper ions from aqueous solution by chitosan–g–poly(acrylic acid)/attapulgite composites, *Journal of Hazardous Materials* 168: 970–977.

Warheit, D.B. 2008. How meaningful are the results of nanotoxicity studies in the absence of adequate material characterization? *Toxicology Science* 101: 183–185.

Weng, X., Lin, S., Zhong, Y., and Chen, Z. 2013. Chitosan stabilized bimetallic Fe/Ni nanoparticles used to remove mixed contaminants-amoxicillin and Cd (II) from aqueous solutions, *Chemical Engineering Journal* 229: 27–34.

Wilson, M.A., Tran, N.H., Milev, A.S., Kannangara, K., Volk, H., and Lu, G.Q.M. 2008. Nanomaterials in soil, *Geoderma* 146: 291–302.

Xiao, Y., Liang, H., Chen, W., and Wang, Z. 2013. Synthesis and adsorption behaviour of chitosan-coated MnFe$_2$O$_4$ nanoparticles for trace heavy metal ions removal, *Applied Surface Science* 285: 498–504.

Xie, Y., Huang, Q., Liu, M., Wang, K., Wan, Q., Deng, F., Lu, L., Zhang, X., and Wei, Y. 2015. Mussel inspired functionalization of carbon nanotubes for heavy metal ion removal, *RSC Advances* 5: 68430–68438.

Xu, D., Zhu, K., Zheng, X., and Xiao, R. 2015. Poly(ethylene-co-vinyl alcohol) functional nanofiber membranes for the removal of Cr(VI) from water, *Industrial Engineering Chemistry Research* 54: 6836–6844.

Xu, Y., Rosa, A., Liu, X., and Su, D.S. 2011. Characterization and use of functionalized carbon nanotubes for the adsorption of heavy metal anions, *New Carbon Materials* 26: 57–62.

Yaghmaeian, K., Mashizi, R.K., Nasseri, S., Mahvi, A.H., Alimohammadi, M., and Nazmara, S. 2015. Removal of inorganic mercury from aquatic environments by multi-walled carbonnanotubes, *Journal of Environmental Health Science and Engineering*, 13: 55.

Yang, W., Ding, P., Zhou, L., Yu, J., Chen, X., and Jiao, F. 2013. Preparation of diamine modified mesoporous silica on multi-walled carbon nanotubes for the adsorption of heavy metals in aqueous solution, *Applied Surface Science* 282: 38–45.

Yang, W.C., Shim, W.G., Lee, J.W., and Moon, H. 2003. Adsorption and desorption dynamics of amino acids in a non-ionic polymeric sorbent XAD-16 column, *Korean Journal of Chemical Engineering* 20: 922–929.

Yao, X.-Z., Guo, Z., Yuan, Q.-H., Liu, Z.-G., Liu, J.-H., and Huang, X.-J. 2014. Exploiting differential electrochemical stripping behaviors of Fe$_3$O$_4$ nanocrystals toward heavy metal ions by crystal cutting, *ACS Applied Materials and Interfaces* 6: 12203–12213.

Yu, Y., Shapter, J.G., Popelka-Filcoff, R., Bennett, J.W., and Ellis, A.V. 2014. Copper removal using bio-inspired polydopamine coated natural zeolites, *Journal of Hazardous Materials* 273: 174–182.

Zeng, G.M., Li, X., Huang, J.H., Zhang, C., Zhou, C.F., and Niu, J. 2011. Micellar-enhanced ultrafiltration of cadmium and methylene blue in synthetic wastewater using SDS, *Journal of Hazardous Materials* 185: 1304–1310.

Zhang, C., Sui, J., Li, J., Tang, Y., and Cai, W. 2012. Efficient removal of heavy metal ions by thiol-functionalized super paramagnetic carbon nanotubes, *Chemical Engineering Journal* 210: 45–52.

Zhang, J., Jin, Y., and Wang, A. 2011. Rapid removal of Pb(II) from aqueous solution by chitosan-g-poly(acrylic acid)/attapulgite/sodium humate composite hydrogels, *Environmental Technology* 32: 523–531.

Zhang, S., Zhang, Y., Bi, G., Liu, J., Wang, Z., Xu, Q., Xu, H., and Li, X. 2014a. Mussel-inspired polydopamine biopolymer decorated with magnetic nanoparticles for multiple pollutants removal, *Journal of Hazardous Materials* 270: 27–34.

Zhang, W., Shi, X., Zhang, Y., Gu, W., Li, B., and Xian, Y. 2013. Synthesis of water-soluble magnetic graphene nanocomposites for recyclable removal of heavy metal ions, *Journal of Materials Chemistry* A 1: 1745–1753.

Zhang, X., Jia, X., Zhang, G., Hua, J., Sheng, W., Ma, Z., Lu, J., and Liu, Z. 2014b. Efficient removal and highly selective adsorption of Hg^{2+} by polydopamine nanospheres with total recycle capacity, *Applied Surface Science* 314: 166–173.

Zhang, Y., Li, Q., Sun, L., Tang, R., and Zhai, J. 2010. High efficient removal of mercury from aqueous solution by polyaniline/humic acid nanocomposite, *Journal of Hazardous Materials* 175: 404–409.

Zhao, G.X., Ren, X.M., Gao, X., Tan, X.L., Li, J.X., and Chen, C.L. 2011. Removal of Pb(II) ions from aqueous solutions on few-layered graphene oxide nanosheets, *Dalton Transactions* 40: 10945–10952.

Zhou, L., Ji, L., Ma, P., Shao, Y., Zhang, H., Gao, W., and Lia, Y. 2014. Development of carbon nanotubes/CoFe$_2$O$_4$ magnetic hybrid material for removal of tetrabromobisphenol A and Pb(II), *Journal of Hazardous Materials* 265: 104–114.

Zhu, J., Wei, S., Gu, H., Rapole, S.B., Wang, Q., Luo, Z., Haldolaarachchige, N., Young, D.P., and Guo, Z. 2012. One-pot synthesis of magnetic graphene nanocomposites decorated with core@double-shell nanoparticles for fast chromium removal, *Environmental Science & Technology* 46: 977–985.

Zhu, Y., Murali, S., Cai, W., Li, X., Suk, J.W., Potts, J.R., and Ruoff, R.S. 2010. Graphene and graphene oxide: Synthesis, properties, and applications, *Advanced Materials* 22: 3906–3924.

Zodrow, K., Brunet, L., Mahendra, S., Li, D., Zhang, A., Li, Q.L., and Alvarez, P.J.J. 2009. Polysulfone ultrafiltration membranes impregnated with silver nanoparticles show improved biofouling resistance and virus removal, *Water Research* 43: 715–723.

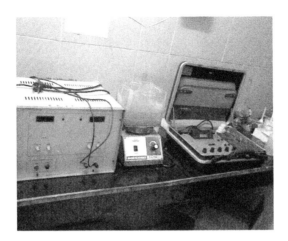

FIGURE 2.1 Picture of electrocoagulation setup.

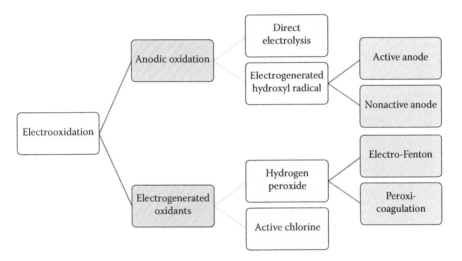

FIGURE 3.1 Classification of EO processes.

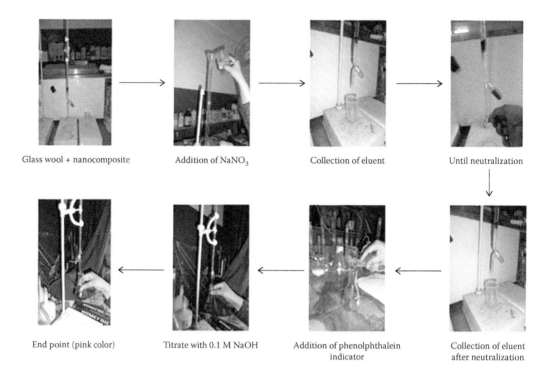

Glass wool + nanocomposite   Addition of NaNO₃   Collection of eluent   Until neutralization

End point (pink color)   Titrate with 0.1 M NaOH   Addition of phenolphthalein indicator   Collection of eluent after neutralization

FIGURE 5.1   Experimental setup for the determination of ion-exchange capacity.

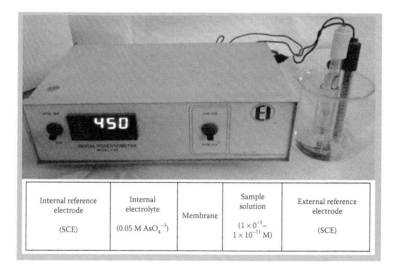

Internal reference electrode (SCE)	Internal electrolyte (0.05 M AsO₄⁻³)	Membrane	Sample solution (1 × 0⁻¹ – 1 × 10⁻¹¹ M)	External reference electrode (SCE)

FIGURE 5.2   A simple set up of an ion-selective electrode.

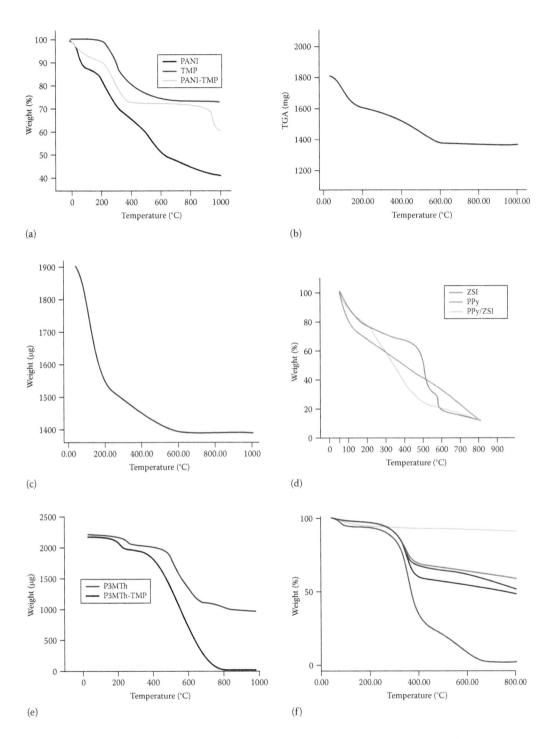

FIGURE 5.7 TGA curves of (a) PANI-TMP, (b) POTST, (c) POAST, (d) Ppy/ZSI, (e) P3MTh-TMP, and (f) PCz/TiO$_2$ nanocomposite cation exchangers.

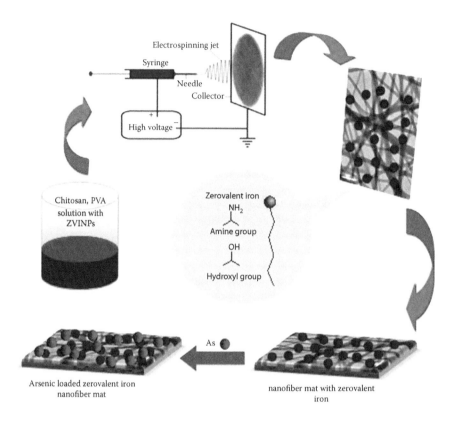

FIGURE 6.2 Schematic representation of the formation of a nonfibrous mat and its application toward arsenic removal.

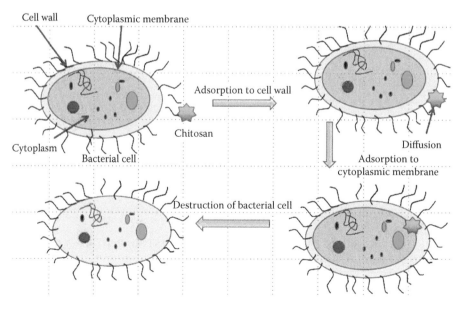

FIGURE 6.5 Schematic representation of antibacterial activity chitosan.

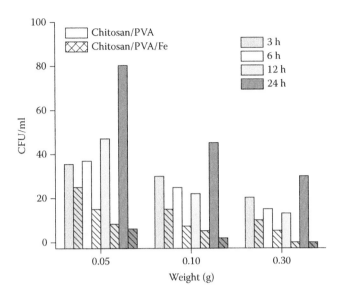

FIGURE 6.6 Antibacterial activity of CPF mats. (Reproduced from Chauhan, D., et al., *RSC Adv.* 4, 54694–54702, 2014.)

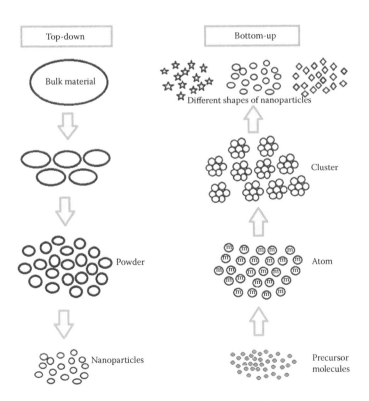

FIGURE 7.1 Schematic presentation of "top-down" and "bottom-up" approaches used for nanomaterial synthesis.

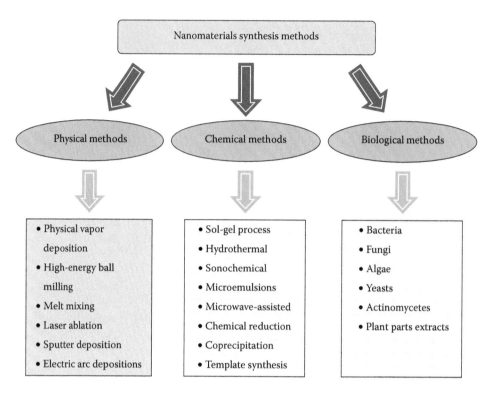

FIGURE 7.2   Flow chart of nanoparticle synthesis methods.

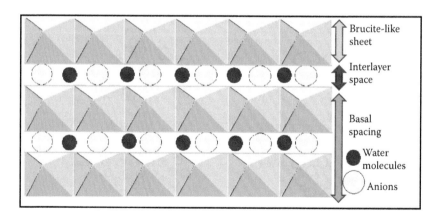

FIGURE 10.1   Schematic representation of LDH structure.

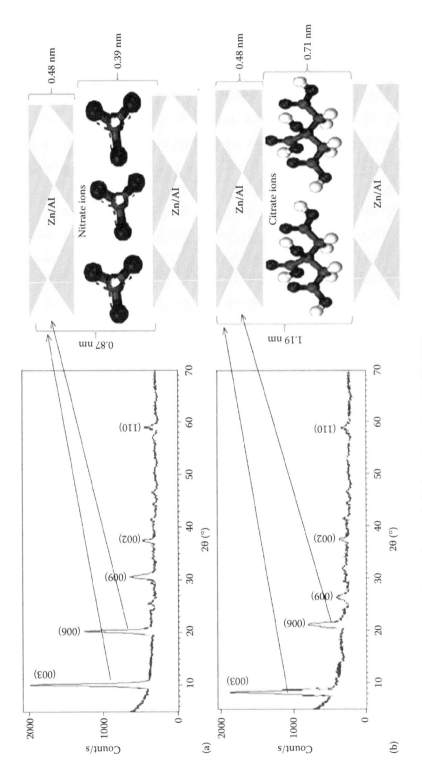

FIGURE 10.2 X-ray diffraction pattern for (a) nitrate and (b) citrate intercalated Zn/Al-LDH.

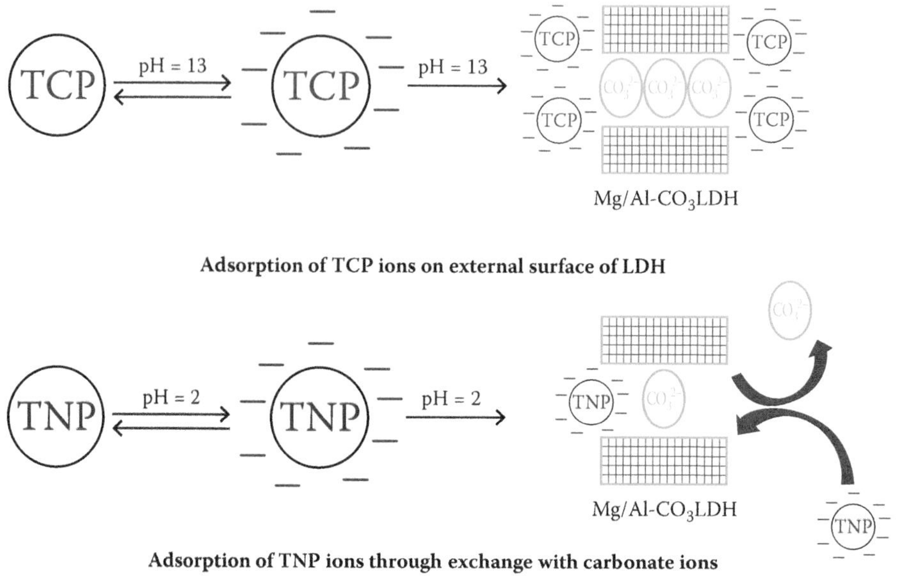

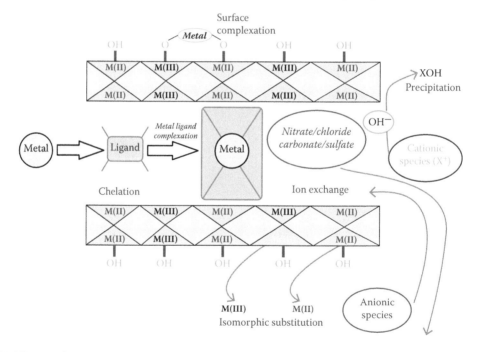

FIGURE 10.5 Schematic representation of adsorption of TCP and TNP at given pH condition.

FIGURE 10.6 Schematic representation of probable mechanism for adsorption of various pollutants.

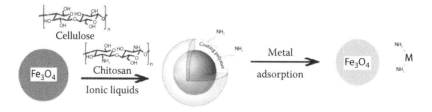

FIGURE 12.10 Preparation of magnetic CL-CS hydrogels and adsorption of heavy metals. (Liu, Z. et al., 2012, Magnetic cellulose–chitosan hydrogels prepared from ionic liquids as reusable adsorbent for removal of heavy metal ions, *Chem. Commun.* 48, 7350–7352, figure 1. Reproduced by permission of The Royal Society of Chemistry.)

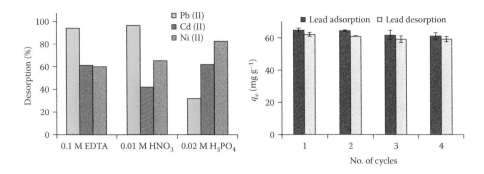

FIGURE 12.12 (a) Percentage recovery of Pb^{2+}, Cd^{2+}, and Ni^{2+} from CDpoly-MNPs using different desorption eluents, and (b) four consecutive adsorption–desorption cycles of CDpoly-MNP adsorbent for Pb^{2+} (initial concentration: 300 mg L^{-1}; pH: 5.5; desorption agent: 10 mL of 0.01 M HNO_3). (Reprinted from figure 4, *Carbohydr. Polym.*, 91, Badruddoza, A.Z.M. et al., Fe_3O_4/cyclodextrin polymer nanocomposites for selective heavy metals removal from industrial wastewater, 322–332, Copyright (2013), with permission from Elsevier.)

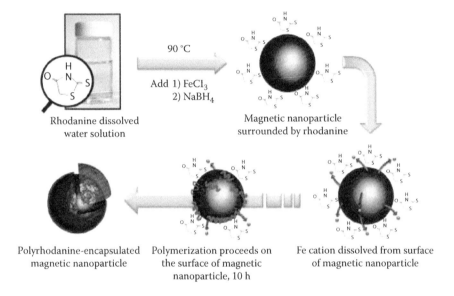

FIGURE 12.14 Fabrication process of polyrhodanine-encapsulated MNPs. (Reprinted from figure 1, *J. Colloid Interface Sci.*, 359, Song, J. et al., Adsorption of heavy metal ions from aqueous solution by polyrhodanine-encapsulated magnetic nanoparticles, 505–511, Copyright (2011), with permission from Elsevier.)

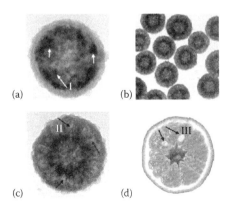

FIGURE 12.23 TEM images of (a) single Fe_3O_4/PPy composite microsphere, (b) porous PPy microspheres, (c) single porous PPy microsphere, and (d) digital photograph of a real orange cross-section. (Wang, Y. et al., 2012, Synthesis of orange-like Fe_3O_4/PPy composite microspheres and their excellent Cr(VI) ion removal properties, *J. Mater. Chem.* 22, 9034–9040, figure 2. Reproduced by permission of The Royal Society of Chemistry.)

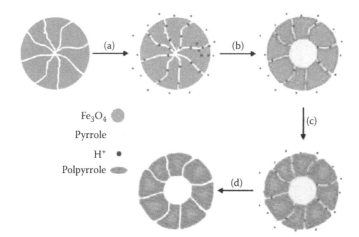

FIGURE 12.24 Synthetic procedure for orange-like Fe_3O_4/PPy composite microspheres: (a) dispersion of Fe_3O_4 microspheres in mixed solutions containing pyrrole and acid, (b) release of Fe^{3+} from the surface and the interior of the Fe_3O_4 microspheres in acidic solution and the polymerization of pyrrole monomers nearby, (c) generation of PPy with decreasing Fe_3O_4 microspheres, and (d) formation of orange-like Fe_3O_4/PPy composite microspheres after washing. (Wang, Y. et al., 2012, Synthesis of orange-like Fe_3O_4/PPy composite microspheres and their excellent Cr(VI) ion removal properties, *J. Mater. Chem.* 22, 9034–9040, figure 6. Reproduced by permission of The Royal Society of Chemistry.)

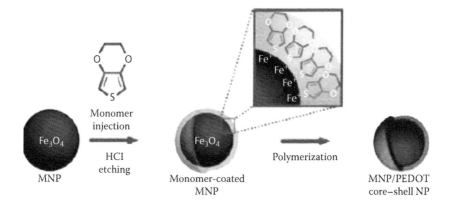

FIGURE 12.26 Fabrication procedure of Fe_3O_4–PEDOT NPs by seeded polymerization mediated with acidic etching. (Shin, S. and Jang, J., 2007, Thiol containing polymer encapsulated magnetic nanoparticles as reusable and efficiently separable adsorbent for heavy metal ions, *Chem. Commun.* 41, 4230–4232, scheme 1. Reproduced by permission of The Royal Society of Chemistry.)

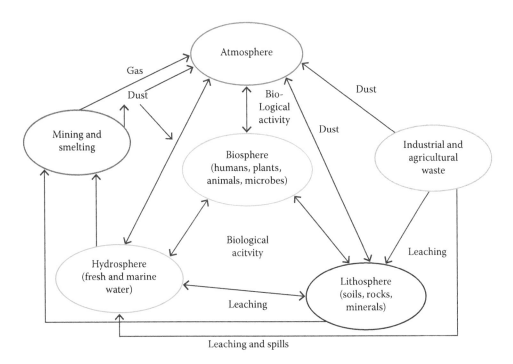

FIGURE 13.1 Arsenic input from various natural and anthropogenic sources. (Redrawn from Mudhoo et al. 2011.)

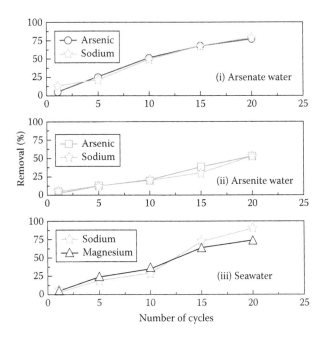

FIGURE 13.6 Simultaneous removal performance of Na and As: (i) sodium arsenate-containing water and (ii) sodium arsenite-containing water; (iii) removal efficiency of sodium and magnesium (desalination) from seawater (From Mishra, A.K. and Ramaprabhu, S., *J. Exp. Nanosci.*, 7, 85–97, 2011.)

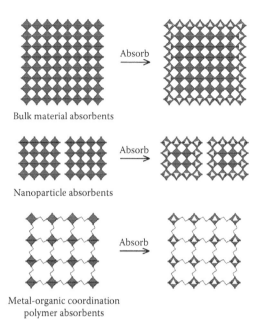

FIGURE 14.1 Schematic illustration of as-proposed new strategy for efficient adsorbent. (Reprinted with permission from scheme 1, Zhu, B.J. et al., 2012, Iron 1,3,5-benzenetricarboxylic metal–organic coordination polymers prepared by solvothermal method and their application in effcient As(V) removal from aqueous solutions, *J. Phys. Chem. C* 116, 8601–8607. Copyright (2012) American Chemical Society.)

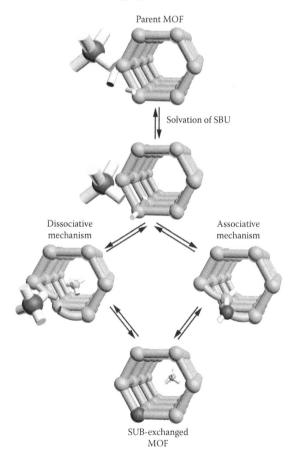

FIGURE 14.2 Simplified mechanistic pathways for cation exchange at MOF SBUs. Green and red spheres represent exiting and entering metal ions, respectively. Organic linkers are shown in gray and solvent is depicted in yellow. (Brozek, C.K. and Dinca, M., 2014, Cation exchange at the secondary building units of metal–organic frameworks, *Chem. Soc. Rev.* 43, 5456–5467, scheme 1. Reproduced by permission of The Royal Society of Chemistry.)

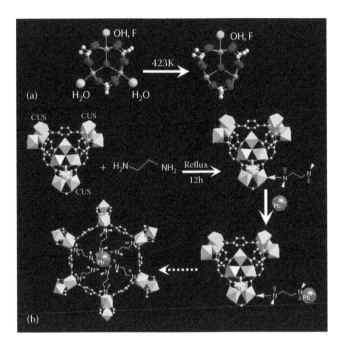

FIGURE 14.3 (a) Generation of coordinated–unsaturated sites from Cr trimers in MIL-101 after vacuum treatment at 423 K for 12 h, and (b) adsorption principle of amino-functionalized MIL-101 for Pb^{2+} ions. (Reprinted with permission from figure 1, Luo, X. et al., 2015, Adsorptive removal of Pb(II) ions from aqueous samples with amino-functionalization of metal–organic frameworks MIL-101(Cr), *J. Chem. Eng. Data* 60, 1732–1743. Copyright (2015) American Chemical Society.)

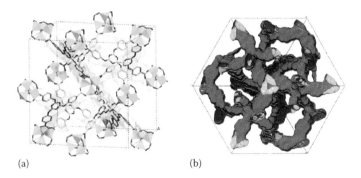

FIGURE 14.9 (a) View of packing of PCN-100, and (b) view of channels and free volume of PCN-100. (Reprinted with permission from figure 3, Fang, Q.-R. et al., 2010, Functional mesoporous metal-organic frameworks for the capture of heavy metal ions and size-selective catalysis, *Inorg. Chem.* 49, 11637–11642. Copyright (2010) American Chemical Society.)

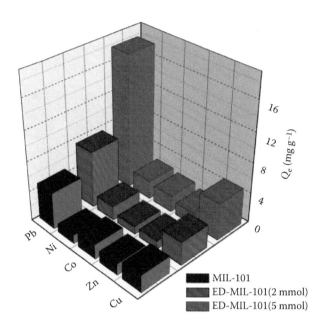

FIGURE 14.12 Selectivity adsorption capacity of MIL-101 and ED-MIL-101 for Pb²⁺ ions in the presence of coions. (Reprinted with permission from figure 14 from Luo, X. et al., 2015, Adsorptive removal of Pb(II) ions from aqueous samples with amino-functionalization of metal–organic frameworks MIL-101(Cr), *J. Chem. Eng. Data* 60, 1732–1743. Copyright (2015) American Chemical Society.)

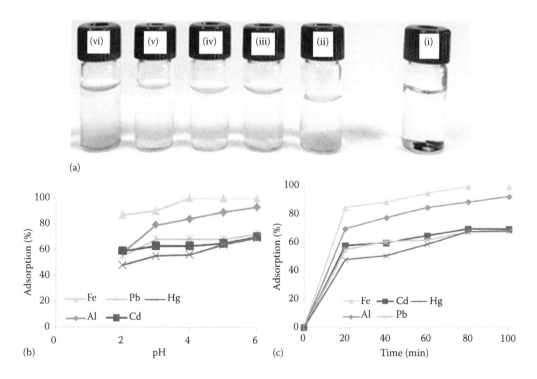

FIGURE 14.13 (a) Photographs of (i) TATAB-based MOF, and (ii) adsorbed heavy metal ions: (ii) Al⁺³, (iii) Hg⁺², (iv) Cd⁺², (v) Pb²⁺, and (vi) Fe³⁺, (b) effect of pH, and (c) effect of exposure time on the adsorption of heavy metal ions by TATAB-MOF nanostructures. (Reprinted from figures 5, 7, and 8, respectively, *Inorg. Chim. Acta*, 430, Abbasi, A. et al., A new 3D cobalt (II) metal–organic framework nanostructure for heavy metal adsorption, 261–267, Copyright (2015), with permission from Elsevier.)

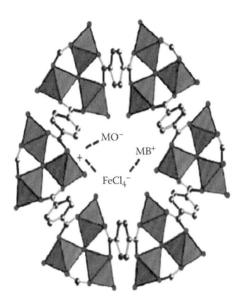

FIGURE 14.26 Proposed electrostatic interaction between dyes and adsorbents. (Reprinted from scheme 2, *J. Hazard. Mater.*, 185, Haque, E. et al., Adsorptive removal of methyl orange and methylene blue from aqueous solution with a metal-organic framework material, iron terephthalate (MOF-235), 507–511, Copyright (2011), with permission from Elsevier.)

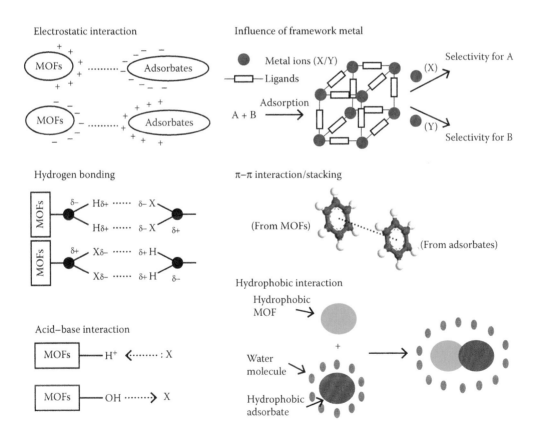

FIGURE 14.27 Possible mechanisms for adsorptive removal of hazardous materials over MOFs. (Reprinted from figure 9, *J. Hazard. Mater.*, 283, Hasan, Z. and Jhung, S.H. Removal of hazardous organics from water using metal-organic frameworks MOFs): Plausible mechanisms for selective adsorptions, 329–339, Copyright (2015), with permission from Elsevier.)

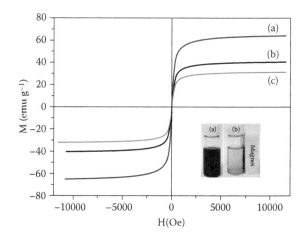

FIGURE 14.31 Magnetic hysteresis loops of (a) Fe_3O_4 and (b) Fe_3O_4-MIL-100(Fe) obtained after 20, and (c) 40 assembly cycles. Inset demonstrates Fe_3O_4-MIL-100(Fe) (a) well dispersed in water and (b) separated easily from water by a magnet. (Reprinted from figure 7, *Chem. Eng. J.*, 283, Shao, Y. et al., Magnetic responsive metal–organic frameworks nanosphere with core–shell structure for highly efficient removal of methylene blue, 1127–1136, Copyright (2016), with permission from Elsevier.)

8 Synthesis, Properties, and Applications of Carbon Nanotubes in Water and Wastewater Treatment

Geoffrey S. Simate and Lubinda F. Walubita

CONTENTS

ABSTRACT

The last two and a half decades have seen the emergence of the most promising materials—carbon nanotubes (CNTs). In recent years, extensive work carried out worldwide has revealed the intriguing mechanical, chemical, thermal, and electrical properties of these novel nanoscale materials. This chapter will discuss the synthesis, properties, and general applications of CNTs in the treatment of water and wastewater.

8.1 INTRODUCTION

The last two and a half decades have seen the emergence of the most promising materials—carbon nanotubes (CNTs). These materials are characterized by having at least a single dimension with the size of a billionth of a meter (10^{-9} m) (Pokropivny 2007). The structure of CNTs can be visualized as rolled hexagonal carbon networks (or graphene layers) that are capped by pentagonal carbon rings (Terrones 2003). There are typically two forms of CNTs, according to the number

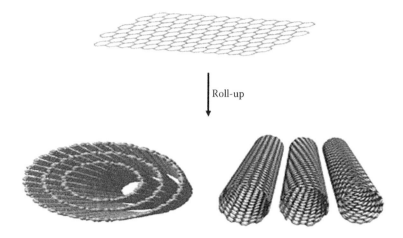

Roll-up

FIGURE 8.1 Models and schematic representation of multiwalled and single-walled CNT. (From Merkoci, A., *Microchimica Acta*, 152, 157–174, 2006; Liu, R., The functionalisation of carbon nanotubes, PhD thesis, 2008.)

of rolled-up graphene layers that form the tube (as illustrated in Figure 8.1); that is, single-walled CNTs (SWCNTs) and multiwalled CNTs (MWCNTs). Depending on the way the graphene layer is wrapped into a cylinder, three different geometries can be formed for the SWCNTs: armchair, chiral, and zigzag (Tom 2003; Eatemadi 2014) as depicted in Figure 8.2 (Grobert 2007). In the zigzag conformation, two opposite C–C bonds of each hexagon are parallel to the tube axis, whereas in the armchair conformation, the C–C bonds are perpendicular to the axis (Grobert 2007). In all other arrangements, the opposite C–C bonds lie at an angle to the tube axis, resulting in a so-called helical nanotube that is chiral. In addition, there are two structural models that can be used to describe the structures of MWCNTs: the Russian doll model and the parchment model (Iyuke and Simate 2011; Eatemadi 2014). In the Russian doll model, a CNT contains another nanotube inside it and the inner nanotube has a smaller diameter than the outer nanotube. In the parchment model, a single graphene sheet is rolled around itself multiple times. In MWCNTs, the nanotubes are typically bound together by strong van der Waals interaction forces and form tight bundles (Dai 2002a).

Extensive work carried out worldwide in recent years has revealed the intriguing mechanical, chemical, thermal, and electrical properties of CNTs. These characteristic properties are what

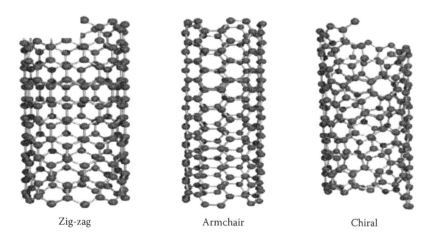

Zig-zag   Armchair   Chiral

FIGURE 8.2 Different forms of SWNTs. (From Grobert, N., *Materials Today*, 10, 28–35, 2007; Eatemadi, A., et al., *Nanoscale Research Letters*, 9, 393, 2014.)

make them versatile in their applications. This chapter will discuss the synthesis, properties, and applications of CNTs in the treatment of water and wastewater. Along with this introductory section, the chapter is divided into four other sections covering the following themes: (1) production of CNTs, (2) properties of CNTs, (3) general applications of CNTs in the treatment of water and wastewater (which include adsorbents, heterogeneous coagulants and flocculants, filtration membranes, antimicrobial materials, catalysts and catalyst supports, sensing and detection), and (4) challenges and concluding remarks.

8.2 PRODUCTION OF CARBON NANOTUBES

Since the "rediscovery" of CNTs (Monthioux and Kuznetsov 2006) by Iijima (1991) as a by-product of fullerene synthesis, several methods have been developed for the production of CNTs (Dresselhaus et al. 2001; Agboola et al. 2007; Simate et al. 2010; Iyuke and Simate 2011). However, the three very useful and widespread methodologies include arc discharge, laser ablation, and chemical vapor deposition (CVD) (Robertson 2004; Agboola et al. 2007; Eatemadi 2014). A review of several techniques show that two key requirements for the synthesis of CNTs include (i) a carbon source, and (ii) a heat source to achieve the desired operating temperature (See and Harris 2007). However, the type of nanotube that is produced also depends strongly on the presence or absence of a catalyst; MWCNTs are most commonly produced via noncatalytic means, whereas SWCNTs are usually the dominant products under catalytic growth conditions (Bernholc et al. 1998; Eatemadi 2014).

The following subsections give a general overview of the three main production techniques of CNTs.

8.2.1 Electric-Arc Discharge

This is the first method that produced CNTs (Iijima 1991). It is the easiest and most widely used technique (Baddour and Briens 2005). In arc discharge, CNTs are produced from carbon vapor generated by an electric-arc discharge between two graphite electrodes (with or without catalysts) under an inert gas atmosphere of helium or argon (Journet et al. 1997; Journet and Bernier 1998; Lee et al. 2002; Agboola et al. 2007). Figure 8.3 shows a typical schematic representation of the arc-discharge apparatus used for producing CNTs (Saito et al. 1996; Harris 1999; Saito and Uemura

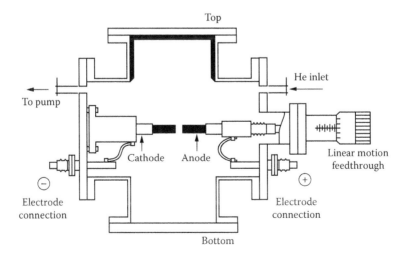

FIGURE 8.3 Arc-discharge setup. (From Harris, P. J. F., *Carbon Nanotubes and Related Structures: New Materials for the Twenty-First Century*, Cambridge University Press, Cambridge, 1999; Saito, Y., et al., *Journal of Applied Physics* 80, 3062–3067, 1996; Saito, Y., et al., *Carbon*, 38, 169–182, 2000.)

2000). The arrangement of the apparatus is such that both electrodes are water-cooled and that the diameter of the anode is usually smaller than the diameter of the cathode (Journet and Bernier 1998; Baddour and Briens 2005).

Typically, the anode is moved toward the cathode until they are less than 1 mm apart and a current of 100 A is passed through the electrodes, creating a plasma between them. Throughout the operation, the anode must be continuously moved to ensure a constant distance between the electrodes.

The average temperature in the interelectrode plasma region is extremely high, of the order of 4000 K (i.e., about 3727°C) (Journet and Bernier 1998). The high temperature occurring between the two rods during the process results in the consumption of the anode and subsequent sublimation of carbon (Journet and Bernier 1998), that is, vaporization of carbon and deposition onto the cathode and walls of the reaction vessel (Baddour and Briens 2005; Varshney 2014). It is the deposit on the cathode that contains the CNTs (Varshney 2014).

The electric-arc process produces MWCNTs in the absence of catalysts, whereas SWCNTs are produced in the presence of catalysts (Baddour and Briens 2005; Grobert 2007). As a rule of thumb, arc-discharge MWCNTs are typically 20 µm long, have a diameter of around 10 nm, and the number of walls is limited to ~20–30 (Grobert 2007). The SWCNTs produced by arc discharge occur in bundles (Journet et al. 1997) with their diameters ranging from 1 to 2 nm (Grobert 2007). However, it remains difficult to measure the SWCNT length accurately, because of the entanglement of the SWCNT bundles (Grobert 2007).

The yield of nanotubes is not very high, but can be in gram quantities (Journet and Bernier 1998). However, some studies have shown that a composite of Ni–Y–graphite can produce high yields (<90%) of SWNTs (Saito et al. 1998) and, nowadays, this mixture is used worldwide for the creation of SWNTs of high yields (Eatemadi 2014). Research studies have also shown that arc-discharge MWCNT and SWCNT samples commonly contain substantial amounts of by-products such as amorphous carbon and polyhedral carbon (Grobert 2007). Furthermore, encapsulated metal catalyst particles are also present in SWCNT samples (Grobert 2007).

Though this technique produces CNTs of reasonable quality with fewer structural defects, its disadvantage is that it uses high temperatures of up to 1500°C, which makes it difficult to scale up for commercial purposes (Iyuke and Simate 2011). The other disadvantage of this method is that there is relatively little control over the alignment (i.e., chirality) of the created nanotubes, a property that is important for their characterization and usage (Eatemadi 2014).

8.2.2 Laser Ablation

Historically, laser ablation was the first technique used to generate fullerene clusters in the gas phase (Kroto et al. 1985). This method is considered as a second, very useful and powerful technique for producing CNTs (Journet and Bernier 1998). Smalley and associates were the first to synthesize CNTs by the laser-ablation method (Guo et al. 1995). In this technique, a piece of graphite target is vaporized by laser irradiation under an inert atmosphere (Journet and Bernier 1998; Paradise and Goswami 2007). Ideally, the laser-ablation technique operates under similar conditions to arc discharge. For example, both methods use the condensation of carbon atoms generated from the vaporization of graphite targets (Grobert 2007), and SWNTs are formed when graphite targets containing catalysts, such as Ni, Co, and Pt, are used (Braidy et al. 2002).

Typically, the process is as follows (Journet and Bernier 1998): A graphite target is placed in the middle of a long quartz tube mounted in a temperature-controlled furnace. After the sealed tube has been evacuated, the furnace temperature is increased to 1200°C. The tube is then filled with a flowing inert gas and a scanning laser beam is focused onto the graphite target by way of a circular lens. The laser beam scans across the target surface to maintain a smooth, uniform face for vaporization. The laser vaporization produces carbon species, which are swept by the flowing inert gas from the high-temperature zone and deposited on a conical water-cooled copper collector at the end of the apparatus. Finally, the soot containing nanotubes is collected from the water-cooled copper collector, the walls of the quartz tube, and the downstream face of the graphite target.

As stated already, this method produces MWCNTs when the vaporized carbon target is pure graphite, whereas the addition of transition metals (e.g., Co, Ni, Fe, or Y) as catalysts to the graphite target results in the production of SWCNTs (Agboola et al. 2007). The SWCNTs formed exist as "ropes" and are bundled together by van der Waals forces (Dresselhaus et al. 1996; Agboola et al. 2007; Grobert 2007). Generally, the CNTs synthesized by the laser vaporization processes have fewer structural defects, in addition to superior mechanical and electrical properties. The major disadvantage with this method is that it is an expensive technique, because it involves the use of high-purity graphite rods, high-power lasers (Iyuke and Simate 2011; Eatemadi 2014), and elaborate configurations (Agboola et al. 2007). In addition, the quantity of CNTs that can be synthesized per day is not as high as in the arc-discharge technique (Eatemadi 2014). However, the quality, length, diameter, and chirality distribution of the nanotubes are believed to be comparable with those of SWCNTs grown by arc discharge (Grobert 2007).

8.2.3 Chemical Vapor Deposition (CVD)

CVD methods have been successful in making carbon fiber, filament, and nanotube materials for more than three decades. The technique of CVD involves the use of an energy source (such as plasma, a resistive or inductive heater, or a furnace) to transfer energy to a gas-phase carbon source so as to produce fullerenes, CNTs, and other sp^2-like nanostructures (Meyyappan 2004). The CVD technique can be applied both in the absence and the presence of a catalyst substrate; the former is a homogeneous gas-phase process where the catalyst is in the gas phase, and the latter is a heterogeneous process that uses a supported catalyst (Corrias et al. 2003).

At the moment, the CVD method (or variations thereof) is the only promising process for the production of CNTs on a reasonably large scale compared with the arc-discharge and laser vaporization methods (Coleman 2008). In addition, the process tends to produce nanotubes with fewer impurities (catalyst particles, amorphous carbon, and nontubular fullerenes) compared with other techniques (Esawi and Farag 2007). The variants of the CVD are as a result of the means by which chemical reactions are initiated, the type of reactor used, and the process conditions (Deshmukh et al. 2010). The CVD is simple, flexible, versatile, and allows high specificity of single-walled or multiwalled nanotubes through the appropriate selection of process parameters; for example, metal catalysts, reaction temperature, and flow rate of feed stock (Nolan et al. 1995; Agboola et al. 2007; Moisala et al. 2006).

Recently, a swirled floating catalyst (or fluid) chemical vapor deposition (SFCCVD) reactor was developed with the aim of upscaling production capacity (Iyuke 2007; Iyuke and Simate 2011). The simplified schematic representation of a SFCCVD is shown in Figure 8.4. Typically, it consists of a vertical quartz or silica plug flow reactor inside a furnace. The upper end of the reactor is connected to a condenser that leads to two delivery cyclones, where the CNTs produced are collected. Feed materials, including carrier gases, are uniformly mixed with the aid of a swirled coiled mixer to give an optimum interaction. The flow of gases into the SFCCVD reactor is aided by a system of valves and rotameters (Yah et al. 2011).

Table 8.1 summarizes and compares the three methods of CNT synthesis. From Table 8.1, it is clear that the CVD method provides the most yields and is also relatively inexpensive.

8.3 PROPERTIES OF CARBON NANOTUBES

Comprehensive analyses of the properties of CNTs have been extensively discussed in various publications in the literature (Sinnott and Andrews 2001; Dai 2002a; Terrones 2003; Popov 2004; Baddour and Briens 2005; Paradise and Goswami 2007). From these publications, this section will focus on and cite some of the most important and distinctive properties that make CNTs superior to most traditional materials, specifically in reference to the publication by Baddour and Briens (2005).

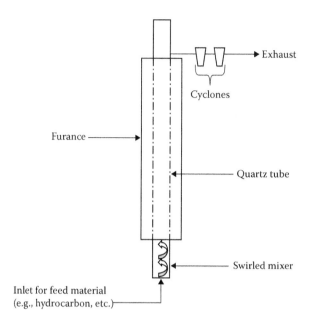

FIGURE 8.4 Simplified schematic presentation of a swirled floating catalytic CVD. (From Iyuke, S. and Simate, G. S., Synthesis of carbon nanomaterials in a swirled floating catalytic chemical vapour deposition reactor for continuous and large scale production. In: Naraghi. M. (Ed.), *Carbon Nanotubes: Growth and Applications*, 35–58, In-Tech, Rijeka, 2011.)

TABLE 8.1
Comparison of CNT Production Methods

Method	Arc Discharge	Laser Ablation	CVD
Pioneer	Iijima (1991)	Guo et al. (1995)	Yacaman et al. (1993)
Description	Arc evaporation of graphite in the presence of inert gas; CNT formed on electrodes during quenching	Vaporization of graphite target by laser; CNT formed on receiver during quenching	Decomposition of hydrocarbons with or without transition metal catalyst to form CNT
Operating temperature	>3000°C	>3000°C	<1200°C
Operating pressure	50–7600 Torr generally under vacuum	200–750 Torr generally under vacuum	760–7600 Torr
Yield (%)	<75	<75	>75
SWCNT or MWCNT	Both	Only SWCNTs	Both
Advantages	Simple; inexpensive; good-quality CNTs	Relatively high-purity CNTs; room temperature synthesis option with continuous laser	Simple; inexpensive; low temperature; high purity and high yields; aligned growth is possible; fluidized bed technique for large scale
Disadvantages	Purification of crude product is required; method cannot be scaled up; must have high temperature	Cannot produce MWCNTs, method only adapted to lab-scale; expensive; crude product purification required	CNTs are usually MWCNTs; parameters must closely be watched to obtain SWCNTs

Source: Baddour, C. E., et al., *International Journal of Chemical Reactor Engineering*, 3, Review R3, 2005; See, C. H., et al., *Industrial and Engineering Chemistry Research*, 46, 997–1012, 2007.

8.3.1 MECHANICAL PROPERTIES

Many studies have been performed on the mechanical properties of CNTs, including those conducted by Treacy et al. (1996), Cornwall and Wille (1997), Lu (1997), Krishnan et al. (1998), and Yu et al. (2000). The research studies have shown that CNTs are far lighter than steel, and are also between 10 and 100 times stronger (Rosso 2001). In fact, they have been described as the strongest fibers known to man (Rosso 2001) and are also the stiffest materials yet discovered (Varshney 2014). The strength of CNTs results from the covalent sp^2 bonds formed between the individual carbon atoms (Varshney 2014). Due to the carbon–carbon sp^2 bonds, CNTs are expected to be extremely strong along their axes, and are also expected to have a very large Young's modulus in their axial direction (Popov 2004; Baddour and Briens 2005; Varshney 2014). Experimental and theoretical results have shown an elastic modulus greater than 1 TPa (that of diamond is 1.2 TPa) (Treacy et al. 1996; Krishnan et al. 1998; Popov 2004); with the elastic modulus of MWCNTs being higher than that of SWCNTs (Yamabe 1995; Rosso 2001; Paradise and Goswami 2007). It has also been predicted that CNTs have the highest Young's modulus of all different types of composite tubes such as BN, BC_3, BC_2N, C_3N_4, CN, and so on (Table 8.2) (Delmotte and Rubio 2002; Paradise and Goswami 2007).

The definition of Young's modulus involves the second derivative of the energy with respect to the applied stress/strain. In general, the strength of the chemical bonds determines the actual value of Young's modulus and smaller diameters result in a smaller Young's modulus. However, tests conducted on CNTs show that little dependence exists on the diameter of the tube with Young's modulus, which does help the hypothesis that CNTs do possess the highest Young's modulus (Paradise and Goswami 2007).

Due to the high in-plane tensile strength of graphite, both single-walled and multiwalled nanotubes are expected to have large bending constants, since these mostly depend on Young's modulus (Paradise and Goswami 2007). Research has shown that nanotubes are very flexible; they can be elongated, twisted, flattened, or bent into circles before fracturing (Iijima et al. 1996; Paradise and Goswami 2007). Simulations conducted by Bernholc and colleagues indicate that, in many cases, the nanotubes are able to regain their original shape, a unique characteristic property that is very beneficial in most engineering applications (Bernholc et al. 1998; Paradise and Goswami 2007). Furthermore, their "kink-like" ridges allow the structure to relax elastically while under compression, unlike carbon fibers that fracture easily (Dresselhaus et al. 1996; Paradise and Goswami 2007).

TABLE 8.2
Mechanical Properties of Carbon Nanotubes and Other Materials

Material	Young's Modulus (GPa)	Tensile Strength (GPa)	Density (g cm^{-3})
Single-walled nanotube	900–1700	75	—
Multiwalled nanotube	1800 average, 690–1870	150	2.6
Steel	208	0.4	7.8
Epoxy	3.5	0.005	1.25
Wood	15	0.008	0.6

Source: Paradise, M., et al., *Materials and Design*, 28, 1477–1489, 2007; Yamabe, T., *Synthetic Metals*, 70, 1511–1518, 1995.

8.3.2 ELECTRICAL PROPERTIES

Not only are CNTs extremely strong, but they have very fascinating electrical properties (Varshney 2014). In fact, individual nanotubes, similarly to macroscopic structures, can be characterized by a set of electrical properties: resistance, capacitance, and inductance (Varshney 2014). Studies have shown that the electronic capabilities possessed by CNTs arise predominately from the intralayer interactions, rather than from interlayer interactions between multilayers within a single CNT or between two different nanotubes (Dresselhaus et al. 1995).

An interest in using CNTs in nanoscale electronic devices has led to a large amount of research being done on their electrical properties in the recent past (Dunlap 1992; Voit 1995; Baddour and Briens 2005, etc.). The theoretical and experimental results show that CNTs have superior electrical properties compared with conventional materials (Paradise and Goswami 2007). For example, CNTs can produce an electric current carrying capacity 1000 times higher than traditional copper wires (Collins and Avouris 2000). Furthermore, depending on their structure, CNTs can be almost perfect one-dimensional conductors, in which various phenomena have been observed at low temperatures (Popov 2004; Baddour and Briens 2005); for example, (1) single electron charging, (2) resonant tunneling through discrete energy levels, and (3) proximity-induced superconductivity. However, at high temperatures, CNTs have been found to behave as one-dimensional Luttinger liquids (liquids where the energy state of their electrons is strongly affected by weak Coulomb interactions) (Popov 2004; Baddour and Briens 2005).

Depending on tubule diameter and/or chiral angle, the nanotube can be either a metal, a semiconductor (Tans et al. 1998; Paradise and Goswami 2007), or a small bandgap semiconductor (Zhou et al. 2000a). In addition, the I-tight-binding model within the zone-folding scheme shows that one-third of CNTs are found to be metallic while two-thirds are semiconducting, depending on their indices (Popov 2004; Paradise and Goswami 2007). Studies have shown that metallic conduction can be achieved without the introduction of doping effects (Paradise and Goswami 2007). For semiconducting nanotubes, the bandgap has been found to be proportional to a fraction of the diameter and is independent of the tubule chirality (Dresselhaus et al. 1995; Paradise and Goswami 2007).

Since the electrical properties of CNTs are dependent on the tube structure, CNTs can be used as junctions between metal and semiconductor, semiconductor and semiconductor, and metal and metal (Popov 2004; Baddour and Briens 2005). Ideally, there are three types of CNT junctions that can be achieved: on-tube, Y, and crossed junctions (Baddour and Briens 2005). An on-tube junction can be attained by joining two tubes of different chirality (Dunlap 1992; Baddour and Briens 2005) or by chemical doping of nanotube segments (Zhou et al. 2000b; Baddour and Briens 2005). The Y and crossed junctions are constructed from Y-branched CNTs (Papapdopulos et al. 2000; Baddour and Briens 2005) and crossed CNTs (Fuhrer et al. 2000). These various CNT junctions can be used to build parts of nanoscale devices (Baddour and Briens 2005).

8.3.3 THERMAL PROPERTIES

Nanotubes are extremely stable at high temperatures, and can withstand about 2800°C in a vacuum and up to 750°C at normal atmospheric pressures (Rosso 2001; Varshney 2014). However, CNTs may be damaged at temperatures lower than 750°C (in air) due to oxidation mechanisms; and above 2800°C (in an inert atmosphere), they transform into regular, polyaromatic solids (phases built with stacked graphenes instead of single graphenes) (Méténier et al. 2002). It is predicted that CNTs can transmit up to 6000 Wm^{-1} K^{-1} at room temperature, compared with copper, a metal well known for its good thermal conductivity, which transmits 385 Wm^{-1} K^{-1} (Varshney 2014). It has also been found that CNTs have a high thermal conductivity that is comparable to diamond crystal (1000–2600 W mK^{-1}) and in-plane graphite sheet (Baddour and Briens 2005). In addition, research studies indicate that all nanotubes are expected to be very good thermal conductors along the tube—exhibiting a property known as *ballistic conduction*—but good insulators laterally to the

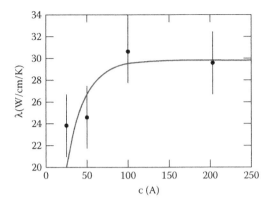

FIGURE 8.5 Thermal conductivity. (From Rosso, M. A., *Origins, Properties, and Applications of Carbon Nanotubes and Fullerenes*. IT 283 Advance Materials and Processes, California State University, Fresno, CA, 2001.)

tube axis (Varshney 2014). In other words, the thermal conductivity of CNTs is expected to be very high in the axial direction, but very low in the lateral direction. Studies have also shown that at temperatures above 1 K and below room temperature, there is a linear dependence of the specific heat and thermal conductivity of CNTs on temperature (Baddour and Briens 2005). Figure 8.5 illustrates this relationship of thermal conductivity to temperature. It is these thermal characteristics, as well as other characteristic factors, that make nanotubes so well suited to serve as electrical conductors.

8.3.4 CHEMICAL PROPERTIES

Just as in the case of a fullerene, a perfect CNT does not have functional groups; therefore, these quasi-one-dimensional cylindrical aromatic macromolecules are chemically inert (Niyogi et al. 2002). However, curvature-induced pyramidalization and misalignment of the π-orbitals of the carbon atoms induces a local strain, and thus, CNTs are expected to be more reactive than a flat graphene sheet. From the chemistry standpoint, it is conceptually useful to divide the CNTs into two regions: the end caps and the side wall (Niyogi et al. 2002). The end caps of the CNTs resemble a hemispherical fullerene, and because it is impossible to reduce the maximum pyramidalization angle (θ_p) of any fullerene below about $\theta_p^{max} = 9.7°$, the end caps will always be quite reactive, irrespective of the diameter of the CNT. In other words, fullerene-like tips of CNTs are known to be more reactive, whereas their walls are not reactive (Lin et al. 2003). The CNT walls are unreactive due to the seamless arrangement of hexagon rings, which lack dangling bonds (Lin et al. 2003).

Research studies have also shown that the CNTs, as produced, are insoluble in all organic solvents and aqueous solutions (Tasis et al. 2006), due to their strong intertube van der Waals and π-π interactions (Liu 2008). In other words, CNTs readily bundle together and it is very difficult to dissolve or disperse them in a solution. This lack of solubility and the difficulty of manipulation in any solvent have led to many limitations to the use of CNTs (Matarredona et al. 2003; Tasis et al. 2006). It must be noted, however, that the successful incorporation of CNTs into practical materials relies on the capability of breaking up the bundles into individual nanotubes and keeping them in homogeneous and stable suspensions (Matarredona et al. 2003). Therefore, CNTs have to be functionalized to obtain optimal performance in various applications (Simate 2012).

Indeed, CNTs can undergo various chemical reactions that make them more soluble so that they can be integrated into inorganic, organic, and biological systems (Tasis et al. 2006). Particularly, the oxidatively introduced carboxylic groups are useful for anchoring functional moieties, through either ionic or covalent linkage (Balasubramanian and Burghard 2005; Liu 2008). For example, dissolution of CNTs in organic solvents requires the introduction of a hydrophobic substituent onto the carboxylic groups (Liu 2008). Major approaches in this regard include amidation or esterification (Liu 2008). Functionalization is, therefore, an important aspect of the chemistry of CNTs (Dai 2002b). In fact, functionalization permits easy manipulation of CNTs for use in diverse technological fields (Rai et al. 2007).

The main approaches for the functionalization or modification of the CNT structures can be grouped into three categories, namely: (1) the covalent attachment of chemical groups onto the π-conjugated skeleton of CNT through reactions, (2) the noncovalent adsorption or wrapping of various functional molecules, and (3) the endohedral filling of their inner empty cavities (Tasis et al. 2006). Readers are referred to Tasis et al. (2006) for more detailed information on different functionalization techniques.

8.4   APPLICATIONS OF CARBON NANOTUBES IN THE TREATMENT OF WATER AND WASTEWATER

The versatility of CNTs in industrial applications has also been witnessed in water and wastewater treatment. In fact, recent advances suggest that many of the recent problems involving water quality could be solved or greatly ameliorated using CNTs (Savage and Diallo 2005). Therefore, this section provides an (obviously not exhaustive) overview on some of the possible applications of CNTs in the treatment of water and wastewater. In particular, the section will discuss the applications of CNTs (1) as adsorbents and heterogeneous coagulants/flocculants, (2) in filtration techniques, (3) as antimicrobial materials, (4) as catalysts/ catalyst supports, and (5) for sensing and detection.

8.4.1   Carbon Nanotubes as Adsorbents and Heterogeneous Coagulants/Flocculants

The use of CNTs as adsorbents for heavy metals (Li et al. 2003a, 2006, 2007), organic pollutants (Lu et al. 2005; Goering and Burghaus 2007; Lu and Su 2007; Li et al. 2007), and inorganic contaminants (Li et al. 2003b) has been studied by several researchers in the recent past. As can be seen from Table 8.3, a variety of contaminants have been studied as target pollutants for CNTs with different physical structures and surface chemistry (Liu et al. 2013). These studies have shown that CNTs have very good and unique adsorption properties. Several researchers have reported that metal-ion adsorption onto CNTs is described either by the Langmuir equation or the Freundlich equation or both (Li et al. 2002, 2005; Di et al. 2006; Lu and Chiu 2006; Lu and Liu 2006). The Langmuir equation is valid for the dynamic equilibrium sorption process on completely homogeneous surfaces, whereas the Freundlich equation is applicable to a heterogeneous surface. Tables 8.4 and 8.5 show the maximum metal-ion sorption capacities of raw and surface oxidized CNTs and other sorbents, respectively, as calculated by the Langmuir equation (Rao et al. 2007). The conditions under which the studies were carried out are also shown in the tables. The superior adsorption capacities of CNTs compared with other adsorbents is mainly attributed to their fibrous shape with high aspect ratio, provision of large external surface area, associated sorption sites that can be easily accessed by many contaminants, and the presence of well-developed mesopores (Upadhyayula et al. 2009; Qu et al. 2013). Research studies have also shown that CNTs experience fast adsorption kinetics due to the highly accessible adsorption sites and the short intraparticle diffusion distance (Qu et al. 2013).

Electrostatic attraction, sorption–precipitation, and chemical interaction between the metal ions and the surface functional groups of CNTs are the mechanisms by which the metal ions are adsorbed onto CNTs (Rao et al. 2007). However, the chemical interaction between the metal ions and the surface functional groups of CNTs, schematically shown in Figure 8.6, is considered to be the major sorption mechanism (Lu and Chiu 2006; Lu and Liu 2006). Essentially, protons in the carboxylic and phenolic groups of CNTs exchange with the metal ions in the aqueous phase, thus leading to a drop in the value of the solution's pH.

TABLE 8.3
Application of Carbon Nanotubes in Removal of Target Pollutants from Aqueous Solutions

Adsorbents	Pollutants	Comments
As-grown CNTs and graphitized CNTs	1,2-Dichlorobenzene (DCB)	As-grown CNTs had rough surface which made adsorption of organics much easier. Graphitized CNTs became smooth and the adsorption of organics decreased. The removal efficiency of DCB by both as-grown CNTs and graphitized CNTs kept stable in the pH range 3–10. When pH exceeded 10, removal dropped suddenly due to adsorption of water molecules onto –COO– groups that could hinder the access of DCB.
CNTs purified by mixed HNO_3 and H_2SO_4	Polycyclic aromatic hydrocarbons (PAHs)	Adsorption capacity of phenanthrene by MWCNTs was related to surface area or micropore volume. SWCNTs exhibited a larger adsorption capacity toward PAHs than MWCNTs.
	PAHs	In low concentration range, adsorption affinity of 13 PAHs was directly related to the solubility of their subcooled liquid.
	Phenolic compounds phenol, pyrogallol, 1-naphthol	Four possible solute–sorbent interactions: hydrophobic effect, electrostatic interaction, hydrogen bonding, and π-π electron donor–acceptor interaction acted simultaneously. CNTs with smaller outer diameters had higher distribution coefficients (K_d). The K_d values of the three polar phenolics tended to increase with increasing pH and then decrease with pH over their pK_a value.
	Triton X-series surfactants	Hydrophobic and π-π interactions between the surfactants and CNTs were the dominant mechanisms. The adsorption of Triton X-series surfactants facilitated suspending CNTs in water. Adsorption remained constant within pH range 2–12, indicating that electrostatic interaction and hydrogen bonding were not the major mechanisms.
	Uranium	Acid treatment increased the surface acidic functional groups of CNTs, therefore increased their colloidal stability and adsorption capacity for uranium.
As-received CNTs	Atrazine and trichloroethylene	Hydrophobic interactions were the dominant adsorption mechanism. The impurities in SWCNTs contributed a significant mass but did not provide strong adsorption sites.
CNTs activated by KOH etching	Pharmaceutical antibiotics: sulfamethoxazole, tetracycline, tylosin	KOH etching is an effective activation method to improve the adsorption affinity and adsorption reversibility of organic pollutants on carbon nanotubes. The activated CNTs showed an increased adsorption capacity toward antibiotics due to more interconnected pore structure and less pore deformation.
As-prepared and oxidized MWCNTs	Ionizable aromatic compounds (IACs): 1-naphthylamine, 1-naphthol, phenol	The adsorption capacity of MWCNTs toward IACs was higher than other common adsorbents such as natural bentonite, apatite, and kaolinite. Oxidation of MWCNTs increased the surface area and added oxygen-containing functional groups to the surfaces of MWCNTs, which depressed the adsorption of IACs on MWCNTs. The adsorption was considerably hindered when pH > pK_a.
MWCNTs grown by CVD	Antibiotic ciprofloxacin	The adsorption of ciprofloxacin was compared between AC, carbon xerogel, and CNT. CNT exhibited the best adsorption performance per unit surface area.
CNTs	Microcystins (MCs)	CNTs showed higher adsorption affinity to MCs as compared with ACs. The pore size of CNTs was a fit for the molecular dimension of microcystins. CNTs with smaller outside diameter could absorb more MCs.
MWCNTs	Natural organic matter (NOM)	The higher molecular weight fraction of NOM was preferentially adsorbed, as proved by size exclusion chromatographic analysis.

(Continued)

TABLE 8.3 (CONTINUED)

Application of Carbon Nanotubes in Removal of Target Pollutants from Aqueous Solutions

Adsorbents	Pollutants	Comments
CNTs opened using hydrothermal opening method	Phenol	The influence of hydrothermal opening conditions on surface chemical composition and adsorption capacity of CNTs were investigated.
Amorphous Al_2O_3 supported by CNTs	Fluoride	The CNT-supported Al_2O_3 had much higher adsorption capacity toward fluoride, which may be attributed to the nanosize Al_2O_3 clusters on CNTs and the intrinsic adsorption capacity of CNTs toward fluoride. The adsorption performs well at pH 5–9, which was a much broader range than that of the activated alumina (pH < 6).
CNTs purified by HNO_3	Lead	The adsorption capacity of acid-refluxed CNTs (11.2 mg g^{-1}) was higher than that of ACs (about 5.5 mg g^{-1}). The surface oxygen-containing functional groups were the most important fact for lead adsorption. The higher adsorption capacity of CNTs at pH 7 may be due to the cooperating role of adsorption and precipitation.
Ceria nanoparticles supported on CNTs	Arsenate	The As(V)-loaded adsorbent could be efficiently regenerated. Ca^{2+} and Mg^{2+} ions in water enhanced the adsorption capacity of CeO_2–CNTs toward arsenate due to the formation of ternary surface complex.
Iron(III) oxide coated ethylenediamine functionalized MWCNTs	Arsenate	Influence of pH, iron oxide loading, and interfering ions were modeled using MINTEQ program.

Source: Liu, X., et al., *Journal of Environmental Sciences*, 25, 1263–1280, 2013.

TABLE 8.4

Maximum Sorption Capacities of Various Divalent Metal Ions with Carbon Nanotubes

Adsorbent	Cd^{2+}	Cu^{2+}	Ni^{2+}	Pb^{2+}	Zn^{2+}	Conditions
			Qm			
CNTs	1.1	—	—	—	—	pH: 5.5, T: 25, S/L: 0.05/100, $C_0 = 9.50$ mg L^{-1}
CNTs	—	—	—	1.0	—	pH: 7.0, T: room, S/L: 0.05/100, $C_0 = 2$–14 mg L^{-1}
CNTs (HNO_3)	—	—	35.6	—	—	pH: 5.0, T: 25, S/L: 0.05/100, $C_0 = 10$–80 mg L^{-1}
SWCNTs	—	—	9.22	—	—	pH: 7.0, T: 25, S/L: 0.05/100
SWCNTs	—	—	—	—	11.23	pH: 7.0, T: 25, S/L: 0.05/100, $C_0 = 10$–80 mg L^{-1}
SWCNTs (NaOCl)	—	—	47.85	—	43.66	—
MWCNTs	—	—	7.53	—	10.21	—
MWCNTs (NaOCl)	—	—	38.46	—	32.68	—
MWCNTs (HNO_3)	7.42	—	6.89	—	—	pH: 8.0, T: room, $C_0 = 1$ (column study)
MWCNTs (HNO_3)	10.86	24.49	—	97.08	—	pH: 5.0, T: room, S/L: 0.05/100, $C_0 = 2$–15 (Cd^{2+}), 5–30 (Cu^{2+}), 10–80 (Pb^{2+})
MWCNTs (HNO_3)	9.80	—	—	—	—	pH: 6.55, T: 60, S/L: 0.15/500, $C_0 = 6$–20 mg L^{-1}
CNTs (H_2O_2)	2.6	—	—	—	—	—
CNTs ($KMnO_4$)	11.0	—	—	—	—	—

Source: Rao, G. P., et al., *Separation and Purification Technology*, 58, 224–231, 2007.

Note: Q_m = maximum sorption capacity (mg g^{-1}), T = temperature (°C), S/L = solid/liquid (g mL^{-1}), C_0 = initial concentration (mg L^{-1}).

TABLE 8.5
Maximum Sorption Capacities of Various Divalent Metal Ions with Other Adsorbents

Adsorbent	Qm					Conditions
	Cd^{2+}	Cu^{2+}	Ni^{2+}	Pb^{2+}	Zn^{2+}	
Fly ash	8.00	8.10	—	—	—	pH: 5.0, T: room, S/L: 0.25/300, $C_0 = 335$ (Cu^{2+}), 320 (Cd^{2+})
Inactivated lichen	—	7.69	—	—	—	pH: 5.0, T: 15, S/L: 0.5/30, $C_0 = 100$
Granulated activated carbon	—	—	20.55	—	—	pH: 7.0, T: 25, S/L: 0.05/100, $C_0 = 60$
Powdered activated carbon	—	—	—	—	13.50	pH: 7.0, T: 25, S/L: 0.05/100, $C_0 = 10$–80 mg L^{-1}
Crab shell	198.97	62.28	—	267.29	—	pH: 5.0, T: 30, S/L: 40/1000, $C_0 = 1000$
Green macroalga	4.70	5.57	—	28.72	2.66	pH: 5.0, S/L: 0.5/100, $C_0 = 100$
Palm shell activated carbon	—	—	—	95.20	—	pH: 5.0, T: 27, S/L: 0.5/100, $C_0 = 10$–700
Activated carbon cloths	15.30	5.80	17.30	—	—	pH: 5.0, T: 20, S/L: 0.5/250, $C_0 = 40$
Kaolinite	—	11.04	2.79	—	—	pH: T: 40, S/L: 1/100
Pinus sylvestris	—	—	—	11.38	—	pH: 4.0, T: room, S/L: 4/100, $C_0 = 50$
Iron slag	—	88.50	—	95.24	—	pH: 5.5, S/L (gm L^{-1}): 2/1000, $C_0 = 20$
Modified chitosan	38.50	109.00	9.60	—	—	pH: 6.0 (Cu^{2+}), 2.0–3.0 (Cd^{2+}, Ni^{2+}), T: 25, S/L: 0.1/100, $C_0 = 100$ (Cu^{2+}), 50 (Cd^{2+}, Ni^{2+})
Granular biomass	60.00	55.00	26.00	255.00	—	pH: 4.0–5.5, T: 21 S/L: 0.5/50, $C_0 = 10$
Sugar beet pulp	24.39	21.16	11.86	73.76	17.79	pH: 4.7, T: 20, $C_0 = 2.5 \times 10^{-3}$ M

Source: Rao, G. P., et al., *Separation and Purification Technology*, 58, 224–231, 2007.

Note: Q_m = maximum sorption capacity (mg g^{-1}), T = temperature (°C), S/L = solid/liquid (g mL^{-1}), C_0 = initial concentration (mg L^{-1}).

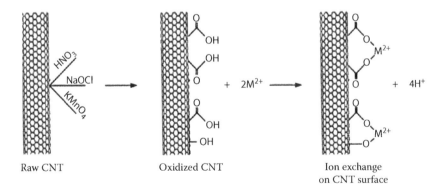

FIGURE 8.6 Schematic diagram of the major mechanism for sorption of divalent metal ions onto carbon nanotube surface. (From Rao, G. P., et al., *Separation and Purification Technology*, 58, 224–231, 2007.)

As for organic ions and organic chemicals, the interactive mechanisms are discussed in a critical review article by Pan and Xing (2008), and are summarized here. In general, adsorption heterogeneity and hysteresis are the two widely recognized features of organic chemical–CNT interactions. However, because different mechanisms may act simultaneously—mainly hydrophobic interactions, π–π bonds, electrostatic interactions, and hydrogen bonds—the prediction of organic chemical adsorption on CNTs is not straightforward. The dominant adsorption mechanism is different for different types of organic chemicals (such as polar and nonpolar); thus, different models may be needed to predict organic chemical–CNT interaction. In addition, each adsorption mechanism is expected to be affected differently by environmental conditions. For example, when H-bonding is the predominant mechanism, increased oxygen-containing functional groups on CNTs would increase the sorption. However, for sorption controlled by hydrophobic interactions, increased functional groups would decrease the accessibility and affinity of CNTs for organic chemicals.

Studies have shown that the kinetic sorption process of metal ions onto CNTs may be well described by the pseudo-second-order rate law, and the rate constants increased with a rise in temperature (Li et al. 2005; Lu et al. 2006) indicating an endothermic reaction (Rao et al. 2007). This could be explained by the increased temperature resulting in a rise in the diffusion rate of metal ions across the external boundary layer and within the pores of CNTs due to decreased solution viscosity (Rao et al. 2007).

Recent studies have also shown that CNTs can be used as heterogeneous coagulants or flocculants (Simate 2012; Simate et al. 2012a). In these studies, both acid-functionalized and pristine CNTs demonstrated the ability to successfully coagulate colloidal particles in brewery wastewater. The results also showed that the heterogeneous coagulation of colloidal particles in brewery wastewater by acid-functionalized CNTs occurs by the mechanism of charge neutralization.

It is noted that the main advantage of employing positively charged CNTs in the treatment of water and wastewaters is that soluble pollutants are expected to be adsorbed; and, at the same time, colloidal particles can be removed through heterogeneous coagulation, which results from surface charge neutralization between the CNTs and colloidal particles (Simate et al. 2012a). The other advantage of CNTs is that their surface chemistry can be tuned to target specific contaminants and, thus, may have unique applications in polishing steps to remove recalcitrant compounds or in the preconcentration of trace organic contaminants for analytical purposes (Qu et al. 2013). These applications require small quantities of materials and, hence, are less sensitive to the material cost.

8.4.2 Carbon Nanotubes in Membrane Filtration

Membrane technologies constitute vital units of many water treatment systems (Liu et al. 2013). However, a major challenge of membrane technology is the inherent trade-off between membrane selectivity and permeability (Qu et al. 2013). Despite these drawbacks, research studies have shown that the incorporation of nanomaterials into membranes can offer a great opportunity to improve membrane permeability, fouling resistance, and mechanical and thermal stability, as well as rendering new functions for contaminant degradation and self-cleaning (Qu et al. 2013). For example, the addition of CNTs to polymeric membranes has been suggested as a possible strategy to reduce membrane breakage and fouling (Brunet et al. 2008). Some studies have also shown that the addition of MWCNTs to polyamide membranes to form a nanocomposite structure improves the mechanical properties of these membranes and their ability to reject key contaminants, with little compromise in membrane permeability (Shawky et al. 2011). Other studies have shown that CNT-blended polysulfone membranes (Choi et al. 2006) and polyethersulfone membranes (Celik et al. 2011) are more hydrophilic and have an enhanced fouling resistance due to the hydrophilic carboxylic groups of functionalized CNTs. Other functional groups such as hydrophilic isophthaloyl-chloride groups (Qiu et al. 2009) and amphilic-polymer groups with protein-resistant ability (Liu et al. 2010) can also be introduced onto CNT surfaces.

Despite combating fouling problems and adding mechanical strength to conventional membranes, studies have shown that there may be a reduction in longer-term membrane performance due to adsorptive fouling, since the addition of CNTs increases membrane hydrophobicity (Shawky et al. 2011). Therefore, some additional factors to consider in the development of these composites should include the need to balance the dimensions of the membrane film cast with those of the reinforcing CNTs, and the effect of derivative CNTs on membrane casting and performance (Shawky et al. 2011).

Some studies have shown that the inner hollow cavities of CNTs or interstices between vertically oriented CNTs can provide a great possibility for filtering water (Simate 2016). For example, Srivastava et al. (2004) successfully fabricated CNT nanofilters that efficiently filtered heavier hydrocarbon species (C_mH_n [m > 12]), from hydrocarboneceous oils such as petroleum (C_mH_n [n = 2m + 2, m = 1–12]). These filtration membranes consisted of hollow cylinders with radially aligned CNT walls. The benefit of using CNT nanofilters is that high flux rates can be obtained with reasonably low pressures (Brady-Estevez et al. 2008; Upadhyayula et al. 2009). In addition, the cytotoxic nature of CNTs prohibits the accumulation of pathogenic biofilms on their surfaces, unlike granulated activated carbon (GAC) filters (Camper et al. 1985; Upadhyayula et al. 2009), which makes cleaning of CNT nanofilters easier than GAC filters. These two benefits of CNT nanofilters greatly offset their high costs. Overall, a major advantage of using nanotube filters over conventional membrane filters lies in that they can be cleaned repeatedly after each filtration process to regain their full filtering efficiency (Srivastava et al. 2004). In fact, a simple process of ultrasonication and autoclaving (~121°C for 30 min) was found to be sufficient for cleaning these filters. In conventional cellulose nitrate/acetate membrane filters used in water filtration, however, strong bacterial adsorption on the membrane surface affects their physical properties and, thus, prevents their reusability as efficient filters; and in addition, most of the typical filters used for virus filtration are not reusable (Srivastava et al. 2004).

Because of the high thermal stability of the nanotubes, CNT nanofilters can also be operated at temperatures of ~400°C, which is several times higher than the highest operating temperatures of conventional polymer membrane filters (~52°C) (Srivastava et al. 2004). The nanotube filters, owing to their high mechanical and thermal stability, may compete with commercially available ceramic filters. Furthermore, in the future, these filters may be tailored to specific needs by controlling nanotube density in the walls and the surface characteristics by chemical functionalization (Srivastava et al. 2004).

8.4.3 CARBON NANOTUBES AS ANTIMICROBIAL MATERIALS

Advances in CNT research have shown that CNTs have antimicrobial mechanisms (Kang et al. 2007, 2008, 2009; Arias and Yang et al. 2009; Yang et al. 2010; Dong et al. 2012; Dizaj et al. 2015; Yah and Simate 2015), with Kang et al. (2007) being the first to show the antimicrobial activities of SWCNTs against *Escherichia coli*. The antimicrobial actions of CNTs and other nanomaterials include destruction of cell membranes, blockage of enzyme pathways, and alterations of microbial cell walls and nucleic material pathways (Yah and Simate 2015). Other antimicrobial mechanisms of nanomaterials include photocatalytic production of reactive oxygen species (ROSs) that can inactivate viruses and cleave DNA; disruption of the structural integrity of the bacterial cell envelope, resulting in leakage of intracellular components; and interruption of energy transduction (Mahendra et al. 2009). In particular, the antibacterial mechanism of CNTs is attributed to a physical interaction in which CNTs pierce cells (Mauter and Elimelech 2008; Li et al. 2008) or oxidative stress that compromises cell membrane integrity (Kang et al. 2007, 2008; Narayan et al. 2005).

Some studies have also shown that CNTs play a significant role in enhancing the activities of other antimicrobial agents (Yah and Simate 2015). For example, the combination of SWCNTs and hydrogen peroxide (H_2O_2) or NaOCl increases the sporicidal effect on the spores of organisms such as *Bacillus* species when compared with treatment with H_2O_2 or NaOCl alone at the same concentrations (Lilly et al. 2012). In such treatments, synergistic mechanisms of efficacy are established, due to the contribution of multiple antimicrobial effects. Further analysis shows that SWCNTs do not only play the role of antimicrobial effect, but also increase the permeability/susceptibility of the *Bacillus* species pathogen to H_2O_2 or NaOCl, thus significantly developing highly effective sporicidal effects (Lilly et al. 2012). Furthermore, findings by Gilbertson et al. (2014) found that oxygen functional groups, when functionalized on MWCNTs, can enhance several MWCNT properties such as redox, electrochemical, and antimicrobial activities. The redox activities include the ability to enhance the oxidation of glutathione and the reduction of surface carboxyl groups that promote the functional performance of MWCNTs' antimicrobial activities for biomedical applications (Gilbertson et al. 2014).

8.4.4 CARBON NANOTUBES AS CATALYSTS/CATALYST SUPPORTS

A significant amount of research has been carried out on the application of CNTs in photocatalytic oxidation (Table 8.6). Basically, photocatalytic oxidation is an advanced oxidation process (AOP) used for the removal of trace contaminants and microbial pathogens (Qu et al. 2013). It is a useful pretreatment technique for hazardous and nonbiodegradable contaminants that enhances their biodegradability. Photocatalysis can also be used as a polishing step for treating recalcitrant organic compounds (Qu et al. 2013). CNTs have found widespread application as ideal building blocks in hybrid photocatalysts, because of their excellent mechanical, electrical, and optical properties (Liu et al. 2013). In addition, CNTs have a large electron-storage capacity and, as already discussed, they can behave either as a semiconductor or a metal. When CNTs are in contact with TiO_2 nanoparticles, for example, they can trigger electron transfer from the conducting band of TiO_2 to the CNT surface due to their lower Fermi level (Liu et al. 2013). The CNTs would then store photogenerated electrons and inhibit the recombination of electrons and holes. Subsequently, the electrons can be transferred to another electron acceptor, such as molecular oxygen, thus forming ROSs (O_2^-, H_2O_2 and ·OH) that can degrade and further mineralize organic pollutants.

Several studies have also utilized CNTs as catalysts and/or catalyst supports in wet air oxidation (WAO) for the treatment of organic and toxic wastewaters (Garcia et al. 2005, 2006; Gomes et al. 2004; Taboada et al. 2009; Yang et al. 2007, 2008). Basically, WAO is a liquid-phase reaction between soluble and suspended organic materials in water and dissolved oxygen from the air (Luck 1999). It is a moderate temperature and high pressure technology (U.S. Army 2003), with typical conditions ranging from 180°C and 2 MPa to 315°C and 15 MPa (Luck 1999).

Although heterogeneous catalysts, such as noble metal and transition metal oxide catalysts, have shown good catalytic activity in the catalytic WAO (CWAO) of organic compounds, there is a shift toward other materials to avoid the leaching of active phases of metal catalysts at acidic operating conditions (Serp et al. 2003; Yang et al. 2015). As a result, several studies have investigated the use of CNTs (particularly MWCNTs) as catalyst supports for Fe (Rodríguez et al. 2009), Cu (Ovejero et al. 2006; Rodríguez et al. 2009), Ru (Ovejero et al. 2006; Garcia et al. 2006), and Pt (Garcia et al. 2006; Ovejero et al. 2006) in CWAO. Recently, there has also been a focus in the study of the catalytic activities of MWCNTs without a metal phase (Yang et al. 2008; Gomes et al. 2008; Rocha et al. 2011; Milone et al. 2011; Yang et al. 2012). In these studies, MWCNTs have exhibited good catalytic performance, and surface functional groups have played an important role in the catalytic performance of the MWCNTs. Figure 8.7 shows a reaction mechanism for producing HO_2· radicals in CWAO of phenol over MWCNTs (Yang et al. 2007). The proposed mechanism is such that dissolved molecular oxygen in the water is first adsorbed on the CNTs and then dissociated on the basal planes of the graphite layers to form the dissociated oxygen atoms (Mestl et al. 2001; Yang et al. 2007). Thereafter, the carboxylic groups (–COOH) on the functionalized MWCNTs and the dissociated oxygen atoms could form HO_2· radicals by hydrogen bonding (Yang et al. 2007; Liu et al. 2013). The radicals may also arouse some radical chain reactions, lead to molecular breakdown, and then decompose phenol to CO_2, H_2O, and low organic compounds in CWAO (Rivas et al. 1998; Yang et al. 2007). The mechanism shows that MWCNTs could effectively improve the formation of the radicals.

8.4.5 CARBON NANOTUBES FOR SENSING AND DETECTION

Nanotechnology is currently being used to develop small and portable sensors with enhanced capabilities for detecting biological and chemical contaminants at very low concentration levels in the environment, including in water. For example, magnetic nanoparticles, quantum dots

TABLE 8.6

Carbon Nanotubes as Photocatalyst Support for the Degradation of Organic Pollutants

Hybrid Photocatalyst	Target Pollutants	Highlights
CNT/TiO$_2$	Azo dye	Enhancement of adsorption of dye; inhibition of charge recombination.
CNT/mesoporous TiO$_2$	Acetone	Inhibition of charge recombination; more hydroxyl groups on the catalyst and more hydroxyl radicals generated.
CNT/TiO$_2$ nanowire film and CNT/P25 film	Methyl orange	CNT/TiO$_2$ nanowire film was more suitable for photocatalytic filtration application owing to its lower pore blockage.
CNT/TiO$_2$	Rhodamine B	Electrospinning method was employed to fabricate CNT/TiO$_2$ composite.
SWCNT/TiO$_2$	Congo red	The addition of silica promoted the coating of TiO$_2$ on CNT. Imitate contact between CNT and TiO$_2$ were needed to achieve enhanced photocatalytic activity.
CNT/TiO$_2$	Phenol	SWCNTs were better support for TiO$_2$ than MWCNTs due to more interfacial contact.
CNT/TiO$_2$ heterojunction array	Phenol	The thickness of TiO$_2$ layer could be controlled by varying the deposition time.
MWCNT/TiO$_2$	2,6-dinitro-p-cresol	No obvious decline in efficiency of the composite photocatalysts was observed after five repeated cycles.
MWCNT/TiO$_2$	2,4-dinitrophenol	The composite was very effective in decolorization and COD reduction of real wastewater from DNP manufacturing.
MWCNT/TiO$_2$	Eosin yellow	Nitrogen was doped to enhance visible photoactivity, and palladium was doped to reduce charge recombination.
MWCNT/TiO$_2$	Acid Blue 92	MWCNT prevented the agglomeration of TiO$_2$ particles.
CNT/TiO$_2$	Aniline, nitrobenzene, benzoic acid	The presence of oxygen-containing functional groups had a positive effect on the photocatalytic activity of the composite.
CNT/TiO$_2$	Methyl orange	CNT also acted as a dispersing template to control the morphology of TiO$_2$.
CNT/titanium silicate	4-Nitrophenol, Rhodamine B	Ball milling removing the physical contact between CNT and titanium silicate greatly reduced photocatalytic activity, indicating the significance of interfacial charge transfer.
MWCNT/TiO$_2$	Atrazine	Microwaves were used to enhance photocatalytic activity; CNT had a beneficial effect on absorbing microwave energy.
TiO$_2$–CeO$_2$/CNT	Phenol derivatives	Presence of mesopores in composite matrix.
Titania nanotube/CNT	Acetaldehyde	Visible light photoactivity was caused by Ti–O–C bond, which was confirmed by XPS and EPR.
CNT/ZnS	Methylene blue	Postrefluxing treatment played a key role in the improvement of the interaction between ZnS nanocrystals and CNTs.
CNT/ZnS	Methyl orange	Microwave-assisted synthesis promoted the dispersion of ZnS and the size of ZnS nanospheres were easily tunable.
CNT/CdS	Azo dye	CNTs hampered the photocorrosion of CdS.
CNT/WO$_3$	Rhodamine B	Mass ratio of CNTs and WO$_3$ were optimized.

Source: Liu, X., et al., *Journal of Environmental Sciences*, 25, 1263–1280, 2013.

(QDs), noble metals, dye-doped nanoparticles, and CNTs have been the commonly used nanomaterials in pathogen detection (Qu et al. 2013). As for CNTs, research studies have shown that the high conductivity along their length makes them outstanding electrode materials (Qu et al. 2013). As a result, CNTs can greatly facilitate electrochemical detection by promoting electron transfer (McCreery 2008; Qu et al. 2013) and electrode–analyte interactions. Besides their

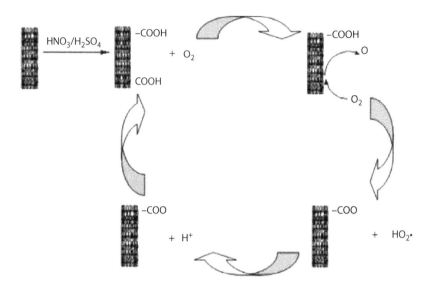

FIGURE 8.7 Mechanism producing the radical (HO_2) in CWAO of phenol over the MWCNTs. (From Yang, S. X., et al., *Catalysis Communications*, 8, 2059–2063, 2007.)

excellent electronic properties, the high adsorption capacity of CNTs also increases their detection sensitivity (Collins et al. 2000).

In trace organic or inorganic contaminant detection, CNTs can be used in both detection and concentration. This is because adsorption of charged species onto CNTs results in changes in conductance, thus providing the basis for the correlation between analyte concentration and current fluctuation (Mauter and Elimelech 2008). CNTs have also been extensively studied regarding the preconcentration of a variety of organic compounds (Cai et al. 2003) and metal ions (Duran et al. 2009). Overall, CNTs have great potential for environmental analysis of trace metal or organic pollutants, as they offer high adsorption capacity as well as fast kinetics (Qu et al. 2013).

Despite the several advantages and applications discussed, the major challenge for CNT-based sensors is the heterogeneity of CNTs (Qu et al. 2013). The separation of metallic and semiconducting SWNTs has been extensively studied but is still far from perfect. The production and purification processes of CNTs often introduce impurities, contaminants, and even degradation of the CNT structure. Therefore, better synthesis, purification, and separation methods are required to produce more homogeneous CNTs (Qu et al. 2013). Furthermore, although most of the nanosensors possess excellent photostability and sensitivity, nonspecific binding is still a major challenge for their application in water and wastewater (Qu et al. 2013).

8.5 CHALLENGES AND CONCLUDING REMARKS

CNTs (and other carbon nanomaterials) have reached the forefront of many industrial research projects since their rediscovery (Iyuke and Simate 2011). These materials sound like a product designer's dream. However, as would be expected with many technologies that offer benefits, every new technology comes with many challenges, starting with the unacceptability of the issues of practicality and cost implications. These issues have to be addressed sensibly so that the full benefits can be realized. Therefore, it would seem incomplete if there is no discussion of some of the issues and challenges that have faced CNT applications from the very beginning. In fact, these challenges have slowed down the pace of nanotube commercialization.

The full potential of applications of nanotubes will not be realized until the growth of nanotubes can be further optimized and controlled (Dai 2002a; Iyuke and Simate 2011). For example, for practical commercial applications of CNTs, one would need quantities in the kilogram range (Subramoney 1999). For applications such as membrane composites, it is desired to obtain high-quality nanotubes at the kilogram or ton level using growth methods that are simple, efficient, and inexpensive (Dai 2002a; Paradise and Goswami 2007). However, as large quantities of CNT materials reach the consumer market, it will also be necessary to establish disposal and/or reuse procedures (de Volder 2013). The CNTs may enter municipal waste streams, where, unless they are incinerated, cross-contamination during recycling is possible (Köller et al. 2008; de Volder 2013). Therefore, broader partnerships among industry, academia, and government are needed to investigate the environmental and societal impact of CNTs throughout their life cycle (de Volder 2013).

The CNTs also need to be synthesized in longer lengths, and improved techniques are required to align and evenly distribute them (Iyuke and Simate 2011). It is also important to grow structurally perfect nanotubes and control the chirality of SWCNTs (Dai 2002a). For devices such as nanotube-based electronics and nanofilters, scale-up will unavoidably rely on self-assembly techniques or controlled growth strategies on surfaces combined with microfabrication techniques (Dai 2002a). The heterogeneity of CNTs is also a challenge. Those CNTs with a wide distribution of diameters, divisions of the individual nanotubes, and the presence of residual metals are not suitable in some applications, such as CNT-based sensors (Eatemadi 2014).

Researchers also agree that one of the major obstacles to using CNTs is cost (Breuer and Sundararaj 2004). Generally, the synthesis techniques used for making nanotubes are expensive. The price of CNTs as of 2003 was from US$200/g for MWCNTs, and the cost is up to 10 times more for SWCNTs (Corrias et al. 2003). However, the cost factor also leads to the use of MWCNTs more than that of SWCNTs (Baugham et al. 2002; Paradise and Goswami 2007). It should be noted that the cost of raw materials used for CNT production (methane, ethylene, propylene, etc.) is low. In addition, the catalysts that are used in the CVD process do not contain rare or noble metals. The cost of catalysts today is relatively high because of the complicated technology needed for the creation of nanodimensional structures, with prescribed size and structure of catalytically active sites. However, the technologies of both the CNT CVD process and the catalyst production process are being continuously improved. So, in the near future, the cost of CNT may be less than US$100/kg.

In the water industry, a lot of challenges arise when attempting to use CNTs in their present state as coagulants or flocculants (Simate et al. 2011, 2012a; Simate 2012). Firstly, CNTs lack dispersion and solubility properties. In recent years, however, there have been several successful attempts to prepare water-soluble CNTs by various techniques (Zhang et al. 2009; Pei et al. 2008) and improvements in their dispersivity through functionalization (Zhang et al. 2002). However, converting CNTs to complete water-soluble form is not recommended, since this poses practical difficulties for using them as adsorbent media (Upadhyayula et al. 2009). Nevertheless, dispersivity of CNTs is important, because the benefit of water-soluble CNTs can be exploited further in the fabrication of CNT–composite membranes (Upadhyayula et al. 2009). On the other hand, semidispersible and partially hydrophobic CNTs exhibit greater affinity toward bacteria than either completely dispersed or weakly dispersed CNTs (Upadhyayula et al. 2009).

The other issues of utmost importance to consider are the safety, health, and environmental aspects related to CNTs. At the moment, there are very few studies on these subjects. Isolated experiments have shown that nanotubes are not toxic (Endo et al. 2008). However, several other studies have shown that CNTs have potential impacts on human health and ecosystems (Moore 2006; Smart et al. 2006; Wiesner 2006; Simate et al. 2012b). Whatever the final word is, potential users and researchers need to be aware of protocols on how to handle these materials, both in the laboratory and in an industrial setting (Endo et al. 2008). Well-defined regulatory controls need to be in place when large quantities of the material are handled routinely, and there need to be standard procedures established on how to package and transport nanotube materials and how to handle them (Endo et al. 2008).

In conclusion, the myriad applications of CNTs in various fields—including water—since their rediscovery are no longer debatable. Due to the great importance of CNTs, it is clear that novel technologies in water will emerge in the future. It is also expected that more applications of CNTs will be found as some of their unknown unique properties are discovered. However, the growth in demand for CNTs in more applications entails the need for the reproducibility of specific CNT materials, and a proper transition from laboratory- to large-scale production (Grobert 2007). Furthermore, the compatibility between the aforementioned nanotechnologies and current water and wastewater treatment processes and infrastructure also needs to be addressed (Qu et al. 2013). Most treatment plants and distribution systems in developed countries are expected to remain in place for decades to come. As a result, it is important to be able to implement nanotechnology with minimal changes to existing infrastructure in the near term.

REFERENCES

Agboola, A. E., Pike, R. W., Hertwig, T. A., and Lou, H. H. (2007). Conceptual design of carbon nanotube processes. *Clean Technologies and Environmental Policy* 9: 289–311.

Arias, L. R. and Yang, L. (2009). Inactivation of bacterial pathogens by carbon nanotubes in suspensions. *Langmuir* 25 (5): 3003–3012.

Baddour, C. E. and Briens, C. (2005). Carbon nanotube synthesis: A review. *International Journal of Chemical Reactor Engineering* 3: Review R3.

Balasubramanian, K. and Burghard, M. (2005). Chemically functionalized carbon nanotubes. *Small* 1 (2): 180–192.

Baugham, R., Zakhidov, A., and Heer, W. (2002). Carbon nanotubes: The route toward applications. *Science* 297 (5582): 787–792.

Bernholc, J., Brabec, C., Buongiorno, N. M., Maiti, A., Roland, C., and Yakobson, B. I. (1998). Theory of growth and mechanical properties of nanotubes. *Applied Physics A* 67: 39–46.

Brady-Estevez, A. S., Kang, S., and Elimelech, M. (2008). A single walled carbon nanotube filter for removal of viral and bacterial pathogens. *Small* 4: 481–484.

Braidy, N., El Khakani, M. A., and Botton, G. A. (2002). Single-wall carbon nanotubes synthesis by means of UV laser vaporization. *Chemical Physics Letters* 354: 88–92.

Breuer, O. and Sundararaj, U. (2004). Big returns from small fibers: A review of polymer/carbon nanotube composites. *Polymer Composites* 25 (6): 630–645.

Brunet, L., Lyon, D. Y., Zodrow, K., Rouch, J. C., Caussat, B., Serp, P., Remigy, J. C., Wiesner, M.R., and Alvarez, P. J. J. (2008). Properties of membranes containing semi-dispersed carbon nanotubes. *Environmental Engineering Science* 25 (4): 1–11.

Cai, Y. Q., Jiang, G. B., Liu, J. F., and Zhou, Q. X. (2003). Multiwalled carbon nanotubes as a solid-phase extraction adsorbent for the determination of bisphenol a, 4-n-nonylphenol, and 4-tertoctylphenol. *Analytical Chemistry* 75 (10): 2517–2521.

Camper, A. K., Lechevallier, M. W., Broadway, S. C., and McFeters, G. A. (1985). Growth and persistence of pathogens on granular activated carbon filters. *Applied and Environmental Microbiology* 50: 1378–1382.

Celik, E., Park, H., Choi, H., and Choi, H. (2011). Carbon nanotube blended polyethersulfone membranes for fouling control in water treatment. *Water Research* 45 (1): 274–282.

Choi, J. H., Jegal, J., and Kim, W. N. (2006). Fabrication and characterization of multi-walled carbon nanotubes/polymer blend membranes. *Journal of Membrane Science* 284 (1–2): 406–415.

Coleman, K. S. (2008). Nanotubes. *Annual Reports on the Progress of Chemistry, Section A: Inorganic Chemistry* 104: 379–393.

Collins, P. G. and Avouris, P. (2000). Nanotubes for electronics. *Scientific American* 283: 62–69.

Collins, P. G., Bradley, K., Ishigami, M., and Zettl, A. (2000). Extreme oxygen sensitivity of electronic properties of carbon nanotubes. *Science* 287 (5459): 1801–1804.

Cornwell, C. F. and Wille, L. T. (1997). Elastic properties of single-walled carbon nanotubes in compression. *Solid State Communications* 101 (8): 555–558.

Corrias, M., Caussat, B., Ayral, A., Durand, J., Kihn, Y., Kalck, P., and Serp, P. (2003). Carbon nanotubes produced in a fluidized bed catalytic CVD: First approach of the process. *Chemical Engineering Science* 58: 4475–4482.

Dai, H. (2002a). Carbon nanotubes: Opportunities and challenges. *Surface Science* 500: 218–241.

Dai, H. (2002b). Carbon nanotubes: Synthesis, integration, and properties. *Accounts of Chemical Research* 35 (12): 1035–1044.

Delmotte, J. P. and Rubio, A. (2002). Mechanical properties of carbon nanotubes: A fiber digest for beginners. *Carbon* 40 (10):1729–1734.

Deshmukh, A. A., Mhlanga, S. D., and Coville, N. J. (2010). Carbon spheres. *Materials Science and Engineering Review* 70 (1–2): 1–28.

de Volder, M. F. L., Tawfick, S. H., Baughman, R. H., and Hart, A. J. (2013). Carbon nanotubes: Present and future commercial applications. *Science* 339: 535–539.

Di, Z. C., Ding, J., Peng, X. J., Li, Y. H., Luan, Z. K., and Liang, J. (2006). Chromium adsorption by aligned carbon nanotubes supported ceria nanoparticles. *Chemosphere* 62: 861–865.

Dizaj, S. M., Mennati, A., Jafari, S., Khezri, K., and Adibkia, K. (2015). Antimicrobial activity of carbon-based nanoparticles. *Advanced Pharmaceutical Bulletin* 5 (1): 19–23.

Dong, L., Henderson, A., and Field, C. (2012). Antimicrobial activity of single-walled carbon nanotubes suspended in different surfactants. *Journal of Nanotechnology* 2012: 1–7.

Dresselhaus, M. S., Dresselhaus, G., and Avouris, P. (Eds.) (2001). *Carbon Nanotubes: Synthesis, Structure, Properties and Applications*. Springer, New York.

Dresselhaus, M. S., Dresselhaus, G., and Eklund, P. C. (1996). *Science of Fullerenes and Carbon Nanotubes*. Academic Press, New York.

Dresselhaus, M. S., Dresselhaus, G., and Saito, R. (1995). Physics of carbon nanotubes. *Carbon* 33 (7): 883–891.

Dunlap, B. I. (1992). Connecting carbon tubules. *Physical Reviews B* 46: 1933.

Duran, A., Tuzen, M., and Soylak, M. (2009). Preconcentration of some trace elements via using multiwalled carbon nanotubes as solid phase extraction adsorbent. *Journal of Hazardous Materials* 169 (1–3), 466–471.

Eatemadi, A., Daraee, H., Karimkhanloo, H., Kouhi, M., Zarghami, N., Akbarzadeh, A., Abasi, M., Hanifehpour, Y., and Joo, S. W. (2014). Carbon nanotubes: Properties, synthesis, purification, and medical applications. *Nanoscale Research Letters* 9: 393.

Endo, M., Strano, M. S., and Ajayan, P. M. (2008). Potential applications of carbon nanotubes. In: Jorio, A., Dresselhaus, G., and Dresselhaus, M. S. (Eds.), *Carbon Nanotubes, Topics in Applied Physics*, Vol. 111, 13–62. Springer-Verlag, Berlin, Heidelberg.

Esawi, A. M. K. and Farag, M. M. (2007). Carbon nanotube reinforced composites: Potential and current challenges. *Materials and Design* 28 (9): 2394–2401.

Fuhrer, M. S., Nygård, J., Shih, L., Forero, M., Yoon, Y. G., Mazzoni, M. S. C., Choi, H. J., et al. (2000). Crossed nanotube junctions. *Science* 288 (5465): 494–497.

Garcia, J., Gomes, H. T., Serp, P., Kalck, P., Figueiredo, J. L., and Faria, J. L. (2005). Platinum catalysts supported on MWNT for catalytic wet air oxidation of nitrogen containing compounds. *Catalysis Today* 102: 101–109.

Garcia, J., Gomes, H. T., Serf, P., Kalck, P., Figueiredo, J. L., and Faria, J. L. (2006). Carbon nanotube supported ruthenium catalysts for the treatment of high strength wastewater with aniline using wet air oxidation. *Carbon* 44 (12): 2384–2391.

Gilbertson, L. M., Goodwin, D. G., Taylor, A. D., Pfefferle, L., and Zimmerman, J. B. (2014). Toward tailored functional design of multi-walled carbon nanotubes (MWNTs): Electrochemical and antimicrobial activity enhancement via oxidation and selective reduction. *Environmental Science and Technology* 48: 5938–5945.

Goering, J. and Burghaus, U. (2007). Adsorption kinetics of thiophene on single-walled carbon nanotubes (CCNTs). *Chemical Physics Letters* 447: 121–126.

Gomes, H. T., Machado, B. F., Ribeiro, A., Moreira, I., Rosário, M., Silva, A. M. T., Figueiredo, J. L., and Faria, J. L. (2008). Catalytic properties of carbon materials for wet oxidation of aniline. *Journal of Hazardous Materials* 159 (2–3): 420–426.

Gomes, H. T., Samant, P. V., Serp, P., Kalck, P., Figueiredo, J. L., and Faria, J. L. (2004). Carbon nanotubes and xerogels as supports of well-dispersed Pt catalysts for environmental applications. *Applied Catalysis, B: Environmental* 54 (3): 175–182.

Grobert, N. (2007). Carbon nanotubes: Becoming clean. *Materials Today* 10 (1): 28–35.

Guo, T., Nikolaev, P., Thess, A., Colbert, D. T., and Smalley, R. E. (1995). Catalytic growth of single-walled nanotubes by laser vaporization. *Chemical Physics Letters* 243: 49–54.

Harris, P. J. F. (1999). *Carbon Nanotubes and Related Structures: New Materials for the Twenty-First Century*. Cambridge University Press, Cambridge.

Iijima, S. (1991). Helical microtubules of graphitic carbon. *Nature* 354: 56–58.

Iijima, S., Brabec, C., Maiti, A., and Bernholc, J. (1996). Structural flexibility of carbon nanotubes. *Journal of Chemical Physics* 104 (5): 2089–2092.

Iyuke, S. E. (2007). A process for production of carbon nanotubes, PCT Application WO2007/026213 (published 2007/03/08) priority data ZA 2005/03438 2005/08/29.

Iyuke, S. E. and Simate, G. S. (2011). Synthesis of carbon nanomaterials in a swirled floating catalytic chemical vapour deposition reactor for continuous and large scale production. In: Naraghi. M. (Ed.), *Carbon Nanotubes: Growth and Applications*, 35–58. In-Tech, Rijeka.

Journet, C. and Bernier, P. (1998). Production of carbon nanotubes. *Applied Physics A* 67: 1–9.

Journet, C., Maser, W. K., Bernier, P., Loiseau, A., de la Chapelle, M. L., Lefrant, S., Deniard, P., Lee, R., and Fischer, J. E. (1997). Large-scale production of single-walled carbon nanotubes by the electric-arc technique. *Nature* 388: 756–758.

Kang, S., Herzberg, M., Rodrigues, D. F., and Elimelech, M. (2008). Antibacterial effects of carbon nanotubes: Size does matter. *Langmuir* 24: 6409–6413.

Kang, S., Mauter, M. S., and Elimelech, M. (2009). Microbial cytotoxicity of carbon-based nanomaterials: Implications for river water and wastewater effluent. *Environmental Science and Technology* 43: 2648–2653.

Kang, S., Pinault, M., Pfefferle, L. D., and Elimelech, M. (2007). Single-walled carbon nanotubes exhibit strong antimicrobial activity. *Langmuir* 23: 8670–8673.

Köller, A. R., Som, C., Helland, A., and Gottschalk, F. (2008). Studying the potential release of carbon nanotubes throughout the application life cycle. *Journal of Cleaner Production* 16: 927–937.

Krishnan, A., Dujardin, E., Ebbesen, T. W., Yianilos, P. N., and Treacy, M. M. J. (1998). Young's modulus of single-walled nanotubes. *Physical Review B* 58 (20): 14013–14019.

Kroto, H. W., Heath, J. R., O'Brien, S. C., Curl, R. F., and Smalley, R. E. (1985). C60: Buckminsterfullerene. *Nature* 318 (6042): 162–163.

Lee, S. J., Baik, H. K., Yoo, J., and Han, J. H. (2002). Large scale synthesis of carbon nanotubes by plasma rotating arc discharge technique. *Diamond and Related Materials* 11: 914–917.

Li, Q., Mahendra, S., Lyon, D. Y., Brunet, L., Liga, M. V., Li, D., and Alvarez, P. J. J. (2008). Antimicrobial nanomaterials for water disinfection and microbial control: Potential applications and implications. *Water Research* 42: 4591–4602.

Li, Y. H., Di, Z., Ding, J., Wu, D., Luan, Z., and Zhu, Y. (2005). Adsorption thermodynamic, kinetic and desorption studies of Pb^{2+} on carbon nanotubes. *Water Research* 39: 605–609.

Li, Y-H., Ding, J., Luan, Z., Di, Z., Zhu, Y., Xu, C., Wu, D., and Wei, B. (2003a). Competitive adsorption of Pb^{2+}, Cu^{2+} and Cd^{2+} ions from aqueous solutions by multiwalled carbon nanotubes. *Carbon* 41: 278–2792.

Li, Y-H., Wang, S., Zhang, X., Wei, J., Xu, C., Luan, Z., and Wu, D. (2003b). Adsorption of fluoride from water by aligned carbon nanotubes. *Materials Research Bulletin* 38 (3): 469–476.

Li, Y. H., Wang, S., Wei, J., Zhang, X., Xu, C., Luan, Z., Wu, D., and Wei, B. (2002). Lead adsorption on carbon nanotubes. *Chemical Physics Letters* 357: 263–266.

Li, Y-H., Zhao, Y. M., Hu, W. B., Ahmad, I., Zhu, Y. Q., Peng, J. X., and Luan, Z. K. (2007). Carbon nanotubes: The promising adsorbent in wastewater treatment. *Journal of Physics: Conference Series* 61: 698–702.

Li, Y-H., Zhu, Y., Zhao, Y., Wu, D., and Luan, Z. (2006). Different morphologies of carbon nanotubes effect on the lead removal from aqueous solution. *Diamond and Related Materials* 15: 90–94.

Lilly, M., Dong, X., McCoy, E., and Yang, L. (2012). Inactivation of *Bacillus anthracis* spores by single-walled carbon nanotubes coupled with oxidizing antimicrobial chemicals. *Environmental Science and Technology* 46 (24): 13417–13424.

Lin, T., Bajpai, V., Ji, T., and Dai, L. (2003). Chemistry of carbon nanotubes. *Australian Journal of Chemistry* 56: 635–651.

Liu, R. (2008). The functionalisation of carbon nanotubes, PhD thesis, University of New South Wales, Australia.

Liu, X., Wang, M., Zhang, S., and Pan, B. (2013). Application potential of carbon nanotubes in water treatment: A review. *Journal of Environmental Sciences* 25 (7): 1263–1280.

Liu, Y. L., Chang, Y., Chang, Y. H., and Shih, Y. J. (2010). Preparation of amphiphilic polymer-functionalized carbon nanotubes for low-protein-adsorption surfaces and protein-resistant membranes. *ACS Applied Materials and Interfaces* 2 (12): 3642–3647.

Lu, C. and Chiu, C. (2006). Adsorption of zinc (II) from water with purified carbon nanotubes. *Chemical Engineering Science* 61: 1138–1145.

Lu, C., Chiu, H., and Liu, C. (2006). Removal of zinc (II) from aqueous solution by purified carbon nanotubes: Kinetics and equilibrium studies. *Industrial and Engineering Chemistry Research* 45: 2850–2855.

Lu, C., Chung, Y-L., and Chang, K-F. (2005). Adsorption of trihalomethanes from water with carbon nanotubes. *Water Research* 39: 1183–1189.

Lu, C. and Liu, C. (2006). Removal of nickel (II) from aqueous solution by carbon nanotubes. *Journal of Chemical Technology and Biotechnology* 81: 1932–1940.

Lu, C. and Su, F. (2007). Adsorption of natural organic matter by carbon nanotubes. *Separation and Purification Technology* 58: 113–121.

Lu, J. P. (1997). Elastic properties of carbon nanotubes and nanoropes. *Physical Review Letters* 79: 1297–1300.

Luck, F. (1999). Wet air oxidation: Past, present and future. *Catalysis Today* 53: 81–91.

Mahendra, S., Li, Q., Lyon, D. Y., Brunet, L., and Alvarez, P. J. J. (2009). Nanotechnology-enabled water disinfection and microbial control: Merits and limitations. In: Nora, S., Mamadou, D., Jeremiah, D., Anita, S., and Richard, S. (Eds), *Nanotechnology Applications for Clean Water*, 157–166. William Andrew, Boston, MA.

Matarredona, O., Rhoads, H., Li, Z., Harwell, J. H., Balzano, L., and Resasco, D. E. (2003). Dispersion of single-walled carbon nanotubes in aqueous solutions of the anionic surfactant NaDDBS. *Journal of Physical Chemistry B* 107: 13357–13367.

Mauter, M. S. and Elimelech, M. (2008). Environmental applications of carbon-based nanomaterials. *Environmental Science and Technology* 42 (16): 5843–5859.

McCreery, R. L. (2008). Advanced carbon electrode materials for molecular electrochemistry. *Chemical Reviews* 108 (7): 2646–2687.

Merkoci, A. (2006). Carbon nanotubes in analytical sciences. *Microchimica Acta* 152 (3–4): 157–174.

Mestl, G., Maksimova, N. I., Keller, N., Roddatis, V. V., and Schlögl, R. (2001). Carbon nanofilaments in heterogeneous catalysis: An industrial application for new carbon materials. *Angewandte Chemie International Edition* 40 (11): 2066–2068.

Méténier, K., Bonnamy, S., Béguin, F., Journet, C., Bernier, P., de la Chapelle, L. M., Chauvet, O., and Lefrant, S. (2002). Coalescence of single walled nanotubes and formation of multi-walled carbon nanotubes under high temperature treatments. *Carbon* 40: 1765–1773.

Meyyappan, M. (2004). Growth: CVD and PECVD. In: Meyyappan, M. (Ed.), *Carbon Nanotubes: Science and Applications.* Chapter 4. CRC Press, Boca Raton, FL.

Milone, C., Hameed, A. R. S., Piperopoulos, E., Santangelo, S., Lanza, M., and Galvagno, S. (2011). Catalytic wet air oxidation of p-coumaric acid over carbon nanotubes and activated carbon. *Industrial and Engineering Chemistry Research* 50 (15): 9043–9053.

Moisala, A., Nasibulin, A. G., Brown, D. P., and Jiang, H. (2006). Single-walled carbon nanotube synthesis using ferrocene and iron pentacarbonyl in a laminar flow reactor. *Chemical Engineering Science* 61: 4393–4402.

Monthioux, M. and Kuznetsov, V. L. (2006). Who should be given the credit for the discovery of carbon nanotubes? *Carbon* 44: 1621–1623.

Moore, M. N. (2006). Do nanoparticles present ecotoxicological risks for the health of the aquatic environment. *Environmental International* 32: 967–976.

Narayan, R. J., Berry, C. J., and Brigmon, R. L. (2005). Structural and biological properties of carbon nanotube composite films. *Material Science and Engineering B* 123: 123–129.

Niyogi, S., Hamon, M. A., Hu, H., Zhao, B., Bhowmik, P., Sen, R., Itkis, M. E., and Haddon, R. C. (2002). Chemistry of single-walled carbon nanotubes. *Accounts of Chemical Research* (16): 3476–3479.

Nolan, P. E., Schabel, M. J., Lynch, D. C., and Cutler, A. H. (1995). Hydrogen control of carbon deposit morphology. *Carbon* 33: 79–85.

Ovejero, G., Sotelo, J. L., Romero, M. D., Rodríguez, A., Ocaña, M. A., Rodríguez, G., and García, J. (2006). *Industrial and Engineering Chemistry Research* 45: 2206–2212.

Pan, B. and Xing, B. (2008). Adsorption mechanisms of organic chemicals on carbon nanotubes. *Environmental Science and Technology* 42 (24): 9005–9013.

Papadoupoulos, C., Rakitin, A., Li, J., Vedeneev, A. S., and Xu, J. M. (2000). Electronic transport in Y-junction carbon nanotubes. *Physical Review Letters* 85 (16): 3476–3479.

Paradise, M. and Goswami, T. (2007). Carbon nanotubes: Production and industrial applications. *Materials and Design* 28: 1477–1489.

Pei, X., Hua, L., Liu, W., and Hao, J. (2008). Synthesis of water-soluble carbon nanotubes via surface initiated redox polymerization and their tribological properties as water-based lubricant additive. *European Polymer Journal* 44: 2458–2464.

Pokropivny, V., Lohmus, R., Hussainova, I., Pokropivny, A., and Vlassov, S. (2007). *Introduction to Nanomaterials and Nanotechnology.* Available at http://www.fi.tartu.ee/~rynno/raamat/Introduction%20in%20nanomaterials-sisu.pdf (Accessed September 2015).

Popov, V. N. (2004). Carbon nanotubes: Properties and applications. *Materials Science and Engineering Reports* 43: 61–102.

Qiu, S., Wu, L. G., Pan, X. J., Zhang, L., Chen, H. L., and Gao, C. J. (2009). Preparation and properties of functionalized carbon nanotube/PSF blend ultrafiltration membranes. *Journal of Membrane Science* 342 (1–2): 165–172.

Qu, X., Alvarez, P. J. J., and Li, Q. (2013). Applications of nanotechnology in water and wastewater treatment. *Water Research* 47: 3931–3946.

Rai, P. K., Parra-Vasquez, A. N. G., Chattopadhyay, J., Pinnick, R. A., Liang, F., Sadana, A. K., Hauge, R. H., Billups, W. E., and Pasqual, M. (2007). Dispersions of functionalized single-walled carbon nanotubes in strong acids: Solubility and rheology. *Journal of Nanoscience and Nanotechnology* 7 (10): 3378–3385.

Rao, G. P., Lu, C., and Su, F. (2007). Sorption of divalent metal ions from aqueous solution by carbon nanotubes: A review. *Separation and Purification Technology* 58: 224–231.

Rivas, F. J., Kolaczkowski, S. T., Beltrán, F. J., and McLurgh, D. B. (1998). Development of a model for the wet air oxidation of phenol based on a free radical mechanism. *Chemical Engineering Science* 53 (14): 2575–2586.

Robertson, J. (2004). Realistic applications of CNTs. *Materials Today* 7 (10): 46–52.

Rocha, R. P., Sousa, J. P. S., Silva, A. M. T., Pereira, M. F. R., and Figueiredo, J. L. (2011). Catalytic activity and stability of multiwalled carbon nanotubes in catalytic wet air oxidation of oxalic acid: The role of the basic nature induced by the surface chemistry. *Applied Catalysis B: Environmental* 104 (3–4): 330–336.

Rodríguez, A., Ovejero, G., Mestanza, M., Callejo, V., and García J. (2009). Degradation of methylene blue by catalytic wet air oxidation with Fe and Cu catalyst supported on multiwalled carbon nanotubes. *Chemical Engineering Transactions* 17: 145–151.

Rosso, M. A. (2001). *Origins, Properties, and Applications of Carbon Nanotubes and Fullerenes.* IT 283 Advance Materials and Processes, California State University, Fresno, CA.

Saito, R., Dresselhaus, G., and Dresselhaus, M. S. (1998). *Physical Properties of Carbon Nanotubes*, 4th edition. World Scientific, Singapore.

Saito, Y., Nishikubo, K., Kawabata, K., and Matsumoto, T. (1996). Carbon nanocapsules and single-layered nanotubes produced with platinum-group metals (Ru, Rh, Pd, Os, Ir, Pt) by arc discharge. *Journal of Applied Physics* 80 (5): 3062–3067.

Saito, Y. and Uemura, S. (2000). Field emission from carbon nanotubes and its application to electron sources. *Carbon* 38: 169–182.

Savage, N. and Diallo, M. S. (2005). Nanomaterials and water purification: Opportunities and challenges. *Journal of Nanoparticle Research* 7: 331–342.

See, C. H. and Harris, A. T. (2007). A review of carbon nanotube synthesis via fluidized-bed chemical vapor deposition. *Industrial and Engineering Chemistry Research* 46: 997–1012.

Serp, P., Corrias, M., and Kalck, P. (2003). Carbon nanotubes and nanofibers in catalysis. *Applied Catalysis A: General* 253 (2): 337–358.

Shawky, H. A., Chae, S., Lin, S., and Wiesner, M. R., 2011. Synthesis and characterization of a carbon nanotube/polymer nanocomposite membrane for water treatment. *Desalination* 272 (1–3): 46–50.

Simate, G. S. (2012). The treatment of brewery wastewater using carbon nanotubes synthesized from carbon dioxide carbon source, PhD thesis, University of the Witwatersrand, Johannesburg, South Africa.

Simate, G. S. (2016). The use of carbon nanotubes in the treatment of water and wastewater. In: Thakur, V. K. and Thakur, M. (Eds), *Chemical Functionalisation of Carbon Nanomaterials*, pp. 705–717. CRC Press, Boca Raton.

Simate, G. S., Cluett, J., Iyuke, S. E., Musapatika, E. T., Ndlovu, S., Walubita, L. F., and Alvarez, A. E. (2011). The treatment of brewery wastewater for reuse: State of the art. *Desalination* 273 (2–3): 235–247.

Simate, G. S., Iyuke, S. E., Ndlovu, S., and Heydenrych, M. (2012a). The heterogeneous coagulation and flocculation of brewery wastewater using carbon nanotubes. *Water Research* 46 (4): 1185–1197.

Simate, G. S., Iyuke, S. E., Ndlovu, S., Heydenrych, M., and Walubita, L. F. (2012b). Human health effects of residual carbon nanotubes and traditional water treatment chemicals in drinking water. *Environment International* 39 (1): 38–49.

Simate, G. S., Iyuke, S. E., Ndlovu, S., Yah, C., and Walubita, L. F. (2010). The production of carbon nanotubes from carbon dioxide: Challenges and opportunities. *Journal of Natural Gas Chemistry* 19 (5): 453–460.

Smart, S. K., Cassady, A. I., Lu, G. Q., and Martin, D. J. (2006). The biocompatibility of carbon nanotubes. *Carbon* 44: 1034–1047.

Sinnott, S. B. and Andrews, R. (2001). Carbon nanotubes: Synthesis, properties, and applications. *Critical Reviews in Solid State and Materials Sciences* 26 (3): 145–249.

Srivastava, A., Srivastava, O. N., Talapatra, S., Vajtai, R., and Ajayan P. M. (2004). Carbon nanotube filters. *Nature Materials* 3: 610–614.

Subramoney, S. (1999). Carbon nanotubes: A status report. *Interface* 8: 34–37.

Taboada, C. D., Batista, J., Pintar, A., and Levec, J. (2009). Preparation, characterization and catalytic properties of carbon nanofiber-supported Pt, Pd, Ru monometallic particles in aqueous-phase reactions. *Applied Catalysis B: Environmental* 89 (3–4): 375–382.

Tans, S. J., Verschueren, A. R. M., and Dekker, C. (1998). Room-temperature transistor based on a single carbon nanotube. *Nature* 393 (6680): 49–52.

Tasis, D., Tagmatarchis, N., Bianco, A., and Prato, M. (2006). Chemistry of carbon nanotubes. *Chemical Reviews* 106 (3): 1105–1136.

Terrones, M. (2003). Science technology of the twenty-first century: Synthesis, properties, and applications of carbon nanotubes. *Annual Review of Materials Research* 33: 419–501.

Tom, G. (2003). An Introduction to Carbon Nanotubes. Available at http://web.stanford.edu/group/cpima/education/nanotube_lesson.pdf [Accessed September 2015].

Treacy, M. M. J., Ebbesen, T. W., and Gibson, J. M. (1996). Exceptionally high Young's modulus observed for individual carbon nanotubes. *Nature* 381: 678–680.

Upadhyayula, V. K. K., Deng, S., Mitchell, M. C., and Smith, G. B. (2009). Application of carbon nanotube technology for removal of contaminants in drinking water: A review. *Science of the Total Environment* 408: 1–13.

U.S. Army (2003). Wet Air Oxidation Technology Assessment. Available at https://clu-in.org/acwaatap/06-wetairoxidationfinal.pdf [Accessed September 2015].

Varshney, K. (2014). Carbon nanotubes: A review on synthesis, properties and applications. *International Journal of Engineering Research and General Science* 2 (4): 660–677.

Voit, J. (1995). One-dimensional fermi liquid. *Reports on Progress in Physics* 58: 977–1116.

Wiesner, M. R., Lowry, G. V., Alvarez, P., Dionysiou, D., and Biswas, P. (2006). Assessing the risks of manufactured nanomaterials. *Environmental Science and Technology* 40 (14): 4336–4337.

Yacaman, M., Yoshidam M., Rendón, L., and Santiesteban, J. G. (1993). Catalytic growth of carbon microtubules with fullerene structure. *Applied Physics Letters* 62: 202–204.

Yah, C. S., Iyuke, S. E., Simate, G. S., Unuabonah, E. I., Bathgate, G., Matthews, G., and Cluett, J. D. (2011). Continuous synthesis of multi-walled carbon nanotubes from xylene using the swirled floating catalyst chemical vapour deposition technique. *Journal of Materials Research* 26 (5): 640–644.

Yah, C. S. and Simate, G. S., 2015. Nanoparticles as potential new generation broad spectrum antimicrobial agents. *DARU Journal of Pharmaceutical Sciences* 23 (1): 43.

Yamabe, T. (1995). Recent development of carbon nanotube. *Synthetic Metals* 70 (1–3): 1511–1518.

Yang, C., Mamouni, J., Tang, Y., and Yang, L. (2010). Antimicrobial activity of single-walled carbon nanotubes: length effect. *Langmuir* 26 (20): 16013–16019.

Yang, S., Sun, Y., Tang, H., and Wan, J. (2015). Catalytic wet air oxidation of phenol, nitrobenzene and aniline over the multi-walled carbon nanotubes (MWCNTs) as catalysts. *Environmental Science and Engineering* 9 (3): 436–443.

Yang, S. X., Li, X., Zhu, W. P., Wang, J. B., and Descorme, C. (2008). Catalytic activity, stability and structure of multi-walled carbon nanotubes in the wet air oxidation of phenol. *Carbon* 46 (3): 445–452.

Yang, S. X., Wang, X. G., Yang, H. W., Sun, Y., and Liu, Y. X. (2012). Influence of the different oxidation treatment on the performance of multi-walled carbon nanotubes in the catalytic wet air oxidation of phenol. *Journal of Hazardous Materials* 233–234: 18–24.

Yang, S. X., Zhu, W. P., Li, X., Wang, H. B., and Zhou, Y. R. (2007). Multi-walled carbon nanotubes (MWNTs) as an efficient catalyst for catalytic wet air oxidation of phenol. *Catalysis Communications* 8 (12): 2059–2063.

Yu, M. F., Lourie, O., Dyer, M. J., Moloni, K., Kelly, T. F., and Ruoff, R. S. (2000). Strength and breaking mechanism of multi-walled nanotubes under tensile load. *Science* 287: 637–640.

Zhang, L., Ni, Q. Q., Fu, Y., and Natsuki, T. (2009). One-step preparation of water-soluble single-walled carbon nanotubes. *Applied Surface Science* 255: 7095–7099.

Zhang, N., Xie, J., and Varadan, V. K. (2002). Functionalization of carbon nanotubes by potassium permanganate assisted with phase transfer catalyst. *Smart Materials and Structures* 11: 962–965.

Zhou, C., Kong, J., Yenilmez, E., and Dai, H. (2000b). Modulated chemical doping of individual carbon nanotubes. *Science* 290 (5496): 1552–1555.

Zhou, C. W., Kong, J., and Dai, H. J. (2000a). Intrinsic electrical properties of individual single-walled carbon nanotubes with small band gaps. *Physical Review Letters* 84 (24): 5604–5607.

9 Carbon- and Graphene-Based Nanocomposites for Wastewater Treatment

Ramasamy Boopathy

CONTENTS

ABSTRACT

The rapid industrialization, the widespread utilization of antibiotics and other pharmaceuticals, the expanded use of chemical products in agriculture, and other human activities have caused the introduction of many new synthetic organic and inorganic contaminants (i.e., dyes, pesticides, heavy metals, etc.) to drinking water treatment systems. Many of these emerging contaminants and the products they form, when in contact with organic matter, pose serious threats to human health. Adsorbent materials present a possible solution to each of these problems. Adsorption-based removal of contaminants has been used successfully for many years in water treatment. Traditional adsorbents, such as activated carbon and polymers, are simple to implement and maintain, but are not always efficient enough to remove all biological and chemical contaminants. Therefore, there is an urgent need to develop a new generation of adsorbent materials that can extend the limits of traditional chemical treatment systems and adsorbents. With increasing interest in nanotechnology, many types of metallic and carbon/graphene-based nanomaterials are emerging in water treatment application. In recent years, there have been many discoveries related to antimicrobial and adsorption properties of carbon/graphene-based nanomaterials for the removal of various biological and organic/inorganic contaminants in aqueous solution. Furthermore, progress in the synthesis of multifunctional nanocomposites paves the way for their application in advanced water treatment system design. Hence, carbon/graphene-based nanomaterials and their respective nanocomposites offer many possibilities for novel applications in water treatment.

Keywords: Carbon, Graphene, Nanocomposites, Adsorption, Organic pollutants, Nanoparticles

9.1 INTRODUCTION

The ever-increasing industrial revolution, rapid improvement of industries, population density, and urbanization severely pollute air, water, and soil. A vast amount of pollutants discharged from industrial processes and households have caused significant effects to the ecoenvironment and human life. These pollutants include toxic gases (NO_x, SO_x, CO, NH_3), heavy metals, organics, biotoxics, and so on. There are many kinds of physical, chemical, and biological technologies being developed to control the pollution load in the water system. Among the various available technologies, the adsorption process is widely used and considered as simple and easy to operate for removal of different types of pollutant from the environment. In addition, adsorption does not result in a secondary pollution load during the process.

For any adsorption process, an adsorbent should have a large surface area, pore volume, and surface functional groups, which are the key factors to be considered for its success. Currently, many different porous natures of materials are being developed, such as activated carbon, pillared clays, zeolites, mesoporous oxides, polymers, and metal organic frameworks, showing varying extents of effectiveness for the removal of toxic pollutants from air, water, and soil. Among them, carbonaceous-based adsorbents such as activated carbon, carbon nanotubes, and fullerenes show high adsorption capacity and thermal stability in the process (Ren et al. 2011; Rao et al. 2007; Seymour et al. 2012; Wang et al. 2012).

In the past few years, graphene oxide (GO) and graphene nanosheets (GNs) have also attracted enormous interest in adsorption and many other applications. Graphene is naturally a two-dimensional carbon nanomaterial with a single layer of sp^2-hybridized carbon atoms arranged in six-membered rings. Graphene possesses strong mechanical, thermal, and electrical properties, with a maximum theoretical specific surface area of 2630 m^2/g (Zhu et al. 2010). GO is also functionalized with various oxygen-containing groups for its various applications. There are many reports on the applications of GO and GNs in different areas such as physics, chemistry, biology, and materials science (Zhu et al. 2010; Huang et al. 2011; Liu et al. 2012; Machado and Serp 2012; Yao et al. 2012). This chapter is devoted to describing the research on carbon/graphene-based nanomaterials as adsorbents for the removal of various types of contaminants such as chemical species, heavy metals, and biological species that are present in the water treatment system.

9.2 EFFECTS OF NANOMATERIAL FUNCTIONAL GROUPS ON ADSORPTION AND ITS WATER CHEMISTRY ON ADSORPTION

Carbon/graphene-based nanomaterials such as graphene and carbon nanotubes are being developed and have been found to occur in functionalized or nonfunctionalized forms. Mostly, graphene or carbon nanotubes can be functionalized with –OH and –COOH groups via chemical oxidation methods to produce GO-functionalized carbon nanotubes, which are highly dispersible in water rather than in their pure form. These nanomaterials are also functionalized with metals or metal oxide frameworks. The most commonly used metals/metal oxide frameworks of carbon/graphene-based nanocomposites for water treatment applications include Fe/Fe$_3$O$_4$, Al/Al$_2$O$_3$, TiO$_2$, Ag, and many others (Guo et al. 2012; Shaari et al. 2012; Pyrzynska and Bystrzejewski 2010; Kim et al. 2013; Gupta et al. 2011a,b; He et al. 2010; Lee and Yang 2012). The incorporation of metals/metal oxides in carbon/graphene-based nanocomposites has been found to improve the disinfection properties along with the adsorption properties of the material.

The incorporation of organic functional groups into carbon and graphene-based nanomaterials is carried out via the introduction of the surface functionalization of organic materials, which plays a significant role in the adsorption process. As mentioned, graphene and carbon nanotubes can be modified with various functional groups on their surfaces via covalent or noncovalent methods for their different field applications. The functionalization of carbon/graphene-based nanomaterials is helpful in two major ways. First, it improves the hydrophilicity of carbon nanotubes and graphene, by ameliorating their dispersion in aqueous media. This dispersion phenomenon sufficiently increases the available surface area of each nanoparticle—and, thus, their exposure—for removal of microbial and chemical contaminants. Second, depending on the available surface charge properties of the material toward the target contaminants, nanomaterial surfaces can be modified to maximize the electrostatic interactions between sorbent and sorbate in aqueous media. The main driving force for the adsorption of positively charged compounds (e.g., heavy metals, organic dye pollutants) in water treatment systems is by electrostatic interaction with negatively charged functionalized carbon/graphene material. In addition, the aggregated forms of carbon nanomaterials contain more mesopores than the conventionally available carbon-based adsorbents, such as granular activated carbon, which have more micropores in their structure. Generally, mesopores provide easy access for the adsorption of large- and small-sized fractions of contaminants present in the water system.

The important characteristic for any carbon/graphene-based nanomaterial is the high surface area-to-volume ratio, which allows the particles to be classified as to whether they have an advantage over macro- or microscale pollutant removal in water treatment applications. This property is essential for carbon/graphene-based nanomaterials to serve as adsorbents or scaffolding materials. However, the use of adsorption is affected by the state of nanomaterial aggregation, which is mostly dependent on solution chemistry (Zhang et al. 2011; Saleh et al. 2008). Aggregation behavior can be related to hydrophobic, electrostatic, and van der Waals interactions between nanomaterials and other compounds in solution, as well as interactions between the nanomaterials themselves.

9.3 EFFECTS OF WATER CHEMISTRY ON ADSORPTION OF CHEMICAL SPECIES BY NANOMATERIALS

The pH and ionic strength are the two most important water chemistry parameters to be considered in any adsorption process. The suspension stability of nanomaterials in aqueous solution is strongly dependent on the pH level. A change in the pH level significantly affects the ionization state of the functional groups present on the surface of nanomaterials and, thus, reduces the specific adsorption efficiency of the contaminants. The solution pH also alters the aggregation state of the nanomaterials in suspension, especially those that are nonfunctionalized carbon/graphene single-walled nanotubes (SWNT) and multiwalled nanotubes (MWNT). It is reported that MWNTs should have a negative charge at pH 6 and above, to prevent aggregates forming; however, this reduces their volume of micro- and nanopores. If the surface charge is greatly reduced, to pH 3, this allows aggregates to form more readily in the MWNT material in suspension. Also, the adsorption of contaminants by the functionalized nanomaterials of carbon/graphene is highly affected by the change in pH level. This effect may be explained by the presence of carboxylic acid groups on the material surface, which have a pH of around 4.0–5.0 (Wu et al. 2011). Hence, the removal of most contaminants in water treatment systems are performed at pH > 5.0, by the deprotonated carboxylic groups with a strong electric charge within the material used. For example, the adsorption of Pb^{2+} has been observed to increase from around 60% for pH values below 4.0 to 80%–90% for values above this level (Musico et al. 2013). But, in nanomaterials and composites functionalized with positively charged functional groups used for the removal of anionic contaminants in water systems, the ideal pH level for electrostatic adsorption should be 5.0 or lower.

Similarly, altering the solution ionic strength can mask charges on the surfaces of carbon/graphene nanomaterials, which affect their aggregation state. There are many reports on increases of ionic strength (Na^+, Ca^{2+}) reducing electrostatic repulsions between nanomaterials and contaminates, thus increasing attachment efficiency. For example, MWNT suspensions with Na^+ at a concentration of 25 mM can yield better dispersion of the nanomaterial via electrostatic repulsion (Saleh et al. 2008). In the same way, the addition of divalent cations of Ca^{2+} (2.6 mM) and Mg^{2+} (1.5 mM) into the MWNT suspension may give good aggregate efficiency in solution. Overall, the addition of salt concentrations in water systems has greatly affected the protein sorption process. Hence, the addition of 0.2 M concentration of Na^+ and Ca^{2+} can increase adsorption of the lysozyme

TABLE 9.1

Removal of Chemical, Biological, and Heavy Metal Species Present in Wastewater by Various Nanomaterials/Nanocomposites

Nanomaterial/ Nanocomposite	Chemical and Biological Species Removal	Heavy Metal Species Removal
GO	Biomolecules (Konicki et al. 2013), bacteria, synthetic organics (Musico et al. 2013; Lu and Su 2007)	Metals (Lu and Su 2007)
G	Bacteria, synthetic organics (Wang et al. 2013; Ydue et al. 2014)	Metals (Wang et al. 2013)
SWNT	Natural organic materials, bacteria, viruses (Li et al. 2005; Brady-Estevez et al. 2008)	Metals
MWNT	Bacteria, synthetic organics (Musico et al. 2013)	Metals (Brady-Estevez et al. 2004)
PVK-SWNT, PVK-G, MWNT-Ag, G-Ag, GO-Ag	Bacteria (Santos et al. 2012; Ydue et al. 2014; Wang et al. 2010; Su et al. 2013; Kim et al. 2012; Bao et al. 2011)	—
PVK-GO	Bacteria (Santos et al. 2012)	Metals (Mejias Carpio et al. 2012)
GO-TiO$_2$, MWNT-Al$_2$O$_3$	—	Metals (Wang et al. 2012; Gupta et al. 2011a)
GO-EDTA	Bacteria (Ydue et al. 2014)	Metals (Mejias Carpio et al. 2012)
CNT/Ce-TiO$_2$, MWNT/ ZnO, PVA-MWNT	Organics (Shaari et al. 2012; Saleh et al. 2011; Wang et al. 2005)	—
CNT/TiO$_2$, CNT/Fe-Ni/ TiO$_2$, G-Fe$_3$O$_4$, PVA/ PAA/TiO$_2$/GO	Synthetic organics (Li et al. 2011; Ma et al. 2014; Moon et al. 2013)	—
CS/MWNT	Synthetic organics (Chatterjee et al. 2010)	Metals (Chatterjee et al. 2010)

protein by SWNTs by about a factor of five over a deionized solution. However, a concentration of ionic strength greater than 0.2 M was found to retard adsorption efficiency for both carbon- and graphene-based nanomaterials. For example, at 0.2 M Na$^+$, the adsorption capacity for lysozyme adsorption by SWNTs was reduced from 300 mg/g to just over 200 mg/g, and at the same concentration of divalent Ca^{2+}, the adsorption capacity was reduced to only 50 mg/g (Saleh et al. 2008b).

These results have been achieved based only on the experiments carried out under controlled laboratory conditions and pure conditions (Table 9.1). The effects of complex water chemistries, organic contaminations, solution pH, and salt concentration have not been examined in combination. Future studies are necessary to explore detailed water chemistries for the effects of carbon/ graphene-based nanomaterials on the treatment of actual water contaminants.

9.4 MECHANISM BEHIND REMOVAL OF ORGANIC CONTAMINANTS FROM AQUEOUS SOLUTION BY CARBON/GRAPHENE-BASED NANOMATERIALS

The removal of organic contaminants from water systems is a great concern for their safe use and application. The most common organic contaminants found in water systems include natural organic matter (NOM), such as humic substances, proteins, lipids, and carbohydrates; and synthetic organic compounds, such as dyes, pharmaceuticals, and pesticides, caused by anthropogenic activity. Some of these synthetic compounds are termed *emerging contaminants* because their consequences to health and removal processes are yet to be established. Many of these synthetic compounds are recalcitrant in molecular structures that are left untreated by available conventional chemical and biological waste treatment processes, and have become significant issues for the collection of water downstream (Seymour et al. 2008; Wang et al. 2012; Zhu et al. 2010; Huang et al. 2011; Liu et al. 2012; Machado and Serp 2012; Yao et al. 2012; Guo et al. 2012; Shaari et al. 2012; Pyrzynska and Bystrzejewski 2010; Kim et al. 2013; Gupta et al. 2011a,b, 2013; He et al. 2010; Lee and Yang 2012; Zhang et al. 2011; Saleh et al. 2008; Wu et al. 2011; Musico et al. 2013; Rossner et al. 2009; Lapworth et al. 2012; Stuart et al. 2012; Turgay et al. 2011; Konicki et al. 2013). Carbon-based nanomaterials display promising prospects for the removal of these contaminants through adsorption processes. However, their adsorption efficiency in dealing with these recalcitrant organic chemicals depends on their physicochemical properties, the chemical characteristics of the organic contaminants, and the water solution chemistry (e.g., pH and ionic strength). The most significant physicochemical properties that affect the removal of organic chemicals are (micro- and meso-) pore size and the surface charges of the functionalized nanomaterials (Figure 9.1). Nonfunctionalized carbon/graphene-based nanomaterials are also found to form aggregates in aqueous solutions. The large aggregates formed create diverse pore sizes, which increases their internal surface areas within the bulk solution through their hydrophobic nature, allowing a higher adsorption capacity for NOM and other organic chemicals present in wastewater. For instance, the adsorption of NOM by SWNTs showed much a higher removal capacity of about 22–26 mg/g than conventional activated carbon (14.7 mg/g). The SWNTs form aggregates in aqueous solution, and thus create a larger fraction of mesopores and cause the aggregates to possess lower surface charges, which enhances their adsorption capacity for NOM (Zhang et al. 2011; Saleh et al. 2008; Wu et al. 2011; Musico et al. 2013; Rossner et al. 2009; Lapworth et al. 2012; Stuart et al. 2012; Turgay et al. 2011; Gupta et al. 2013; Konicki et al. 2013; Lu and Su 2007). NOM is adsorbed on the outer surface of the nanomaterials, which facilitates their easy regeneration as compared with activated carbon. The larger surface area of the nanomaterials enables the removal of competitive adsorbates such as trichloroethane and other synthetic organic compounds.

In addition to pore size and surface area, the development of functional groups in nanomaterials will also impact adsorption capacity. It is reported that functionalized MWNTs have a reduced adsorptive capacity for the removal of some aromatic compounds, but they have been observed to have an increased adsorption compared with nonfunctionalized MWNTs for ringed compounds with amino side groups. This clarifies that the hydrophobic effect plays little role in the adsorption

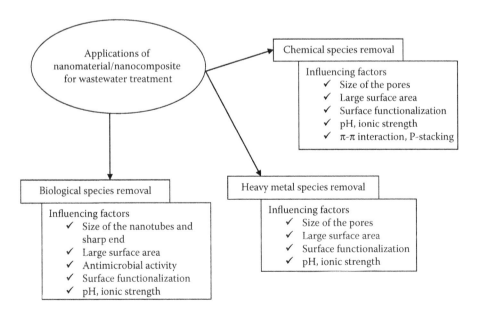

FIGURE 9.1 Schematic representation of factors influencing the effective treatment of wastewater by nanomaterials/nanocomposites.

kinetics of the compounds to the nanomaterials. For example, thiophene is more hydrophobic than 4,6-diaminopyridine, which adsorbed less to graphene and pristine MWNT material. This study reveals that this could be due to Lewis adduct formation between the lone pairs of the amino groups and the oxygen-bearing sites within the nanomaterial. Conversely, aromatic compounds lacking these side groups are adsorbed significantly less (up to 30%) by the functionalized nanotubes than by those in their pure form. However, a similar study reported that benzene, toluene, and similar compounds are adsorbed two to seven times better by functionalized MWNTs than nonfunctionalized nanomaterials (Su et al. 2010). Hence, the functional groups and surface area are the essential criteria for the adsorption of any organic chemicals by carbon/graphene materials. Other factors that influence adsorption of organic chemicals are their chemical composition and molecular structure.

Typically, the higher molecular weight organic compounds will be adsorbed on the outer surface of the nanomaterial, while smaller molecules will be sorbed by the internal micropores. The aromatic rings in any organic contaminant will also affect its removal by carbon/graphene nanomaterials. For instance, contaminants with aromatic rings can form π–π interactions with carbon/graphene-based nanomaterials, which facilitates the removal of these organic compounds from aqueous solution. The aliphatic hydrocarbons are also known to be adsorbed by carbon/graphene nanomaterials, by the interaction of π-stacking. In addition to these effects, the physicochemical properties of the nanomaterials, the chemical characteristics of the organic contaminants, and water chemistry—especially the pH and ionic strength—will also contribute an important role in the removal of organic contaminants from the water system. For instance, most of the protein adsorption is strongly dependent on solution pH and ionic strength. The adsorption capacity of GO for lysozymes is reduced from 350 mg/g to (effectively) zero when the solution pH is increased above its isoelectric point (Smith et al. 2014). Adsorption of dyes has also shown that the solution pH greatly influences the ionization behavior of both the nanoadsorbent and the dye chemicals in the water body. At higher pH levels, the surfaces of the GOs have more negative charges, which significantly increases their electrostatic interaction with positively charged dye molecules. In these conditions, the increase of the solution pH from 4 to 12 increases the methylene blue adsorption capacity of the GOs (Yang et al. 2011). In contrast to the pH level, the effects of ionic strength have contradictory results for the adsorption of organic contaminants. An increase of Na^+ concentration from 0 to 100 mmol/L decreases the adsorption of tetracycline compounds by about 50% for GO adsorbents (Gao et al. 2012). In the presence of Na^+, electrostatic interaction between the charged amino groups in tetracycline and the deprotonated carboxyl groups decreases and, consequently, adsorption is reduced. However, an increase in the ionic strength of Na^+ from 0.0 to 80 mmol/L significantly increases dye adsorption for MWNT adsorbents. This could be explained by the presence of Na^+ dye molecules aggregating and, thereby, their adsorption rate and adsorbed mass increase. However, more extensive investigation needs to be carried out with different organic chemicals for a better understanding to arrive at the mechanism behind the role of ionic strength for any adsorption process.

It is important to point out that, despite nanomaterials being used efficiently for the removal of organic contaminants from aqueous solution, the incorporation of carbon/graphene-based nanomaterials as adsorbents is limited at present, due to their high cost and low rate of recovery after adsorption. Several studies are investigating how to overcome these issues; also, the development of polymeric and metal/metal oxide nanocomposites is being undertaken to address them. For example, chitosan is mostly used as a base material to prepare carbon-based nanocomposites and create different morphologies because of its high adsorption capacity and biocompatibility. Impregnation of chitosan with 0.01 wt% of MWNTs shows significant improvement in the adsorption of Congo Red dye, with a maximum of about 225 mg/g (Chatterjee et al. 2010). The improved adsorption was attributed to the increased surface area and the successful interaction between the aromatic rings in the dye molecules and the MWNTs. Such impregnation enables a homogeneous distribution of nanostructures on chitosan with denser and more porous-like beads with higher mechanical strength.

Similarly, the development of polymeric hydrogel nanocomposites with carbon base has achieved very efficient adsorption and degradation processes of organic contaminants. A hydrogel composed of poly(vinyl alcohol)/poly(acrylic acid)/TiO_2/GO (PVA/PAA/TiO_2/GO) has been commercially employed for two-step adsorption and degradation processes. The first step involves highly dispersed GO in the polymer matrix, which enables the high adsorption of organic contaminants; this is followed by the presence of TiO_2, which helps the photodegradation of the adsorbed organic contaminants. Overall, the incorporation of nanomaterials in polymer matrices seems to be a promising direction for the production of advanced nanocomposites for water treatment. However, further studies should investigate their degradation and persistence in the environment, as well as their implications for human health and safety, before they are widely applied to water treatment systems.

9.5 MECHANISMS OF HEAVY METAL REMOVAL BY NANOCOMPOSITES

In conventional treatment plants, granular activated carbon (GAC) filters are frequently used for the removal of heavy metal pollutants. However, this method often fails to meet regulatory limits set by the authorities, because of poor removal efficiency at low concentrations of heavy metals and slow adsorption rates. Recent studies have shown that carbon nanotubes and GO, in particular, are capable of removing significant amounts of heavy metal pollutants from aqueous solutions. In most cases, nanotubes/SWNTs were capable of removing Pb^{2+}, Zn^{2+}, Cd^{2+}, and Cu^{2+} from aqueous solutions (Lu et al. 2006, 2008). However, the results were found to be contradictory regarding the affinity order of the heavy metal ions in the solution. Furthermore, these results demonstrated that the adsorption capacities of carbon/graphene materials are highly dependent on the pH of the solution. A common aspect of all these cases was that these nanomaterials exhibit a significantly higher adsorption capacity than any other conventional adsorbents in the same conditions. This may be due to the higher surface areas of nanomaterials as compared with conventional adsorbents; surface fictionalization is also considered to be an important factor in heavy metal removal by carbon nanoparticles (Figure 9.1). A comprehensive study showed that the mechanisms of Pb^{2+} adsorption by functionalized MWNTs demonstrated that Pb^{2+} interacted with oxygen-bearing functional groups on the carbon nanoparticle surfaces. Other adsorption mechanisms identified adsorption of Pb^{2+} by physical phenomena on the nonfunctionalized areas of the surfaces, such as the defect sites on the walls of the nanotubes and the open ends. Carbon nanotube material has also been demonstrated to adsorb metal ions via cation–π interaction.

GOs are also considered to be promising materials for heavy metal remediation. Since most heavy metal pollutants are cations (e.g., Cu^{2+}, Pb^{2+}, Cd^{2+}), the negatively charged functional groups of GOs make them excellent adsorbents for heavy metals. Nonfunctionalized graphene is also found to serve as a good adsorbent for both cationic and anionic heavy metal ions to some extent, although it removes less than 10% of Pb^{2+} ions as compared with GO under similar conditions. Most of the cases have indicated that GO is a better adsorbent for many types of heavy metal removal, adsorbing 1.4–10 times as many Cd^{2+} ions than conventional adsorbents, and performing 2.75–88.6 times better on Co^{2+} adsorption, as well as adsorbing approximately 10 times Pb^{2+} than any other material (Zhao et al. 2011). Electrostatic attractions between positively charged metal ions and negatively charged functional groups on GO surfaces are the predominant reason for the adsorption mechanism. In all illustrated functionalized nanomaterials, the negative functional groups on the surfaces of the carbon/graphene-based material will enhance heavy metal ion capture. This surface modification is carried out by activation mechanisms using various oxidizing agents such as $KMnO_4$, H_2O_2, HNO_3, and H_2SO_4. Zeta potential analyses have shown that, depending on the pH, the oxidized nanomaterials may frequently tend to form negatively charged particles in solution due to the presence of –COOH and –OH functional groups on external sites. This greatly enhances their interactions with the positively charged metal ions during water treatment.

Similarly to organic pollutants, the removal of heavy metals is also much influenced by the solution pH and the ionic strength of carbon/graphene-based nanomaterials. If the solution pH is higher than the pKa of the nanomaterial, then the negative surface charge facilitates adsorption of cationic heavy metal pollutants via electrostatic attraction. Conversely, lowering the pH will decrease the adsorption of cationic heavy metals due to ionic repulsion. The solution pH also affects the selective speciation of heavy metals and introduces competing reactions among ionic species in the solution, which can affect the adsorption process. Additionally, if the ionic strength of the solution is optimal, some metal ions will be adsorbed by the nanomaterials for limited adsorption sites. To facilitate the use of these nanomaterials for application to water treatment, the researchers synthesized polymeric nanocomposites of carbon-based nanomaterials. These nanomaterials deliver enhanced heavy metal removal capacity, via the functional fixation of carbon/graphene-based nanomaterials into polymer matrices to improve their availability for adsorption of heavy metals. The results revealed that the introduction of only 10 wt% of GO in poly-N-vinyl carbazole (PVK-GO) nanocomposite could remove about 887.98 mg/g (maximum) of Pb^{2+}, which is considerably higher than any PVK-like adsorbents (Musico et al. 2013). Due to its ease of preparation and use of low GO load, this nanocomposite has a promising outlook for use as an adsorbent in commercial applications.

Similarly, a surface-functionalized MWNT (1 wt%) was blended with polysulfone material for heavy metal removal in water treatment. The incorporation of a functionalized MWNT improved the hydrophilic behavior of the polymer, which thereby increased its metal adsorption capacity by 78% for Cd^{2+} when compared with polysulfone alone. Chitosan is also considered to be an attractive base polymer due to its hydroxyl and amino functional groups, which facilitate heavy metal adsorption. Dispersion of GO sheets in chitosan is able to remove more Pb^{2+} (77 mg/g) from solution than pure chitosan. After desorption of the adsorbed Pb^{2+}, this nanocomposite can retain 93% of its initial capacity, which means it is a promising material for practical applications (Fan et al. 2013). A polyaniline/GO nanocomposite has also been synthesized via *in situ* polymerization using an aniline monomer with GO. This nanocomposite was found to possess a highly porous nanostructure with a higher surface area than polyaniline alone and, with 15 wt% GO loading, enabled 94% greater adsorption of Hg(II) from an aqueous solution over polyaniline. As mentioned, the adsorption capacity of Hg(II) depends on solution pH and the physicochemical properties of the adsorbent. In general, desorption of the adsorbent is mostly achieved via an acidic or alkaline solution, which regenerates approximately 70% of the initial adsorption capacity of this nanocomposite.

Applications of carbon/graphene-based nanomaterials for heavy metal removal from aqueous solutions are often limited, due to the difficulties of separating the nanomaterials after the adsorption process. However, this can be overcome by use of magnetic metal/metal oxide nanocomposites, which are considered to be convenient to separate the nanomaterials by simple application of a magnetic field. There are now more studies on the synthesis of MWNT/nano-Fe_3O_4 nanocomposites, in which nanoclusters of Fe_3O_4 are deposited on the MWNT nanoparticles (Gupta et al. 2011a). The nanocomposite was found to have a higher adsorption capacity for Cr^{3+} removal in aqueous solution compared with pure MWNT or activated carbon. The adsorption of Cr^{3+} was also dependent on the solution pH, contact time, and mixing rate. The nanocomposite was also used as packed bed column material in the continuous filtration process (Wu et al. 2011). Similarly, alumina (Al_2O_3) was deposited on the surface of MWNTs to form an MWNT/Al_2O_3 nanocomposite for the removal of Pb^{2+}. Photocatalytic compounds of TiO_2 were also found to be effective for the application of heavy metal removal. TiO_2 is well known for its ability to remove several heavy metals by surface complexation mechanisms, and GO is known to remove heavy metals by electrostatic interaction. The individual heavy metal removal capacity of these two nanomaterials is further enhanced via formation of flowerlike TiO_2 structures on GO nanosheets, which can effectively remove Cd^{2+}, Pb^{2+}, and Zn^{2+} from aqueous solutions. Overall, many of these reports have shown that metal oxides can be grown in a controlled way on the extensive surfaces of carbon/graphene-based nanomaterials to maximize their adsorption capacities.

Not all toxic heavy metals have always been found as elemental cations in aqueous solutions. Arsenic and chromium, in particular, are often present in a highly oxidized anion state. Recent progress is focused on making use of carbon/graphene-based nanomaterials for the sorption of toxic anions. GO with iron oxide surface groups was found to serve the removal of 54.18 mg/g arsenite and 26.76 mg/g arsenate, according to a Langmuir adsorption model, with very similar adsorption values (53.15 and 39.08 mg/g), even in the case of MWNTs coated with iron oxide nanoparticles (Mishra and Ramaprabhu 2010; Yu et al. 2015). Various nanomaterial adsorbents are being used to remove chromate [Cr(VI)] anions as well, although the removal mechanism is not as clear because of the variety of nanocomposites being used. Pristine, nonfunctionalized MWNTs may be able to completely absorb Cr(VI) present in trace concentrations of aqueous solution. A mixture of GO with the liquid cation Aliquat-336 demonstrated an excellent Cr(VI) sorption capacity (285.71 mg/g), and was able to remove all Cr(VI) from tannery effluent (Kumar and Rajesh 2013). Chromium adsorption onto GO was also found to be strongly improvised by functionalizing the surface with 2,6-diaminopyridine, which was able to absorb high concentrations of Cr(VI) over a relatively low contact time. As with arsenic anions, chromate adsorption is found to be strongly dependent on pH levels; at low pH (1–5), higher rates of adsorption resulted; however, this was not wholly attributable to electrostatic interaction. Still, there are many research activities in progress to explore the use of carbon/graphene-based nanomaterials for the sorption of nonmetal anionic contaminants, such as selenium, nitrate, and dyes.

Overall, the removal of heavy metals by carbon/graphene-based nanomaterials and nanocomposites are largely dependent on the water chemistry (especially pH level), the functional groups on the nanomaterials, the exposure time, and the extent of the mixing rate. In the present scenario, the results of the use of different nanomaterials and nanocomposites for the removal of heavy metals are highly promising. These advanced materials are more efficient than the currently available adsorbents used in water treatment for the removal of heavy metals.

9.6 REMOVAL OF BIOLOGICAL CONTAMINANTS IN WASTEWATER TREATMENT

9.6.1 Antimicrobial Activity of Carbon/Graphene-Based Nanomaterials

Carbon/graphene-based nanomaterials possess excellent bactericidal efficiency, as well as improved viral and bacterial spore adsorption capacity, when compared with conventional adsorbents, due to their greater surface areas. High surface area and greater affinity for certain groups of nano-composite bacteria are the main reasons for their bactericidal activity. This removal performance is important, since it has the potential for the selective removal of pathogens over nonpathogenic microorganisms in the water system. For example, in mixed bacterial cultures of *Staphylococcus aureus* and *Escherichia coli*, *S. aureus* is removed about 100 times more than *E. coli* within 5–30 min of exposure time (Deng et al. 2008). This fast rate of microbial removal and selectivity of pathogens is a good characteristic for the development of filters, point-of-use (POU) water treatment devices, and microbial sensors. In traditional drinking water treatment, the removal of pathogenic bacteria by chemical disinfection or adsorption is minimal. In aqueous solution, SWNT particles can serve to adsorb 27–37 times more *Bacillus subtilis* spores than powdered activated carbon (Upadhyayula et al. 2009). This high adsorption capacity is due to the high adsorption ratio and the fibrous structure of SWNTs, which significantly enhance the active surface availability for spore attachment. In addition to bacteria, viruses have also been successfully removed by carbon/graphene-based nanomaterials through adsorption processes. Commercially available membrane filters coated with carbon nanotubes are also able to remove all viruses from contaminated water containing 107–108 plaque-forming units (PFU)/mL (Brady-Estevez et al. 2008; Mostafavi et al. 2009). The degree of virus particle removal depends linearly on the thickness of the nanotube layer formation on the structure. Removal of viruses in traditional drinking water treatment is a highly significant challenge, as nanometer-sized virus particles can easily escape through regular membranes and activated carbon filtration systems. In addition to the pristine forms of nanomaterials, polymeric and metal/metal oxide nanocomposites of carbon/graphene-based nanomaterials have shown great promise for microbial disinfection in water treatment.

Nanocomposites of poly-N-vinyl carbazole (PVK) with SWNTs (PVK-SWNTs) and GO (PVK-GO) serve to inactivate over 80% of pathogenic cells of both gram-positive and gram-negative bacteria in aqueous solutions. PVK-SWNTs act as a surface coating material, which is highly effective against bacterial biofilm formation (Ahmed et al. 2013). With the addition of only 3 wt% of SWNTs or GO, these polymeric nanocomposites will exhibit the same disinfection capacity as 100 wt% of pure nanomaterials. Nanocomposites of reduced GO in polyoxyethylene sorbitan laureate can be molded into a paperlike structure, which possesses high antimicrobial activity toward gram-positive bacteria. Similarly, nanocomposites of SWNTs (2 wt%) in poly (lactic-co-glycolic acid) and polylysine polymers successfully inactivate bacteria (90%) in aqueous solutions (Aslan et al. 2010, 2012). The microbial inactivation potential of metal/metal oxide nanocomposites, especially metallic silver (Ag) and silver nanoparticles (AgNP), have also received significant attention regarding the removal of biological pollutants in water systems. In aqueous solutions, silver is released as ionic silver (Ag^+) species, which attack bacterial cell membranes and cause cell death. Well-dispersed AgNP with an average diameter of 20–25 nm on graphene nanosheets (GNS) with 0.05 mg/mL of GNS/AgNP can achieve 99.5% disinfection of *Colibacillus* bacteria cultures, as well as of the yeast *Candida albicans* (Yuan et al. 2012). Similarly, a graphene doped with AgNP nanocomposites inhibits gram-negative (*E. coli*) and gram-positive (*S. aureus*) bacteria at concentrations of 0.04 and 0.16 mg/L, respectively (Nguyen et al. 2012). A comparative antimicrobial study of GO and AgNP/GO indicates that the addition of AgNP to GO significantly improves antimicrobial effects by 45% and 25%, respectively, for the removal of *E. coli* and *S. aureus* bacteria in water treatment (Bao et al. 2011). Similarly, much improved disinfection capacity was observed for MWNTs decorated with silver nanoparticle composites. Furthermore, reports also describe that carbon/graphene-based nanocomposites retain 50% of their antibacterial activity after 20 cycles of usage, which significantly lowers the cost of their application to water disinfection. Overall, the enhanced antibacterial effects of metal/carbon nanocomposites were attributed to the good distribution of silver nanoparticles on the carbon/graphene-based nanomaterial scaffold, and synergistic antimicrobial activity of the individual components.

9.6.2 Removal Mechanisms of Microbial Species in Water Treatment by Nanomaterials

Most investigations concerning adsorption and inactivation have examined diverse groups of planktonic microorganisms, including bacteria (both gram-positive and gram-negative), protozoa, and viruses. A few studies have shown that SWNT- and GO-coated surfaces inhibited the formation bacterial biofilm such as *E. coli* and *B. subtilis*. The physicochemical properties of carbon/graphene-based nanomaterials have been linked to antimicrobial activity. The most important phenomena for nanomaterial toxicity to microbes is their nanoscale size; but, their shape, specific surface area, chemical composition, and surface structure also play important roles (Figure 9.1). It has been established that the larger the surface area of an individual nanomaterial, the higher the contact area for the adsorption of microorganisms, which increases antimicrobial capacity. For instance, dispersed nanostructures have been reported to have significantly more toxic effects on cells than the aggregates. The inactivation of *S. aureus*, *E. coli*, and *B. subtilis* was increased from 30%–50% to 90%–100% on the addition of functional groups that increased the dispersibility of the nanomaterials (Murugan and Vimala 2011). Additionally, carbon/graphene-based nanomaterials are synthesized in different geometries (e.g., tubes, sheets, spheres), which can affect their deposition and their interactions with microorganisms. For example, MWNTs are found to be less bactericidal than SWNTs; this is due to the larger diameters of their pores, which reduces their antimicrobial activity. The diameters of SWNTs were found to be much smaller than those of MWNTs, and much shorter; they have more ends per length of nanomaterial, and consequently more adsorption between cells and these ends. When the sharp ends of the nanotubes come into contact with microbes, they damage membranes by puncturing them, causing irreversible damage to the cells. In the case of rod-shaped nanomaterials, such as SWNTs, it was found that the diameter of the nanotubes was a key factor regarding their antibacterial properties. In the presence of nanotubes of small diameter, bacteria cells generate more stress-related gene products than in the presence of those of larger diameter (Kang et al. 2008). The gene expression experiments confirmed that fatty acid beta-oxidation, glycolysis, and fatty acid biosynthesis pathways involved in membrane damage, repair, and lipid recycling were highly significant in the presence of SWNTs, but there is no report that they are significant in the presence of MWNTs.

The studies concluded that the fatty acid beta-oxidation pathways are responsible for fatty acid uptake from the surrounding media, which suggests that more lipids will be released to the media due to membrane damage/cell death in the presence of SWNTs. In the case of the fatty acid biosynthesis pathway, the researchers suggest that the upregulation of this pathway involves the synthesis of new lipid molecules to be incorporated in cells, either for cell growth or repair of membranes damaged by nanomaterials. The high amount of glycolysis in the presence of SWNTs indicates that the cells require more energy to survive in stress-related conditions, and the by-products of this mechanism (acetyl-CoA) are typically used for the synthesis of fatty acids to repair cell membranes. Similarly to carbon nanotubes, the sharp edges of graphene or GO sheets are also found to physically damage and inactivate bacterial cells; another finding explains that the improved toxicity of smaller particles, due to their higher edge-to-weight ratios, are also a reason for bacterial cell death. Contrary to these results, however, several studies have independently discovered that graphene sheets will wrap around bacterial cells, preventing the diffusion of nutrients into the cells and thereby controlling their growth (Liu et al. 2011; Chen et al. 2014).

The other major effect of carbon nanomaterial toxicity that is extensively described in many studies is the production of harmful reactive oxygen species (ROS) (Vecitis et al. 2010; Shvedova et al. 2003, 2012; Lyon and Alvarez 2008; Mauter and Elimelech 2008). In the presence of carbon/graphene-based nanomaterials, ROS such as O^{2-} and H_2O_2 are produced, which cause oxidation reactions in bacterial cells. Excessive ROS can oxidize the fatty acids in the cell membrane and damage the cell permeability, affecting other essential cell functions (Kang et al. 2008). There is also a report of a DNA microarray investigation of *E. coli* in the presence of MWNTs and SWNTs, which showed that, in the presence of both nanomaterials, genes were detoxified by ROS such as superoxide dismutase and catalase and were highly expressed in this microorganism. Also, pathways of DNA damage and repair systems were affected in the presence of these nanomaterials. This pattern of gene expression is also observed in *E. coli* cells exposed to hydrogen peroxide and superoxide. In conclusion, the principal mechanisms caused by the toxicity of all carbon/graphene-based nanomaterials are physical cell membrane damage and oxidative stress. Besides cellular inactivation, carbon nanomaterials are found to be good adsorbents for the removal of bacteria and viruses. In addition, the rate of bacterial inactivation on the membrane was enormously increased, from 10% on the control to 90% on the nanotube-coated filters. A few other studies have examined the virus-adsorbing capacity of carbon nanomaterials but, despite their overall positive findings, most nanomaterials are not examined for their virus-removing capabilities. In addition to microbial removal by carbon nanotubes, there are studies which compare the capacity for removal of gram-positive and gram-negative bacteria using graphene, GO, and PVK-nanocomposites with uncoated nitrocellulose filters. Studies on the removal of viruses by GO or graphene nanocomposites are almost nonexistent.

9.7 APPLICATION OF CARBON/GRAPHENE-BASED NANOCOMPOSITES FOR WATER TREATMENT

In general, carbon/graphene-based nanomaterials in pure form cannot be applied directly to water treatment, due to their extremely small size, which makes their recovery from water treatment systems difficult. Therefore, many research activities are in progress concerning mechanisms to incorporate these nanomaterials into membranes, to embed them into polymeric bases, or to modify them with magnetic metal oxides to facilitate their recovery from water. In the case of nanomaterials in membranes, very frequently these carbon/graphene-based materials are embedded in polymer membranes during synthesis or are bound onto the membrane surfaces by physiosorption or covalent bonds between the functional groups present on the nanomaterials and the membranes. These procedures permit the successful development of nanocomposite water treatment membranes. In the literature, there are many successful examples of the incorporation of carbon nanotubes into membranes, with a resulting improvement in membrane permeability. These membranes can be broadly divided into two categories: (1) carbon nanotubes are aligned on a nonporous polymeric support, where the nanotubes act as pores to allow water to pass through; (2) carbon nanomaterials are blended with the membrane polymer to change the physicochemical properties of the membrane. In the first group of membranes, straw-shaped nanotubes, which may be functionalized or modified to remove specific contaminants, allow water to pass through their hollow pores, while contaminants are diverted from the pore openings via adsorption (Cong et al. 2007). These membranes are highly uniform in their porous structure, which eliminates any concentration irregularities due to asymmetry, as observed in traditional membranes. Most importantly, the permeability of this type of membrane is four to five times higher than theoretical values; this may be attributed to the nanoscale channels, hydrophobicity, and the minimal friction of the nanotubes' smooth inner channels.

In the second group, where carbon nanotubes are blended with the membrane polymer to change the physicochemical properties of the membrane, it shows that the incorporation of the nanomaterials inside the polymer matrix can improve its mechanical properties, and calibrate the hydrophilic or hydrophobic nature of the membrane (Vatanpour et al. 2011). The embedded nanomaterials are closely packed inside the polymer matrix and, thus, provide a lower risk of detachment. The incorporation of multiwalled nanotubes in PVA increases the stiffness and permeability of the membrane. An increased carbon nanotube load (1–4 wt% of the functionalized nanotubes) was found to improve water permeability and rejection of solute. Interfacial polymerization with polyamide and nanotubes produces highly solvent-resistant nanofiltration membranes. In these nanocomposites, the porous structure was reported to enhance the solvent flux by one order of magnitude. Functionalized nanotubes are often used to increase the hydrophilic properties of the membrane with the most common functional groups, including $-COOH$, $-OH$, and $-NH_2$. Due to the functionalization of the carbon nanotubes, the impregnated membrane becomes hydrophilic, rejecting more hydrophobic pollutants, and increases water permeability. Hydrogen bonding between functional groups and water is attributed to the higher hydrophilicity of the functionalized nanotubes on the membranes. Nanocomposite membranes are also formed by adding amine-functionalized MWNTs to an aqueous solution of 1,3-phenylendiamine. The results suggest that the incorporation of NH_2 nanotubes creates nanochannels on the surface; in addition, $-NH_2$ increases the hydrophilicity of the membrane. These significantly improved properties led to a 160% increase in water flux across the membrane.

In addition to the impregnation of nanomaterials inside the polymer matrix, modifications of the surfaces of filters with nanomaterials/nanocomposites are carried out by physiosorption or covalent reactions between the functional groups on the nanomaterial and the membrane surface. These types of nanocomposite membranes are shown to be extremely effective at removing microbes from water. For example, PVK-GO coated onto a nitrocellulose membrane with a pore size of 8.0 nm was found to remove 4.5 logs of *B. subtilis* from wastewater, as compared with the removal of less than 2.0 logs by unmodified membranes. This could be explained by the reduced pore size caused by the membrane coating, as well as bacterial adsorption by the nanomaterial/nanocomposite. Additionally, the coated filters were found to possess significant antimicrobial properties toward both gram-negative and gram-positive bacteria, thus reducing the viability of the captured bacteria, indicating that these materials prevented biofilm formation on the filters and consequent biofouling

(Joshi et al. 2014; Chen et al. 2008; Hu et al. 2010). Despite high biocompatibility, antimicrobial properties, and extreme selectivity, these membranes have been given little attention regarding water treatment application.

In addition to the use of modified filters with nanomaterials, researchers have also impregnated polymeric materials with nanomaterials to create macroscale base structures. This process allows the removal of contaminants by the nanomaterials embedded in the beads by sieving rather than microfiltration. Earlier, chitosan beads were the object of much of this interest. This polymer is, itself, a good adsorbent for diverse contaminants, due to its variety of functional groups, while also being biocompatible and biodegradable. However, it lacks mechanical strength. This limitation may be enhanced by the addition of carbon nanomaterials into the matrix, which can also improve adsorption capacity. For example, it was found that the impregnation of chitosan beads with SWNTs increased the adsorption rate from 223 to 450 mg/g for heavy metal removal. Another study found that a chitosan/GO blend successfully removed Pb^{2+} ions from solution with a maximum adsorption capability of 77 mg/g (Fan et al. 2013). In addition to the competitive sorption rate and easy separation properties, it was also found to be recyclable, with up to 90% of the original capacity after four generations of washing, although this diminishes in subsequent washings. Related hydrogels are able to remove gram-positive and gram-negative bacterial populations, and gels formed with biopolymers other than chitosan (bovine serum albumin [BSA], DNA) display similarly enhanced biocompatibilities and adsorption capacities; however, chitosan's greater resistance to degradation makes its application to environmental systems much easier. A three-dimensional GO nanosponge synthesized by centrifugation was able to rapidly remove almost all methylene blue (99.1%) and methyl violet (98.8%) from aqueous solution (Liu et al. 2012), and a similar study described that the affinity of reduced GO (rGO) aerogels could be optimized for the removal of particular pollutants by interspersing the polymer matrix of the gel with hydrophilic or hydrophobic amino acids, which greatly increased the gel's attraction for heavy metals (cysteine) or oils (lysine). Similar materials are receiving much attention as ultrahigh storage capacitors, and are being investigated for water treatment applications.

Another alternative method under investigation is the coupling of carbon/graphene-based nanomaterials with magnetic nanoparticles for water treatment. Metal oxides with magnetic properties are often used as the base metal for nanocomposites, which enables separation from solution at low cost. In a recent study, magnetic G/Fe_3O_4 was synthesized through an *in situ* coprecipitation method for the removal of synthetic dye in aqueous solutions. Homogeneous incorporation of Fe_3O_4 on graphene sheets was observed, and the resultant nanocomposite could easily be separated from the aqueous solution by an applied magnetic field. At an adsorbent dose of 0.5 g/L, almost 90% of dye was quickly removed via $\pi-\pi$ adsorption. To achieve the success and long-term operation of adsorption, it needs to be regenerated in a cost-effective manner. The G/Fe_3O_4 nanocomposite was found to regain 94% of its original adsorption capacity after collection and washing with ethanol solution. In a similar study, the G/Fe_3O_4 nanocomposite was able to remove a higher concentration than traditional adsorbents of each of five different carbamate pesticide compounds from environmental water samples. The nanocomposite did not lose its adsorption capacity or magnetic separation properties, even after 12 sequential adsorption–desorption cycles. Several other reports confirmed the successful synthesis of TiO_2, ZnO, and Al_2O_3 with carbon/graphene-based nanomaterials for the separation of organic contaminants from aqueous solutions. These studies elucidated that nanomaterials, in combination with polymers or other nanomaterials such as magnetic particles, have great potential for the development of advanced functionalized materials for water treatment applications. These nanohybrid materials will allow the safer use of these nanomaterials in water treatment processes, due to their 100% capacity for recovery, and also avoid other adverse environmental impacts that may occur from the release of these materials into the environment. However, it is necessary to further investigate the removal efficiency of these nanocomposites for diverse biological and chemical contaminants, as well as their potential for reusability prior to any use in water treatment.

9.8 LIMITATIONS OF CARBON/GRAPHENE-BASED NANOMATERIALS IN WATER TREATMENT

Several cytotoxicity studies are being carried out on carbon/graphene-based nanomaterials concerning their problematic effects on human and other mammalian cells. The research reveals that, in antimicrobial studies, nanomaterial size, shape, dose, and exposure time are found to play an important role regarding their cytotoxic properties. A comprehensive study investigating the correlation of the size of GO with its cytotoxicity showed that smaller nanosheets were capable of causing higher cytotoxicity than larger sheets. In agreement with studies of other nanosized carbon particles, *in vivo* toxicological studies of GO using mice report the accumulation of nanomaterials in kidney and lung tissues. In this study, doses of GO of only 0.4 mg were found to be a lethal concentration (Wang et al. 2011; Warheit 2006). However, the cytotoxic effects of these materials are greatly improved when they are dispersed in biocompatible polymers to form nanocomposites. Studies comparing graphene and GO in poly-N-vinyl carbazole show that the composite material reduces cytotoxic activity by a factor of two in NIH 3T3 fibroblast cells. The impregnation of these nanomaterials in polymers reduced cell inactivation from 40% to 20% in the case of graphene, and from 15% to 7.5% for GO. However, the antimicrobial effects of these nanocomposites actually improved over pure nanomaterials.

Similar results have been reported for carbon-based nanomaterials coated with polyethylene glycol (PEG) and BSA. PEGylation of SWNTs reduced the cytotoxicity effect from 80% to 20% in the study of rat neuronal PC12 cells (Li et al. 2014). These results are very promising, suggesting to researchers that the toxicity of the nanomaterials would be dramatically reduced by combining them with biocompatible polymers. Additionally, studies have shown that the cytotoxic effects of nanomaterials are highly dependent on dose and contact time. The first published cytotoxicity study was performed using human epidermal keratinocyte cells exposed to different concentrations of SWNTs. The results illustrated that significant loss of cell viability and morphological changes in the cells are caused by oxidative stress. Similarly, effects were reported even for the addition of MWNTs with human epidermal keratinocyte cells. Adsorbent dose- and time-dependent cytotoxic effects were observed, and the MWNTs were found to accumulate inside the cells. The aggregation state of the nanotubes is also a factor of interest with respect to cytotoxicity. The MWNTs were homogenized by a grinding process to increase their dispersion property. This higher nanotube dispersion increased cytotoxicity and cell inflammation significantly. GO nanosheets exhibited cytotoxicity to human fibroblast cells at concentrations as low as 20 µg/mL. In the presence of GO, cells are morphologically changed and subsequent cell death will occur (Pinto et al. 2013). More detailed toxicological studies of GO have discovered that not only are the nanomaterials in the cytoplasm, but they are also found in lysosomes, mitochondria, and even in the nucleus. Overall, cytotoxicity data universally indicate the adverse effects of SWNTs, MWNTs, and GO on mammalian cells at various degrees, among which SWNTs are the most hazardous (Zhang et al. 2010).

Recent research demonstrates the great potential of carbon/graphene-based nanomaterials and their nanocomposites for water treatment applications. However, the increasing use of these substances will inevitably lead to their increased release into the environment, which raises questions about the long-term effects of these nanomaterials in ecosystems. As far as work concerning antimicrobial activity and cytotoxicity to humans is concerned, the number of studies on the ecotoxicological effects of carbon/graphene-based nanomaterials is very limited. Nonetheless, some of the impacts of these nanomaterials in bacterial communities in soil and aquatic systems are documented. The reports claim that both SWNTs and MWNTs inhibit the respiration of activated sludge microorganisms at high concentrations. Acute exposure to GO, even at concentrations of 10 mg/L, also causes significant loss of metabolic activity in wastewater bacterial communities. The biogeochemical cycles of nutrients such as carbon, nitrogen, and phosphorus are also slowed by the presence of GO (Ahmed and Rodrigues 2013). Similarly, a study investigating the impact of functionalized SWNTs on soil microbial communities found significant impacts on the communities' metabolism activity. However, after reaching a minimum after 3 days, the bacterial community began to recover, and cell counts returned to normal after 14 days. This study demonstrates that SWNTs can cause only temporary adverse effects under acute exposure, even at high concentrations. The effects of chronic exposure on the addition of smaller concentrations, on the other hand, are largely unknown. In the case of other carbon/graphene-based nanomaterials, there is a lack of systematic investigation of chronic and acute exposure to different microbial communities.

However, there is a report of great discoveries on the biodegradation of carbon/graphene-based nanomaterials. Recent research discovered that *Shewanella* sp. is able to reduce GO to graphene (Wang et al. 2011; Jiao et al. 2011). These findings suggest that the impact of these nanomaterials in the environment may be self-attenuating, with little care for the handling of these materials. While these results are promising for the future of the materials, many questions remain to be answered by future research on the topic of their environmental impact, especially in long-term applications. Also, these studies were performed at very high concentrations and in pure conditions. Hence, little is known about the environmental and human effects of these nanomaterials when embedded in polymer matrices. Even though some of these results are promising, vast amounts of work remain to be done to determine the applications of these carbon/graphene-based nanomaterials to water treatment.

REFERENCES

Ahmed, F., Rodrigues, D.F. 2013. Investigation of acute effects of graphene oxide on wastewater microbial community: A case study. *J. Hazard. Mater.* 256:33–39.

Ahmed, F., Santos, C.M., Mangadlao, J., Advincula, R., Rodrigues, D.F. 2013. Antimicrobial PVK: SWNT nanocomposite coated membrane for water purification: Performance and toxicity testing. *Water Res.* 47: 3966–3975.

Aslan, S., Deneufchatel, M., Hashmi, S., et al. 2012. Carbon nanotube-based antimicrobial biomaterials formed via layer-by-layer assembly with polypeptides. *J. Colloid Interface Sci.* 388:268–273.

Aslan, S., Loebick, C.Z., Kang, S., Elimelech, M., Pfefferle, L.D., Van Tassel, P.R. 2010. Antimicrobial biomaterials based on carbon nanotubes dispersed in poly (lactic-co-glycolic acid). *Nanoscale.* 2:1789–1794.

Bao, Q., Zhang, D., Qi, P. 2011. Synthesis and characterization of silver nanoparticle and graphene oxide nanosheet composites as a bactericidal agent for water disinfection. *J. Colloid Interface Sci.* 360:463–470.

Brady-Estevez, A.S., Kang, S., Elimelech, M. 2008. A single-walled carbon-nanotube filter for removal of viral and bacterial pathogens. *Small.* 4(4):481–484.

Chatterjee, S., Lee, M.W., Woo, S.H. 2010. Adsorption of Congo Red by chitosan hydrogel beads impregnated with carbon nanotubes. *Bioresour. Technol.* 101:1800–1806.

Chen, H., Muller, M.B., Gilmore, K.J., Wallace, G.G., Li, D. 2008. Mechanically strong, electrically conductive, and biocompatible graphene paper. *Adv. Mater.* 20:3557–3561.

Chen, J., Peng, H., Wang, X., Shao, F., Yuan, Z., Han, H. 2014. Graphene oxide exhibits broad-spectrum antimicrobial activity against bacterial phytopathogens and fungal conidia by intertwining and membrane perturbation. *Nanoscale.* 6:1879–1889.

Cong, H., Zhang, J., Radosz, M., Shen, Y. 2007. Carbon nanotube composite membranes of brominated poly (2,6-diphenyl-1,4-phenylene oxide) for gas separation. *J. Membr. Sci.* 294:178–185.

Deng, S., Upadhyayula, V.K.K., Smith, G.B., Mitchell, M.C. 2008. Adsorption equilibrium and kinetics of microorganisms on single-wall carbon nanotubes. *Sens. J. IEEE.* 8:954–962.

Fan, L., Luo, C., Sun, M., Li, X., Qiu, H. 2013. Highly selective adsorption of lead ions by water-dispersible magnetic chitosan/graphene oxide composites. *Colloids Surf. B.* 103:523–529.

Gao, Y., Li, Y., Zhang, L., et al. 2012. Adsorption and removal of tetracycline antibiotics from aqueous solution by graphene oxide. *J. Colloid Interface Sci.* 368:540–546.

Guo, J., Wang, R., Tjiu, W.W., Pan, J., Liu, T. 2012. Synthesis of Fe nanoparticles at graphene composites for environmental applications. *J. Hazard. Mater.* 225:63–73.

Gupta, V.K., Agarwal, S., Saleh, T.A. 2011a. Chromium removal by combining the magnetic properties of iron oxide with adsorption properties of carbon nanotubes. *Water Res.* 45:2207–2212.

Gupta, V.K., Agarwal, S., Saleh, T.A. 2011b. Synthesis and characterization of alumina-coated carbon nanotubes and their application for lead removal. *J. Hazard. Mater.* 85:17–23.

Gupta, V.K., Kumar, R., Nayak, A., Saleh, T.A., Barakat, M.A. 2013. Adsorptive removal of dyes from aqueous solution onto carbon nanotubes: A review. *Adv. Colloid Interface Sci.* 193:24–34.

He, F., Fan, J., Ma, D., Zhang, L., Leung, C., Chan, H.L. 2010. The attachment of Fe_3O_4 nanoparticles to graphene oxide by covalent bonding. *Carbon.* 48:3139–3144.

Hu, W., Peng, C., Luo, W., et al. 2010. Graphene-based antibacterial paper. *ACS Nano.* 4:4317–4323.

Huang, X., Yin, Z., Wu, S., et al. 2011. Graphene-based materials: Synthesis, characterization, properties, and applications. *Small.* 7:1876–1902.

Jiao, Y., Qian, F., Li, Y., Wang, G., Saltikov, C.W., Gralnick, J.A. 2011. Deciphering the electron transport pathway for graphene oxide reduction by *Shewanella oneidensis* MR-1. *J. Bacteriol.* 193:3662–3665.

Joshi, R.K., Carbone, P., Wang, F.C., et al. 2014. Precise and ultrafast molecular sieving through graphene oxide membranes. *Science.* 343:752–754.

Kang, S., Herzberg, M., Rodrigues, D.F., Elimelech, M. 2008. Antibacterial effects of carbon nanotubes: Size does matter. *Langmuir.* 24:6409–6413.

Kim, E.S., Hwang, G., Gamal, E.M., Liu, Y. 2012. Development of nanosilver and multi-walled carbon nanotubes thin-film nanocomposite membrane for enhanced water treatment. *J. Membr. Sci.* 394:37–48.

Kim, J.D., Yun, H., Kim, G.C., Lee, C.W., Choi, H.C. 2013. Antibacterial activity and reusability of CNT-Ag and GO-Ag nanocomposites. *Appl. Surf. Sci.* 283:227–233.

Konicki, W., Cendrowski, K., Chen, X., Mijowska, E. 2013. Application of hollow mesoporous carbon nanospheres as an high effective adsorbent for the fast removal of acid dyes from aqueous solutions. *Chem. Eng. J.* 228:824–833.

Kumar, A.S.K., Rajesh, N. 2013. Exploring the interesting interaction between graphene oxide, Aliquat-336 (a room temperature ionic liquid) and chromium (VI) for wastewater treatment. *RSC Adv.* 3:2697–2709.

Lapworth, D.J., Baran, N., Stuart, M.E., Ward, R.S. 2012. Emerging organic contaminants in groundwater: A review of sources, fate and occurrence. *Environ. Pollut.* 163:287–303.

Lee, Y.C., Yang, J.W. 2012. Self-assembled flower-like TiO_2 on exfoliated graphite oxide for heavy metal removal. *J. Ind. Eng. Chem.* 18:1178–1185.

Li, Y., Feng, L., Shi, X., et al. 2014. Surface coating-dependent cytotoxicity and degradation of graphene derivatives: Towards the design of nontoxic, degradable nano-graphene. *Small.* 10:1544–1554.

Li, Y.H., Di, Z., Ding, J., Wu, D., Luan, Z., Zhu, Y. 2005. Adsorption thermodynamic, kinetic and desorption studies of Pb^{2+} on carbon nanotubes. *Water Res.* 39(4):605–609.

Li, Z., Gao, B., Chen, G.Z., Mokaya, R., Sotiropoulos, S., Li Puma, G. 2011. Carbon nanotube/titanium dioxide (CNT/TiO_2) core–shell nanocomposites with tailored shell thickness, CNT content and photocatalytic/photoelectrocatalytic properties. *Appl. Catal. B.* 110:50–57.

Liu, F., Chung, S., Oh, G., Seo, T.S. 2012. Three-dimensional graphene oxide nanostructure for fast and efficient water-soluble dye removal. *ACS Appl. Mater. Interfaces.* 4:922–927.

Liu, S., Zeng, T.H., Hofmann, M., et al. 2011. Antibacterial activity of graphite, graphite oxide, graphene oxide, and reduced graphene oxide: Membrane and oxidative stress. *ACS Nano.* 5:6971–6980.

Liu, Y., Dong, X., Chen, P. 2012. Biological and chemical sensors based on graphene materials. *Chem. Soc. Rev.* 41:2283–2307.

Lu, C., Chiu, H., Liu, C. 2006. Removal of zinc (II) from aqueous solution by purified carbon nanotubes: Kinetics and equilibrium studies. *Ind. Eng. Chem. Res.* 45:2850–2855.

Lu, C., Liu, C., Rao, G.P. 2008. Comparisons of sorbent cost for the removal of Ni^{2+} from aqueous solution by carbon nanotubes and granular activated carbon. *J. Hazard Mater.* 151:239–246.

Lu, C., Su, F. 2007. Adsorption of natural organic matter by carbon nanotubes. *Sep. Purif. Technol.* 58:113–121.

Lyon, D.Y., Alvarez, P.J.J. 2008. Fullerene water suspension (nC60) exerts antibacterial effects via ROS-independent protein oxidation. *Environ. Sci. Technol.* 42:8127–8132.

Ma, L., Chen, A., Lu, J., Zhang, Z., He, H., Li, C. 2014. *In situ* synthesis of CNTs/Fe–Ni/TiO_2 nanocomposite by fluidized bed chemical vapor deposition and the synergistic effect in photocatalysis. *Particuology.* 14:24–32.

Machado, B.F., Serp, P. 2012. Graphene-based materials for catalysis. *Catal. Sci. Technol.* 2:54–75.

Mauter, M.S., Elimelech, M. 2008. Environmental applications of carbon-based nanomaterials. *Environ. Sci. Technol.* 42:5843–5859.

Mejias Carpio, I.E., Santos, C.M., Wei, X., Rodrigues, D.F. 2012. Toxicity of a polymer-graphene oxide composite against bacterial planktonic cells, biofilms, and mammalian cells. *Nanoscale.* 4(15):4746–4756.

Mishra, A.K., Ramaprabhu, S. 2010. Magnetite decorated multiwalled carbon nanotube based supercapacitor for arsenic removal and desalination of seawater. *J. Phys. Chem. C.* 114:2583–2590.

Moon, Y.E., Jung, G., Yun, J., Kim, H.I. 2013. Poly (vinyl alcohol)/poly (acrylic acid)/TiO_2 graphene oxide nanocomposite hydrogels for pH-sensitive photocatalytic degradation of organic pollutants. *Mater. Sci. Eng. B.* 178(17):1097–1103.

Mostafavi, S.T., Mehrnia, M.R., Rashidi, A.M. 2009. Preparation of nanofilter from carbon nanotubes for application in virus removal from water. *Desalination*. 238:271–280.

Murugan, E., Vimala, G. 2011. Effective functionalization of multiwalled carbon nanotube with amphiphilic poly (propyleneimine) dendrimer carrying silver nanoparticles for better dispersability and antimicrobial activity. *J. Colloid Interface Sci.* 357:354–365.

Musico, Y.L.F., Santos, C.M., Dalida, M.L.P., Rodrigues, D.F. 2013. Improved removal of lead (II) from water using a polymer-based graphene oxide nanocomposite. *J. Mater. Chem. A*. 1:3789–3796.

Nguyen, V.H., Kim, B.K., Jo, Y.L., Shim, J.J. 2012. Preparation and antibacterial activity of silver nanoparticles-decorated graphene composites. *J. Supercrit. Fluids*. 72:28–35.

Pinto, A.M., Goncalves, I.C., Magalhaes, F.D. 2013. Graphene-based materials biocompatibility: A review. *Colloids Surf. B*. 111:188–202.

Pyrzynska, K., Bystrzejewski, M. 2010. Comparative study of heavy metal ions sorption onto activated carbon, carbon nanotubes, and carbon-encapsulated magnetic nanoparticles. *Colloids Surf. A*. 362:102–109.

Rao, G.P., Lu, C., Su, F. 2007. Sorption of divalent metal ions from aqueous solution by carbon nanotubes: A review. *Sep. Purif. Technol.* 58:224–231.

Ren, X., Chen, C., Nagatsu, M., Wang, X. 2011. Carbon nanotubes as adsorbents in environmental pollution management: A review. *Chem. Eng. J.* 170:395–410.

Rossner, A., Snyder, S.A., Knappe, D.R.U. 2009. Removal of emerging contaminants of concern by alternative adsorbents. *Water Res.* 43:3787–3796.

Saleh, N.B., Pfefferle, L.D., Elimelech, M. 2008. Aggregation kinetics of multiwalled carbon nanotubes in aquatic systems: Measurements and environmental implications. *Environ. Sci. Technol.* 42:7963–7969.

Saleh, T.A., Gondal, M.A., Drmosh, Q.A., Amani, Z.H., Al-Yamani, A. 2011. Enhancement in photocatalytic activity for acetaldehyde removal by embedding ZnO nano particles on multiwall carbon nanotubes. *Chem. Eng. J.* 166(1):407–412.

Santos, C.M., Mangadlao, J., Ahmed, F., et al. 2012. Graphene nanocomposite for biomedical applications: Fabrication, antimicrobial and cytotoxic investigations. *Nanotechnology*. 23(39):395.

Seymour, M.B., Su, C.M., Gao, Y., Lu, Y.F., Li, Y.S. 2012. Characterization of carbon nanoonions for heavy metal ion remediation. *J. Nanopart. Res.* 14:1087.

Shaari, N., Tan, S.H., Mohamed, A.R. 2012. Synthesis and characterization of CNT/Ce- TiO_2 nanocomposite for phenol degradation. *J. Rare Earths*. 7:007.

Shvedova, A., Castranova, V., Kisin, E., et al. 2003. Exposure to carbon nanotube material: Assessment of nanotube cytotoxicity using human keratinocyte cells. *J. Toxicol. Environ. Health A*. 66:1909–1926.

Shvedova, A.A., Pietroiusti, A., Fadeel, B., Kagan, V.E. 2012. Mechanisms of carbon nanotube-induced toxicity: Focus on oxidative stress. *Toxicol. Appl. Pharmacol.* 261:121–133.

Smith, S.C., Ahmed, F., Gutierrez, K.M., Frigi Rodrigues, D. 2014. A comparative study of lysozyme adsorption with graphene, graphene oxide, and single-walled carbon nanotubes: Potential environmental applications. *Chem. Eng. J.* 240:147–154.

Stuart, M., Lapworth, D., Crane, E., Hart, A. 2012. Review of risk from potential emerging contaminants in UK groundwater. *Sci. Total Environ.* 416:1–21.

Su, F., Lu, C., Hu, S. 2010. Adsorption of benzene, toluene, ethylbenzene and p-xylene by NaOCl-oxidized carbon nanotubes. *Colloids Surf. A*. 353:83–91.

Su, R., Jin, Y., Liu, Y., Tong, M., Kim, H. 2013. Bactericidal activity of Ag doped multi-walled carbon nanotubes and the effects of extracellular polymeric substances and natural organic matter. *Colloids Surf. B*. 104:133–139.

Turgay, O., Ersoz, G., Atalay, S., Forss, J., Welander, U. 2011. The treatment of azo dyes found in textile industry wastewater by anaerobic biological method and chemical oxidation. *Sep. Purif. Technol.* 79:26–33.

Upadhyayula, V.K.K., Deng, S., Smith, G.B., Mitchell, M.C. 2009. Adsorption of *Bacillus subtilis* on single-walled carbon nanotube aggregates, activated carbon and NanoCeramTM. *Water Res.* 43:148–156.

Vatanpour, V., Madaeni, S.S., Moradian, R., Zinadini, S., Astinchap, B. 2011. Fabrication and characterization of novel antifouling nanofiltration membrane prepared from oxidized multiwalled carbon nanotube/polyethersulfone nanocomposite. *J. Membr. Sci.* 375:284–294.

Vecitis, C.D., Zodrow, K.R., Kang, S., Elimelech, M. 2010. Electronic structure-dependent bacterial cytotoxicity of single-walled carbon nanotubes. *ACS Nano*. 4:5471–5479.

Wang, G., Qian, F., Saltikov, C.W., Jiao, Y., Li, Y. 2011. Microbial reduction of graphene oxide by Shewanella. *Nano Res.* 4:563–570.

Wang, K., Ruan, J., Song, H., et al. 2011. Biocompatibility of graphene oxide. *Nanoscale Res. Lett.* 6:1–8.

Wang, L., Zhu, D., Duan, L., Chen, W. 2010. Adsorption of single-ringed N-and S-heterocyclic aromatics on carbon nanotubes. *Carbon*. 48(13):3906–3915.

Wang, S., Ng, C.W., Wang, W., Li, Q., Li, L. 2012. A comparative study on the adsorption of acid and reactive dyes on multiwall carbon nanotubes in single and binary dye systems. *J. Chem. Eng. Data.* 57:1563–1569.

Wang, S., Sun, H., Ang, H.M., Tade, M.O. 2013. Adsorptive remediation of environmental pollutants using novel graphene-based nanomaterials. *Chem. Eng. J.* 226:336–347.

Wang, X., Chen, X., Yoon, K., Fang, D., Hsiao, B.S., Chu, B. 2005. High flux filtration medium based on nano-fibrous substrate with hydrophilic nanocomposite coating. *Environ. Sci. Technol.* 39(19):7684–7691.

Warheit, D.B. 2006. What is currently known about the health risks related to carbon nanotube exposures. *Carbon.* 44:1064–1069.

Wu, M., Kempaiah, R., Huang, P.J.J., Maheshwari, V., Liu, J. 2011. Adsorption and desorption of DNA on graphene oxide studied by fluorescently labeled oligonucleotides. *Langmuir.* 27:2731–2738.

Yang, S.T., Chen, S., Chang, Y., Cao, A., Liu, Y., Wang, H. 2011. Removal of methylene blue from aqueous solution by graphene oxide. *J. Colloid Interface Sci.* 359:24–29.

Yao, J., Sun, Y., Yang, M., Duan, Y. 2012. Chemistry, physics and biology of graphene-based nanomaterials: New horizons for sensing, imaging and medicine. *J. Mater. Chem.* 22:14313–14329.

Ydue, M., Santos, C., Dalida, M., Rodrigues, D.F. 2014. Surface modification of membrane filters using graphene and graphene oxide-based nanomaterials for bacterial inactivation and removal. *ACS Sustain. Chem. Eng.* 2:1559–1565.

Yu, F., Sun, S., Ma, J., Han, S. 2015. Enhanced removal performance of arsenate and arsenite by magnetic graphene oxide with high iron oxide loading. *Phys. Chem. Chem. Phys.* 17:4388–4397.

Yuan, W., Gu, Y., Li, L. 2012. Green synthesis of graphene/Ag nanocomposites. *Appl. Surf. Sci.* 261:753–758.

Zhang, S., Shao, T., Karanfil, T. 2011. The effects of dissolved natural organic matter on the adsorption of synthetic organic chemicals by activated carbons and carbon nanotubes. *Water Res.* 45:1378–1386.

Zhang, Y., Ali, S.F., Dervishi, E., et al. 2010. Cytotoxicity effects of graphene and single-wall carbon nanotubes in neural phaeochromocytoma-derived PC12 cells. *ACS Nano.* 4:3181–3186.

Zhao, G., Li, J., Ren, X., Chen, C., Wang, X. 2011. Few-layered graphene oxide nanosheets as superior sorbents for heavy metal ion pollution management. *Environ. Sci. Technol.* 45:10454–10462.

Zhu, Y., Murali, S., Cai, W., Li, X., Suk, J.W., Potts, J.R., Ruoff, R.S. 2010. Graphene and graphene oxide: Synthesis, properties, and applications. *Adv. Mater.* 22:3906–3924.

10 Nanoscale Layered Double Hydroxides for Wastewater Remediation
Recent Advances and Perspectives

Sushmita Banerjee and Ravindra Kumar Gautam

CONTENTS

ABSTRACT

Layered double hydroxides (LDHs), well known as anionic clays or hydrotalcite-like compounds, are an important class of materials consisting of positively charged metal oxide/hydroxide sheets with intercalated anions and water molecules. LDHs have earned great interest because of potential application in a wide variety of fields, such as additives in polymers, adsorption materials, precursors for functional materials, in pharmaceutics, photochemistry, and electrochemistry. In recent years, there has been considerable interest in utilizing LDHs for the remediation of environmental contaminants. In this regard, significant progress has been achieved in the research and development of LDHs and their application as potential adsorbents in the removal of organic and inorganic wastes from aqueous solutions by the process of adsorption. In this chapter, we have tried to compile the scattered information available on LDHs and its use in water treatment processes in an organized manner. The chapter entails several aspects of LDHs such as their composition, structure, synthesis, and characterization, with special focus on the sorption mechanism of LDHs for various organic and inorganic contaminants from aqueous solution. Moreover, a few recommendations related to the development and potential applications of LDHs for future research are also proposed.

Keywords: Layered double hydroxides (LDHs), Adsorption, Dyes, Heavy metal, Pesticides, Sorption mechanism, Regeneration

10.1   INTRODUCTION

In recent decades, a class of anionic clays, also known as layered double hydroxides (LDHs), has gained substantial attention from environmental researchers owing to its enormous capability to solve environmentally related issues (Seftel et al. 2008; Cheng et al. 2009; Zhang et al. 2012a,b; Theiss et al. 2012; Chen et al. 2013; Elkhattabi et al. 2013). These unique materials are available in naturally occurring form but are also synthesized easily in laboratory conditions. Historically, the synthesis of these substances was first reported by a German scientist, W. Feitknecht, in 1942, who prepared the material by reacting a dilute solution of metal salts with a base; however, the interpretation of the structural aspect of the LDHs were carried out principally by Allmann and Taylor in the early 1970s (Reichle 1986). These materials are also referred to as hydrotalcite-like compounds, due to their structural similarity with magnesium–aluminum hydroxycarbonate, formulated as $Mg_6Al_2(OH)_{16}(CO_3).4H_2O$. Structurally, LDHs are composed of piles of positively charged mixed metal hydroxide layers, also termed *brucite sheets*, which are electrically compensated by the presence of intercalated anions. However, these interlayered anions exhibit weak bonding with brucite sheets, thus offering tremendous ability in the exchange of organic and inorganic anions from the external surroundings. Synthetic LDHs are quite promising, due to their low cost, high surface area, chemical versatility, thermal stability, and high anion-exchange capacity in the range of 200–400 meq/100 g; this facilitates the tuning of their composition and properties in a wide range. Thus, LDHs are found to have extensive applications in diverse fields, including adsorption of gases (Yong et al. 2001; Ishihara et al. 2013), adsorption of solids (Cheng et al. 2009; Chen and Song 2013; Elkhattabi et al. 2013), as antacids (Holtmeier et al. 2007; Parashar et al. 2012), catalysts (Ping et al. 2005; Wang et al. 2011), drug delivery (Rojas et al. 2012; Lv et al. 2015), electrochemistry (Mousty and Prévot 2013; Nejati and Zeynali 2014), flame retardants (Gao et al. 2014), magnetism (Coronado et al. 2008; Abellan et al. 2015), photochemistry (Chen et al. 2012; Parida and Mohapatra 2012), polymerization (Costa et al. 2007; Kuila et al. 2008), and so on.

Recently, LDH compounds have been investigated worldwide due to their remarkable ability in the elimination of various toxic materials from the environment. In this period to date, a variety of LDHs were synthesized by fine tuning of synthetic parameters. LDHs with varied chemical combinations were attempted by using divalent and trivalent metal cation ratios, along with suitably intercalated anions that result in the formation of a diverse range of inorganic–organic assemblies with desirable physical and chemical properties. Significant advancements have been evident in the research and development of LDHs and their application in resolving environmental problems. For instance, the use of LDHs in membrane fabrication has been extensively studied (Liu et al. 2014; Dong et al. 2015; Liu et al. 2015). Recently, Liu et al. (2014) employed hydrotalcite material for the preparation of a nanofiltration membrane that efficiently separates organic matter from aqueous media. The admirable performance of LDHs in photocatalyst-assisted degradation of pollutants was examined by several researchers (Yuan et al. 2009; Parida and Mohapatra 2012; Alanis et al. 2013; Prince et al. 2015). Parida and Mohapatra (2012) have reported enhanced photocatalytic activity of a novel carbonate-intercalated LDH catalyst for the degradation of the azo dyes methyl violet and malachite green under solar light, with a high decolorization efficiency of 99% within 2 h. Li and Duan (2006), in their extensive review, described potential applications of LDHs in various ion-exchange and adsorption studies. The rich interlayered chemistry of LDHs suggests great scope for anion-exchange reactions with great vigor. LDHs offer tremendous capacity as anion exchangers, due to the presence of highly labile anions in their interlayer region. This characteristic property further assists in the elimination of undesirable or toxic anions from aqueous phase through exchange with other nonhazardous anions. Miyata (1983) and Yamaoka et al. (1989) gave a comparative list of ion selectivities for monovalent and divalent anions, respectively, in the following order, given as

$$NO_3^- < Br^- < Cl^- < OH^- \tag{10.1}$$

$$SO_4^{2-} < CrO_4^{2-} < HAsO_4^{2-} < HPO_4^{2-} < CO_3^{2-} \tag{10.2}$$

This list further helps in the synthesis of typical LDH compounds intercalated with the desired anions, which can easily replace targeted anions from effluents. Radha et al. (2005) have attempted the exchange of nitrate anions with chloride ions of NaCl solution from the gallery region of LDH at an initial pH solution value of 12. An interesting investigation was carried out by Costantino and coworkers (Costantino et al. 2014) in which the ion-exchange properties of LDH toward halide ions were investigated. The selectivity of carbonate-intercalated LDHs toward the halides were found to decrease with an increase of the halide–ionic radius, and the selectivity order was analyzed as $F^- > Cl^- > Br^- > I^-$. Thus, LDHs play a momentous role as ion exchangers. Additionally, appealing textural characteristics, such as a high surface area, good porosity, and the presence of a large number of active sites, impelled researchers to explore the potential of LDHs as adsorbents. LDH-based adsorbents exhibited enormous potential to remove a wide range of organic, inorganic, and microbial contaminants from water and wastewater. Undeniably, LDH compounds have proved to be of great importance in dealing with pollution problems; more specifically, their role in dealing with water pollution is highly commendable. In this chapter, attempts have been made to address the vital aspects of LDHs, including their composition, structure, different methods of synthesis, and characterization; moreover, special focus has been directed toward the interpretation of the interaction behavior of LDHs with various pollutants present in the liquid phase.

10.2 COMPOSITION AND STRUCTURE OF LDHs

LDH compounds represent two-dimensional structures. The typical arrangement follows a brucite-like [Mg (OH)$_2$] pattern, where hydroxyl groups octahedrally surround the divalent metal cations—in some cases, monovalent cations. Besserguenev et al. (1997) also reported the formation of an octahedral unit, with units sharing edges to form a sheet-like structure. These octahedral units, which contain divalent metal cations, become isomorphously substituted by trivalent metallic cations; therefore, the sheets acquire a positive charge and, thus, the charge density is proportional to the molar ratio of metal cations, such as $x = M^{3+}/(M^{2+} + M^{3+})$ (Tao et al. 2009). The whole structure is, therefore, organized as a pile of such positively charged sheets, which further becomes electrically balanced by the organic and inorganic anions and some water molecules residing in the interlayer region between these brucite-like sheets, as illustrated in Figure 10.1. LDHs can be described by the general formula, given as

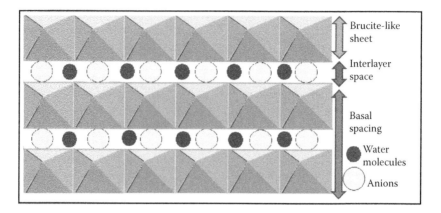

FIGURE 10.1   **(See color insert)** Schematic representation of LDH structure.

$$\left[M_{1-x}^{2+} M_x^{3+} (OH)_2 \right]^{z+} \left[A^{n-} \right]_{z/n} \cdot yH_2O \qquad (10.3)$$

where:

M^{2+} and M^{3+} are divalent and trivalent cations respectively

A^{n-} are interlayer anions

The layer charge is $z = x$ when the M^{2+} is a divalent cation and $z = 2x-1$ when M^{2+} is a monovalent cation. The maximum molar amount of interlayer water that can be structurally incorporated as a monolayer in an LDH is $1 - (z/n)$. The structural cation ratio for M^{2+}/M^{3+} mostly varies from 1 to 5 (You et al. 2001), and the value of x generally exists between 0.16 and 0.45 (Whilton et al. 1997). The value of x plays a crucial role in LDH synthesis; for values of x beyond this range the compounds are formed with unreliable hydrotalcite-like structures. At a high value of x, the formation of Al(OH)$_3$ occurs due to an increased number of aluminum octahedrals, while at a low value of x, there is a high density of Mg octahedrals, thus leading to the formation of Mg(OH)$_2$ (Brindley and Kikkawa 1979).

It has been also analyzed that the ionic radii contributes significantly in the formation of LDH structures. For divalent and trivalent cations, the ion radii values between 0.65–0.80 Å and 0.62–0.69 Å, respectively, are considered as the most preferable for the formation of brucite-like layers. However, higher ionic radii, such as those of Ba, Cd, Sc, and La, drastically discourage the formation of true brucite-like layers. Hence, the nature of the cations in the brucite-like sheets are certainly not limited to +2 and +3 combinations; along with interlayer anions, together with the coefficient, the value of x may be varied in a wide range that generates a large class of isostructural materials (Cavani et al. 1991). Usually the M^{2+} and M^{3+} ions have ionic radii comparable to that of Mg^{2+} (ionic radius = 0.65 Å), which can easily lodge in the holes of the close, compact packing of hydroxyl anions in the brucite sheets to form LDHs. The most commonly used M^{2+} and M^{3+} species include Mg^{2+}, Zn^{2+}, Fe^{2+}, Cu^{2+}, Ni^{2+}, Co^{2+}, or Ca$^+$, and Al^{3+}, Fe^{3+}, Cr^{3+}, Mn^{3+}, or V^{3+}, respectively. Practically all divalent metal ions from Mg^{2+} to Mn^{2+} are known to form hydrotalcites except Cu^{2+}, as due to the Jahn–Teller effect, Cu^{2+} can form hydrotalcites with a distorted octahedral shape only in the presence of other dipositive metal cations, on condition that the ratio of Cu^{2+} to the other divalent cation must be greater than 1 (Khan and O'Hare 2002). LDHs can also be obtained using cations of monovalent–trivalent (Li-Al) and divalent–tetravalent (Co-Ti) combinations. In terms of anion selection, there is no such major limitation; a wide range of charge compensatory anions, either organic (carboxylates, humic acids, aminoacids) or inorganic (halides, oxyanions, silicates) are known to be successfully used as intercalated anions.

In addition to the availability of diverse chemical combinations, LDHs also exhibit polymorphisms due to different stacking arrangement patterns of two-dimensional sheets in various manners. The stacking is known to be accomplished in two possible ways, either with a rhombohedral or a hexagonal symmetry. In the rhombohedral symmetry (3R) one unit cell is equal to three times the interlayer separation, while for the hexagonal form (2H), the interlayer is repeated twice (Rives and Ulibarri 1999). Of these, LDHs with rhombohedral symmetry exist commonly.

10.3 SYNTHESIS OF LDHs

For any water treatment process, the two most important things that make a treatment system successful are its cost-effectiveness and decontamination efficiency. Therefore, to achieve these, it is important to focus on the synthesis techniques of the adsorbent (LDHs). In general, LDHs can be effortlessly synthesized in the laboratory using various methods. However, the most simple and trouble-free technique is recognized as the coprecipitation method. This method facilitates the production of LDHs in copious quantities. However, this method suffers from various setbacks, including reduced reproducibility, lack of control in particle size, and poor crystallinity. As a result, to overcome such problems, other preparatory routes were explored, such as precipitation at low and high supersaturation, urea hydrolysis, separate nucleation and aging steps (SNAS), ion exchange, rehydration using the memory effect, hydrothermal process, secondary intercalation, dissolution and recoprecipitation (D-R), sol-gel, and the mechanochemical process. A common problem with all these methods is that during preparations of LDHs using anions other than CO_3^{2-}, contamination from CO_2 easily takes place; consequently, CO_3^{2-} anions enter and are firmly held in the interlayer. The various methods used for LDH synthesis are summarized below in Table 10.1. Table 10.2 represents LDHs synthesized using metal combinations and intercalated anions of diverse range.

10.4 CHARACTERIZATION OF LDHs

LDH material, with its characteristic layered structure, can be substantiated by analyzing the synthesized material using various instrumentation techniques. Some of the most frequently used analytical techniques are Fourier transform infrared spectroscopy (FTIR), powdered x-ray diffraction (PXRD), thermogravimetric analysis (TGA), differential thermal analysis (DTA), scanning electron microscopy (SEM), transmission electron microscopy (TEM), and surface area pore volume analysis by the Brunauer–Emmett–Teller (BET) method.

The assessment of various interlayered anions participate in the formation of LDHs; their bonding and orientation can be easily analyzed by interpreting the resultant absorption peaks of the compound, by means of a specialized technique, known as infrared spectroscopy. FTIR spectroscopy is an advanced form of infrared spectroscopy and is widely used these days, due to an increased level of sensitivity and improved signal-to-noise ratio. The energy level of $12–2 \times 10^{13}$ Hz, emitted by the infrared radiation, generates vibrations in molecules by absorbing the specific frequency of infrared radiation, as each functional group has its own frequency. This specific frequency helps in determining the presence of the exact functional groups in the sample. Some of the most important characteristic absorption bands exhibited by LDH compounds are as follows: peaks at 3600–3200 cm^{-1}, which are indicative of the stretching mode of hydrogen-bonded hydroxyl groups from the hydroxide layers of brucite sheets and interlayered water molecules (Nicola et al. 1997). The peak intensity between 1600 and 1650 cm^{-1} corresponds to the bending mode of water molecules (Hernandez-Moreno et al. 1985). The presence of interlayer anions such as carbonate, nitrate, sulfate, chloride, and phosphate can be ascertained from the peaks located at 1364–1390 cm^{-1}, 1380–1385 cm^{-1}, 1000–1250 cm^{-1}, 1350–1370 cm^{-1}, and 1050 cm^{-1}, respectively (Gaini et al. 2009; Halajania et al. 2013). The absorbance peaks in the low-frequency region, that is, below 1000 cm^{-1}, corresponds to the lattice vibration of M–O and O–M–O in the octahedral sheets, which strongly supports the formation of the distinctive phases of the LDHs (Zhang et al. 2012a).

TABLE 10.1

Summary of Various Synthesis Techniques for Layered Double Hydroxides

S. No	Synthesis Technique	Synthesis Procedure	References
1	Coprecipitation	• In this method, nitrate or chloride salts of M^{2+} and M^{3+} metal cations are taken in an appropriate stoichiometric ratio and dissolved together to form a mixed metal solution. • This is followed by simultaneous addition of an alkali solution in dropwise manner into the mixed metal solution under vigorous stirring, keeping pH value in the range of 7–10. • Commonly, nitrate and chloride salts of the metal ion are preferred due to their low affinity for the positively charged layers, which further eases modification of the interlayer anions according to specific requirements. • In the case where carbonate intercalation is not desired, it is necessary to perform synthesis under N_2 atmosphere. • Synthesis by means of coprecipitation can be carried out in four different ways; (a) precipitation at low supersaturation (b) precipitation at high supersaturation (c) SNAS (d) urea hydrolysis.	Cheng et al. 2009; Elkhattabi et al. 2013
1(a)	Precipitation at low supersaturation	• The method consists of slow addition of mixed cations metal solution containing desirable intercalated anions. • The pH value of the metal solution is kept constant by precipitating it with the simultaneous addition of alkaline solution at an appropriate rate. • Monitoring of the solution pH is essential in this process. • The LDHs derived from this method possess high crystallinity. • Highly controlled and desired ratio of M^{2+} to M^{3+} in the LDH is obtained.	Rodrigues et al. 2012
1(b)	Precipitation at high supersaturation	• The pH level during synthesis fluctuates greatly. • The divalent and trivalent metal cations are precipitated with an alkaline solution that contains desirable anions that are supposed to be intercalated. • The variable pH condition during synthesis leads to the formation of secondary products, such as $M(OH)_2$ or $M(OH)_3$. • Relatively inappropriate and undesired ratios of M^{2+} to M^{3+} in the LDH were obtained. • The method gives poorly developed crystallites because of the formation of large-sized crystallization nuclei. • Thermal treatment is desirable for improved crystallinity and to increase percentage yields.	Zhang et al. 2008
1(c)	Separate nucleation and aging steps (SNAS)	• This is the modified form of the coprecipitation process. • The process involves very rapid mixing of the components in a special apparatus known as a colloid mill, which assists in accelerating the nucleation process. • In subsequent steps, aging is required, which takes place separately. • The method offers great control of the size distribution and a relatively higher degree of crystallinity. • The drawback of this method is the high cost required for instrumental setup.	Zhao et al. 2002; Lin et al. 2006

TABLE 10.1 (CONTINUED)
Summary of Various Synthesis Techniques for Layered Double Hydroxides

S. No	Synthesis Technique	Synthesis Procedure	References
1(d)	Urea hydrolysis	• The urea hydrolysis process advances slowly, which results in a low degree of supersaturation during precipitation. • Exceptional characteristic features of urea, such as weak Brønsted base, high solubility in water, and high rate of hydrolysis can be easily managed by controlling reaction temperature, thus rendering it an attractive precipitating agent for metal cations. • The resultant products exhibit highly homogenous particle size distribution. • The crystals develop having high crystallinity with well-defined hexagonal shaped crystallites. • The method is not widely applicable, as the constant presence of carbonate released from the urea decomposition leads to synthesis of carbonate-intercalated LDHs and carbonate anions.	Costantino et al. 1998; Inayat et al. 2011.
2.	Ion exchange	• The technique involves exchange of intercalated anions of LDH with that of desirable anions. • For efficient anion exchange, LDHs with nitrate as interlayered ions may be a good choice. • In general, the exchange medium is aqueous, but the use of organic solvent as a medium favors the exchange to organic anions. • Various factors interfere with the ion-exchange process, such as charge and ionic radius of the incoming anions, medium of exchange, and pH value. • The method mostly favors carbonate ion intercalation; therefore, preparation must be carried out under inert atmosphere. • Employing this method, intercalation of various inorganic and organic anions can be easily accomplished.	Carriazo et al. 2006; Costantino et al. 2014
3.	Rehydration using structural memory effect	• The method entails synthesis of LDHs by rehydration and reconstruction using the structural memory effect. • The presence of interlayer anions and water molecules are easily eliminated during calcination and result in the formation of mixed metal oxides. • The calcined sample can be regenerated by rehydrating it with the desired anions dissolved in an aqueous organic solvent. • Temperature and relative humidity are the important factors that influence the synthesis process.	Mascolo and Mascolo 2015
4.	Hydrothermal synthesis	• The method is considered to be highly applicable in the case when organic anions, having low affinity for LDHs, are attempted for intercalation, while other synthesis methods, namely, coprecipitation and ion exchange, do not work efficiently. • The synthesis involves two possible routes. • In the first case, mixed metal cation solutions are required to be thermally treated at high temperatures of >300°C in a pressured autoclave for a fixed period of time.	Kovanda et al. 2005

(Continued)

TABLE 10.1 (CONTINUED)
Summary of Various Synthesis Techniques for Layered Double Hydroxides

S. No	Synthesis Technique	Synthesis Procedure	References
		• In the second case, LDHs are synthesized at relatively low temperatures and retain the components for aging by refluxing them at a constant temperature for 18 h.	
		• The resultant products are of small and uniform sized and display large surface areas.	
5.	Sol-gel process	• In this process, the metallic precursors undergo hydrolysis and partial condensation followed by gel formation.	Prince et al. 2009
		• Metallic precursors such as metal acetates, acetylacetonates, and alkoxides are mostly used.	
		• The process is usually very sensitive to experimental parameters such as temperature, pH, chemical composition, precursor concentrations, and the nature of the solvent.	
		• This process presents several advantages such as high phase purity, good compositional homogeneity, and a high specific surface area of the particles.	
6.	Salt-oxide method	• The method simply involves reaction between solid and liquid components.	Roussel et al. (2001)
		• Accordingly, a solution of trivalent metal salts is slowly added into a divalent metal oxide suspension under vigorous stirring conditions.	
		• The mixture is required to be aged for a longer time period of 24–48 h, which promotes a reaction between the precursors.	
		• Recording of pH level is important during the addition of trivalent metal solution, which further indicates the progress of reaction.	
		• The method has been specifically applied to the synthesis of Zn/Cr-Cl LDHs.	
7.	Miscellaneous methods	Secondary intercalation (prepillaring) method; intercalation method involving dissolution and recoprecipitation; surface synthesis; templated synthesis, electrochemical synthesis, mechano-chemical synthesis.	Pang et al. 2013; Valente et al. 2009; Li et al. 2012; Zhao et al. 2009; Nejati and Zeynali 2014; Iwasaki et al. 2012.

The analysis of the x-ray diffraction pattern helps in the identification of the characteristic layered patterns as well as the crystalline phases of the sample. The identification of d-spacing values from the resultant peaks, with characteristic diffraction planes of (003), (006), and (009), gives an idea of the thickness of the brucite sheets, and the gallery height of the LDH compound (Chen and Song 2013). The d-spacing value of the corresponding reflection can be used for the determination of the lattice structure of LDH material. Seftel et al. (2008) recommended some formulas that help in the calculation of unit cell parameters such as a, c, and c' ($a = 2d_{110}$, $c = 3d_{003}$ and $c = 3c'$). The ratio of divalent to trivalent cation can be predicted using parameter a. Parameter c can be estimated as three times the value of the basal spacing (c') due to rhombohedral symmetry. The gallery spacing can be calculated as the difference between c' and the thickness of the brucite-like sheet, assumed as 4.8 Å.

Moreover, the peak obtained also helps in differentiating between the crystalline and amorphous phases; in the former case the peaks are sharp in nature, while the latter exhibits typical broad and diffuse peaks. Figure 10.2 represents the diffraction pattern for nitrate- and citrate-intercalated Zn/

TABLE 10.2
Summary of Commonly Used LDHs

M^{2+}	M^{3+}	Intercalated Anion	Synthesis Technique	References
Zn	Al	Nitrate	Electrochemical	Yarger et al. 2008
Mg	Al	Nitrate	Coprecipitation	Vreysen and Maes 2008
Li	Al	Nitrate	Coprecipitation	Zhang et al. 2012a
Ni	Al	Nitrate	Coprecipitation	Zadeh and Sadeghi 2012
Zn	Fe	Carbonate	Coprecipitation	Parida and Mohapatra 2012
Zn	Cr	Nitrilotriacetate	Ion exchange	Gutmann et al. 2000
Cu	Cr	Nitrate	Separate nucleation and aging	Tian et al. 2012
Co	Bi	Nitrate	Urea hydrolysis	Jaiswal et al. 2015a
Ca	Al	Dodecyl sulfate	Hydrothermal	Dutta and Pramanik 2013
Ni	Fe	Chloride	Mechanochemical	Iwasaki et al. 2012
Cu	Fe	Nitrate	Coprecipitation	Nejati et al. 2013
Co	Al	Chloride	Hydrothermal	Arai and Ogawa et al. 2009
Co	Ni	Ethylene glycol	Coprecipitation	Wang et al. 2015
Ca	Fe	Phosphate	Ion exchange	Woo et al. 2011
Cu	Al	Carbonate	Hydrothermal	Britto and Kamath 2009
Cu-Mg	Fe-La	Nitrate	Coprecipitation	Guo et al. 2012
Cu-Mg	Mn	Carbonate	Coprecipitation	Kovanda et al. 2005
Ni-Mg	Mn	Carbonate	Coprecipitation	Kovanda et al. 2005
Cu	Al-Fe	Carbonate	Coprecipitation	Trujillano et al. 2005
Zn-Mg	Al	Nitrate	Coprecipitation	Zheng et al. 2012
Zn	Ga-Al	Carbonate	Coprecipitation	Prince et al. 2014
Pd-Mg	Al	Carbonate	Coprecipitation	Shen et al. 2011
Pd-Mg	Al-Zr	Carbonate	Coprecipitation	Shen et al. 2011
Mg	Al-La	Carbonate	Coprecipitation	Rodrigues et al. 2012
Mg	Al-Zr	Chloride	Coprecipitation	Curtius et al. 2009
Mg	Fe-Zr	Carbonate	Coprecipitation	Chitrakar et al. 2010
Mg	Al	Tetraborate	Hydrothermal	Li et al. 1996
Zn	Al	Ethylenediaminetetraacetic acid	Ion exchange	Rojas et al. 2009
Mg	Al	Oxalate, malate, tartrate	Ion exchange	Zhang et al. 2004
Mg	Al	Arsenate	Coprecipitation	Prasanna and Kamath 2009
Mg	Al	Diethylenetriaminepentaacetic acid	Coprecipitation	Liang et al. 2010
Mg	Al	Citrate	Ion exchange	Zhang et al. 2012
Mg	Al	Dodecylsulfate	Coprecipitation	Otero et al. 2012
Mg	Al	Tetradecanedioic acid	Coprecipitation	Otero et al. 2012
Mg	Al	Polysulfides	Ion exchange	Ma et al. 2014
Mg	Al	Amino acids	Reconstruction	Nakayama et al. 2004
Mg	Al	Dibenzoyl-D-tartaric acid	Ion exchange	Jiao et al. 2009
Mg	Al	Carboxymethyl chitosan	Hydrothermal	Wang and Zhang 2014
Zn	Al	Hexamethylenediaminetetracetic acid	Ion-exchange	Kosorukov et al. 2013

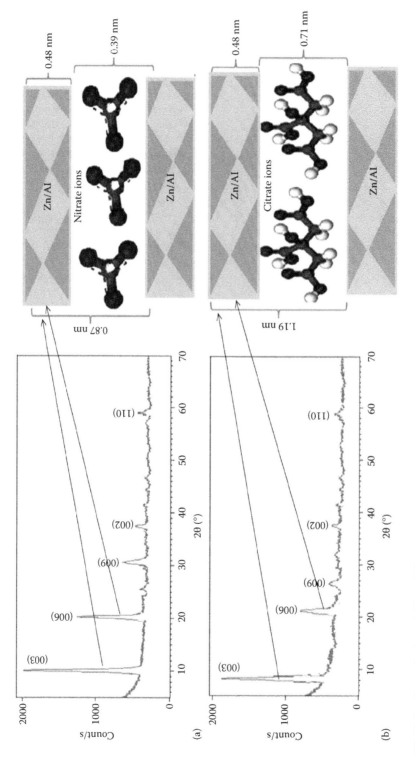

FIGURE 10.2 (See color insert) X-ray diffraction pattern for (a) nitrate and (b) citrate intercalated Zn/Al-LDH.

Al-LDH. The diffraction pattern (Figure 10.2a) displays characteristic basal peaks for (003) and (006) planes at a low 2θ angle and nonbasal peaks at reflection planes of (009), (002), and (110) at a high 2θ angle. The peaks at 10° and 20.2°, with d-spacing values of 8.72 Å and 4.77 Å, indicate basal height and brucite layer thickness, respectively, and the gallery height is estimated as ~3.90 Å (8.72–4.8 Å). In Figure 10.2b, it can be observed that when citrate ions replace nitrate ions, the diffraction peak at a basal plane of (003) shifts toward the left, with a basal height of 11.9 Å, and the gallery height is evaluated as 7.1 Å. This suggests that intercalation of citrate ions increases the gallery space between the brucite layers.

The thermal characteristics of LDH material can be investigated through various thermoanalytical techniques, namely, TGA and DTA. Through TGA analysis, the thermal stability of the material under investigation can be interpreted; moreover, through this technique, one can also predict the presence of certain chemical compounds through their decomposition temperature, as the study involves changes in weight with respect to temperature. DTA analysis addresses the thermal property of the LDH materials. It has been demonstrated by several researchers (Zhang et al. 2004; Wang et al. 2007; Prasanna and Kamath 2009; Chen and Song 2013) that, in most cases, decomposition of LDHs usually occurs in four steps. The first weight loss mostly takes place at temperatures between 65°C and 100°C, and is mainly attributed to loss of surface water or moisture. The second weight loss can be observed at 200°C–250°C due to loss of interlayered water. The third weight loss represents the decomposition of interlayered anions such as carbonate, chloride, nitrate, sulfate and hydroxyl in the temperature range of 250°C–600°C. The fourth weight loss, within the temperature range 650°C–800°C, indicates the complete collapse of the layered structure and formation of metal oxides. Moreover, the weight losses at corresponding temperatures may exhibit typical endothermic peaks, which further substantiate the decomposition process of LDHs. Figure 10.3 represents the TG/DTA curve of Zn/Al-NO$_3$ synthesized by our group, exhibiting characteristics of four major weight losses with three typical endothermic peaks; the material completely decomposed and was transformed into stable metal oxides at 845°C.

SEM and TEM analysis facilitate morphological assessment of the LDH particles; furthermore, the techniques also help in determining particle size. The SEM image of LDH crystals shown in

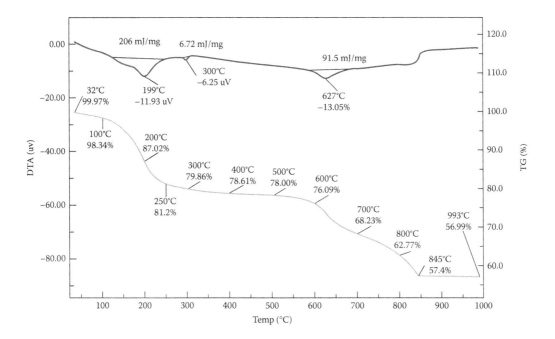

FIGURE 10.3 TG/DTA profile for Zn/Al-NO$_3$ LDH synthesized by the authors.

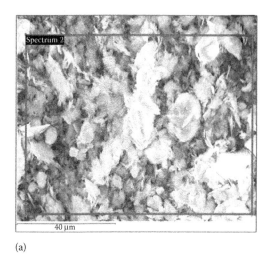

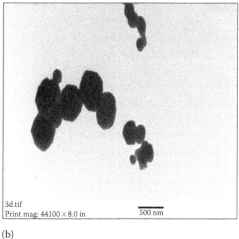

(a)   (b)

FIGURE 10.4   (a) SEM image; (b) TEM image of Zn/Al-NO$_3$ synthesized by the authors.

Figure 10.4a exhibits highly agglomerated conditions; each crystal appears to possess flat, flaky, plate-like features with hexagonal morphology. The TEM image (Figure 10.4b) clearly reveals the smooth and hexagonal geometry of the crystals with a diameter of 0.5 µm. LDHs with similar morphologies were reported by Zheng and Zhang (2012) for Zn/Mg/Al/NO$_3$-LDHs, Chen and Song (2013) for chromotropic acid-intercalated Zn/Al-LDHs, and de Sá et al. (2013) for Ca/Al/NO$_3$-LDHs.

The surface area and porosity play a crucial role that determines the applicability of the LDHs for specific purposes. The specific surface area of the LDH particles are determined through an N$_2$ adsorption–desorption isotherm using the BET method (1938). The pore size and pore volume analysis are also determined from a similar N$_2$ adsorption–desorption isotherm, by means of the computational method given by Barrett, Joyner, and Halenda (BJH), which assumes that, for microporosity, the pore size should be <2 nm; for mesoporosity, it is between 2 and 50 nm; and sizes >50 nm indicate the presence of macropores. The N$_2$ isotherm for LDH particles mostly follows the type IV adsorption isotherm (Table 10.3) according to the International Union of Pure and Applied Chemistry (IUPAC) classification with a characteristic H-3 type hysteresis loop that indicates the presence of slit-like pores of mesoporous character.

TABLE 10.3

Textural Details of Some LDH Materials

LDH	N$_2$ Adsorption Isotherm Type	BET Surface Area (m^2/g)	Pore Size Range (nm)[a]	References
Zn/Ti	IV	91.96–107.88	15–100	Shao et al. 2011
Fe$_3$O$_4$/MgAl	—	110.6	3.59*	Shou et al. 2015
Mg/Fe	II	53	>100	Asouhidou et al. 2011
Mg/Al	II & IV	72	40*	Asouhidou et al. 2011
Mg-Ni/Al	IV	137.4	7–23	Zaghouane-Boudiaf et al. 2012
Cu/Mg/Fe/La	IV	134	16.6*	Guo et al. 2012
Zn/Al	IV	253	3–8	Lin et al. 2015
Ni/Al	IV	108.7	15*	Sun et al. 2015
Cu/Mg/Fe	IV	70	34.6*	Guo et al. 2013b

[a] Average pore size.

Other analytical techniques include vibrating sample magnetometry (VSM); Chen et al. (2012) reported ferromagnetic behavior of a magnetic LDH composite and obtained a high saturation magnetization value of 22.6 emu/g, thus offering good scope for complete separation of the composite material using a simple hand-held external magnet. An energy-dispersive x-ray analyzer (EDAX) determines the elemental composition of the LDHs quantitatively. Ultraviolet visible (UV–vis) diffuse reflection spectroscopy (DRS) is mostly used to analyze the performance of LDHs in photocatalytic activity. Shao et al. (2011) predicted high photocatalytic activity of a Zn/Cr-LDH through DRS spectra with a low bandgap of 3.06 eV. X-ray photoelectron spectroscopy (XPS) also verifies the elemental composition of the LDH material, but qualitatively. Wu et al. (2013) reported XPS spectra of a Mg/Al/Cl-LDH, which clearly reveals peaks for Mg, Al, O, C, and Cl. The pH_{ZPC} (the pH value representing electrical neutrality of the material) of the LDHs have been also reported by several researchers. The technique helps to ascertain the probable interaction mechanism between LDHs and other species in the aqueous phase.

10.5 ADSORPTION OF WATER POLLUTANT SPECIES ON LDHs

The adsorption process is considered to be one of the most suitable and proven technologies, having wide prospective applications in both water and wastewater treatment. The adsorbents that are economically feasible, synthesized in large scales, nontoxic, and have the potential to treat a wide range of pollutants are highly preferable. In this perspective, LDHs are considered as one of the most appropriate adsorbent materials. In the sorption process, the interaction between LDH and pollutants essentially depends on the surface chemistry of the adsorbent and adsorbate and the prevailing system conditions. The surface chemistry of the adsorbate and adsorbent determines the nature of the adsorption process, which may be controlled either by physical or chemical forces. On the other hand, the prevailing system conditions comprise various process parameters that control the rate of the adsorption process. It has been demonstrated by numerous researchers that by monitoring the process variables, the sorption performance of LDHs in scavenging of water pollutants is significantly enhanced. Therefore, in the following section the influence of process variables affecting the sorption of LDHs are briefly discussed.

10.5.1 Factors Affecting Sorption Performance of LDHs

10.5.1.1 pH

The solution pH is regarded as one of the most important parameters, which significantly controls the rate of the sorption process. Therefore, in most cases, pH is termed the *master variable*. As it is already known that the solution pH monitors the degree of ionization of the adsorbate species, as well as altering the surface chemistry of the adsorbent through the dissociation of various functional groups present on the adsorbent surfaces, pH-assisted adsorption by LDHs therefore depends mostly on the types of pollutants and the types of LDHs. The mechanisms of pollutant removal as a function of pH were mostly reported in the literature in terms of the pKa value of the adsorbate and the pH_{ZPC} value of the adsorbents. Ulibarri et al. (1995) demonstrated that phenolic compounds, namely, trichlorophenol (TCP) and trinitrophenol (TNP), were found to adsorb mainly at pH values close to and higher than that of the respective pKa values (TCP = 6.9; TNP = 3.8) of the phenols, as phenol dissociation promotes the sorption process. The maximum adsorption of 14% was observed for TCP at a pH value of 13.0 and 55% for TNP at a pH value of 2.0. Figure 10.5 displays the adsorption mechanism of TCP and TNP. In the case of TCP, a high pH (13.0) favors sorption, as when pH > pKa, TCP completely dissociates to phenolate form and becomes adsorbed on the external surfaces of the LDH. However, for TNP, due to its low pKa value, it almost exists in anionic form at a low pH value (2.0) and thus acts a Brønsted acid that results in the exchange of carbonate anions with TNP anions, which is also responsible for high adsorption. It has been observed that the pH

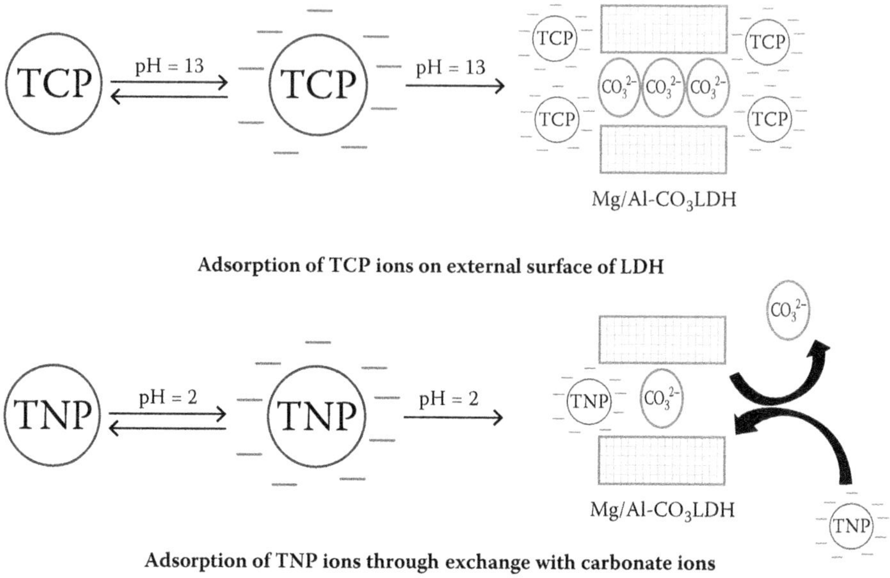

FIGURE 10.5   (See color insert) Schematic representation of adsorption of TCP and TNP at given pH condition.

of the solution increases with increases in TNP adsorption, which further substantiates the replacement of carbonate with TNP anions.

The removal of pesticides such as carbetamide and metamitron by dodecyl sulfate-intercalated Mg/Al-LDHs reflects an interesting trend. The lowest adsorption was observed at the pH values where the pesticides exist in their ionic form, according to their pKa values of 10.9 and 2.9 for carbetamide and metamitron, respectively (Bruna et al. 2006).

Li et al. (2009) reported that, depending on the total concentration of Cr(IV) and the solution pH, the Cr(IV) ions exist in different ionic states. Maximum removal of more than 90% was observed at pH values between 2.0 and 4.0. The ion-exchange and surface-complexation reactions were investigated as the main mechanisms that better described the pH-dependent removal of chromate ions.

The adsorption of arsenate on Cu/Mg/Fe/La-LDHs is reportedly favored in low pH conditions. Guo et al. (2010) demonstrated that in the pH range of 3–10, the dominant species are $H_2AsO_4^-$ and $HAsO_4^{2-}$. The authors explained that at low pH values, the adsorbent surface becomes positively charged, as when $pH < pH_{ZPC}$, the surface of the LDH becomes protonated; as the pH_{ZPC} of the LDH is evaluated as 9.7, low pH values therefore favor the uptake of arsenate anions as a result of electrostatic attraction.

de Sá et al. (2013) investigated the adsorption of Sunset Yellow FCF food dye onto Ca/Al-NO₃, which was greatly influenced by the solution pH. The authors reported a pH value of 4.0 to be the optimum pH value at which maximum adsorption of the dye (~50 mg/g at initial dye concentration of 50 mg/L) occurred, and between pH 4.0 and 10, the dye uptake capacity decreased considerably. The removal mechanism was interpreted in terms of the pH_{ZPC} of the adsorbent.

Rojas (2014) demonstrated that, in alkaline conditions, the adsorption of Cu^{2+}, Cd^{2+}, and Pb^{2+} by Ca/Al-LDHs is highly beneficial, as more than 90% of removal was observed at pH 9.0. However, Guo et al. (2013a) reported that adsorption of the dye Acid Brown 14 on calcined Mg/Fe-LDHs was almost independent of the solution pH conditions. Nearly 100% of dye removal was achieved in the pH range 4–11.

Thus, the inconsistency observed in the results for the adsorption of different pollutant species as a function of pH may be due to the involvement of various factors, such as negligence in the proper monitoring of pH while conducting sorption experiments, or adjustments to pH without using pH buffers. Moreover, in several studies, the interference of competing ions remains ignored.

10.5.1.2 Competitive Ions

Wastewater, in general, is a concoction of several organic and inorganic components; therefore, it is quite expected that these chemical components may interfere with the water treatment process. Many studies have been reported (Yang et al. 2005; Lv et al. 2007; Zhao et al. 2011; Zhang et al. 2012; Guo et al. 2013a,b) that determine the effect of competitive ions on the adsorption of targeted ions by LDHs.

Yang et al. (2005) observed that the competing ions have a stronger effect on the adsorption of As(V) over uncalcined LDHs than on that over calcined LDHs. Moreover, they also demonstrated that competitive anions have greater influence on Se(IV) uptake than on As(V) uptake. The effect of competing anions on the adsorption of As(V) reduces in the order: $HPO_4^{2-} > CO_3^{2-} > SO_4^{2-} > NO_3^-$; for Se(IV), the order was demonstrated as: $HPO_4^{2-} > SO_4^{2-} > CO_3^{2-} > NO_3^-$.

Lv et al. (2007) investigated the effect of existing coanions on the removal of fluoride ions by Mg/Al-LDHs. The removal of fluoride ions declined significantly in the presence of anions, in the order $HCO_3^- < Cl- < H_2PO_4^- < SO_4^{2-}$. However, another adsorptive study of fluoride ions onto Li/Al-LDHs exhibits a quite different trend to that reported by Lv et al. (2007), where, in the presence of HCO_{3-}, fluoride adsorption was slightly affected, but in the presence of SO_4^{2-}, the lowest removal percentage for fluoride ions was observed. However, Zhang et al. (2012a) showed that SO_4^{2-} and HCO_3^- have similar effects on fluoride sorption capacity, and the results were obtained as: $NO_3^- \approx Cl^- < SO_4^{2-} \approx HCO_3^- < CO_3^{2-} < HPO_4^{2-} < PO_4^{3-}$. The trend reveals that divalent and trivalent anions exhibit a prominent effect on fluoride sorption as a result of high negative charge density (Zhang et al. 2012a). The adsorption behavior of Pb(II) was also found to be affected in the presence of various electrolyte cations and anions (Zhao et al. 2011). The removal percentage of Pb(II) considerably declined in the presence of K^+, followed by Na^+ and Li^+, and this behavior was mainly attributed to ionic radii, as K^+ has comparable ionic radii to those of Pb(II), and thus actively competes for adsorption sites. Likewise, in the presence of inorganic anionic radicals, the adsorption percentage of Pb(II) followed the order: $ClO_4^- > NO_3^- > Cl^-$. The lowest removal percentage was observed in the presence of Cl−, due to its smaller size, which easily takes up more ion-exchange sites and decreases the removal percentage. The reduction in uptake capacity of Mg/Fe-LDHs for Acid Brown-14 dye in the presence of coexisting ions has been demonstrated by Guo et al. (2013a). The dye adsorption percentage was minimally affected in the presence of NO_3^- and SO_4^{2-}; however, the effect was more pronounced in the cases of PO_4^{3-} and HPO_4^{2-}, where the adsorption percentages dropped below 40%.

Thus, it can be interpreted from the foregoing results that the charge density and ionic radii of the interfering ions plays a significant role in the sorption process of various organic and inorganic ions. However, in most studies, it has been reported that the effect of charge density is more pronounced than that of ionic radii. The greater the charge density of the existing coions, the more actively they will compete with the ionic species under investigation. However, in many studies it has been found that the pH level was not taken into consideration; in such cases, the results obtained may be unreliable to some extent.

10.5.1.3 LDH Dosage

The investigation of the optimum adsorbent dose helps to determine the precise amount of adsorbent to achieve maximum removal of pollutants from the batch adsorption system; further, the parameter also determines the economic perspective of the treatment process. In general, in most of the studies it has been reported that the adsorption capacity is enhanced with the increase in adsorbent dosage; after a certain dosage value, the adsorption curve reaches a flattened position, which indicates that with increased dosage, the number of active sorption sites of adsorbent also increases, and are thus readily available for binding with the adsorbate species. Abdelkader et al. (2011) reported a similar trend for the adsorption of anionic dye Orange G over noncalcined and calcined LDHs. The dye uptake amount increased from 5 to 50 mg/g for noncalcined LDHs, and for calcined LDHs the amount of uptake rose significantly from 78 to 200 mg/g with an increasing sorbent dose from

0.1 to 1.4 g/L. A similar finding was also demonstrated by Zhang et al. (2012b), who reported that the uptake capacity of Mg/Al-citrate-LDHs for uranium(VI) spectacularly increases from 134 to 184 mg/g, with an increase in adsorbent dose from 0.1 to 30 g/L at an initial uranium(IV) concentration of 200 mg/L. However, in some studies, contrary findings were reported; for instance, in a sorption study of bromide ions, with an increase in sorbent dose (0.2–4.0 g/L) the adsorption capacity of calcined LDHs steadily decreases from 180 to 17 mg/g at a fixed anion concentration of 100 mg/L (Lv et al. 2008). In the case of a chloride sorption study, the author experienced similar sorption behavior; as the dosage of the Zn/Al-LDH increased from 1.5 to 7.5 g/L, the uptake capacity slumped to 10 mg/g from 44 mg/g, using an initial chloride concentration of 100 mg/L (Lv et al. 2009).

10.5.1.4 Temperature

Temperature plays an important role in the sorption process by altering the rate of the sorption reaction. Moreover, it is also presumed that changing temperature amends the equilibrium capacity of the adsorbent for a particular adsorbate, due to the formation or weakening of hydrogen bonds or van der Waals interactions between the adsorbate and adsorbent species. In the case where uptake efficiency enhances with increasing temperature, the adsorption process is then regarded as endothermic in nature. Several endothermic phenomena have been reported when using LDHs as adsorbents. Li et al. (2009) observed that the adsorptive capacity of Cr(VI) rose from 105 to 112 mg/g as the adsorption temperature increased from 293 K to 313 K. The authors verified the findings by examining thermodynamics parameters, where a positive value of ΔH° also indicates the endothermic behavior of the sorption process. A related condition has also been observed for adsorption of radionickel on Mg_2Al-LDH (Zhao et al. 2013). The sorption capacity significantly rose from 46.5 mg/g to 51 mg/g as the temperature increased from 298 K to 338 K. Arsenate adsorption on an Mg/Al-polycinnamide-LDH is also enhanced with rising temperature, and the endothermic nature of the process is represented by a positive ΔH° value, which further indicates structural modification of the adsorbent and affinity of the adsorbent for arsenate ions (Islam et al. 2013). Likewise, sorption of Pb(II) and F⁻ ions are favored at higher temperatures (Zhao et al. 2011; Zhang et al. 2012).

However, in several studies, it has been observed that rising temperature depresses the uptake capacity, while lower temperature favors the adsorption process, thus indicating the exothermic nature of the sorption process. The adsorptive removal of phosphate ions by an Mg/Al-LDH was found to increase with rising temperature, and a negative value of ΔH° further confirmed the exothermic behavior of the adsorption process (Das et al. 2006). Similarly, Sui et al. (2012) also demonstrated that removal of norfloxacin from aqueous solution follows an exothermic sorption process. The uptake capacity of 2,4-dichlorophenoxyacetic acid by a Cu/Fe-LDH was analyzed as an exothermic process by Nejati et al. (2013). The maximum uptake capacity was evaluated as 1667 mg/g at 298 K, while at 333 K the adsorbed amount fell to 1420 mg/g.

In some studies, it was demonstrated that the adsorption process was independent of temperature. Yang et al. (2005) reported that the adsorption of Se(IV) on uncalcined LDHs was not affected by changes in the adsorption temperature. The adsorption of boron on calcined LDHs was demonstrated to remain unaffected when temperature the was increased from 288 K to 298 K; however, as the temperature was increased further, from 298 K to 308 K, the amount of boron uptake slightly decreased (Paez et al. 2014).

10.6 ADSORPTION ISOTHERMS AND ADSORPTION KINETICS

Adsorption isotherm studies have been extensively reported on regarding LDH-water pollutants-based systems. The analysis of adsorption data through adsorption isotherms is important, as it helps in ascertaining the nature of the possible interaction that may occur between LDH and pollutant species. Moreover, this study also helps in predicting the surface properties as well as the efficacy of the LDHs by evaluating their maximum sorption capacity. Importantly, the study determines the effect of initial pollutant concentrations on adsorption over LDHs at constant temperature. It has been demonstrated that the uptake capacity of the sorbent greatly depends on the initial adsorbate concentration. For instance, Sui et al. (2012) reported that, at 298 K, the sorption capacity of Mg/Al-LDH increased rapidly with increasing norfloxacin concentration. However, the uptake amount of methyl orange by Zn/Mg/Al-LDHs reportedly decreased from 1121 to 680 mg/g as the initial dye concentration increased from 70 to 110 mg/L (Zheng and Zhang 2012). Studies on adsorption isotherms are, in general, carried out through the fit of the adsorption data to various mathematical expression–based isotherm models. The most appropriate model that represents the linearized relationship is used for the design of the preferred water treatment system. To date, varieties of equilibrium isotherm models have been applied for the interpretation of sorption behavior of LDHs; among them, the Langmuir, Freundlich, and modified Langmuir models have been extensively investigated. Table 10.3 lists isotherm models of various types that are suitable for different adsorbate–adsorbent systems. From Table 10.3, it is seen that most of the sorption studies follow either the Langmuir or Freundlich isotherm model, which indicates that adsorption of adsorbate over LDHs occurred in monolayered and multilayered manners, respectively. Furthermore, adsorption isotherm studies also help to investigate the dependence of the sorption system on temperature. Several authors (Lv et al. 2007, 2009; Halajnia et al. 2013; Nejati et al. 2013; Guo et al. 2013a,b) investigated the thermodynamics parameters of the sorption system by using the K_L or b parameter of the Langmuir isotherm model.

The investigation of the sorption process as a function of time is better represented as adsorption kinetics. In its usual course, it has been reported that the adsorption rate in the initial period of time was very rapid; however, as time proceeded, the adsorption rate slowed down and attained the equilibrium condition. Ni et al. (2007) noticed similar behavior for the adsorption of methyl orange over calcined LDHs; the uptake amount increased rapidly in the initial 60 min, then adsorption occurred slowly and reached a state of equilibrium at 120 min. A similar trend has been experienced by Boudiaf et al. (2011) for the adsorption of 2,4,5 trichlorophenol by Mg/Al-organo-LDHs, where rapid adsorption was observed within 30–40 min; afterward, the sorption rate became sluggish and the system reached equilibrium after 60 min. The rapid adsorption that was observed during the initial contact period was mainly attributed to the availability of a large number of vacant adsorption sites, but as the time proceeds, the sorption process becomes impeded, due to the slow pore diffusion and electrostatic repulsion between adsorbate molecules in the bulk phase and on solid surfaces. The mechanism of adsorbate–adsorbent interaction can be addressed through various kinetic models. In LDH-based sorption studies, the most widely explored kinetic models were the pseudo-first-order, pseudo-second-order, intraparticle diffusion, and Boyd models. The pseudo-first-order and pseudo-second-order kinetic models determine the rate of the sorption process; on the other hand, the diffusion-controlled adsorption process can be efficiently investigated through the intraparticle diffusion and Boyd models, which also facilitate in-depth understanding of the mechanistic path of the sorption process. It was also investigated that those sorption processes that were very rapid were usually dictated by surface diffusion mechanis; whereas sorptions that take a long time to complete are controlled by the pore diffusion mechanism. For instance, adsorption of As(V) on an Mg/Fe-LDH occurred rapidly and was almost completed within 15 min of contact time (Turk et al. 2009). Rapid adsorption within a short period of contact time was attributed to surface diffusion; in addition, it was also presumed that the LDH uptake capacity was proportional to the number of the active sites on its surface. Similar observations were reported by Chen et al. (2011); it was observed that about 98% of Congo red removal by Fe_3O_4-Mg/Al-LDH was completed within 15 min. Further, it was also recommended that those adsorbents that exhibit quick adsorption kinetics offer cost-effective industrial wastewater treatment. In some adsorption systems, it was also demonstrated that the sorption process proceeded slowly and required a long time period to accomplish, thus indicating that adsorption occurs through the diffusion of adsorbates into the pores of the sorbent. A similar phenomenon was observed by Liu et al. (2007) for the adsorption of 1-phenyl-1,2-ethanediol (PED) by carboxymethyl-β-cyclodextrin intercalated Zn/Al-LDHs (CMCD LDHs). The adsorption of PED was found to advance slowly and continued up to 18,000 s; pore diffusivity was suspected to be rate limiting, which was finally confirmed from the experimental and theoretical uptake curves plotted between F (fractional attainment of equilibrium) and t (time). In another study, Liu et al. (2013) reported similar adsorption behavior for adsorption of the amino acid phenylalanine on carboxymethyl-α/β-cyclodextrin. However, in some studies, it was also reported that both surface and pore diffusion occurred concurrently. Khenifi et al. (2010) demonstrated that the intraparticle diffusion of glyphosate and glucosinate uptake on Ni/Al-LDH plots exhibited two separate linear regions. It was further verified that the initial curved portion represented the boundary layer effect and the second linear segment was ascribed to the pore diffusion. Thus, both processes were simultaneously responsible for the adsorption of glyphosate and glucosinate. Table 10.4 summarizes detailed information of adsorption-related characteristics for the sorption of various water pollutants on LDHs, with the aim of offering immediate understanding to readers, so that they can easily interpret the relation between adsorbate–adsorbent and process parameters for various adsorption systems.

10.7 SORPTION MECHANISMS

It was demonstrated that the interaction of adsorbates with LDHs was mechanistically controlled by various physicochemical process including ion exchange, memory effect, chelation, surface complexation, isomorphic substitution, and precipitation (Figure 10.6).

10.7.1 Ion Exchange

The ion-exchange process is considered as one of the prominent sorption mechanisms for anionic species. In this process, the interlayered anions of LDHs are replaced with targeted anionic species. However, the extent of anionic exchange depends on the anions present in the interlayered spaces of the LDHs. Adsorption of 4-chloro-2-methylphenoxyacetic acid (MCPA) through ion exchange was investigated by Inacio et al. (2001). The confirmation of the ion-exchange process was made after examination of the x-ray diffraction peaks of LDH-loaded MCPA. From the analysis of the diffraction peaks, it was seen that the basal spacing became enlarged after adsorption, thus indicating the exchange of MCPA with chloride and nitrate ions.

Li et al. (2009) reported removal of $HCrO_4^-$ by an Mg/Al/Cl-LDH through exchange with Cl^- ions. However, the anionic exchange was found to be intense in acidic conditions, in the pH range 2–5. The exchange mechanism was represented as (Li et al. 2009)

$$Sur - OH_2^+/Cl^- + HCrO_4^- \rightarrow Sur - OH_2^+/HCrO_4^- + Cl^- \tag{10.4}$$

Likewise, removal of arsenate ions from aqueous solution also occurred via ion exchange and the mechanism demonstrated by Xu et al. (2010) is given as

$$Mg_2Al - Cl - LDH + HAsO_4^{2-} \rightarrow Mg_2Al - (HAsO_4)_y (Cl)_z + Cl^- \tag{10.5}$$

Zhang et al. (2012) demonstrated that the removal of fluoride ions through the ion-exchange process beomes significantly affected in presence of other competing ions. The uptake capacity of Li/Al-LDH for fluoride ions significantly decreases in the presence of other anions, in the order as follows:

$$NO_3^- \approx Cl^- < SO_4^{2-} \approx HCO_3^- < CO_3^{2-} < HPO_4^{2-} < PO_4^{3-} \tag{10.6}$$

TABLE 10.4

Adsorption-Related Characteristics for Removal of Various Water Pollutants by LDHs

Pollutant type	Pollutant	Type of LDH	Pollutant Concentration and LDH Dosage	Initial pH; Temp (K)	Adsorption Capacity	Adsorption Isotherm	Equilibrium Time	Kinetic Model	References
Herbicide	MCPA (4-chloro-2-methylphenoxyacetic acid)	Uncalcined chloride Mg/Al-LDHs	0.05–2.0 mmol/L and 20–100 mg/100 mL	7.0; 298	0.50 mmol/g	Freundlich	24 h	—	Inacio et al. 2001
Herbicide	Glyphosate	Uncalcined nitrate Ni/Al-LDHs	200 mg/g and 1 g/L	7.0; 298	172.4 mg/g	Langmuir	300 min	Pseudo-second-order	Khenifi et al. 2010
Herbicide	2,4-D (2,4-dichlorophenoxyacetic acid)	Uncalcined nitrate Cu/Fe-LDHs	1.5 mmol/L and 0.05 g/L	4.0; 298	1336 mg/g	Langmuir and Freundlich	248 min	Pseudo-second-order	Nejati et al. 2013
Pesticide	Carbetamide	Dodecyl sulfate-intercalated Mg/Al-LDHs	1.0 mmol/L and 0.125 g/100 mL	6.0; room temp.	—	Freundlich	2 h	—	Bruna et al. 2006
Pesticide	S-Metolachlor	Tetradecane-dioate intercalated Mg/Al-LDHs	0.35 mmol/L and 20 mg/30 mL	6.6; 303	—	Freundlich	24 h	—	Otero et al. 2012
Surfactant	DBS (dodecylbenzene sulfonate)	Calcined carbonate Mg/Al-LDHs	0.05 g/30 mL	6.0; room temp.	5263 µmol/g	Langmuir	24 h	—	Ulibarri et al. 2001
Surfactant	SDBS (sodiumdodecylbenzene sulfonate)	Uncalcined carbonate Mg/Al-LDHs	2.50×10^{-4}– 8.0×10^{-3} mol/dm^3 and 200 mg/0.05 dm^3	7–9; 298	—	—	72 h	—	Reis et al. 2004
Phenol	TNT (trinitrophenol)	Calcined carbonate Mg/Al-LDHs	0.05 g/30 mL	2.0; room temp.	4545 µmol/g	Langmuir	24 h	—	Ulibarri et al. 2001
Phenol	TCP (2,4,6-trichlorophenol)	Dodecylbenzenesulfonate intercalated Zn/Al-LDHs	300 mg/L and 30 mg/50 mL	3.0; 298	166.8 mg/g	Freundlich	2 h	Pseudo-second-order	Zhao et al. 2015
Dye	Orange G	Calcined carbonate Mg/Fe-LDHs	50–800 mg/L and 1.0–1.8 g/L	3.0–13.0; 323	378.8 mg/g	Langmuir	9 h	Pseudo-second-order	Abdelkader et al. 2011
Dye	Remazol Red 3BS	Calcined Mg/Al-LDHs	0.092 mmol/L and 0.05 g/50 mL	6.0; 338	134.37 mg/g	Langmuir	3 h	Pseudo-second-order	Asouhidou et al. 2011
Dye	Methyl orange	Uncalcined Zn/Mg/Al-LDHs	0.07–0.12 g/L and 0.05–0.25 g/L	3.0; 298	883.24 mg/g	Langmuir	90 min	Pseudo-second-order	Zheng et al. 2012
Dye	Congo red	Uncalcined carbonate Mg/Fe-LDHs	100 mg/10 mL and 30 mg/25 mL	1.0–4.0; 323	104.6 mg/g	Langmuir	180 min	Pseudo-second-order	Ahmed and Gasser 2012
Dye	Acid Brown 14	Calcined carbonate Mg/Fe-LDHs	0.5 to 50 mg/L and 0.2 g/L	4.0; 323	370 mg/g	Langmuir	60 min	Pseudo-second-order	Guo et al. 2013a

(Continued)

TABLE 10.4 (CONTINUED)

Adsorption-Related Characteristics for Removal of Various Water Pollutants by LDHs

Pollutant type	Pollutant	Type of LDH	Pollutant Concentration and LDH Dosage	Initial pH; Temp (K)	Adsorption Capacity	Adsorption Isotherm	Equilibrium Time	Kinetic Model	References
Dye	Reactive Blue 4	Unicalcined nitrate Mg/Al-LDHs	95 mg/L and 0.015 g/20 mL	2.0; 295	328.9 mg/g	Langmuir-Freundlich	—	—	Aguiar et al. 2013
Anion	Phosphate	Calcined carbonate Mg/Al-LDHs	50 mg/L and 0.4 g/L	6.0; 303	44 mg/g	Langmuir and Freundlich	4 h	Pseudo-first-order	Das et al. 2006
Anion	Bromide	Calcined carbonate Mg/Al-LDHs	100 mg/L and 0.05 g/50 mL	7.0; 303	362.3 mg/g	Langmuir	24 h	Pseudo-second-order	Lv et al. 2008
Anion	Selenite	Uncalcined chloride Zn/Al-LDHs	110.5 mg/L and 0.1 g/100 mL	3.4–7.4; 298	129.3 mg/g	Langmuir	60 min	Pseudo-second-order	Mandal et al. 2009
Anion	Cr(IV)	Uncalcined chloride Mg/Al-LDHs	20–200 mg/L and 2.0 g/L	2.5–5.0; 303	112 mg/g	Langmuir	150 min	Pseudo-second-order	Li et al. 2009
Anion	Arsenate	Uncalcined Cu/Mg/Fe/La-LDHs	1–15 mg/L and 20 mg/100 mL	3.0; 323	43.5 mg/g	Langmuir	8 h	Pseudo-second-order	Guo et al. 2012
Anion	Boron	Calcined Mg/Al-LDHs	10–200 mg/L and 0.2 g/40 mL	8.0; 288	25.52 mg/g	Langmuir and Freundlich	5 days	—	Paez et al. 2014
Anion	Fluoride	Calcined Li/Al-LDHs	20 mg/L and 0.1 g/50 mL	6.0–7.0; 313	47.24 mg/g	Freundlich	24 h	Pseudo-second-order	Zhang et al. 2012a
Heavy metal	Cd(II)	Uncalcined Fe_3O_4 carbonate Mg/Al-LDHs	80–500 mg/L and 0.08 g/L	8.0; 323	54.7 mg/g	Langmuir	300 min	Pseudo-second-order	Shan et al. 2015
Heavy metal	Pb(II)	Uncalcined diethylenetriaminepentaacetic acid-intercalated Mg/Al-LDHs	300 mg/L and 1.0 g/L	3.5–5.5; 298	170 mg/g	Langmuir	300 min	Pseudo-second-order	Liang et al. 2010
Heavy metal	Cu(II)	Calcined carbonate Ni/Al-LDHs	5–25 mg/L and 10 g/L	7.0; 303	87.6 mg/g	Langmuir	120 min	Pseudo-second-order	Jaiswal et al. 2015b
Radionuclide	[63]Ni(II)	Uncalcined carbonate Mg/Al-LDHs	10 mg/L and 0.4 g/L	8.6; 338	51.02 mg/g	Langmuir	24 h	Pseudo-second-order	Zhao et al. 2013
Radionuclide	U(VI)	Uncalcined magnetic citrate intercalated Mg/Al-LDHs	200 mg/L and 0.05 g/50 mL	6.0; 328	180 mg/g	Freundlich	4 h	Pseudo-second-order	Zhang et al. 2012
Radionuclide	[152+154]Eu	Uncalcined sodium lauryl sulfate-intercalated Mg/Al-LDHs	60 mg/L and 0.01 g/5 mL	4.0–7.0; 303	156.45 mg/g	Langmuir	90 min	Pseudo-second-order	Mahmoud and Someda 2012
Organic matter	Humic acid	Uncalcined chloride Mg/Fe-LDHs	200 mg/L and 0.05/10 mL	7.0; 298	76.70 mg/g	Langmuir	120 min	—	Gasser et al. 2008
Pharmaceutical	Nofloxacin	Uncalcined chloride Mg/Al-LDHs	20 mg/L and 1 g/500 mL	3.0–10.2; 298	—	Freundlich	50 min	Pseudo-second-order	Sui et al. 2012

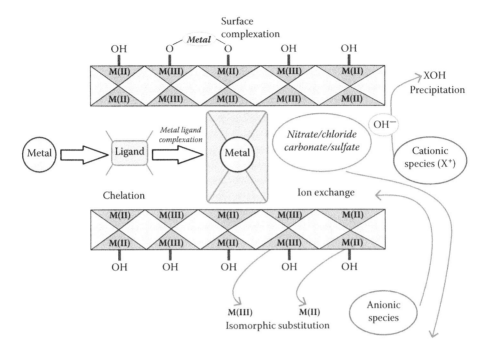

FIGURE 10.6 **(See color insert)** Schematic representation of probable mechanism for adsorption of various pollutants.

It was observed that the divalent and trivalent anions exhibited a greater effect on uptake efficiency than monovalent anions, as the high negative charge density ions were easily replaced with the interlayered ions.

10.7.2 Memory Effect

In this process, calcined LDHs or layered double oxides (LDOs) are employed to scavenge the pollutants from aqueous solution by the reconstruction method. In general, when LDOs are exposed to a hydric environment, they start to rearrange themselves into lamellar structures, which they have lost due to calcination by capturing hydroxyl ions and water molecules from the nearby environment; meanwhile, during the reconstruction process, they also capture pollutant species into their interlayered space. This process is also favorable for anion removal. The removal of acidic pesticides, namely, 2,4-D, clopyralid, and picloram, was reportedly accomplished using Mg/Al mixed metal oxides (Pavlovic et al. 2005). It was determined that the pesticide adsorption was as a result of the reconstruction of mixed metal oxides with pesticides. It was also stated that during intercalation the anionic pesticides arranged themselves in flat positions in the interlayer. A similar phenomenon was observed by Lv et al. (2006) concerning the adsorption of chloride ions from aqueous solution by a calcined Mg/Al-CO$_3$-LDH. The synthesis of the calcined LDH (Equation 10.7) and the mechanism of chloride uptake through reconstruction (Equation 10.8) can be expressed as

$$Mg_{1-x}Al_x (OH)_2 (CO_3)_{x/2} \cdot mH_2O \xrightarrow{500°C} Mg_{1-x}Al_xO_{1+x/2} + x/2CO_2 + (m+1)H_2O \quad (10.7)$$

$$Mg_{1-x}Al_xO_{1+x/2} + (x/n)A^{n-} + (m+1+(x/2)+y)H_2O \rightarrow Mg_{1-x}Al_x(OH)_2(A^{n-})_{x/n} \cdot mH_2O + xOH^- \quad (10.8)$$

where A^{n-} in Equation 10.8 represents chloride ions. The reconstructed LDH consists of chloride ions substantiated through various analytical techniques such as XRD, FTIR, and thermo-gravimetry mass spectroscopy (TG-MS).

Adsorption of Acid Brown-14 dye on a calcined Mg/Fe-LDH by the memory effect was reported by Guo et al. (2013a). It was further determined that the dye anions were intercalated in between the brucite sheets by the chemisorption process. The $-SO_3^-$ groups of the dye AB-14 interacted with the metal ions and the $-OH$ group of the LDH brucite sheets through electrostatic attraction and hydrogen bonding, respectively.

10.7.3 CHELATION

The removal process consists of the intercalation of LDHs with chelating agents, which, in the presence of metallic species, form metal–ligand complexes within the interlayered spaces of LDHs. This removal mechanism is highly preferred for the uptake of metallic species. Gutmann et al. (2000) investigated the interaction of Cu(II) and Ni(II) with nitrilotriacetate intercalated in the Zn/Cr-LDH. The metal–ligand complexes were arranged in vertical and tilted positions within the LDHs. Moreover, analysis of the x-ray diffraction pattern suggested that the arrangement of complexes varied with temperature and metal ions. An interesting study reported by Rojas et al. (2009) was related to the elimination of Cu(II) from aqueous solution using EDTA-modified LDHs as scavengers. In the study, it was observed that during EDTA intercalation within the Zn/Al-LDH, the brucite layer became partially eroded. As a result, Zn-EDTA complexation took place and the brucite layer became intercalated. Thus, the uptake of Cu(II) occurred through the exchange reaction with the Zn-EDTA complex. The tentative reaction suggested by Rojas et al. (2009) was given as

$$Zn/Al - LDH + \left[Zn\left(EDTA\right) \right]^{2-} + \left[Cu\left(H_2O_6\right) \right]^{2+} \rightarrow Cu - LDH + \left[Cu\left(EDTA\right) \right]^{2-} + \left[Zn\left(H_2O_6\right) \right]^{2+}$$

(10.9)

The sorption of Pb(II) ions by an LDH intercalated with diethylenetriaminepentaacetic acid (DPTA) was reported by Liang et al. (2010). The complexation of Pb(II) with a Mg/Al/DPTA-LDH was confirmed through XPS spectra analysis that exhibited typical binding energy peaks for Pb and significant chemical shifts as a result of Pb-DPTA complexation.

The uptake of rare earth ions such as La(III), Sc(III), and Y(III) from the aqueous phase by metal–ligand complex formation was also successfully tested by Kameda et al. (2013). Cu/Al/EDTA-LDHs were employed for the uptake of rare earth ions and the order of removal was observed as Sc(EDTA)$^-$ > La(EDTA)$^-$ > Y(EDTA)$^-$. This order was based on the formation of stable metal–chelate complexes.

10.7.4 SURFACE COMPLEXATION

In this process, the hydroxyl group present on the LDH surface participated in complex formation with the targeted pollutants. The mechanism was found to be effective for the uptake of both cationic and anionic species. This removal phenomenon was observed by Li et al. (2009) for the removal of Cr(VI) ions. It was observed that $HCrO_4^-$ anions interacted electrostatically with the positively charged surface of Mg/Al-LDH. The mechanism of interaction was given as

$$Sur - OH_2^+ + HCrO_4^- \rightarrow Sur - OH_2^+/HCrO_4^-$$

(10.10)

The removal of molybate ions by surface complexation was suggested by Ardau et al. (2012) as an important uptake mechanism. It was also reported that this removal mechanism was prominent when the metal cation ratio of the Zn(II)/Al(III)-LDH was low. The adsorption of boron by Paez et al. (2014) also occurred via a similar surface-complexation mechanism. However, for boron species such as $B(OH)_4^-$ and H_3BO_3, two different surface complexations were investigated. It was observed that the adsorption of $B(OH)_4^-$ was as a result of the formation of outer-sphere complexes of $B(OH)_4^-$ with OH, as given in Equation 10.11. For boric acid (H_3BO_3), the adsorption was presumed to be an acid–base interaction with the OH layer that leads to the formation of inner-sphere complexes, as represented in Equation 10.12.

$$M-OH_2^+ + B(OH)_4^- \rightarrow M-OH_2^+B(OH)_4^- \tag{10.11}$$

$$M-OH + H_3BO_3 \rightarrow M-H_2BO_3 + H_2O \tag{10.12}$$

Shan et al. (2015) also observed the removal of Cd(II) by the formation of outer-sphere surface complexes of Cd(II) with the OH layer.

10.7.5 Isomorphic Substitution

This adsorption mechanism was typically observed for the removal of metal cations. In this process, the divalent and trivalent metal cations of LDHs were readily substituted with the targeted metal cations in the aqueous phase. This very phenomenon was reported by Pavlovic et al. (2009) for the removal of Cu(II), Pb(II), and Cd(II) ions using Zn/Al-LDH. It was supposed that the targeted metal ions might be substituted with the Zn(II) of the brucite sheets. The tentative reaction mechanism was represented as

$$\left[Zn_3Al(OH)_8\right]^+ + X^{2+} \rightarrow \left[Zn_3Al(OX)(OH)_7\right]^{2+} + H^+ \tag{10.13}$$

where X^{2+} is the targeted metallic species. Shan et al. (2015) also suggested a similar mechanism for Cd(II) removal.

10.7.6 Precipitation

In this process, pollutants were sequestered through the formation of their precipitates. Xu et al. (2010) investigated the removal of arsenate and phosphate through the precipitation mechanism. It was reported that, in the presence of Ca-containing LDHs, the precipitation reaction was initiated. When phosphate ions were present, the Ca(II) of the LDH reacted to form hydroxyapatite, as represented by Equation 10.14:

$$Ca_3Al(OH)_8Cl \rightarrow 3Ca^{2+} + Al(OH)_3 + 5OH^- + Cl^- \tag{10.14}$$

$$5Ca^{2+} + 3PO_4^{3-} + OH^- + Ca_5(PO_4)_3(OH) \tag{10.15}$$

In a similar manner, arsenate ions reacted with Ca(II) to form johnbaumite. The reaction was given as

$$5Ca^{2+} + 3AsO_4^{3-} + OH^- + Ca_5(AsO_4)_3(OH) \tag{10.16}$$

Rojas et al. (2014) described similar mechanisms for removal of Cu(II), Cd(II), and Pb(II). The removal mechanisms of these cations, as suggested by Rojas et al. (2014), were given as

$$Ca_3Al(OH)_8 NO_3 \cdot 3H_2O + 3Cu^{2+} \rightarrow Cu_3Al(OH)_8 NO_3 \cdot nH_2O + 3Ca^{2+} \tag{10.17}$$

$$Ca_3Al(OH)_8 NO_3 \cdot 3H_2O + 2Cd^{2+} \rightarrow Cd_2Al(OH)_6 NO_3 \cdot nH_2O + 3Ca^{2+} + 2OH^- \tag{10.18}$$

$$Ca_3Al(OH)_8 NO_3 \cdot 3H_2O + 3Pb^{2+} \rightarrow Pb_3(OH)_5 NO_3 + Al(OH)_3 + 3H_2O + 3Ca^{2+} \tag{10.19}$$

González et al. (2015) also interpreted the precipitation mechanism as a means of removal of Cu(II), Cd(II), and Pb(II) from aqueous solution. In the removal process, three types of LDHs were employed; out of these three, LDH-Cl exhibited maximum metal sorption capacity. The buffer property of LDHs plays an important role, as slight dissolution of the LDH induced precipitation of metal cations into metal hydroxides.

10.8 REGENERATION OF LDHs

To make the treatment processes more effective and economical, it is highly desirable to explore the regeneration prospects of the adsorbent material. Investigations into the regeneration of LDHs were mostly carried out by employing two promising methods, namely, thermal and chemical. In the case of thermal treatment, the loaded adsorbent was calcined at temperatures higher than 523 K, which probably eliminated organic pollutants, and the calcined material was ready for further use. The reversibility of adsorption was also tested via the chemical treatment process. Moreover, by means of various analytical techniques such as FTIR, XRD, and thermo-gravimetric differential thermal analysis (TG-DTA), it was possible to ensure that regeneration was accomplished. However, regeneration through chemical treatment was found to be successful when physisorption and ion exchange controlled the sorption process; whereas, in the case of chemical adsorption, regeneration is not feasible. Ma et al. (2013) reported that the regeneration of polysulfide (S_x)-intercalated LDHs loaded with Hg^0 was not possible, as the adsorption of Hg^0 mainly occurred via the chemisorption mechanism. A detailed list of the regeneration of LDHs exhausted with various kinds of pollutants is presented in Table 10.5.

10.9 REMOVAL OF BIOLOGICAL POLLUTANTS BY LDHs

The efficacies of LDHs toward the removal of biological contaminants from wastewater have also been investigated by various researchers. Jin et al. (2007) employed Mg/Al and Zn/Al-LDH nanocomposites for the removal of bacteria (two different strains of *E. coli*) and viruses (MS2 and ϕX174) from synthetic groundwater and raw river water. The results indicated that the LDHs exhibit a high adsorption efficiency of >99% for viruses and bacteria from synthetic groundwater at viral concentrations between 5.9×10^6 and 9.1×10^6 plaque-forming units (pfu)/L and bacterial concentrations between 1.6×10^{10} and 2.6×10^{10} colony-forming units (cfu)/L. The loading capacities for the bacteria and viruses were evaluated as 10^{13} cfu/kg and 10^{10} pfu/kg, respectively. However, when the LDHs were exposed to raw river water, the removal efficiency was not very different—between 87% and 99%—but the uptake capacity of the LDHs reduced sharply to four to five orders of magnitude lower with respect to the uptake capacity of LDHs when exposed to synthetic groundwater. Park et al. (2012) investigated the removal of a bacteriophage (MS2) by an Ni/Al-LDH using batch mode and by Ni/Al-LDH-coated sand via the column experimentation mode. The maximum loading capacity of the LDH was evaluated as 5.5×10^7 pfu/g when the initial MS2 concentration was 1.35×10^5 pfu/mL. Removal as a function of pH was also examined, and it was found that at pH levels between 4.3 and 8.2, the removal percentage was between 98% and 99%, and decreased slightly to 96% at pH 9.4. This suggested that the MS2 uptake by the Ni/Al-LDH was not very sensitive to variations in pH levels. Moreover, it was also demonstrated that the effect of competitive divalent anions such as SO_4^{2-}, CO_3^{2-}, and HPO_4^- significantly affected the removal process. Column

TABLE 10.5
Regeneration Conditions Adopted to Investigate the Reversibility of Adsorption

LDHs	Pollutants	Regeneration Condition	Regeneration/Desorption Efficiency	References
Mg/Al	Dicamba	Chemically treated with 5.0 mmol/dm^3 of Na$_2$CO$_3$, Na$_2$SO$_4$, Na$_2$HPO$_4$, NaNO$_3$, NaCl	~100 % of desorption observed	You et al. 2002
Li/Al	Cr(VI)	Heated deionized water (333 K)	99% of desorption observed	Wang et al. 2006
Zn/Al	Methyl orange	Thermally treated at 723–773 K	Regenerated LDH shows adsorption efficiency of >80% up to two cycles	Ni et al. 2007
Mg/Al	Bromide	Chemically treated with Na$_2$CO$_3$ 0.1 M followed by calcination at 773 K	Regenerated LDHs exhibit almost the same adsorption capacity even after fifth regeneration cycle	Lv et al. 2008
Ni/Al-aluminosilicate	Cr(III)	Chemically treated with HNO$_3$ 0.2 mmol/mL	90% desorption observed	Carnizello et al. 2009
Zn/Al	Phosphate	Chemically treated with NaOH of 5 wt%	Desorption rate more than 80% up to seventh cycle	Cheng et al. 2009
Mg/Al	Cr(VI)	Chemically treated with FeCl$_2$ 0.2 M at pH 2.0	97% of desorption observed	Li et al. 2009
Zn/Al	Se(IV)	Water	8% of desorption observed	Mandal et al. 2009
Mg/Al-dodecyl sulfate	S-Metolachlor	Chemically treated with ethanol	Possibility of regeneration limited to one regeneration step	Otero et al. 2012
Mg/Fe	Congo red	Thermally treated at 673–773 K	Regenerated LDH loses adsorption efficiency from 95% to <50% after third regeneration cycle	Ahmed and Gasser 2012
Mg/Fe	Acid Brown 14	Thermally treated at 773 K	Regenerated LDH shows adsorption efficiency of ~60% up to one cycle	Guo et al. 2013
Mg/Al-polycinnamamide	Arsenate	Chemically treated with alkali at pH 8, 10, and 12	Regenerated LDH shows adsorption efficiencies of 2.44%, 3.21%, and 3.62 % at given pH conditions	Islam et al. 2013
Mg/Al	^{63}Ni(II)	Washing with acetic acid	Regenerated LDH used efficiently up to seventh cycle	Zhao et al. 2013
Mg/Al	Boron	Deionized water at pH 4.0	76.5% of desorption observed	Paez et al. 2014
Ni/Al	Remazol brilliant violet	Chemically treated with NaNO$_3$ followed by calcined at 523 K	Adsorption performance remains the same after three cycles	Pahalagedara et al. 2014
Co/Bi	Cr(VI)	Chemically treated with 0.1 M NaHCO$_3$	Adsorption competency remains more than 60% after fourth regeneration cycle	Jaiswal et al. 2015
Zn/Al-dodecylbenzene-sulfonate	2,4,6-trichlorophenol	Chemically treated with acetone	Removal percentage remains >60% even after tenth adsorption cycle	Zhao et al. 2015
Fe$_3$O$_4$-Mg/Al	Co(II)	Chemically treated with NaNO$_3$ 0.01 M	Adsorption efficiency remains stable even after eighth regeneration cycle	Shou et al. 2015

experiments also revealed that Ni/Al-LDH-coated sand was highly effective in scavenging MS2 from the liquid phase. To make the biological wastewater treatment more effective, the biogranulation method was employed in several treatment studies. The method involves agglutination of suspended solids with bacteria. It was demonstrated that LDHs act as nuclei that promote granulation through microbial adhesion. This concept was utilized by Liu et al. (2013) for the removal of bacteria (*Bacillus subtilis*) using an Ni/Fe-LDH. The adsorption process was influenced by temperature, pH, and ionic strength. The maximum adsorption capacity was evaluated as 45 mg/g. The removal mechanism was interpreted as electrostatic interaction between the positively charged LDH surface and the negatively charged bacterium surface, because of the presence of the $-PO_4^-$ group leading to the formation of biogranule-like aggregates.

10.10 PROBLEMS

10.10.1 STRUCTURAL CONSTANCY

For any adsorbent-based water treatment system, it is important to ensure that the adsorbent should maintain its integrity until and unless the exhausted material is properly disposed of. However, for LDHs, material structural stability is an important concern, as in several studies it was reported that the lamellar structure started to erode in acidic conditions. In such cases, it is anticipated that the degradation of the LDH structure would release harmful metal cations into the treated water. Therefore, it is desirable that harmful metal precursors are avoided during LDH synthesis, especially when LDHs are employed for the treatment of drinking water. Moreover, the problems related to LDH dissolution can be overcome if metal cations, whose hydrolysis product is less soluble, are given preference for LDH synthesis.

10.10.2 THERMAL TREATMENT

When LDHs are calcined at temperatures above 773 K, their layered structure is transformed into mixed metal oxides that possess a high surface area, thermal stability, and basic properties. When these calcined products were released into aqueous media for treatment purposes, it was demonstrated that the material was restructured by incorporating targeting pollutant species into its interlayered space, due to the memory effect. But, during reconstruction, the carbonate anions also became intercalated due to the presence of atmospheric CO_2; as a result, the sorption of anionic species is significantly affected.

10.11 FUTURE RECOMMENDATIONS

In most of the removal studies, the efficiency of LDHs was limited to the batch adsorption mode. It has been demonstrated from the literature survey that very little attention was paid to employing LDHs in column-related studies. Therefore, it would be highly desirable if we focus our research interest on the exploration of the performance of LDHs in column-based adsorption studies. Moreover, it would certainly be advantageous if efforts are directed toward the development, optimization, and application of LDHs for water treatment at pilot as well as industrial levels by the use of column operations.

10.12 CONCLUDING REMARKS

LDHs exhibit enormous potential in the sequestration of various organics, inorganics, and biological contaminants from wastewater. LDHs can be synthesized effortlessly in laboratory conditions and these materials exhibit great compositional flexibility. Thus, LDHs possess great potential for practical applications in a wide variety of fields. In this chapter, the applications of LDHs were focused on environmental cleanup studies; more specifically, attention was given to the use of LDHs as an adsorbent in water treatment studies. The LDH-based sorption studies were greatly influenced by various operational parameters such as pH, effect of competing ions, adsorbent dosage, and temperature. Furthermore, LDHs displayed varying interaction mechanisms with different kinds of pollutant species; for instance, anionic pollutants were mainly scavenged by LDHs through the ion-exchange and memory effect processes, whereas chelation and isomorphic substitution were the preferred sorption mechanism exhibited by LDHs for cationic species. However, the materials do not receive attention at commercial levels due to certain issues which need to be addressed in further research. In a nutshell, LDHs offer tremendous scope in wastewater treatment; hence, collaborative efforts of research in this specific direction are earnestly required to achieve fast, highly efficient, and economical treatment technology.

ACKNOWLEDGMENTS

SB and RKG are thankful to the Council of Scientific and Industrial Research (CSIR) and the University Grant Commission (UGC), New Delhi for providing financial assistance as a senior research fellowship (SRF).

REFERENCES

Abdelkader, N.B., A. Bentouami, Z. Derriche, N. Bettahar, L.C. de Menorval, Synthesis and characterization of Mg–Fe layer double hydroxides and its application on adsorption of Orange G from aqueous solution, *Chem. Eng. J.* 169 (2011) 231–238.

Abellan, G., C. Marti-Gastaldo, A. Ribera, E. Coronado, Hybrid materials based on magnetic layered double hydroxides: A molecular perspective, *Acc. Chem. Res.* 48 (2015) 1601–1611.

Aguiar, J.E., B.T.C. Bezerra, B. de M. Braga, P.D. da Silva Lima, R.E.F.Q. Nogueira, S.M. Pereira de Lucena, I. José da Silva Jr., Adsorption of anionic and cationic dyes from aqueous solution on non-calcined Mg-Al layered double hydroxide: Experimental and theoretical study, *Sep. Sci. Technol.* 48 (2013) 2307–2316.

Ahmed, I.M., M.S. Gasser, Adsorption study of anionic reactive dye from aqueous solution to Mg–Fe–CO_3 layered double hydroxide (LDH), *Appl. Surf. Sci.* 259 (2012) 650–656.

Alanis, C., R. Natividad, C. Barrera-Diaz, V. Martinez-Mirand, J. Prince, J.S. Valente, Photocatalytically enhanced Cr(VI) removal by mixed oxides derived from MeAl (Me:Mg and/or Zn) layered double hydroxides, *Appl. Catal. B* 140–141 (2013) 546–551.

Arai, Y., M. Ogawa, Preparation of Co–Al layered double hydroxides by the hydrothermal urea method for controlled particle size. *Appl. Clay Sci.* 42 (2009) 601–604.

Ardau, C., F. Frau, E. Dore, P. Lattanzi, Molybdate sorption by Zn–Al sulphate layered double hydroxides, *Appl. Clay Sci.* 65–66 (2012) 128–133.

Asouhidou, D.D., K.S. Triantafyllidis, N.K. Lazaridis, K.A. Matis, Adsorption of reactive dyes from aqueous solutions by layered double hydroxides, *J. Chem. Technol. Biotechnol.* 87 (2011) 575–582.

Besserguenev, A.V., A.M. Fogg, R.J. Francis, S.J. Price, D. O'Hare, Synthesis and structure of the gibbsite intercalation compounds [$LiAl_2(OH)_6$]X {X = Cl, Br, NO_3} and [$LiAl_2(OH)_6$]Cl·H_2O using synchrotron x-ray and neutron powder diffraction, *Chem. Mater.* 9 (1997) 241–247.

Brindley, G.W., S. Kikkawa, A crystal-chemical study of Mg,Al and Ni,Al hydroxy-perchlorates and hydroxy-carbonates, *Am. Mineral.* 64 (1979) 836–843.

Bruna, F., I. Pavlovic, C. Barriga, J. Cornejo, M.A. Ulibarri, Adsorption of pesticides carbetamide and meta-mitron on organohydrotalcite, *Appl. Clay Sci.* 33 (2006) 116–124.

Brunauer, S., P.H. Emmett, E. Teller, Adsorption of gases in multimolecular layers, *J. Am. Chem. Soc.* 60 (1938) 309–319.

Carriazo, D., C. Domingo, C. Martin and V. Rives, Structural and texture evolution with temperature of layered double hydroxides intercalated with paramolybdate anions, *Inorg. Chem.*, 45 (2006) 1243–1251.

Cavani, F., F. Trifiro, A. Vaccari, Hydrotalcite-type anionic clays: Preparation, properties and applications, *Catal. Today 11* (1991) 173–301.

Chen, C., P. Gunawan, R. Xu, Self-assembled Fe_3O_4-layered double hydroxide colloidal nanohybrids with excellent performance for treatment of organic dyes in water, *J. Mater. Chem.* 21 (2011) 1218–1225.

Chen, D., Y. Li, J. Zhang, J. Zhou, Y. Guo, H. Liu, Magnetic Fe_3O_4/ZnCr-layered double hydroxide composite with enhanced adsorption and photocatalytic activity, *Chem. Eng. J.* 185–186 (2012) 120–126.

Chen, Y., Y.F. Song, Highly selective and efficient removal of Cr (VI) and Cu(II) by the chromotropic acid-intercalated Zn−Al layered double hydroxides, *Ind. Eng. Chem. Res.* 52 (2013) 4436−4442.

Cheng, X., X. Huang, X. Wang, B. Zhao, A. Chen, D. Sun, Phosphate adsorption from sewage sludge filtrate using zinc–aluminum layered double hydroxides, *J. Hazard. Mater.* 169 (2009) 958–964.

Coronado, E., J.R. Galan-Mascaros, C. Marti-Gastaldo, A. Ribera, E. Palacios, M. Castro, R. Burriel, Spontaneous magnetization in Ni-Al and Ni-Fe layered double hydroxides, *Inorg. Chem.* 47 (2008) 9103–9110.

Costa, F.R., U. Wagenknecht, G. Heinrich, LDPE/Mg–Al layered double hydroxide nanocomposite: thermal and flammability properties, *Polym. Degrad. Stab.* 92 (2007) 1813–1823.

Costantino, U., F. Marmottini, M. Nocchetti, R. Vivani, New synthetic routes to hydrotalcite-like compounds: Characterisation and properties of the obtained materials, *Eur. J. Inorg. Chem.* (1998) 1439–1446.

Costantino, U., R. Vivani, M. Bastianini, F. Costantino, M. Nocchetti, Ion exchange and intercalation properties of layered double hydroxides towards halide anions, *Dalton Trans.* 43 (2014) 11587–11596.

Das, J., B.S. Patra, N. Baliarsingh, K.M. Parida, Adsorption of phosphate by layered double hydroxides in aqueous solutions, *Appl. Clay Sci.* 32 (2006) 252–260.

Dong, H., L. Wu, L. Zhang, H. Chen, C. Gao, Clay nanosheets as charged filler materials for high-performance and fouling-resistant thin film nanocomposite membranes, *J. Membrane Sci.* 494 (2015) 92–103.

dos Reis, M.J., F. Silverio, J. Tronto, J.B. Valim, Effects of pH, temperature, and ionic strength on adsorption of sodium dodecylbenzenesulfonate into Mg–Al–CO_3 layered double hydroxides, *J. Phys. Chem. Solids* 65 (2004) 487–492.

Elkhattabi, E.H., M. Lakraimi, M. Badreddine, A. Legrouri, O. Cherkaoui, M. Berraho, Removal of Remazol Blue 19 from wastewater by zinc–aluminium–chloride-layered double hydroxides, *Appl. Water Sci.* 3 (2013) 431–438.

Gao, Y., J. Wu, Q. Wang, C.A. Wilkieb, D. O'Hare, Flame retardant polymer/layered double hydroxide nanocomposites, *J. Mater. Chem. A* 2 (2014) 10996–11016.

González, M.A., I. Pavlovic, C. Barriga, Cu(II), Pb(II) and Cd(II) sorption on different layered double hydroxides. A kinetic and thermodynamic study and competing factors, *Chem. Eng. J.* 269 (2015) 221–228.

Guo, Y., Z. Zhu, Y. Qiu, J. Zhao, Adsorption of arsenate on Cu/Mg/Fe/La layered double hydroxide from aqueous solutions, *J. Hazard. Mater.* 239–240 (2012) 279–288.

Guo, Y., Z. Zhu, Y. Qiu, J. Zhao, Enhanced adsorption of acid brown 14 dye on calcined Mg/Fe layered double hydroxide with memory effect, *Chem. Eng. J.* 219 (2013a) 69–77.

Guo, Y., Z. Zhu, Y. Qiu, J. Zhao, Synthesis of mesoporous Cu/Mg/Fe layered double hydroxide and its adsorption performance for arsenate in aqueous solutions, *J. Environ. Sci.* 25 (2013b) 944–953.

Gutmann, N.H., L. Spiccia, T.W. Turney, Complexation of Cu(II) and Ni(II) by nitrilotriacetate intercalated in Zn-Cr layered double hydroxides, *J. Mater. Chem.* 10 (2000) 1219–1224.

Halajnia, A., S. Oustan, N. Najafi, A.R. Khataee, A. Lakzian, Adsorption–desorption characteristics of nitrate, phosphate and sulfate on Mg–Al layered double hydroxide, *Appl. Clay Sci.* 80–81 (2013) 305–312.

Hernandez-Moreno, J.M., M.A. Ulibarri, J.L. Rendon, C.J. Serna, IR characteristics of hydrotalcite-like compounds, *Phys. Chem. Minerals* 12 (1985) 34–38.

Holtmeier, W., G. Holtmann, W.F. Caspary, U. Weinqartner, On-demand treatment of acute heartburn with the antacid hydrotalcite compared with famotidine and placebo: Randomized double-blind cross-over study, *J. Clin. Gastroenterol.* 41 (2007) 564–70.

Inacio, J., C. Taviot-Gueho, C. Forano, J.P. Besse, Adsorption of MCPA pesticide by MgAl-layered double hydroxides, *Appl. Clay Sci.* 18 (2001) 255–264.

Inayat, A., M. Klumpp, W. Schwieger, The urea method for the direct synthesis of ZnAl layered double hydroxides with nitrate as the interlayer anion, *Appl. Clay Sci.* 51 (2011) 452–459.

Ishihara, S., P. Sahoo, K. Deguchi, S. Ohki, M. Tansho, T. Shimizu, J. Labuta, J.P. Hill, K. Ariga, K. Watanabe, Y. Yamauchi, S. Suehara, N. Iyi, Dynamic breathing of CO_2 by hydrotalcite, *J. Am. Chem. Soc.* 135 (2013) 18040–18043.

Islam, M., P.C. Mishra, R. Patel, Microwave assisted synthesis of polycinnamamide Mg/Al mixed oxide nanocomposite and its application towards the removal of arsenate from aqueous medium, *Chem. Eng. J.* 230 (2013) 48–58.

Iwasaki, T., H. Yoshii, H. Nakamura, S. Watano, Simple and rapid synthesis of Ni−Fe layered double hydroxide by a new mechanochemical method, *Appl. Clay Sci.* 58 (2012) 120–124.

Jaiswal, A., R. Mani, S. Banerjee, R.K. Gautam, M.C. Chattopadhyaya, Synthesis of novel nano-layered double hydroxide by urea hydrolysis method and their application in removal of chromium(VI) from aqueous solution: Kinetic, thermodynamic and equilibrium studies. *J. Mol. Liq.* 202 (2015a) 52–61.

Jaiswal, A., S. Banerjee, R.K. Gautam, M.C. Chattopadhyaya, Synthesis of microporous takovite and its environmental application. *J. Mol. Liq.* 209 (2015b) 759–766.

Kameda, T., K. Hoshi, T. Yoshioka, Preparation of Cu-Al layered double hydroxide intercalated with ethylenediaminetetraacetate by coprecipitation and its uptake of rare earth ions from aqueous solution, *Solid State Sci.* 17 (2013) 28–34.

Khan, A.I., D. O'Hare, Intercalation chemistry of layered double hydroxides: Recent developments and applications, *J. Mater. Chem.* 12 (2002) 3191–3198.

Khenifi, A., Z. Derriche, C. Mousty, V. Prévot, C. Forano, Adsorption of glyphosate and glufosinate by Ni_2AlNO_3 layered double hydroxide, *Appl. Clay Sci.* 47 (2010) 362–371.

Kovanda, F., D. Kolousek, Z. Cilova, V. Hulinsky, Crystallization of synthetic hydrotalcite under hydrothermal conditions, *Appl. Clay Sci.* 28 (2005) 101–109.

Kuila, T., S.K. Srivastava, A.K. Bhowmick, A.K. Saxena, Thermoplastic polyolefin based polymer: Blend-layered double hydroxide nanocomposites, *Composites Sci. Technol.* 68 (2008) 3234–3239.

Li, F., X. Duan, Applications of layered double hydroxides, in *Layered Double Hydroxides* 2006. Eds. X. Duan, D.G. Evans, 193–223, Springer, Berlin.

Li, W., K.J.T. Livi, W. Xu, M.G. Siebecker, Y. Wang, B.L. Phillips, D.L. Sparks, Formation of crystalline Zn−Al layered double hydroxide precipitates on γ-Alumina: The role of mineral dissolution, *Environ. Sci. Technol.* 46 (2012) 11670−11677.

Li, Y., B. Gao, T. Wu, D. Sun, X. Li, B. Wang, F. Lu, Hexavalent chromium removal from aqueous solution by adsorption on aluminum magnesium mixed hydroxide, *Water Res.* 43 (2009) 3067–3075.

Liang, X., W. Hou, Y. Xu, G. Sun, L. Wang, Y. Sun, X. Qin, Sorption of lead ion by layered double hydroxide intercalated with diethylenetriaminepentaacetic acid, *Colloid. Surfaces A* 366 (2010) 50–57.

Lin, Y., J. Wang, D.G. Evans, D. Li, Layered and intercalated hydrotalcite-like materials as thermal stabilizers in PVC resin, *J. Phys. Chem. Solids* 67 (2006) 998–1001.

Lin, Y., Z. Zeng, J. Zhu, Y. Wei, S. Chen, X. Yuan, L. Liu, Facile synthesis of ZnAl-layered double hydroxide microspheres with core−shell structure and their enhanced adsorption capability, *Mater. Lett.* 156 (2015) 169–172.

Liu, J., L. Yu, Y. Zhang, Fabrication and characterization of positively charged hybrid ultrafiltration and nanofiltration membranes via the in-situ exfoliation of Mg/Al hydrotalcite, *Desalination* 335 (2014) 78–86.

Liu, X., L. Ge, W. Li, X. Wang, F. Li, Layered double hydroxide functionalized textile for effective oil/water separation and selective oil adsorption, *Appl. Mater. Interfaces* 7 (2015) 791–800.

Liu, X., M. Wei, F. Li, X. Duan, Intraparticle diffusion of 1-phenyl-1,2-ethanediol in layered double hydroxides, *Mater. Interfaces Electrochem. Phenom.* 53 (2007) 1591–1600.

Lv, F., L. Xu, Y. Zhang, Z. Meng, Layered double hydroxide assemblies with controllable drug loading capacity and release behavior as well as stabilized layer by layer polymer multilayers, *ACS Appl. Mater. Interfaces* 7 (2015) 19104–19111.

Lv, L., J. He, M. Wei, D.G. Evansa, X. Duan, Uptake of chloride ion from aqueous solution by calcined layered double hydroxides: Equilibrium and kinetic studies, *Water Res.* 40 (2006) 735–743.

Lv, L., Y. Wang, M. Wei, J. Cheng, Bromide ion removal from contaminated water by calcined and uncalcined $MgAl-CO_3$ layered double hydroxides, *J. Hazard. Mater.* 152 (2008) 1130–1137.

Ma, S., Y. Shim, S.M. Islam, K.S. Subrahmanyam, P. Wang, H. Li, S. Wang, X. Yang, M.G. Kanatzidis, Efficient Hg vapor capture with polysulfide intercalated layered double hydroxides, *Chem. Mater.* 26 (2014) 5004−5011.

Mandal, S., S. Mayadevi, B.D. Kulkarni, Adsorption of aqueous selenite [Se(IV)] species on synthetic layered double hydroxide materials, *Ind. Eng. Chem. Res.* 48 (2009) 7893–7898.

Mascolo, G., M.C. Mascolo, On the synthesis of layered double hydroxides (LDHs) by reconstruction method based on the "memory effect," *Microporous Mesoporous Mater.* 214 (2015) 246–248.

Miyata, S., Anion exchange properties of hydrotalcite-like compounds, *Clay Clay Miner.* 31 (1983) 305–311.

Mousty, C., V. Prévot, Hybrid and biohybrid layered double hydroxides for electrochemical analysis, *Anal. Bioanal. Chem.* 405 (2013) 3513–3523.

Nejati, K., K. Asadpour-Zeynali, Electrochemical synthesis of nickel–iron layered double hydroxide: Application as a novel modified electrode in electrocatalytic reduction of metronidazole, *Mater. Sci. Eng. C* 35 (2014) 179–184.

Nejati, K., S. Davary, M. Saati, Study of 2,4-dichlorophenoxyacetic acid (2,4-D) removal by Cu-Fe-layered double hydroxide from aqueous solution, *Appl. Surf. Sci.* 280 (2013) 67–73.

Otero, R., J.M. Fernández, M.A. Ulibarri, R. Celis, F. Bruna, Adsorption of non-ionic pesticide S-Metolachlor on layered double hydroxides intercalated with dodecylsulfate and tetradecanedioate anions, *Appl. Clay Sci.* 65–66 (2012) 72–79.

Pang, X., X. Ma, D. Li, W. Hou, Synthesis and characterization of 10-hydroxycamptothecin-sebacate: Layered double hydroxide nanocomposites, *Solid State Sci.* 16 (2013) 71–75.

Parashar, P., V. Sharma, D.D. Agarwal, N. Richhariya, Rapid synthesis of hydrotalcite with high antacid activity, *Mater. Lett.* 74 (2012) 93–95.

Parida, K.M., L. Mohapatra, Carbonate intercalated Zn/Fe layered double hydroxide: A novel photocatalyst for the enhanced photo degradation of azo dyes, *Chem. Eng. J.* 179 (2012) 131–139.

Pavlovic, I., C. Barriga, M.C. Hermosin, J. Cornejo, M.A. Ulibarri, Adsorption of acidic pesticides 2,4-D, Clopyralid and Picloram on calcined hydrotalcite, *Appl. Clay Sci.* 30 (2005) 125–133.

Pereira de Sá, F., B.N. Cunha, L.M. Nunes, Effect of pH on the adsorption of Sunset Yellow FCF food dye into a layered double hydroxide (CaAl-LDH-NO$_3$), *Chem. Eng. J.* 215–216 (2013) 122–127.

Prince, J., A. Montoya, G. Ferrat, Jaime S. Valente, Proposed general sol-gel method to prepare multimetallic layered double hydroxides: Synthesis, characterization, and envisaged application, *Chem. Mater.* 21 (2009) 5826–5835.

Prince, J., F. Tzompantzi, G. Mendoza-Damian, F. Hernandez-Beltran, J.S. Valente, Photocatalytic degradation of phenol by semiconducting mixed oxides derived from Zn(Ga)Al layered double hydroxides, *Appl. Catal. B* 163 (2015) 352–360.

Radha, A.V., P.V. Kamath, C. Shivakumara, Mechanism of the anion exchange reactions of the layered double hydroxides (LDHs) of Ca and Mg with Al, *Solid State Sci.* 7 (2005) 1180–1187.

Reichle, W.T., Synthesis of anionic clay minerals (mixed metal hydroxides, hydrotalcite), *Solid States Ionics* 22 (1986) 135–141.

Rives, V., M.A. Ulibarri, Layered double hydroxides (LDH) intercalated with metal coordination compounds and oxometalates, *Coord. Chem. Rev.* 181 (1999) 61–120.

Rodrigues, E., P. Pereira, T. Martins, F. Vargas, T. Scheller, J. Correa, J. Del Nero, S.G.C. Moreira, W. Ertel-Ingrisch, C.P. De Campos, A. Gigle, Novel rare earth (Ce and La) hydrotalcite like material: Synthesis and characterization, *Materials Lett.* 78 (2012) 195–198.

Rojas, R., Copper, lead and cadmium removal by Ca Al layered double hydroxides, *Appl. Clay Sci.* 87 (2014) 254–259.

Rojas, R., M.C. Palena, A.F. Jimenez-Kairuz, R.H. Manzo, C.E. Giacomelli, Modeling drug release from a layered double hydroxide–ibuprofen complex, *Appl. Clay Sci.* 62–63 (2012) 15–20.

Rojas, R., M.R. Perez, E.M. Erro, P.I. Ortiz, M.A. Ulibarri, C.E. Giacomelli, EDTA modified LDHs as Cu^{2+} scavengers: Removal kinetics and sorbent stability, *J. Colloid Interface Sci.* 331 (2009) 425–431.

Roussel, H., V. Briois, E. Elkaim,, A. De Roy, J.P. Besse, J.P. Jolivet, Study of the formation of the layered double hydroxide [Zn–Cr–Cl], *Chem. Mater.* 13 (2001) 329–337.

Seftel, E.M., E. Popovici, M. Mertens, K. De Witte, G. Van Tendeloo, P. Cool, E.F. Vansant, Zn–Al layered double hydroxides: Synthesis, characterization and photocatalytic application, *Microporous Mesoporous Mater.* 113 (2008) 296–304.

Shou, J., C. Jiang, F. Wang, M. Qiu, Q. Xu, Fabrication of Fe$_3$O$_4$/MgAl-layered double hydroxide magnetic composites for the effective decontamination of Co(II) from synthetic wastewater, *J. Mol. Liq.* 207 (2015) 216–223.

Sui, M., Y. Zhou, L. Sheng, B. Duan, Adsorption of norfloxacin in aqueous solution by Mg–Al layered double hydroxides with variable metal composition and interlayer anions, *Chem. Eng. J.* 210 (2012) 451–460.

Sun, Y., J. Zhou, W. Cai, R. Zhao, J. Yuan, Hierarchically porous NiAl-LDH nanoparticles as highly efficient adsorbent for p-nitrophenol from water, *Appl. Surface Sci.* 349 (2015) 897–903.

Tao, Q., H. He, R.L. Frost, P. Yuan, J. Zhu, Nanomaterials based upon silylated layered double hydroxides, *Appl. Surf. Sci.* 255 (2009) 4334–4340.

Theiss, F.L., M.J. Sear-Hall, S.J. Palmer, R.L. Frost, Zinc aluminium layered double hydroxides for the removal of iodine and iodide from aqueous solutions, *Desalin. Water Treat.* 39 (2012) 166–175.

Turk, T., I. Alp, H. Deveci, Adsorption of As(V) from water using Mg–Fe-based hydrotalcite (FeHT), *J. Hazard. Mater.* 171 (2009) 665–670.

Ulibarri, M.A., I. Pavlovic, C. Barriga, M.C. Hermosin, J. Cornejo, Adsorption of anionic species on hydrotalcite-like compounds: Effect of interlayer anion and crystallinity, *Appl. Clay Sci.* 18 (2001) 17–27.

Ulibarri, M.A., I. Pavlovic, M.C. Hermosin, J. Cornejo, Hydrotalcite-like compounds as potential sorbents of phenols from water, *Appl. Clay Sci.* 10 (1995) 131–145.

Valente, J.S., M. Sanchez-Cantu, E. Lima, F. Figueras, Method for large-scale production of multimetallic layered double hydroxides: Formation mechanism discernment, *Chem. Mater.* 21 (2009) 5809–5818.

Wang, J., Z. Lei, H. Qin, L. Zhang, F. Li, Structure and catalytic property of Li-Al metal oxides from layered double hydroxide precursors prepared via a facile solution route, *Ind. Eng. Chem. Res.* 50 (2011) 7120–7128.

Whilton, N.T., P.J. Vickers, S. Mann, Bioinorganic clays: Synthesis and characterization of amino- and poly-amino acid intercalated layered double hydroxides, *J. Mater. Chem.* 7 (1997) 1623–1629.

Xu, Y., Y. Dai, J. Zhou, Z.P. Xu, G. Qian, G.Q.M. Lu, Removal efficiency of arsenate and phosphate from aqueous solution using layered double hydroxide materials: Intercalation vs. precipitation, *J. Mater. Chem.* 20 (2010) 4684–4691.

Yamaoka, T., M. Abe, M. Tsuji, Synthesis of Cu-Al hydrotalcite like compound and its ion exchange property. *Mat. Res. Bull.* 24 (1989) 1183–1199.

Yang, L., Z. Shahrivari, P.K.T. Liu, M. Sahimi, T.T. Tsotsis, Removal of trace levels of arsenic and selenium from aqueous solutions by calcined and uncalcined layered double hydroxides (LDH), *Ind. Eng. Chem. Res.* 44 (2005) 6804–6815.

Yong, Z., V. Mata, A.E. Rodrigues, Adsorption of carbon dioxide onto hydrotalcite-like compounds (HTlcs) at high temperatures, *Ind. Eng. Chem. Res.* 40 (2001) 204–209.

You, Y., G. F. Vance, H. Zhao, Selenium adsorption on Mg–Al and Zn–Al layered double hydroxides, *Appl. Clay Sci.* 20 (2001) 13–25.

Yuan, S., Y. Li, Q. Zhang, H. Wang, ZnO nanorods decorated calcined Mg–Al layered double hydroxides as photocatalysts with a high adsorptive capacity, *Colloids Surfaces A* 348 (2009) 76–81.

Zaghouane-Boudiaf, H., M. Boutahala, L. Arab, Removal of methyl orange from aqueous solution by uncalcined and calcined MgNiAl layered double hydroxides (LDHs), *Chem. Eng. J.* 187 (2012) 142–149.

Zhang, W.H., X.D. Guo, J. He, Z.Y. Qian, Preparation of Ni(II)/Ti(IV) layered double hydroxide at high supersaturation, *J. European Ceramic Soc.* 28 (2008) 1623–1629.

Zhang, T., Q. Li, H. Xiao, H. Lu, Y. Zhou, Synthesis of Li−Al layered double hydroxides (LDHs) for efficient fluoride removal, *Ind. Eng. Chem. Res.* 51 (2012a) 11490−11498.

Zhang, X., L. Ji, J. Wang, R. Li, Q. Liu, M. Zhang, L. Liu, Removal of uranium(VI) from aqueous solutions by magnetic Mg–Al layered double hydroxide intercalated with citrate: Kinetic and thermodynamic investigation, *Colloid. Surfaces A* 414 (2012b) 220–227.

Zhao, D., G. Sheng, J. Hu, C. Chen, X. Wang, The adsorption of Pb(II) on Mg_2Al layered double hydroxide, *Chem. Eng. J.* 171 (2011) 167–174.

Zhao, D., Y. Ding, S. Chen, H. Xuan, Y. Chen, Effect of environmental parameters on the sequestration of radionickel by Mg_2Al layered double hydroxide, *J. Radioanal. Nucl. Chem.* 298 (2013) 1197–1206.

Zhao, P., X. Liu, W. Tian, D. Yan, X. Sun, X. Lei, Adsolubilization of 2,4,6-trichlorophenol from aqueous solution by surfactant intercalated ZnAl layered double hydroxides, *Chem. Eng. J.* 279 (2015) 597–604.

Zhao, Y., F. Li, R. Zhang, D.G. Evans, X. Duan, Preparation of layered double-hydroxide nanomaterials with a uniform crystallite size using a new method involving separate nucleation and aging steps, *Chem. Mater.* 14 (2002) 4286–4291.

Zhao, Y., M. Wei, J. Lu, Z.L. Wang, X. Duan, Biotemplated hierarchical nanostructure of layered double hydroxides with improved photocatalysis performance, *ACS Nano* 3 (2009) 4009–4016.

Zheng, Y.M., N. Li, W.D. Zhang, Preparation of nanostructured microspheres of Zn–Mg–Al layered double hydroxides with high adsorption property, *Colloid. Surfaces A* 415 (2012) 195–201.

11 Activated Carbon-Doped Magnetic Nanoparticles for Wastewater Treatment

Muhammad Abbas Ahmad Zaini, Lee Lin Zhi, and Tang Shu Hui

CONTENTS

ABSTRACT

There is burgeoning concern over the presence of persistent pollutants in water bodies. Effluent from factories and domestic activities are among the major contributors to water pollution. Polluted water has negative implications for aquatic creatures, public health, and the ecosystem. Owing to its textural and surface chemistry properties, activated carbon has been widely used in wastewater treatment. Recent advances in nanotechnology offer considerable advantages in wastewater treatment via the introduction of activated carbon decorated magnetic nanoparticles. This chapter highlights the current research trend on the applications of magnetic activated carbon-doped nanoparticles over the past 10 years. The rationale and advantages of using novel adsorbents with magnetic properties in the wastewater treatment technologies are discussed. Attention is also directed toward the synthesis and characterization of magnetic activated carbons, target pollutants, and removal mechanisms. Some limitations within recent studies are highlighted to offer directions for future research.

Keywords: Magnetic activated carbon, Nanoparticles, Characterization, Wastewater treatment

11.1 INTRODUCTION

Activated carbon-doped magnetic nanoparticles have gained considerable attention in the field of water and wastewater treatment. Cobalt, iron, and nickel are among the magnetic elements that can be doped into activated carbon. However, iron is widely used in the manufacture of activated carbon decorated magnetic nanoparticles, because nickel and cobalt are highly toxic and susceptible to oxidation, and may jeopardize the performance of the adsorbent (Vatta et al. 2006). Nanoparticles in the size from 1 to 100 nm exhibit a variety of unique magnetic and adsorptive characteristics to activated carbon (Vatta et al. 2006; Zhu et al. 2013). Furthermore, the advantage of using the smaller-sized particles lies in a higher effective surface area to accommodate various contaminants in water (Zhu et al. 2013).

The use of magnetic nanoparticles allows for the use of the so-called ferromagnetism mechanism, by which the host material forms permanent magnets, is attracted to magnets, and remains magnetized after the external field is removed (Vatta et al. 2006). With ferromagnetic materials, the spinning electrically charged particles create magnetic dipoles (also known as magnetrons) that will align in the same direction through the exchange of forces. Because ferromagnetism is size dependent, the magnetic nanoparticles are assumed to possess uniform magnetization and negligible interparticle magnetic dipole–dipole interactions, which could result in particles coming into contact and agglomerating with one another (Vatta et al. 2006; Zhu et al. 2013).

The behavior of magnetic nanoparticles is usually described by a hysteresis loop that can be obtained using a magnetometer. Two parameters of interest are remanence (M_R) and coercivity (H_C). Remanence (measured in electromagnetic units per gram, emu/g) is a measure of magnetization, and is defined as the leftover magnetization after the external magnetic field is removed, while coercivity (measured in oersteds, Oe) is a measure of the reverse field needed to bring the magnetization to zero after attaining the magnetic saturation moment (M_S). Coercivity indicates the resistance and the ability of magnetic nanoparticles to withstand an external magnetic field without becoming demagnetized. The coercivity of the magnetic nanoparticles is highly size dependent; it increases with decreasing particle size. However, decreasing the particle size below a critical diameter will lead to zero coercivity with no hysteresis. At this point, the magnetic particles are said to become superparamagnetic, that is, they become magnetic only in the presence of an external field (Akbarzadeh et al. 2012).

Carbon and activated carbon have been used as the hosting matrix for magnetic nanoparticles, owing to their unique mechanical, physical, and chemical properties. Magnetic activated carbon is a nanocomposite material composed of an inorganic magnetic component in the form of nanosized particles embedded in organic activated carbon. The material integrates the properties of both individual magnetic nanoparticles and activated carbon, resulting in dual attributes within a single material. Consequently, this material may offer new features that cannot be achieved by its individual components (Zhu et al. 2013).

This chapter provides an overview on the latest developments of activated carbon-doped magnetic nanoparticles. The preparation methods and the characteristics of the resultant materials are extensively reviewed. Focus has also been given to the applications of magnetic activated carbon in removing water contaminants. Finally, the limitations in the present studies are highlighted, with an intention to shed some light for future research.

11.2 PREPARATION AND CHARACTERIZATION OF MAGNETIC ACTIVATED CARBONS

The methods most often used for magnetic modification of activated carbon are impregnation and chemical coprecipitation. Normally, impregnating the activation with magnetic nanoparticles requires treatment at a higher temperature. The impregnation of iron and nickel salts onto activated carbon is usually carried out at temperatures ranging between 300°C and 750°C (Kakavandi et al. 2013; Quinones et al. 2014; Tolga et al. 2014). Some researchers reported the preparation of magnetic activated carbon nanocomposites through chemical coprecipitation followed by thermal treatment at temperatures from 150°C to 300°C (Wan et al. 2014; Oh et al. 2015; Wang et al. 2015). Most recently, Zhang and coworkers (2015) introduced a single-step doping of magnetic nanoparticles by mixing together K_2CO_3 (activating agent) and Fe_2O_3 during the activation of peanut shell at 800°C. Generally, the magnetic nanoparticles formed depend on the types of inorganic magnetic salt, operating conditions, and atmosphere used. Table 11.1 summarizes the preparation methods and the characteristics of magnetic activated carbon-doped nanoparticles.

Iron(III) chloride (ferric chloride) and iron(II) chloride (ferrous chloride) are among the commonly used magnetic elements in the preparation of magnetic activated carbon-doped nanoparticles. The chemical coprecipitation process includes alkaline precipitation of ferrous and ferric salts

TABLE 11.1

Preparation and Characterization of Magnetic Activated Carbon-Doped Nanoparticles

Precursor	Operating Conditions		Magnetic Properties			Size of Magnetic Particle (nm)	Specific Surface Area (m²/g)		Microporous Volume (cm³/g)		Mesopore Volume (cm³/g)		References
	AC	MAC	M_S (emu/g)	H_C (Oe)	M_R (emu/g)		AC	MAC	AC	MAC	AC	MAC	
PAC	—	1:2 ratio Fe:AC;[a] 70°C	—	—	—	—	1040	868.0	0.308	0.070	0.202	0.065	Zahoor and Mahramanlioglu (2011)
CAC	—	3.35% metallic ratio Ni[b]	0.120	33.33	0.12×10^{-3}	—	942.9	879.0	—	—	—	—	Jia et al. (2011)
		6% metallic ratio Ni[b]	0.250	20.31	0.03×10^{-3}	—		—	—	—	—	—	
		12.53% metallic ratio Ni[b]	0.330	22.51	0.05×10^{-3}	—		—	—	—	—	—	
		16.37% metallic ratio Ni[b]	—	—	—	—		859.1	—	—	—	—	
Almond shell	—	1:2 ratio Fe:AC; 60–70°C	—	—	—	—	733.0	527.0	0.410	0.300	0.100	0.070	Mohan et al. (2011)
PAC	1:1 ratio H₃PO₄; 450°C, 1 h	1:1 ratio Fe:AC[a]	—	—	—	—	1020	551.0	—	—	—	—	Faulconer et al. (2012)
		1:2 ratio Fe:AC[a]	—	—	—	—		709.0		—		—	
		1:3 ratio Fe:AC[a]	—	—	—	—		790.0		—		—	
		1:3 ratio Fe:AC;[a] 450°C, 3 h	—	—	—	—		46.86		—		—	
CAC	—	1:3 ratio SrFe₁₂O₁₉:AC;[a] 1000°C, 2 h	19.60	239.7	4.900	36.6	1050	570.9	0.200	0.110	0.430	0.130	Xie et al. (2012)
Coconut shell	0.5:1 ratio MgCl₂; 600°C, 2 h	2.5% Fe content;[c] 450°C, 2 h	—	—	—	23	408.0	426.0	0.149	0.154	0.087	0.076	Lima et al. (2013)
		5.0% Fe content;[c] 450°C, 2 h	—	—	—	28		388.0		0.138		0.079	
		10.0% Fe content;[c] 450°C, 2 h	—	—	—	32		220.0		0.074		0.085	
		15.0% Fe content;[c] 450°C, 2 h	—	—	—	34		210.0		0.069		0.086	
CAC	—	1:1 ratio Fe:AC;[a] 70°C	—	—	—	—	1043	551.5	0.472	0.487	—	—	Chomehoey et al. (2013)
		1:2 ratio Fe:AC;[a] 70°C	—	—	—	—		720.7		0.508		—	
		1:3 ratio Fe:AC;[a] 70°C	—	—	—	—		750.3		0.461		—	

(Continued)

TABLE 11.1 (CONTINUED)
Preparation and Characterization of Magnetic Activated Carbon-Doped Nanoparticles

Precursor	Operating Conditions AC	Operating Conditions MAC	Magnetic Properties M_S (emu/g)	H_C (Oe)	M_R (emu/g)	Size of Magnetic Particle (nm)	Specific Surface Area (m²/g) AC	MAC	Microporous Volume (cm³/g) AC	MAC	Mesopore Volume (cm³/g) AC	MAC	References
As-prepared carbon nanotube	Fe: modified by KOH activation		27.20	0.870	48.00	—	132.4	662.1	—	—	—	—	Ma et al. (2013)
PAC		1:1 ratio Fe:AC;[c] 750°C, 1 h	—	—	—	30–80	1301	671.2	—	—	—	—	Kakavandi et al. (2013)
Coal	1%–3% FeCl₃, pH 6.5;[a] modified by KOH activation: 550°C, 0.75 h + 820°C, 1.5 h		2.000	0.020	35.20	—	1000	925.0	0.370	0.270	0.140	0.200	Liu et al. (2014)
	2%–5% FeCl₃, pH 6.5;[a] modified by KOH activation: 550°C, 0.75 h + 820°C, 1.5 h		22.00	0.400	42.00	—		1328		0.330		0.340	
	3%–9% FeCl₃, pH 6.5;[a] modified by KOH activation: 550°C, 0.75 h + 820°C, 1.5 h		35.56	0.540	45.00	—		909.6		0.240		0.230	
	4%–5% FeCl₃, pH 3.5;[a] modified by KOH activation: 550°C, 0.75 h + 820°C, 1.5 h		3.000	0.020	35.00	—		987.4		0.290		0.190	
	4%–5% FeCl₃, pH 8.0;[a] modified by KOH activation: 550°C, 0.75 h + 820°C, 1.5 h		23.32	0.410	43.00	—		1224		0.340		0.260	
GAC		0.11:1 ratio Fe:AC;[c] 350°C, 2 h	1.600	—	—	9.0	640.0	331.0	0.299	0.163	—	—	Quinones et al. (2014)
		0.43:1 ratio Fe:AC;[c] 450°C, 3 h	3.200	—	—	16.5		285.0		0.152	—	—	
		0.43:1 ratio Fe:AC;[c] 550°C, 4 h	4.400	—	—	17.5		263.0		0.147	—	—	
Lignite	4:1 ratio KOH; 800°C, 1 h	2:1 ratio Fe(NO₃)₃:AC;[c] 750°C, 3 h	—	—	—	—	947.0	667.0	0.430	0.210	0.150	0.060	Tolga et al. (2014)
Coconut shell-based AC		1.4:1 ratio FeSO₄:AC;[a] 25°C;[d] 12 h	1.950	41.51	0.220	20–70	875.0	535.0	0.340	0.220	0.260	0.090	Wan et al. (2014)
		1.4:1 ratio FeSO₄:AC;[a] 70°C;[d] 12 h	9.900	80.75	1.790	8, 40		568.0		0.230		0.070	
		1.4:1 ratio FeSO₄:AC;[a] 150°C;[d] 12 h	14.76	76.07	2.720	50		652.0		0.260		0.130	

Precursor		Synthesis conditions									References
CAC	—	1:0.5 ratio Fe:AC[e]	18.57	208.5	1.100	13.8	—	—	—	—	Pal and Sen (2014)
		1:2.5 ratio Fe:AC[e]	11.42	220.4	0.810	19.83	—	—	—	—	
		1:3 ratio Fe:AC[e]	11.71	257.6	0.970	20.18	—	—	—	—	
PAC	—	1:1 ratio Fe:AC[a] 300°C, 1 h	—	—	—	—	786.0	607.0	—	—	Oh et al. (2015)
		1:1.5 ratio Fe:AC[a] 300°C, 1 h	—	—	—	—	—	556.0	—	—	
		1:2 ratio Fe:AC[a] 300°C, 1 h	—	—	—	—	—	429.0	—	—	
CAC	—	0.1:1 molar ratio Ni:AC[a] 150°C[d] 12 h	40.10	188.7	—	—	1243	936.8	—	—	Wang et al. (2015)
		0.04:1 molar ratio Ni:AC[a] 150°C[d] 12 h	8.700	167.8	—	—	—	1103	—	—	
		0.02:1 molar ratio Ni:AC[a] 150°C[d] 12 h	3.500	106.9	—	—	—	1157	—	—	
		0.01:1 molar ratio Ni:AC[a] 150°C[d] 12 h	0.900	125.6	—	—	—	1192	—	—	
Coal-based AC	—	1:1 ratio Fe:AC[a] 65°C	—	—	—	—	974.0	659.0	—	—	Han et al. (2015)
Coconut-based AC	—	1:1 ratio Fe:AC[a] 65°C	—	—	—	—	975.0	643.0	—	—	
Biochar	—	1:1 ratio Fe:AC[c] 65°C	—	—	—	—	261.0	219.0	—	—	
Biochar	—	1:1 ratio Fe:AC[c] 65°C	—	—	—	—	6.100	68.00	—	—	

Note: For the impregnation ratio in the preparation of activated carbon, X:Y indicates the mass ratio of activating agent to precursor. AC: Activated carbon; MAC: magnetic activated carbon; PAC: powdered activated carbon; CAC: commercial activated carbon; GAC: granular activated carbon; M_S: saturation magnetization; M_R: remanent magnetization; H_C: coercivity.

a   Chemical coprecipitation.
b   Electroless plating.
c   Impregnation with high-temperature treatment.
d   Hydrothermally treated.
e   Electrical explosion of wires.

in the presence of activated carbon, followed by heating of the aqueous suspension. Alternatively, hot water and steam can be used for thermal treatment (hydrothermal treatment). Magnetic iron oxides such as magnetite (Fe_3O_4), maghemite (γ-Fe_2O_3), or different types of ferrites are usually formed through coprecipitation. The coprecipitation of Fe^{2+} and Fe^{3+} ions often results in a broad size distribution of nanoparticles (Zhu et al. 2013).

Kahani et al. (2007) synthesized magnetic activated carbon by adding NH_4OH into a mixture containing the desired molar ratio of $FeCl_3$ and $FeCl_2$ at 70°C in the presence of an N_2 flow to prevent possible oxidation. The reaction can be described as $Fe^{2+} + 2Fe^{3+} + 8OH^- \rightarrow Fe_3O_4 + 4H_2O$ (Kahani et al. 2007; Zhu et al. 2011). The doping procedure involves mixing the magnetite (Fe_3O_4) precipitates with activated carbon in distilled water for 5 h (Kahani et al. 2007). Alternatively, chemical coprecipitation can be carried out by one-step mixing in the presence of an N_2 flow. For this procedure, a mixture of goethite (FeO_2H), activated carbon, and iron(II) chloride in water is added to concentrated NaOH. Then, the reaction mixture is boiled and refluxed for 2 h. The formation of magnetite can be described as $Fe^{2+} + 2FeO_2H + 2OH^- \rightarrow Fe_3O_4 + 2H_2O$ (Kahani et al. 2007). The one-step mixing coprecipitation process gives a better nanoparticle distribution on the activated carbon surface, even though the magnetic saturation (H_S) of the resultant activated carbon is somewhat lower (Kahani et al. 2007).

Electrical explosion of wires (EEW) is a method to produce iron nanopowders using the pulse of a high-density current. The iron nanoparticles obtained are ground and blended with activated carbon using a mortar and pestle (Pal and Sen 2014). Jia and coworkers (2011) described the preparation of magnetic Ni-based activated carbon through the electroless plating method. In this process, chitosan is loaded onto the activated carbon surface to immobilize the divalent nickel cations. The adsorbed cations are then reduced to Ni^0, using $NaBH_4$ (Jia et al. 2011).

Activated carbon-based magnetic nanocomposites can also be prepared via the thermodecomposition method. In this method, $Fe(CO)_5$ is added to a mixture of sodium dodecylbenzenesulfonate containing graphene and dimethylformamide, and the suspension is heated at 153°C and refluxed for 4 h (Zhu et al. 2013 and references therein). During this process, the iron precursor is transformed into iron nanoparticles that adhere to the graphene sheets. The residual organic solvents are removed from the adsorbent surface by heating at 500°C for 2 h under an inert atmosphere. The process yields uniformly dispersed nanoparticles on the graphene sheet with an average diameter of 22 nm (Zhu et al. 2013 and references therein).

In general, the specific surface area and pore volume decrease after magnetization. A higher metallic ratio can cause a smaller surface area; the nanosized magnetic particles tend to enter and be deposited onto the inner wall of the activated carbon pores, thereby partially blocking them (Zahoor and Mahramanlioglu 2011; Faulconer et al. 2012; Lima et al. 2013; Quinones et al. 2014; Oh et al. 2015). Mohan et al. (2011) suggested that the decrease in surface area is due to the smaller carbon proportion, because of the presence of iron particles in the structure of magnetic activated carbon. Liu et al. (2014) reported an increase in surface area with an increase in the concentration of $FeCl_3$ up to 5%, while further increasing the nanoparticle loading decreases the surface area. It is suggested that an iron salt could act as the activating agent to break down light molecules at high temperature and create new pores, while too high an iron concentration may enlarge the pores and consequently decrease the surface area of the magnetic activated carbon.

From Table 11.1, it is obvious that a higher temperature treatment commonly results in a decrease of the specific surface area of activated carbon. For the same Fe:C ratio, Zahoor and Mahramanlioglu (2011), Mohan et al. (2011), Faulconer et al. (2012), and Chomehoey et al. (2013) reported a 16.5%–30.9% decrease in surface area after coprecipitation at temperatures below 70°C. By using the same Fe:C ratio, but at a higher temperature (300°C), Oh et al. (2015) reported a 45.4% deterioration in the surface area of magnetic activated carbon. Faulconer et al. (2012) also demonstrated a similar result; a decrease in surface area from 790 to 47 m^2/g when the magnetic activated carbon is heat treated at 450°C. On the other hand, nanoparticle impregnation at 450°C and 550°C was reported to produce magnetic activated carbons with specific surface areas of 258 and

263 m^2/g, respectively (Quinones et al. 2014). The evolution of volatiles at higher temperatures usually results in the merging of micropores into mesopores, and consequently leads to a lower specific surface area of the magnetic activated carbon. This is true for activated carbon with a less developed graphitic structure that is susceptible to decomposition with heat.

Liu and coworkers (2014) studied the effect of the solution pH on the surface area of magnetic activated carbon. The variation in surface area is possibly due to the nature of iron particles at different pH levels—pH 2 (Fe^{3+}), pH 8 ($Fe(OH)_3$)—whereby excessive mobile divalent cations at lower pH levels can act as an additional activating template to the precursor that is responsible for the etching of micropores into mesopores, thus slightly decreasing the surface area (Liu et al. 2014). Han et al. (2015) demonstrated a decrease in surface area of char and activated carbon after magnetization, except for char containing $CaCO_3$; dissolution of pore-deposited $CaCO_3$ in an acidic iron-salt solution could be the reason for the increase in surface area.

Basically, a material with higher metallic content possesses higher saturation magnetization (M_S) and coercivity (H_C) values (Pal and Sen 2014; Jia et al. 2011; Liu et al. 2013; Wang et al. 2015). However, for a material with higher carbon content, a more ordered arrangement of carbon layers on the entire surface impedes the attraction of an external magnetic field, hence reducing the contribution of magnetic components to the magnetic activated carbon (Xie et al. 2012). H_C is a measure of the ability of magnetic nanoparticles to withstand an external magnetic field without becoming demagnetized. In other words, a high H_C indicates that the material is magnetically hard and is suitable to be used as a permanent magnet. The increase in H_C may be due to complex interactions, which can create strong pinning centers for core moments during demagnetization (Sajitha et al. 2007). In Table 11.1, nearly all magnetic activated carbons have high M_S and low H_C values—a ferromagnetic behavior. This is a form of magnetic mechanism that allows the host material to be attracted by a magnetic field. Moreover, a low remanent magnetization value (M_R) demonstrates that the host material can be redispersed for regeneration after an external magnetic field is removed (Jia et al. 2011; Quinones et al. 2014).

Based on the information given in Table 11.1, a lower fraction of iron nanoparticles used in the doping process generally produces a lower magnetic saturation moment (magnetic content) (M_S) of activated carbon, and vice versa (Lima et al. 2013; Liu et al. 2014; Quinones et al. 2014). Basically, magnetic properties are determined through the formation of magnetite particles on the activated carbon. With high-temperature thermal treatment, carbon monoxide is released by partial combustion, and reduces Fe^{3+} to Fe^{2+}. Hence, magnetite particles are formed on the surface and in the pores of activated carbon. Any inaccessibility of the carbon monoxide diffusion to the pores could result in a lower reduction of Fe^{3+} to Fe^{2+} (Lima et al. 2013).

One of the factors that affect the M_S value is the magnetic particle size. The size of magnetic nanoparticles on the activated carbon surface is calculated as the anatase crystallite size (d_A) from the (101) diffraction peak using the Scherrer equation:

$$d_A = 0.9\lambda / \beta \cos\theta$$

where:
- λ is the wavelength of the Cu K α radiation
- β is the peak width, corrected for instrumental broadening
- θ is the Bragg angle (Lima et al. 2013; Quinones et al. 2014; Raj and Joy 2015)

The existence of nanoparticles is usually confirmed using an electron microscope and energy-dispersive x-ray spectroscopy (Pal and Sen 2014; Lima et al. 2013). From Table 11.1, the size of magnetic nanoparticles doped in activated carbon ranges from 9 to 80 nm. Although it is important to establish that the synthesized and doped magnetic nanoparticles are in the nanometer scale, some authors are silent on the particle size. According to Pal and Sen (2014), particles with a size smaller than 20 nm are said to possess superparamagnetic behavior; the particles become magnetized only when exposed to an external magnetic field, and are without remanent magnetization (M_R) when the field is removed. However, most of the magnetic nanoparticles doped in activated carbon exhibit ferromagnetic behavior, due to having a particle size larger than the minimum critical value (Vatta et al. 2006).

The M_S value is small for smaller particles (Kahani et al. 2007). On the other hand, the particle size is large when the activated carbon weight is higher, when van der Waals interactions form clusters of nanoparticles (Pal and Sen 2014). The solution's pH level can also affect the M_S value. Liu et al. (2014) reported a maximum M_S when the pH level increases from 2 to 8, as abundant OH^- combines with Fe^{3+} to form $Fe(OH)_3$ precipitates. The effect of temperature on the M_S value had been reported by Wan et al. (2014). A higher temperature (150°C) contributes to a higher M_S value (14.8 emu/g), as compared with the M_S value of 1.95 emu/g at 25°C. A higher temperature can enhance the aggregation of magnetic particles on the activated carbon surface through the orientated attachment mechanism and, hence, subsequently increase the crystallinity.

11.3 MAGNETIC ACTIVATED CARBONS FOR WASTEWATER TREATMENT

Magnetic activated carbon-doped nanoparticles have a great potential to be applied in wastewater treatment, because of the large surface-to-volume ratio of the adsorbent to accommodate water contaminants and easy adsorbent separation from the liquid media by applying a magnetic field. The rationale of using magnetic nanoparticles not only relies on their magnetic properties. There is also evidence that magnetic materials can satisfactorily remove various heavy metals, namely, As(III), As(V), Co(II), Cr(VI), and Pd(II). Earlier reviews suggest that the removal of heavy metals from water by magnetic nanoparticles is pH-sensitive, and is mainly driven through reduction, adsorption, and electrostatic interaction in a magnetic flocculation-precipitation system (Vatta et al. 2006; Zhu et al. 2013). The removal of Cr(VI) is elevated when the graphene is decorated with magnetic nanoparticles, and selective toward $HCrO_4^-$, as removal is favored at lower pH levels (Zhu et al. 2013 and references therein). Therefore, the integration of magnetic nanoparticles and activated carbon is believed to enhance the removal performance of the adsorbent toward the target pollutants. This section will give an overview on the use of magnetic activated carbon-doped nanoparticles in removing water contaminants.

The ability of magnetic nanoparticles to enhance the removal of certain water contaminants is pH dependent. Depci (2012) reported a 30%–50% increase in cyanide removal at neutral pH by an iron-impregnated lignite activated carbon as compared with an unmagnetized one. The inclusion of iron nanoparticles introduces additional active sites (–Fe–OH) to the activated carbon. It is suggested that the removal mechanisms are electrostatic attraction: $Fe-OH_2^+ + CN^- \leftrightarrow Fe-OH_2^+ \cdots CN^-$, direct exchange: $Fe-OH_2^+ + CN^- \leftrightarrow Fe-CN + H_2O$, and the formation of iron-cyanide complexes on the activated carbon surface: $Fe(CN)_6^{-3}$, $Fe(CN)_6^{-4}$ (Depci 2012).

Some studies revealed a slightly improved or comparable uptake of water contaminants such as stable oil (Raj and Joy 2015), imidacloprid (Zahoor and Mahramanlioglu 2011), and arsenic (Yao et al. 2014), even after the doping of magnetic nanoparticles, which usually decreases the specific surface area. This is due to the magnetic nanoparticles lodging in the micropores (Depci 2012; Zahoor and Mahramanlioglu 2011; Yao et al. 2014). As a consequence, the mesopore volume is fairly unaltered, suggesting that the removal of the aforementioned pollutants is mesopore-sensitive.

Gholamvaisi and coworkers (2014) reported similar Bismarck brown dye removal by an Fe_3O_4-decorated activated carbon and an unmodified one, but the former revealed a slower rate of adsorption. In a related study, Mohan et al. (2011) demonstrated a decrease of nearly half in the adsorption intensity of trinitrophenol for magnetic activated carbon from almond shells, despite a similar uptake as compared with the unmodified material. In these works, the main advantage of doping the magnetic nanoparticles is the convenience of recycling the adsorbent using a permanent magnet that is fast and economical.

Table 11.2 summarizes some of the recent studies on the applications of magnetic activated carbons for the removal of water contaminants.

The solution pH, surface charge, and surface functional groups of magnetic activated carbons are among the parameters that have significant influence on the removal of dyes and metal ions. The surface functional groups that are commonly found in most magnetic activated carbon-doped nanoparticles include O–H, C=C, C–O, C=O, and Fe–O in iron oxides. For cationic or positively charged adsorbates (e.g., methylene blue, Bismarck brown, aniline, and metal ions), the removal favors negatively charged surfaces—that is, when the solution pH > pH_{pzc}, through which the electrostatic attraction between the adsorbates and the surface leads to a better adsorption performance. If the solution pH < pH_{pzc}, the carbon surface is protonated and becomes positively charged owing to the protonated functional groups, that is, carboxylic groups ($-CO-OH_2^+$), phenolic groups ($-OH_2^+$), and Fe–OH_2^+ (Depci 2012). The same charge repulsion between the cationic adsorbates and the positively charged surface often results in a decrease in adsorption capacity. The positively charged surface favors the adsorption of anionic (negatively charged) adsorbates such as Alizarin Red S, Congo red, phenol, methyl orange, acid yellow dye, and trinitrophenol. Therefore, the adsorption of cationic adsorbates is more favorable at a higher pH, whereas the adsorption of anionic adsorbates is enhanced at a lower pH.

From Table 11.2, the adsorption process is mainly carried out in an acidic, slightly acidic, or neutral environment, because the pH_{pzc} values of magnetic activated carbons are between 5.5 and 8.3. Ranjithkumar et al. (2014) showed that the removal of Acid Yellow 17 dye decreases with a decreasing pH as the dye molecules compete with the OH^- ions for the active sites. Fayazi et al. (2015), Kakavandi et al. (2013), Mohan et al. (2011), and Jiang et al. (2015) have also reported similar findings.

In addition, some water pollutants may exist in three different forms, namely, cationic, neutral/nonionized, and anionic (Arvand and Alirezanejad 2011), due to the protonation or deprotonation of the functional groups, depending on the solution pH. Wan and coworkers (2014) stated that sulfamethoxazole has two ionizable functional groups, that is, $-NH_2$ and NH (sulfonamide). At a pH value greater than 5.81, the anionic form of sulfamethoxazole predominates, while at pH < 5.81, it is in a neutral nonionized state. The adsorption performance of sulfamethoxazole deteriorates at a higher pH (pH > pH_{pzc} = 6.23) due to a greater repulsion between the anionic molecules and the negatively charged adsorbent surface. According to Jiang et al. (2015), methyl orange exists in quinoid form under acidic conditions (pH < pH_{pzc}), while it alters into an azo structure (anionic form) in basic conditions (pH > pH_{pzc}). On the other hand, Fayazi and coworkers (2015) reported a decrease in Alizarin Red S (anionic dye) dissociation at pH < 1, thereby decreasing the concentration of anionic dye accessible to the active sites of magnetic activated carbons.

The solution pH is a parameter of great influence not only to the adsorption process, but also to the performance of the activated carbon–iron oxide composite. The effect of the solution pH on the adsorption capacity and magnetization properties of magnetic carbon composites has been studied by several researchers (Oliveira et al. 2002, 2003; Zahoor and Mahramanlioglu 2011). Zahoor and Mahramanlioglu (2011) reported a notable change in the iron oxide content of magnetic activated carbon in the pH range of 1–4.8, while no major changes were observed in the pH range of 4.8–8. The loss in magnetization increases with decreasing pH, and no magnetization was recorded at a pH of 1. At low pH levels, the H^+ ions attack the iron oxides on the activated carbon surface, leading to the dissolution of iron oxide and the loss of magnetization properties. Therefore, the effect of pH on the adsorption of imidacloprid was investigated in the pH range of 4.8–8, and the adsorption performance did not differ greatly with differences in pH level (Zahoor and Mahramanlioglu 2011). Oliveira et al. (2002, 2003) also reported similar findings, where the iron oxide doped on activated carbon was totally dissolved at a pH of 1. In the pH range of 5–11, the magnetic activated carbon composite exhibited good resistance and retained its magnetic properties.

Generally, an adsorbent with high specific surface area tends to give a better adsorption capacity and fast equilibrium. From Table 11.2, it can be seen that the magnetic activated carbons with high surface areas, ranging from 652 to 1236 m^2/g, show maximum adsorption capacities of 20.24–710 mg/g, with

TABLE 11.2

Removal Performance of Activated Carbon-Based Magnetic Nanocomposites

Magnetic Adsorbent	Specific Surface Area (m²/g)	Functional Groups (FTIR)	Targeted Pollutant	Max Removal (mg/g)	pH	Time (h)	References
Almond-shell MAC	527	O–H, C≡C, C–O, C=O, Fe–O	2,4,6-trinitrophenol (TNP)	73.96	2.0; pH < pH$_{pzc}$ = 6.78	48	Mohan et al. (2011)
Coconut-shell MAC	652	O–H, C–OH, C–O, Fe–O, Mn–O	Sulfamethoxazole (antibiotic)	159	7; pH$_{pzc}$ = 6.23	24	Wan et al. (2014)
MAC nanocomposite	347.8	C–H, O–H, Fe–O, C=O, C–O	Alizarin Red S (ARS)	108.69	2.0; pH$_{pzc}$ ~ 6.8	1	Frayazi et al. (2015)
AC–NiFe$_2$O$_4$ composite	157.1	O–H, N–N, Fe–O, C–O–C, Ni–O, C=O, C–O	Methyl orange	180.18	3.0; pH$_{pzc}$ = 8.30	0.5	Jiang et al. (2015)
Coconut-shell AC–iron oxide magnetic nanocomposite	—	Fe–O, O–H, C–O, C=O	Premium oil	12,930	—	0.33	Raj and Joy (2015)
			Used oil	7650	—	0.13	
Activated carbon–Fe$_3$O$_4$ composite	—	C–C, Fe–O, O–H, C=O, C–O	Acid Yellow 17 dye	100	5	—	Ranjithkumar et al. (2014)
Magnetic PAC	671.2	—	Aniline	90.91	6.0	5	Kakavandi et al. (2013)
Magnetic PAC	—	—	Bismarck brown (B.B.) dye	224.0	7.0	1	Gholamvaisi et al. (2014)
Magnetic iron oxide–CNT	662.1	—	Arsenic	9.74	—	0.17	Ma et al. (2013)
Magnetic PAC	868	—	Imidacloprid	94.89	7.4	0.67	Zahoor and Mahramanlioglu (2011)
Peanut-shell MAC	1236	—	Methylene blue	405	—	—	Zhang et al. (2015)
Magnetic cellulose–Fe$_3$O$_4$–AC composite	11.87	—	Congo red	66.09	5	10	Zhu et al. (2011)
Iron oxide–AC nanocomposite	678.3	—	Arsenic	20.24	6; pH$_{pzc}$ = 7.9	1	Yao et al. (2014)
MAC (Fe$_3$O$_4$–AC)	671.2	—	Lead (Pb^{2+})	58.82	6; pH$_{pzc}$ = 5.5	1	Kakavandi et al. (2015)
AC–iron oxide magnetic composite	658	—	Phenol	117	5	—	Oliveira et al. (2002)
			Chloroform	710			
			Chlorobenzene	305			
Clay–iron oxide magnetic composites	58	—	Nickel (Ni^{2+})	2465.1	5	—	Oliveira et al. (2003)
			Copper (Cu^{2+})	3177.3			
			Cadmium (Cd^{2+})	8318.4			
			Zinc (Zn^{2+})	4903.5			
Coconut MAC	—	—	Gold (Au)	15.5	—	64	Wang et al. (1994)

Note: AC: activated carbon; MAC: magnetic activated carbon; PAC: powdered activated carbon.

equilibrium times between 10 min and 24 h. A higher surface area provides more active sites, which are accessible to the adsorbate molecules. Initially, the adsorption occurs at a faster rate as adsorbate molecules diffuse rapidly from the bulk solution to the external surface of magnetic activated carbon. When the sites are almost occupied, the adsorbate molecules must overcome the adsorbent mass transfer resistance to pass from the solution into the inner pores. From this point, adsorption occurs at a slower rate, as the sites are nearly saturated and equilibrium is almost achieved.

Few studies demonstrated reasonably good regeneration efficiency of spent magnetic activated carbons after several cycles. Ray and Joy (2015) achieved a better regeneration of coconut shell–based activated carbon–iron oxide magnetic nanocomposite for stable oil removal using petroleum ether than that of heating at 500°C under an N_2 flow. Wan and coworkers (2014) reported the use of H_2O_2 to regenerate magnetic manganese ferrite activated carbon that was previously used to remove sulfamethoxazole. In a related study, Oh et al. (2015) suggested the use of peroxymonosulfate to regenerate methylene blue-loaded $CuFe_2O_4$-activated carbon. Generally, the performance of the spent activated carbons decreases upon regeneration, due to the collapse of pores, decomposition of surface active sites, and leaching of magnetic nanoparticles.

11.4   LIMITATIONS LEADING TO FUTURE DIRECTIONS

Magnetic activated carbon nanoparticles hold much potential for wastewater treatment because of their excellent magnetic properties, which simplify the separation of activated carbon from treated water. Although it has been reported that magnetic nanoparticles could in some way enhance the removal of water contaminants, such evidence is not clearly observed in most of the applications of magnetic activated carbon-doped nanoparticles (Zhu et al. 2013; Vatta et al. 2006). This is partly due to the selection of target adsorbates—mainly dyes and organic pollutants—which may not be affected by magnetic nanoparticles (Zhu et al. 2013; Vatta et al. 2006). Recent studies aimed only to exploit a permanent magnet so that the recovery of activated carbon would become fast and economical (Kakavandi et al. 2013; Gholamvaisi et al. 2014; Jiang et al. 2015). It would be of great importance if the role of magnetic nanoparticles was fully exploited to increase removal of selected water contaminants.

Previous studies have normally demonstrated a decrease in the specific surface area of activated carbons due to the blockage of pore channels by the nanoparticles. This may compromise the removal performance of certain water contaminants when the separation is textural sensitive. As the introduction of magnetic nanoparticles is mainly to produce magnetic behavior in the activated carbon, the trade-off between the magnetic moment, the textural characteristics of the activated carbon, and the removal of the target water contaminants should, therefore, be taken into full consideration. In other words, the amount of doped magnetic nanoparticles should be kept to a minimum to only just provide sufficient magnetic behavior, because unnecessary loading onto the activated carbon will only jeopardize removal performance.

The regeneration of spent activated carbon is another area of considerable interest to be explored. Repetitive use of the same load of adsorbent is essential from the viewpoint of an economic and sustainable environment. It is directly associated with the selection of a regeneration method. The economic and environmental implications of regeneration methods should also be carefully screened and examined prior to being implemented. The use of expensive, toxic, and destructive techniques could be permitted if necessary cautions are taken to minimize losses. Similarly, the regeneration of magnetic activated carbon should consider minimal changes to the textural characteristics and active sites, preserve the adsorbent's magnetic behavior, and prevent the release of nanoparticles into the environment.

Among the major challenges when using magnetic nanoparticles are the recovery of nanosized particles and dissolution in acidic environments. These are of environmental concern, and may occur in the preparation of magnetic activated carbons or in the removal of water contaminants. The release of magnetic nanoparticles to the environment should be avoided, as most of them are very toxic. Therefore, screening and minimum usage of magnetic materials should be carefully observed at the expense of removal performance. Dissolution of magnetic elements is a key barrier in applying these materials, especially in acidic solutions. Magnetite, for example, is soluble at pH values lower than 4 (Vatta et al. 2006; Depci 2012). It should be noted that the doping process does not completely protect magnetic nanoparticles from leaching and dissolution. Therefore, the use of magnetic activated carbons should be cautiously monitored for possible secondary pollution.

11.5 CONCLUSION

In this chapter, we presented recent progress in magnetic activated carbon-doped nanoparticles. Preparation techniques and the characteristics of magnetic activated carbons have been highlighted. Novel adsorbents have a great potential to be used in the removal of various water contaminants. The magnetic behavior of the adsorbent simplifies the separation of the adsorbent from water, using a permanent magnet in a fast and economical way. However, the roles of magnetic nanoparticles to enhance the removal of certain water contaminants are yet to be explored. The present gaps and limitations could serve as a stepping stone for future research.

REFERENCES

Akbarzadeh, A., Samiei, M., and Davaran, S. Magnetic nanoparticles: Preparation, physical properties, and applications in biomedicine. *Nanoscale Research Letters* 7 no.1 (2012): 144.

Arvand, M., and Alirezanejad, F. Sulfamethoxazole-imprinted polymeric receptor as ionophore for potentiometric transduction. *Electroanalysis* 23 no.8 (2011): 1948–57.

Chomehoey, N., Bhongsuwan, D., and Bhongsuwan, T. Arsenic removal from synthetic wastewater by activated carbon-magnetic nanoparticles composite. *ASEAN++ 2013 Moving Forward: The 11th International Conference on Mining, Materials and Petroleum Engineering, The 7th International Conference on Earth Resource Technology,* Chiang Mai, November, 2013.

Depci, T. Comparison of activated carbon and iron impregnated activated carbon derived from Gölbaşı lignite to remove cyanide from water. *Chemical Engineering Journal* 181 (2012): 467–78.

Faulconer, E.K., Hoogesteijn von Reitzenstein, N., and Mazyck, D.W. Optimization of magnetic powdered activated carbon for aqueous Hg(II) removal and magnetic recovery. *Journal of Hazardous Materials* 199–200 (2012): 9–14.

Fayazi, M., Ghanei-Motlagh, M., and Taher, M.A. The adsorption of basic dye (alizarin red S) from aqueous solution onto activated carbon/γ-Fe$_2$O$_3$ nano-composite: Kinetic and equilibrium studies. *Materials Science in Semiconductor Processing* 40 (2015): 35–43.

Gholamvaisi, D., Azizian, S., and Cheraghi, M. Preparation of magnetic-activated carbon nanocomposite and its application for dye removal from aqueous solution. *Journal of Dispersion Science and Technology* 35 no.9 (2014): 1264–69.

Han, Z., Sani, B., Akkanen, J., et al. A critical evaluation of magnetic activated carbon's potential for the remediation of sediment impacted by polycyclic aromatic hydrocarbons. *Journal of Hazardous Materials* 286 (2015): 41–47.

Jia, B., Su, L., Han, G., Wang, G., Zhang, J., and Wang, L. Adsorption properties of nickel-based magnetic activated carbon prepared by Pd-free electroless plating. *BioResources* 6 no.1 (2011): 70–80.

Jia, M., Zhu, Z., Chen, B., et al. One-pot, large-scale synthesis of magnetic activated carbon nanotubes and their applications for arsenic removal. *Journal of Materials Chemistry A* 1 (2013): 4662–66.

Jiang, T., Liang, Y.D., He, Y.J., and Wang, Q. Activated carbon/NiFe$_2$O$_4$ magnetic composite: A magnetic adsorbent for the adsorption of methyl orange. *Journal of Environmental Chemical Engineering* 3 no.3 (2015): 1740–51.

Kahani, S.A., Hamadanian, M., and Vandadi, O. Deposition of magnetite nanoparticles in activated carbons and preparation of magnetic activated carbons. *First Sharjah International Conference on Nanotechnology and its Applications,* Sharjah, Septermber, 2007.

Kakavandi, B., Jafari, A.J., Kalantary, R.R., Nasseri, S., Ameri, A., and Esrafily, A. Synthesis and properties of Fe$_3$O$_4$-activated carbon magnetic nanoparticles for removal of aniline from aqueous solution: Equilibrium, kinetic and thermodynamic studies. *Iranian Journal of Environmental Health Sciences & Engineering* 10 no.19 (2013): 1–9.

Kakavandi, B., Kalantary, R.R., Jafari, A.J., Nasseri, S., Ameri, A., Esrafili, A., and Azari, A. Pb (II) adsorption onto a magnetic composite of activated carbon and superparamagnetic Fe_3O_4 nanoparticles: Experimental and modeling study. *CLEAN–Soil, Air, Water* 43 no.80 (2015): 1157–66.

Lima, S.B., Borges, S.M.S., Rangel, M.C., and Marchtti, S.G. Effect of iron content on the catalytic properties of activated carbon-supported magnetite derived from biomass. *Journal of the Brazilian Chemical Society* 24 no.2 (2013): 344–54.

Liu, K., Xu, L., and Zhang, F. A new preparation process of coal-based magnetically activated carbon. *Chinese Journal of Geochemistry* 33 (2014): 173–77.

Ma, J., Zhu, Z., Chen, B., Yang, M., Zhou, H., Li, C., Yu, F., and Chen, J. One-pot, large-scale synthesis of magnetic activated carbon nanotubes and their applications for arsenic removal. *Journal of Materials Chemistry A* 1 no.15 (2013): 4662–66.

Mohan, D., Sarswat, A., Singh, V.K., Alexandre-Franco, M., and Pittman Jr., C.U. Development of magnetic activated carbon from almond shells for trinitrophenol removal from water. *Chemical Engineering Journal* 172 (2011): 1111–25.

Oh, W.D., Lua, S.K., Dong, Z., and Lim, T.T. Performance of magnetic activated carbon composite as peroxymonosulfate activator and regenerable adsorbent via sulfate radical-mediated oxidation processes. *Journal of Hazardous Materials* 284 (2015): 1–9.

Oliveira, L.C., Rios, R.V., Fabris, J.D., Garg, V., Sapag, K., and Lago, R.M. Activated carbon/iron oxide magnetic composites for the adsorption of contaminants in water. *Carbon* 40 no.12 (2002): 2177–83.

Oliveira, L.C., Rios, R.V., Fabris, J.D., Sapag, K., Garg, V.K., and Lago, R.M. Clay–iron oxide magnetic composites for the adsorption of contaminants in water. *Applied Clay Science* 22 no.4 (2003): 169–77.

Pal, S.P., and Sen, P. Synthesis and characterization of nanocomposites of Fe nanoparticles and activated carbon. arXiv preprint arXiv:1404.2665 (2014).

Quinones, D.H., Rey, A., Alvarez, P.M., Beltran, F.J., and Plucinski, P.K. Enhanced activity and reusability of TiO_2 loaded magnetic activated carbon for solar photocatalytic ozonation. *Applied Catalysis B: Environmental* 144 (2014): 96–106.

Raj, K.G., and Joy, P.A. Coconut shell based activated carbon–iron oxide magnetic nanocomposite for fast and efficient removal of oil spills. *Journal of Environmental Chemical Engineering* 3 no.3 (2015): 2068–75.

Ranjithkumar, V., Hazeen, A.N., Thamilselvan, M., and Vairam, S. Magnetic activated carbon-Fe_3O_4 nanocomposites: Synthesis and applications in the removal of acid yellow dye 17 from water. *Journal of Nanoscience and Nanotechnology* 14 no.7 (2014): 4949–59.

Sajitha, E.P., Prasad, V., Subramanyam, S.V., Mishra, A.K., Sarkar, S., and Bansal, C. Size-dependent magnetic properties of iron carbide nanoparticles embedded in a carbon matrix. *Journal of Physics: Condensed Matter* 19 no.4 (2007): 046214.

Tolga, D., Busetty, S., and Yunus, O. Investigation of the potential of activated and magnetic activated carbon produced from Turkish lignite as gold adsorbents. *Asian Journal of Applied Sciences* 7 no.6 (2014): 486–98.

Vatta, L.L., Sanderson, R.D., and Koch, K.R. Magnetic nanoparticles: Properties and potential applications. *Pure and Applied Chemistry* 78, no.9 (2006): 1793–801.

Wan, J., Deng, H., Shi, J., Zhou, L., and Su, T. Synthesized magnetic manganese ferrite nanoparticles on activated carbon for sulfamethoxazole removal. *CLEAN–Soil, Air, Water* 42 no.9 (2014): 1199–207.

Wang, C., Liu, Q., Cheng, X., and Shen, Z. Adsorption and desorption of gold on the magnetic activated carbon. *Journal of Materials Science & Technology* 10 no.2 (1994): 151–3.

Wang, P., Xu, J., Zhang, B., et al. Adsorption performance for methylene blue of magnetic Ni@activated carbon nanocomposites. *Functional Materials Letters* 8 no.2 (2015): 1550024.

Xie, T.P., Xu, L.J., Liu, C.L., He, C.L., Xiao, Z.M., and Wang, J.C. Preparation of magnetic activated carbon and adsorption properties for $KMnO_4$. *Science China Press* 42 no.8 (2012): 1152–60.

Yao, S., Liu, Z., and Shi, Z. Arsenic removal from aqueous solutions by adsorption onto iron oxide/activated carbon magnetic composite. *Journal of Environmental Health Science and Engineering* 12 (2014): 1–8.

Zahoor, M., and Mahramanlioglu, M. Adsorption of imidacloprid on powdered activated carbon and magnetic activated carbon. *Chemical and Biochemical Engineering Quarterly* 25 no.1 (2011): 55–63.

Zhang, S., Tao, L., Jiang, M., Gou, G., and Zhou, Z. Single-step synthesis of magnetic activated carbon from peanut shell. *Materials Letters* 157 (2015): 281–4.

Zhu, H.Y., Fu, Y.Q., Jiang, R., Jiang, J.H., Xiao, L., Zcng, G.M., Zhao, S.L., and Wang, Y. Adsorption removal of Congo red onto magnetic cellulose/Fe_3O_4/activated carbon composite: Equilibrium, kinetic and thermodynamic studies. *Chemical Engineering Journal* 173 no.2 (2011): 494–502.

Zhu, J., Wei, S., Chen, M., Gu, H., Rapole, S.B., Pallavkar, S., Ho, T.C., Hopper, J., and Guo, Z. Magnetic nanocomposites for environmental remediation. *Advanced Powder Technology* 24 (2013): 459–67.

12 Functionalized Fe₃O₄ Nanoparticles for the Removal and Remediation of Heavy Metals in Wastewater

Functionalized Fe_3O_4 Nanoparticles for the Removal and Remediation of Heavy Metals in Wastewater

*Annamalai Sivaraman, Dhevagoti Manjula Dhevi,
Arun Anand Prabu, and Kap Jin Kim*

CONTENTS

12.1 Introduction 292
12.2 Heavy Metal Removal Using Adsorption..292
    12.2.1 Magnetic Separation Technology 293
    12.2.2 MNPs-Based Adsorbents..293
12.3 Polymer-Fe_3O_4-Based Adsorbents..294
    12.3.1 Natural Polymer-Fe_3O_4-Based Adsorbents...296
        12.3.1.1 Chitosan 297
        12.3.1.2 Cellulose–Chitosan..300
        12.3.1.3 Cyclodextrin–Chitosan 302
        12.3.1.4 Carbon Disulfide–Chitosan 304
        12.3.1.5 Polyrhodanine 304
    12.3.2 Amine-Fe_3O_4-Based Adsorbents...307
        12.3.2.1 Tetraethylenepentamine and Its Related Amines 307
        12.3.2.2 APTEs..309
    12.3.3 Polyvinyl Acetate-Fe_3O_4-Based Adsorbents 311
    12.3.4 Polyacrylate-Fe_3O_4-Based Adsorbents 313
    12.3.5 Polypyrrole-Fe_3O_4-Based Adsorbents 314
    12.3.6 PEDOT-Fe_3O_4-Based Adsorbents 315
    12.3.7 Dendrimer-Fe_3O_4-Based Adsorbents..317
12.4 Future Trends 318
References..318

ABSTRACT

This chapter focuses on recent developments in the utilization of polymer functionalized-metal oxide nanosorbents, particularly magnetite (Fe_3O_4) nanoparticles, for the removal of toxic heavy metals from aqueous media and their limitations for use in real-time field applications. Though ceramics-based metal oxide nanoparticles have been used for the remediation of heavy metals, they bind nonspecifically and, hence, their reversible recovery from treated effluents becomes extremely challenging. In recent times, many researchers have used polymer functionalized-Fe_3O_4 nanoparticles for the remediation of heavy metals and tested their sorption effectiveness under varying conditions such as pH, contact time, and dosage. This modification has the following advantages: (a) increase in number of active sites due to the presence of magnetite and polymer functional groups, (b) high specific surface area due to the presence of nanoparticles, and (c) prevention of self-aggregation of nanoparticles, which increases accessibility to active sites. This chapter will mainly focus on the different types of polymeric materials used to modify Fe_3O_4 as an adsorbent, recovery of heavy metals from the adsorbents and the reuse of adsorbents for many adsorption/desorption cycles. The issues dealt with in this chapter will be useful to researchers working on polymer–nanomaterial-based adsorbents for the remediation of heavy metals from industrial effluents.

12.1 INTRODUCTION

Due to the rapid urbanization in the last decade, most water bodies are becoming contaminated with untreated industrial effluents containing heavy metals and other toxic materials. On the other hand, heavy metals are important strategic resources, as well as essential raw materials in the field of advanced functional materials. For example, copper compounds are widely used in the manufacturing of fungicides, metal plating, wood pulp production, antifouling paints, and electrical and semiconductor devices. Nickel compounds are used as high-performance catalysts. Living organisms also require trace amounts of essential heavy metals (Co, Cu, Fe, Mn, Mo, Sr, and Zn), but their excessive levels can be detrimental when they get passed up the human food chain. Also, nonessential heavy metals (Hg, Cr, Cd, As, Pb, and Sr) originating from natural sources, mining, industrialization, and so on are considered to be a greater threat to public health when mixed into fresh water supplies. For instance, excess copper would cause gastrointestinal distress in short-term exposure, and liver or kidney damage in long-term exposure. The World Health Organization (WHO) reported that drinking water containing zinc at levels above 3 mg L^{-1} can cause an undesirable astringent taste, and the U.S. Environmental Protection Agency has set a maximum allowable level of zinc at 5 mg L^{-1}. In general, the toxicity of a heavy metal is attributed to its inhibition and reduction of various enzymes, complexation with certain amino acid ligands, and substitution of essential metal ions from enzymes (Hall et al. 2002). Toxicological studies have found that the degree of heavy metal toxicity depends on its oxidation state (Babel and Kurniawan 2003). In spite of government regulations, many industries still allow their effluents into the public sewers or drinking water sources like rivers and lakes without adequate treatment to save their operational costs. Contamination of drinking water with toxic metal ions such as Ag^+, Hg^{2+}, Pb^{2+}, Ni^{2+}, Co^{2+}, Cu^{2+}, Cd^{2+}, As^{3+}, Cr^{3+}, Cr^{4+}, and As^{5+} is fast becoming a severe environmental and public health problem, and there is an urgent need to find efficient and cost-effective methods for their removal from water resources.

12.2 HEAVY METAL REMOVAL USING ADSORPTION

Though various techniques like evaporation, flotation, and oxidation (Fu and Wang 2011), chemical reduction–precipitation (Pang et al. 2011; Fu et al. 2012), membrane separation (Song et al. 2011a), ion exchange (Dabrowski et al. 2004), bio/electrokinetic remediation, and adsorption (Badruddoza et al. 2011, 2013; Yang et al. 2013) are available in laboratory scale for the removal of heavy metals, very few methods are effectively established in the industrial scale. Among the aforementioned methods, adsorption is a conventional but efficient technique to remove toxic metal ions and bacterial pathogens from water. Most common absorbents such as activated carbon, silicon dioxide, activated alumina, zeolite, and so on are typically porous with larger specific surface areas. For example, the Brunauer–Emmett–Teller (BET) specific surface area of activated carbon (Calgon F-400) of mesh size 12–40 is as large as 1026 m^2 g^{-1}. However, its specific area of external surface (S_E) is as small as 0.00576 m^2 g^{-1}. Adsorption sites of porous adsorbents are mainly located inside the pores, which leads to some disadvantages: (a) Due to the resistance of pore diffusion, it takes a longer time for the porous adsorbents to reach adsorption equilibrium. (b) The regeneration process of porous adsorbents is also difficult, with a longer regeneration time. This drawback could be overcome by decreasing the particle diameter to a few micrometers or less. (c) Resistance of pore diffusion decreases with reducing particle size. For example, CP-5 with mass mean particle diameter (d_w) of 9.1 μm required only about 3 h for equilibrium, while CP-100 with d_w of 73.3 μm required several days (Papelis et al. 1995). In addition, microsize polymer adsorbents have good regeneration ability and it is also easier to modify the adsorbent's surface to change its chemical characteristics. Therefore, surface-modified absorbents can have higher affinity to some specific metals.

12.2.1 MAGNETIC SEPARATION TECHNOLOGY

Magnetic separation technology (MST) was found to be a conventional but efficient technique to remove metal ions from wastewater (Hu et al. 2006). Magnetic nanoparticles (MNPs) possess the advantages of a large surface area, a high number of surface active sites, and favorable magnetic properties, which lead to higher adsorption efficiency and removal rate of contaminants. Rapid separation of the adsorbent from the effluent solution can easily be realized via a magnetic field and subsequently reused many times, thereby overcoming the separation difficulties in industrial wastewater treatment. Another advantage of high-gradient magnetic separation is that the nonmagnetic impurities can be excluded during the recovery of magnetic adsorbents. Synthesis of magnetic adsorbents of microsize and a high-gradient magnetic force field can be applied to separate, recover, and recycle them from water after the adsorption process. The removal of heavy metal ions, sulfate, nitrate, and dissolved organic carbon (DOC) from wastewater using MST has already been put into practical applications such as the magnetic ion exchange resin (MIEX®) technique developed by Orica Watercare and two leading Australian research organizations (Zhang et al. 2006).

12.2.2 MNPs-BASED ADSORBENTS

The utilization of iron oxides, namely magnetite (Fe$_3$O$_4$)- and maghemite (γ-Fe$_2$O$_3$)-based MNPs as adsorbents for the removal of heavy metals from industrial effluents, has received much attention due to their unique properties, such as high surface area-to-volume ratio, surface modifiability, excellent magnetic properties, greater biocompatibility, ease of separation using an external magnetic field, reusability, and comparatively lower cost (Shen et al. 2009; Dave and Chopda 2014). To improve the practical application prospects of magnetite, it is essential to lessen the separation time of the magnetite from the solution. However, compared with pure bulk magnetic materials, the magnetism of smaller MNPs or nonmagnetic coated nanostructures is weaker, which can prolong the separation of the magnetite from the solution. When the Fe core in the synthesized MNPs was increased, Zhu et al. (2012) reported a large M_S (96.3 emu g^{-1}) for the faster removal of chromium.

Creation of secondary nanostructures with microsphere characteristics can retain a M_S as high as the counterpart bulk material, although the nanostructure does not lead to improvements in magnetization. Magnetization depends on the magnetic moments of the particles per volume unit, and there is no relation with the final magnetic collective system. To combine the individual NPs with the possibility of tuning collective properties and control them, manipulation of the secondary structures of NPs is desired. In particular, the approach of forming large complex structures of NPs appears more attractive, because the proportion of atoms located in the surface increases; these atoms hardly reverse with the magnetic field and the final M_S is as high as its bulk counterpart. Superparamagnetic behavior is exhibited by MNPs having individual spherical cores of <20 ∼ 30 nm diameter; that is, they become easily magnetized/unmagnetized to/from a magnetic field once the field is turned on/off. Extensive reviews have addressed the issues pertaining to the synthesis of MNPs, controlling their size, shape, morphology, and magnetic properties (Laurent et al. 2008; Qiao et al. 2009). Magnetic hybrid hydrogels can be used for removing heavy metals, and the adsorption, desorption, and adsorbent recycling processes are schematically shown in Figure 12.1.

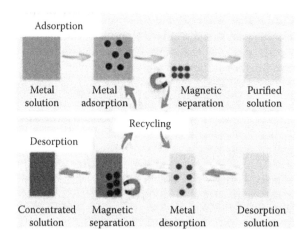

FIGURE 12.1 The overall process of metal adsorption, desorption, and adsorbent recycling. (Liu, Z. et al., 2012, Magnetic cellulose–chitosan hydrogels prepared from ionic liquids as reusable adsorbent for removal of heavy metal ions, *Chem. Commun.* 48, 7350–7352, figure 3. Reproduced by permission of The Royal Society of Chemistry.)

12.3 POLYMER-FE₃O₄-BASED ADSORBENTS

MNPs tend to agglomerate and create heterogeneous size distribution patterns due to their hydrophobic surfaces and large surface area-to-volume ratio. Hence, it is imperative to disperse MNPs in suitable solvents or to coat them with certain molecules or polymers. In general, coating leads to the creation of more hydrophilic nanostructures via end-grafting, encapsulation, hyperbranching, or hydrophobic interactions, which opens up the possibility of creating multimodal and multifunctional MNPs (Fang and Zhang 2009). Coating of MNPs can be achieved through different methods: during or after synthesis; *in situ* or by encapsulation (Figure 12.2). Preparation and application of magnetically active polymeric adsorbents (MAPAs) with several functional groups have been examined, in which the balance between steric and electrostatic repulsive forces is crucial. Surface iron atoms on iron oxide can coordinate with polymeric molecules to donate lone-pair electrons, therefore acting as Lewis acids (Dias et al. 2011). If Fe_3O_4 nanocrystals could be irreversibly dispersed within the polymeric beads, the nonmagnetic polymeric sorbents would then be magnetically active and respond to the magnetic field.

The challenge in magnetization of polymeric beads lies in controlling the process conditions so that the formation of Fe_3O_4 is preferred over nonmagnetic $Fe(OH)_3$ and $Fe(OH)_2$. Under a reducing environment (i.e., absence of O_2), $Fe(OH)_2$ is the predominant solid phase. Under a moderate to highly oxidizing environment, $Fe(OH)_3$ predominates. The redox and pH boundaries within which the formation of magnetite is favored are very narrow. Since the MNP isoelectric point is ~pH 6.8, the surface of magnetite can become positive or negative depending on the actual pH level of the solution. Consequently, protonation of the particle surface below the isoelectric point leads to the formation of $\equiv Fe-OH^{2+}$ moieties. Deprotonation occurs above the isoelectric point, leading to the formation of $\equiv Fe-O^-$ surface moieties, which affects the electrostatic attachments of the polymers on the surface of MNPs. Figure 12.3 shows the stability diagram at 25°C (E_h – pH) depicting the pertinent solid phases; note that Fe_3O_4 formation is thermodynamically favored only within a narrow envelope. Considering oxygen to be the sole electron acceptor, the solid-phase transition between Fe_3O_4 and $Fe(OH)_3(s)$ can be presented as follows:

$$4Fe_3O_4(s) + O_2(aq) + 18H_2O \leftrightarrow 12Fe(OH)_3(s) \tag{12.1}$$

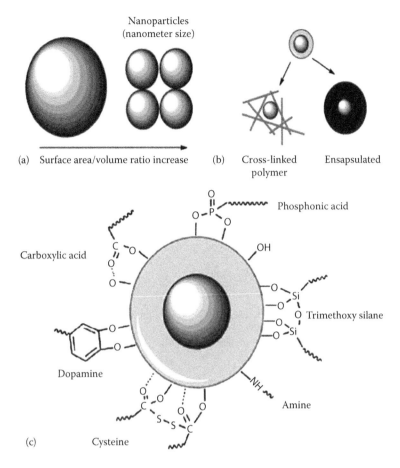

FIGURE 12.2 Properties of iron oxide magnetic nanoparticles: (a) high surface to volume ratios due to their nanosize, (b) versatility in synthesis and coating procedures with polymers, and (c) common chemical moieties for the anchoring of polymeric functional groups at the surface of iron oxide nanoparticles. (Reprinted from figures 1 and 3, *Biotechnol. Adv.*, 29, Dias, A.M.G.C. et al., A biotechnological perspective on the application of iron oxide magnetic colloids modified with polysaccharides.142–155, Copyright (2011), with permission from Elsevier.)

The magnetization process demands the presence of an extremely low concentration of dissolved oxygen that will oxidize Fe^{2+} to Fe$_3$O$_4$ without forming essentially nonmagnetic Fe(OH)$_3$(s). For a cation exchange resin with sulfonic acid functional groups, the primary steps of forming magnetite crystals within the polymer beads can be depicted as shown in Figure 12.4. The entire process consists of three major steps: (a) introducing/loading Fe^{2+} within the sorbent at slightly acidic pH; (b) slow oxidation of Fe^{2+} into crystalline magnetite (Fe$_3$O$_4$) at alkaline pH under a controlled redox environment; and (c) drying of resulting material. The acquired magnetic susceptibility decreases with a decrease in the acidity of the functional groups of the polymer. MAPAs are not permanent magnets, which is why they do not exhibit any clumping; that is, they are well dispersed in water. Since magnetic activity of MAPAs is derived solely from the fine magnetite crystals dispersed in the polymer phase, no loss of magnetite occurred even after 15 cycles (Cumbal et al. 2003).

A number of natural and synthetic adsorbents incorporating various functional groups (polyamine, carboxylate, sulfate, iminodiacetate, thiol, aminophosphonate, etc.) on the surface of Fe$_3$O$_4$ have been extensively explored, and are discussed extensively in the following sections.

O_2
H_2O
$Fe(OH)_{3\,(S)}$
Fe^{2+}
H_2O
H_2
$Fe_3O_{4\,(S)}$
$Fe(OH)_{2\,(S)}$
$Fe_{(S)}$
E_h
pH

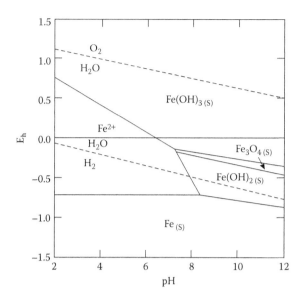

FIGURE 12.3 E–pH diagram for iron oxides highlighting a narrow predominance region for Fe_3O_4(s). (Reprinted from figure 2, *React. Funct. Polym.*, 54, Cumbal, L. et al., Polymer supported inorganic nanoparticles: Characterization and environmental applications, 167–180, Copyright (2003), With permission from Elsevier.)

Step I: Loading of Fe^{2+} at acidic pH
Step II: Magnetite formation mechanisms
Desorption
$3\,Fe^{2+} + 1/2\,O_2 + 3\,H_2O \longrightarrow Fe_3O_4(S) + 6H^+$
$Fe^{2+} + 2\,OH^- \longrightarrow Fe(OH)_2(S)$
$3\,Fe(OH)_2(S) + 1/2\,O_2 \longrightarrow Fe_3O_4(S) + 3\,H_2O$
Step III: Washing with an alcohol or a solvent with low dielectric constant

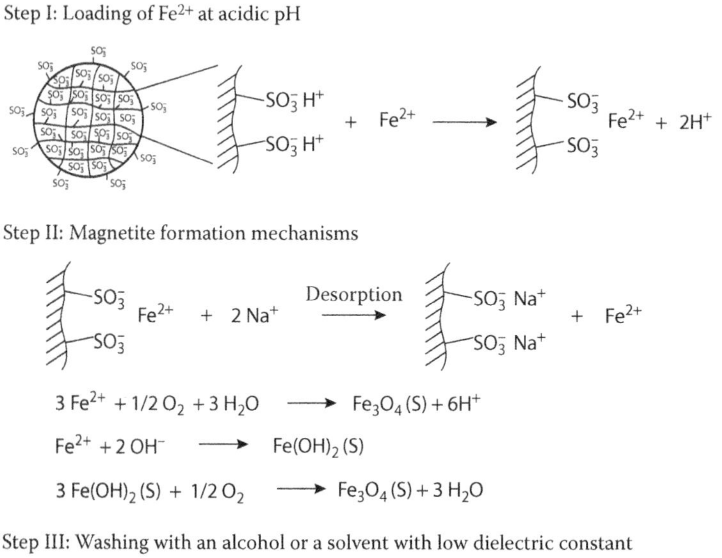

FIGURE 12.4 Illustration of stepwise procedure for magnetization of cation exchange resins. (Reprinted from figure 3, *React. Funct. Polym.*, 54, Cumbal, L. et al., Polymer supported inorganic nanoparticles: Characterization and environmental applications, 167–180, Copyright (2003), with permission from Elsevier.)

12.3.1 Natural Polymer-Fe_3O_4-Based Adsorbents

Among the various biomaterials available, the polysaccharides most commonly used for coating MNPs include agarose, alginate, carrageenan, chitosan, dextran, gum arabic, heparin, pullulan, and starch. Among them, cellulose (CL) and chitosan (CS) are well known for their biocompatibility, biodegradability, thermal, and chemical stability characteristics. CL is the most abundant natural renewable polymer, and beads, films, and resins based on CL have been widely used in adsorption of heavy metals. CS is reported to have the highest sorption capacity for Cu^{2+} ions, aided by plentiful NH$_2$ and OH groups in CS acting as the chelation sites (Lv et al. 2014; Wen et al. 2015). Its excellent adsorption characteristics can be attributed to (i) high hydrophilicity due to large number of OH groups; (ii) the presence of a large number of functional groups (acetamido, NH$_2$, OH) of high chemical reactivity; and (iii) the flexible structure of the polymer chain. However, weaker mechanical strength, poor chemical resistance, and high crystallinity are some of the disadvantages that have limited the usage of CS as an effective adsorbent.

12.3.1.1   Chitosan

Tran et al. (2010) prepared CS/magnetite composite beads by chemical coprecipitation of Fe^{2+} and Fe^{3+} ions by NaOH in the presence of CS, followed by hydrothermal treatment. Briefly, 0.5 g of CS was dissolved into 5 mL of CH$_3$COOH (99.5% wt, d = 1.05 g mL^{-1}) and 45 mL of distilled water (pH = 2–3) FeCl$_2$ and FeCl$_3$ were dissolved in 1:2 molar ratio and then the resulting solution was dropped slowly into NaOH (30% wt solution) to obtain CS/magnetite beads with different mass ratios of CS:magnetite (0:1 [pure Fe$_3$O$_4$], 1:2, 4:1). The suspension was kept at room temperature for 24 h without stirring and was separated by washing several times in water to remove alkalis. The particles were finally dried in vacuum at 70°C for 24 h to obtain CS/magnetite composite beads as an adsorbent. The CS on the surface of the magnetic nanoparticles is available for coordination with heavy metal ions, removing those ions with the assistance of external magnets. The synthetic procedure of composite beads is schematically presented in Figure 12.5. These data indicate the possible binding of iron ions to the NH$_2$ group of CS. In addition, electrostatic interactions occurred between surface negatively charged Fe$_3$O$_4$ and positively protonated CS. This means that the Fe$_3$O$_4$ was coated by the CS and no chemical bonding between the CS and Fe$_3$O$_4$ was formed.

Fan et al. (2011) investigated the performance of a cross-linked magnetic modified CS, coated with magnetic fluids and cross-linked with glutaraldehyde for the adsorption of Zn^{2+} from aqueous solution. A highly efficient adsorption equilibrium was achieved within 30 min and followed the Langmuir model. The maximum adsorption capacity for Zn^{2+} was estimated to be 32.16 mg g^{-1} with a Langmuir adsorption equilibrium constant of 0.01 L mg^{-1} at 298 K. Moreover, the adsorption rate was about 90% of the initial saturation adsorption capacity after being used five times.

Yuwei and Jianlong (2011) prepared magnetic CS nanoparticles by the two-step *in situ* coprecipitation method. The first step was to synthesize Fe$_3$O$_4$ particles and the second was to bind them with CS, using either the verse-phase suspension cross-linking method or the precipitation method. Compared with the former, the precipitation method is quite simple and easy to carry out, especially in posttreatment. Considering that CS can be precipitated under alkaline conditions, the formation of CS–magnetite nanoparticles can be achieved when the acidic solution containing CS, Fe^{2+}, and Fe^{3+} ions are added dropwise into NaOH under constant stirring. The particles (8–40 nm) were superparamagnetic with a saturation magnetization value of about 36 emu g^{-1}, which is higher than those of 17.6 and 16.3 emu g^{-1} obtained for other CS-based Fe$_3$O$_4$ beads (Ma et al. 2005; Ren et al. 2008). The maximum sorption capacity was calculated to be 35.5 mg g^{-1} using the Langmuir isotherm model, depending on the temperature. Acids such as HCl and HNO$_3$ may react with Fe$_3$O$_4$, and are therefore not used for the desorption process. Using ethylenediaminetetraacetic acid (EDTA) at a concentration of 0.02 M with a contact time of 3 h, it was possible to desorb 90% of Cu^{2+} ions adsorbed on magnetic CS nanoparticles into the solution. Increasing the EDTA concentration to 0.1 M resulted in 96% desorption of Cu^{2+}, and the adsorption capacity of the nanoparticles could still be maintained at a level greater than 90% after four cycles.

Yan et al. (2012) prepared MAPAs by coprecipitation of Fe$_3$O$_4$ nanoparticles and poly(acrylic acid) (PAA)-blended CS in alkaline solution. Figure 12.6 shows the saturation magnetization (σ_s) of CS and CS/PAA at about 14.2 and 13.0 emu g^{-1}, respectively. The slightly higher value of σ_s for CS may be due to the smaller size of Fe$_3$O$_4$ particles in CS/PAA bearing smaller magnetization. Moreover, the σ_s of the two kinds of CS-based MAPAs were both much lower than that of pure Fe$_3$O$_4$

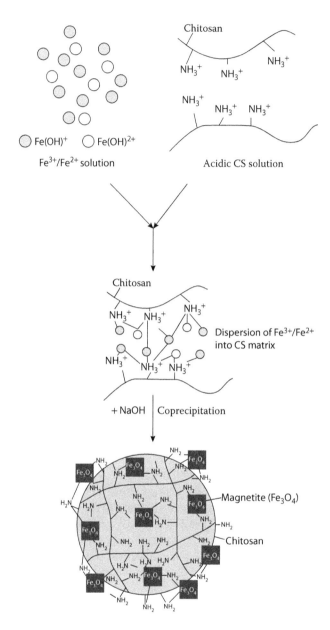

FIGURE 12.5 Schematic formation process of chitosan/magnetite composite beads. (Reprinted from figure 6, *Mater. Sci. Eng. C*, 30, Tran, H.V. et al., Preparation of chitosan/magnetite composite beads and their application for removal of Pb(II) and Ni(II) from aqueous solution 304–310, Copyright (2010), with permission from Elsevier.)

particles (62.8 emu g^{-1}). This trend is attributed to the decrease in Fe$_3$O$_4$ content in MAPAs and the diamagnetic shell of CS. Also, electron exchange between the ligands in the CS shell and surface atoms on the Fe$_3$O$_4$ core could suppress the magnetic moment. Hysteresis of all the investigated species was very weak and showed superparamagnetic properties, indicating that the Fe$_3$O$_4$ particles dispersed in the composites may be nanometer-scale clusters. The very weak hysteresis ensured that the MAPAs would not aggregate due to the remaining magnetism after magnetic separation. Despite a significant decrease of σ_s in comparison with pure Fe$_3$O$_4$, the micrometer-scale CS and CS/PAA particles could be easily separated from aqueous solution under an external magnetic field.

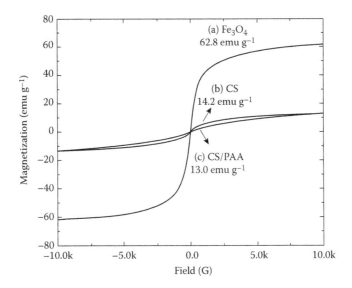

FIGURE 12.6 Magnetization curves of (a) Fe$_3$O$_4$, (b) CS-MAPA, and (c) CS/PAA-MAPA at RT. (Reprinted from figure 4, *J. Hazard. Mater.*, 229–230, Yan, H. et al., Preparation of chitosan/poly(acrylic acid) magnetic composite microspheres and applications in the removal of copper(II) ions from aqueous solutions, 371–380, Copyright (2012), with permission from Elsevier.)

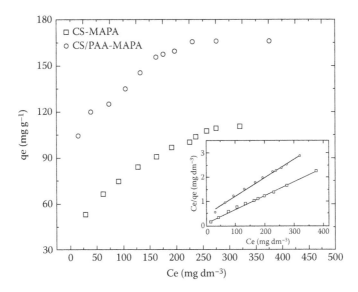

FIGURE 12.7 Adsorption isotherms of Cu^{2+} on CS-MAPA (□) and CS/PAA-MAPA (o) at 30°C and pH 5.5. The insets show the results of fit of the Langmuir isothermal model. (Reprinted from figure 7, *J. Hazard. Mater.*, 229–230, Yan, H. et al., Preparation of chitosan/poly(acrylic acid) magnetic composite microspheres and applications in the removal of copper(II) ions from aqueous solutions 371–380, Copyright (2012), with permission from Elsevier.)

Experimental results revealed that the CS/PAA had greater adsorption capacity than CS. Moreover, the adsorption isotherms were all well described by the Langmuir model (Figure 12.7), while the adsorption kinetics followed the pseudo-second-order equation. Cu^{2+} uptakes of both MAPAs increased linearly with the initial concentrations of Cu^{2+} increasing at the beginning, then reached surface saturation at high concentrations around 250.0 mg dm^{-3}. At lower initial concentrations

TABLE 12.1

Comparison of Maximum Adsorption Capacity of Cu²⁺ Ions Using Various Adsorbents

Adsorbents	Adsorbent Capacity (mg g⁻¹)
CS/PAA-magnetically active polymeric adsorbent (MAPA)	174.0
CS-MAPA	108.0
CS/PAA-GLA hydrogel beads	120.0
CS-GLA hydrogel beads	66.0
Surface carboxymethylated CS beads	130.0
Drying bed activated sludge	62.5
Manganese-coated activated carbon	39.5
Aspergillus terreus	15.2
Magnetic porous ferrospinel MnFe₂O₄	60.5
PAA hydrogels	330.0

Source: Extracted from table 1, *J. Hazard. Mater.* 229–230, Yan, H. et al., Preparation of chitosan/poly(acrylic acid) magnetic composite microspheres and applications in the removal of copper(II) ions from aqueous solutions, 371–380, Copyright (2012), with permission from Elsevier.

of Cu²⁺ ions, the adsorption sites on the adsorbents were sufficient and the Cu²⁺ uptakes relied on the amount of Cu²⁺ ions transported from the bulk solution to the adsorbent surface. However, the adsorption sites on the surface saturated at higher initial concentrations of Cu²⁺ ions. Adsorption equilibrium was achieved after 35 min for CS and 50 min for CS/PAA, respectively. The adsorbent could also easily be regenerated at lower pH levels and reused without almost any loss of adsorption capacity. In contrast, the Cu²⁺ ions loaded with CS and CS/PAA were sufficiently stable at pH levels of higher than 4.0, and both exhibited efficient phosphate removal with maximal uptakes of around 63.0 and 108.0 mgP g⁻¹, respectively. Table 12.1 shows the comparison of maximum adsorption capacity of Cu²⁺ ions using various adsorbents.

12.3.1.2 Cellulose–Chitosan

Peng et al. (2014) prepared cellulose–chitosan (CL–CS) nanoporous magnetic composite microspheres (NMCMs) by the sol-gel transition method using ionic liquids as solvent. These NMCMs have a high adsorption capacity for the effective removal of Cu²⁺ and have many advantages such as a large surface area, good mechanical strength and chemical stability, minimized secondary wastes, and easy recovery from aqueous solution using a magnet. Preparation of NMCMs and their application for Cu²⁺ adsorption is shown in Figure 12.8.

At first, Fe₃O₄ nanoparticles were synthesized by the chemical coprecipitation method under alkaline conditions. CL and CS (1:2) were dissolved in 1-butyl-3-methylimidazolium chloride (IL) at 100°C for 30 min to obtain a 7% wt solution. Magnetic fluid was immediately added to the solution under vigorous stirring for 15 min. An amount of 20 mL of the well-mixed solution was emulsified with a mixture solution containing vacuum pump oil (80 mL) and Tween80 (4 mL) under continuous stirring at 1000 rpm in an oil bath maintained at 100°C. A separate mixture solution containing vacuum pump oil (100 mL) and Tween80 (5 mL) was emulsified with 80 mL of sodium sulfate solution (0.4 mol L⁻¹), and this mixture was added to the previous CL–CS emulsion. Composite microspheres were obtained under slowly decreasing temperature and the final NMCMs (>95% yield) were washed three times with deionized (DI) water followed by ethanol, and stored in DI water for further use. Batch adsorption of Cu²⁺ on NMCMs was carried out as follows: 10 mL Cu²⁺ solutions and 50 mg NMCMs contained in a series of 25 mL conical flasks were shaken at room temperature (RT) for a period of time. The pH of the solution was adjusted with 1 mol L⁻¹ HCl solution. Initial and final concentration of Cu²⁺ after adsorption were analyzed using an atomic absorption spectrophotometer (AAS). The amount of adsorption (q) is calculated using the equation:

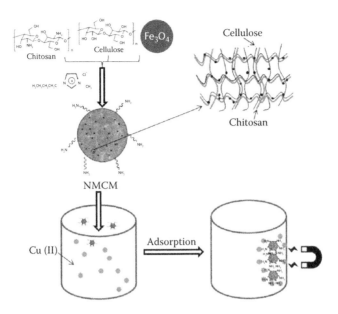

FIGURE 12.8   Schematic illustration of the preparation for NMCMs and application for Cu^{2+} adsorption. (Reprinted with permission from scheme 1, Peng, S. et al., 2014, Nanoporous magnetic cellulose–chitosan composite microspheres: Preparation, characterization, and application for Cu(II) adsorption, *Ind. Eng. Chem. Res.* 53, 2106–2113. Copyright (2014) American Chemical Society.)

$$q = \left(C_0 - C_e\right)\frac{V}{M} \tag{12.2}$$

where:

q is the amount of Cu^{2+} adsorbed onto the bioadsorbents (mg g^{-1})
C_0 and C_e are the initial and equilibrium concentrations of Cu^{2+} (mg L^{-1}), respectively
V is the volume of Cu^{2+} solution (L)
M is the weight of the bioadsorbents (g)

The composite microspheres exhibited efficient adsorption capacity of Cu^{2+} from aqueous solution due to favorable chelating groups in their structure. The adsorption process was best described by a pseudo-second-order kinetic model. Compared with the Freundlich model, the adsorption of Cu^{2+} on NMCMs is better described using the Langmuir equation. NMCMs showed faster Cu^{2+} removal (61% removal efficiency) within 1.5 h. The maximum adsorption capacity for Cu^{2+} reached 65.8 mg g^{-1} within 20 h (initial Cu^{2+} conc. 150 mg L^{-1}). Common coexisting ions almost had no negative effect on the Cu^{2+} adsorption of NMCMs. Solution with pH 1 can efficiently remove the adsorbed Cu^{2+}, which suggest that the NMCMs exhibit poor adsorption capacity in an acidic environment (Figure 12.9a).

For desorption studies, the bioadsorbents loaded with Cu^{2+} were placed in HCl with varying pH values and shaken at RT for 24 h. Unabsorbed Cu^{2+} after magnetic separation was removed by gentle washing with DI water. Desorption ratio (D_s) was defined as the following equation:

$$D_S = \frac{C_e'}{\left(C_0 - C_e\right)}100\% \tag{12.3}$$

where:

C_0 and C_e are the initial and equilibrium concentrations of Cu^{2+} (mg L^{-1})
C_e' is the equilibrium concentration of Cu^{2+} (mg L^{-1}) in the elution medium

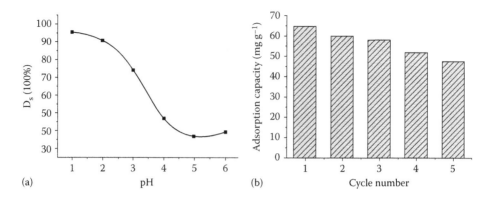

FIGURE 12.9 (a) Desorption of Cu^{2+} by HCl and (b) effect of recycling NMCMs on Cu^{2+} adsorption. (Reprinted with permission from figures 10 and 11 from Peng, S. et al., 2014, Nanoporous magnetic cellulose–chitosan composite microspheres: Preparation, characterization, and application for Cu(II) adsorption, *Ind. Eng. Chem. Res.* 53, 2106–2113. Copyright (2014) American Chemical Society.)

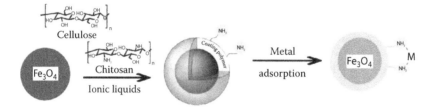

FIGURE 12.10 (See color insert) Preparation of magnetic CL-CS hydrogels and adsorption of heavy metals. (Liu, Z. et al., 2012, Magnetic cellulose–chitosan hydrogels prepared from ionic liquids as reusable adsorbent for removal of heavy metal ions, *Chem. Commun.* 48, 7350–7352, figure 1. Reproduced by permission of The Royal Society of Chemistry.)

The adsorption capacities for Cu^{2+} did not show any decrease even after five cycles (Figure 12.9b).

Liu et al. (2012) prepared magnetic CL–CS hydrogels in which Fe_3O_4 particles were successfully coated by CL and CS in ILs (Figure 12.10). CL is used as a blending polymer for CS, and could provide strong mechanical strength and improve the chemical stabilities of the hydrogel beads under acidic conditions.

12.3.1.3 Cyclodextrin–Chitosan

Renewable and biodegradable cyclodextrins (CDs) are cyclic oligosaccharides consisting of six (α), seven (β), or eight (γ) glucopyranose units linked together via α (1–4) linkages. Their torus-shaped ring structure contains an apolar cavity with $R-NH_2$ groups lying on the outside and secondary hydroxyl groups in the inside. CDs have the ability to complex with various metal ions; their performance can be further improved by modifying them with suitable functional groups through esterification, oxidation reactions, and cross-linking of OH groups outside the interior cavity (Norkus 2009). Complexation of β-cyclodextrin (CD)/metal has been used in the removal of heavy metals (Mahlambi et al. 2010). In a recent study, Badruddoza et al. (2013) synthesized a carboxymethyl-β-cyclodextrin (CM-β-CD) polymer by simple coprecipitation and grafted it onto Fe_3O_4 nanoparticle surfaces (Figure 12.11). The polymer-grafted magnetic nanoparticles (CDpoly-MNPs) were used for the selective removal of Pb^{2+}, Cd^{2+}, Ni^{2+} ions from aqueous solution. Batch adsorption equilibrium was reached after 45 min, and maximum uptakes at 25°C for Pb^{2+}, Cd^{2+}, and Ni^{2+} in noncompetitive adsorption mode were 64.5, 27.7, and 13.2 mg g^{-1}, respectively. Adsorption capacity in polymer-grafted MNPs is enhanced because of the complexing abilities of the multiple −OH and

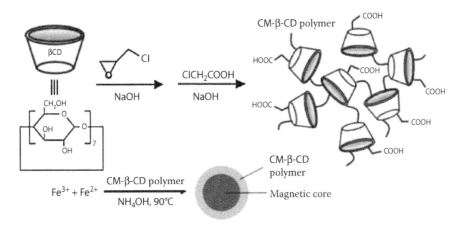

FIGURE 12.11   Grafting mechanism of CM-β-CD polymer on Fe$_3$O$_4$ NPs. (Reprinted from scheme 1, *Carbohydr. Polym.*, 91, Badruddoza, A.Z.M. et al., Fe$_3$O$_4$/cyclodextrin polymer nanocomposites for selective heavy metals removal from industrial wastewater, 322–332, Copyright (2013), with permission from Elsevier.)

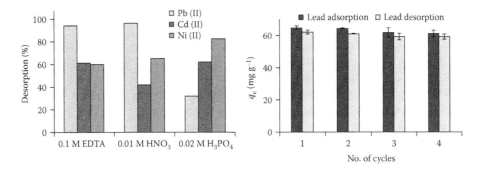

FIGURE 12.12   **(See color insert)** (a) Percentage recovery of Pb^{2+}, Cd^{2+}, and Ni^{2+} from CDpoly-MNPs using different desorption eluents, and (b) four consecutive adsorption–desorption cycles of CDpoly-MNP adsorbent for Pb^{2+} (initial concentration: 300 mg L^{-1}; pH: 5.5; desorption agent: 10 mL of 0.01 M HNO$_3$). (Reprinted from figure 4, *Carbohydr. Polym.*, 91, Badruddoza, A.Z.M. et al., Fe$_3$O$_4$/cyclodextrin polymer nanocomposites for selective heavy metals removal from industrial wastewater, 322–332, Copyright (2013), with permission from Elsevier.)

−COO groups in the polymer backbone with metal ions. In competitive adsorption experiments, CDpoly-MNPs can adsorb with an affinity order of Pb^{2+} ≫ Cd^{2+} > Ni^{2+}, which can be explained by hard and soft acids and bases (HSAB) theory. The results of the desorption of Pb^{2+}, Cd^{2+}, and Ni^{2+} using the three buffers (0.01 M nitric acid, 0.1 M Na$_2$EDTA, and 0.02 M phosphoric acid) are summarized in Figure 12.12a. HNO$_3$ and Na$_2$EDTA solutions show excellent desorption efficiency for Pb^{2+} (96.0% and 94.2% recovery), whereas H$_3$PO$_4$ is a better eluent for Cd^{2+} and Ni^{2+} desorption, with recovery percentages of 61.8% and 82.7%, respectively. The bonding between the active sites of magnetic nanoadsorbent and metal ion is not sufficiently strong to be held in acidic conditions. Under acidic conditions, H$^+$ ions protonate the adsorbent surface; that is, the carboxyl group regeneration is more favorable, thereby reflecting the metal ions from the adsorbent surface leading to desorption of positively charged metal ions. Moreover, Na$_2$EDTA solution can desorb Pb^{2+} ions more efficiently from the surface than Cd^{2+} or Ni^{2+} ions. When Na$_2$EDTA solution was added to the metal adsorbed-CDpoly-MNPs, stronger coordination ligands in the Na$_2$EDTA would form a stronger bonding with Pb^{2+} ions, making the metal ions more easily desorb from the CDpoly-MNPs (Yu et al. 2011). Following the adsorption–desorption process, reusability was checked for four cycles for Pb^{2+} ions (Figure 12.12b). The CDpoly-MNPs adsorbent retained its adsorption capability after repeated adsorption–regeneration cycles with negligible changes, indicating that there are almost no irreversible sites on the surface of CDpoly-MNPs. Hence, these nanoadsorbents can be repeatedly used as efficient adsorbents in wastewater treatment.

12.3.1.4 Carbon Disulfide–Chitosan

CS is soluble in acidic-to-neutral solution and, hence, its chemical stability is reinforced using cross-linking agents such as glutaraldehyde (GLA), epichlorohydrin, tripolyphosphate, ethylene glycol diglycidyl ether, and so on. In the case of GLA, the reaction (Schiff's base reaction) occurs between aldehyde groups of GLA and some amine groups of CS. Cross-linking may reduce adsorption capacity, but the chemical modification introduces a variety of functional groups which can enhance sorption performance. For example, grafting of sulfur compounds can create new chelating groups on the CS backbone. Xu et al. (2015) prepared novel modified-CS encapsulated magnetic Fe_3O_4 nanoparticles [carbon disulfide modified chitosan (CMCS)-Fe_3O_4] for the removal of Cd^{2+} from aqueous solution using the batch method. A ferromagnetic nature is indicated from the magnetic hysteresis loop of CMCS-Fe_3O_4 at RT (Figure 12.13a), its high coercive field (H_c) of 100 Oe, and remnant magnetization (M_r) of 5 emu g^{-1}. The separation test by a magnet confirmed that CMCS-Fe_3O_4 can be easily separated, despite its low M_s. CMCS-Fe_3O_4 showed superior adsorption capacity for Cd^{2+} (Figure 12.13b). The adsorption capacity increased with the equilibrium concentration of Cd^{2+} in solution, progressively saturating the adsorbent for the CMCS-Fe_3O_4. The maximum adsorption capacity for Cd^{2+} was 200 mg g^{-1} from the Langmuir isotherm. The adsorption kinetics followed the pseudo-second-order, and the equilibrium data were well described by the Langmuir isotherm with a high correlation coefficient ($R^2 = 0.9999$) over the Freundlich isotherm ($R^2 = 0.8812$). From Figure 12.13c, the experimental data of Cd^{2+} were in good agreement with the Langmuir mode, suggesting a monolayer adsorption. The common coexisting ions had a negligible impact on Cd^{2+} adsorption. Based on Fourier transform infrared (FTIR) spectral analysis, the –NH_2 and –SH groups on the CMCS-Fe_3O_4 participated mainly in the Cd^{2+} adsorption process. It can be concluded that the adsorption sites for Cd^{2+} are the nitrogen atoms of the amino group in CS and the sulfur atoms of the attached mercapto group. As may be seen in Figure 12.13d, using selected eluents to desorb Cd^{2+} from CMCS-Fe_3O_4 (at 25°C, 0.1 mol L^{-1}), the regeneration efficiency for Cd^{2+} was below 50%, which suggested that the adsorbed Cd^{2+} remained almost stable on the CMCS-Fe_3O_4, and that chemisorption may be the major mode of Cd^{2+} removal by the adsorbent.

12.3.1.5 Polyrhodanine

Supported by the HSAB theory introduced by Pearson (1963), polyrhodanine (PR), containing oxygen, nitrogen, and sulfur atoms in its monomeric structure, can serve as an efficient adsorbent of heavy metal ions (Rastegarzadeh 2010). In a recent study, Song et al. (2011b) synthesized PR-coated γ-Fe_2O_3 nanoparticles by one-step chemical-oxidation polymerization (Figure 12.14) and used them in the removal of heavy metal ions from aqueous solution. Rhodanine molecules coordinated with Fe ions when $FeCl_3$ was added into rhodanine aqueous solution. After $NaBH_4$ was injected, MNPs were formed instantaneously, with coordinated rhodanine monomers acting as a stabilizer. As the reaction proceeded, Fe ions were redissolved from the surface of the MNPs. Chemical-oxidation polymerization was initiated when Fe ions induced the oxidation of rhodanine monomers. The Fe ions accept electrons from the rhodanine monomers, which triggers rhodamine polymerization to form PR-MNPs.

Figure 12.15 a displays the time dependence of heavy metal ions' (Hg^{2+}, Cd^{2+}, Mn^{2+}, and Cr^{3+}) adsorption capacity onto PR-MNPs. A state of equilibrium was reached within 2 h for all four metal ions, in the order $Hg^{2+} \gg Cr^{3+} > Cd^{2+} \sim Mn^{2+}$ (Figure 12.15b). Sulfur groups of PR are classified as Lewis soft bases and have strong affinity toward Lewis soft acids (Hg^{2+} and Cd^{2+} ions). In contrast, the PR shell of the fabricated nanoparticles also contains oxygen and quaternary nitrogen groups (Lewis hard bases), which have affinity to Lewis hard acids (Mn^{2+} and Cr^{3+} ions). The different preferred binding sites of the four metal ions caused different uptake capabilities onto

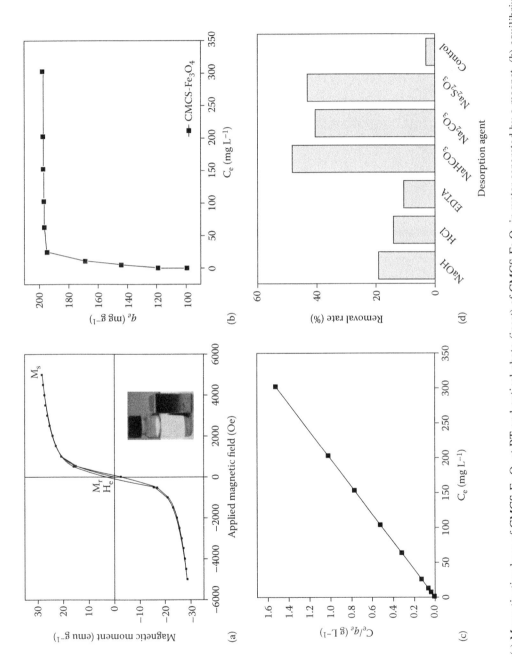

FIGURE 12.13 (a) Magnetization loop of CMCS-Fe₃O₄ at RT and optical photo (inset) of CMCS-Fe₃O₄ in water separated by a magnet, (b) equilibrium adsorption isotherm of Cd²⁺ by CMCS-Fe₃O₄, (c) curve of fit of the Langmuir isotherm model for Cd²⁺ adsorption on CMCS-Fe₃O₄, and (d) reuse of CMCS-Fe₃O₄ regenerated by different eluents. (Reprinted from figures 2, 5, and 8, Xu, L. et al., 2015, *Desalin. Water Treat.* 57:8540–8548, 2015. With permission from Taylor & Francis Ltd.)

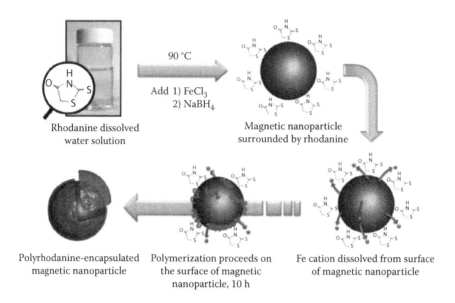

FIGURE 12.14 **(See color insert)** Fabrication process of polyrhodanine-encapsulated MNPs. (Reprinted from figure 1, *J. Colloid Interface Sci.*, 359, Song, J. et al., Adsorption of heavy metal ions from aqueous solution by polyrhodanine-encapsulated magnetic nanoparticles, 505–511, Copyright (2011), with permission from Elsevier.)

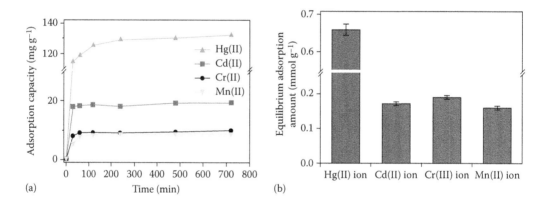

FIGURE 12.15 (a) Time-dependence of adsorption capacity of Hg^{2+}, Cd^{2+}, Cr^{3+}, and Mn^{2+} ions onto PR-MNPs, and (b) equilibrium adsorption amount of heavy metal ions (pH value: 4.0 at 25°C; initial concentration of each metal ion: 80 mg/L). (Reprinted from figure 6, *J. Colloid Interface Sci.*, 359, Song, J. et al., Adsorption of heavy metal ions from aqueous solution by polyrhodanine-encapsulated magnetic nanoparticles, 505–511, Copyright (2011), with permission from Elsevier.)

the PR-MNPs. The disparity in metal ion radii, interaction energies, and oxidation states of the heavy metal ions could also affect the different adsorption capabilities. An adsorption equilibrium study of PR-coated MNPs was carried out using Hg^{2+} ions, which followed a Freundlich isotherm model, and their kinetic data were best described by a pseudo-second-order equation, indicating their chemical adsorption. After the adsorption test, PR-MNPs were harvested using an external magnetic field, and sequentially treated with 0.1 M HCl solution and DI water to regenerate their metal ion binding ability. The variation of adsorption efficiency with respect to the cycle number is displayed in Figure 12.16b. The initial Hg^{2+} ion concentration was 80 mg L^{-1} and the pH value was 6.0. PR-MNPs (5 mg) were placed in contact with Hg^{2+} ions for 4 h at 25°C. Even after five cycles, the metal ion binding capability of the PR-MNPs remained >96%.

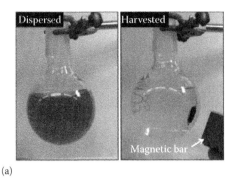
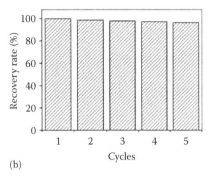

FIGURE 12.16   (a) Dispersed and harvested PR-MNPs, and (b) adsorption efficiency of Hg^{2+} in adsorption–desorption cycles by the PR-MNPs. (Reprinted from figure 8, *J. Colloid Interface Sci.*, 359, Song, J. et al., Adsorption of heavy metal ions from aqueous solution by polyrhodanine-encapsulated magnetic nanoparticles, 505–511, Copyright (2011), with permission from Elsevier.)

12.3.2   AMINE-FE₃O₄-BASED ADSORBENTS

The exceptional binding ability and multiple chelation sites in multiamine grafted adsorbents have attracted considerable interest regarding the efficient removal of trace heavy metals from wastewater. Wang et al. (2015) synthesized core–shell magnetic Fe_3O_4-poly(m-phenylenediamine) particles for chromium reduction and adsorption. According to HSAB theory, the N atom, which is a middle partial hard base, favors the formation of inner-sphere surface complexes with intermediate partial hard acids, therefore contributing to the unique selectivity of multiamine functionalized adsorbents. In saline systems, multiamine adsorbents will preferentially bind with heavy metal ions (partial hard acids), rather than Na^+, K^+, Ca^{2+}, and Mg^{2+} ions (hard acids). As a result, multiamine adsorbents are capable of removing and exclusively recovering trace heavy metals from saline wastewaters or other complex wastewaters. Fe_3O_4 magnetic nanoparticles modified with γ-aminopropyltrimethoxysilane (APTMS) and either the copolymers of acrylic acid and crotonic acid (Ge et al. 2012), polystyrene bead-supported nanoiron oxides (Qui et al. 2012), carboxymethyl-β-CD conjugated Fe_3O_4 nanoparticles (Badruddoza et al. 2011, 2013), monodisperse Fe_3O_4-silica core–shell microspheres (Hu et al. 2010), ascorbic acid-coated Fe_3O_4 nanoparticles (Feng et al. 2012), or magnetic Fe_2O_3-loaded carbonaceous adsorbents (Ohno et al. 2011) could also be used as good adsorbents. Functionalized materials based on polyaniline and amine derivatives have also been developed as absorbents (Wang et al. 2009; Kumar and Chakraborty 2009; McManamon et al. 2012).

12.3.2.1   Tetraethylenepentamine and Its Related Amines

Zhao et al. (2010a) prepared four types of NH_2-functionalized NMPs coupled with different diamino groups, that is to say, tetraethylenepentamine (TEPA), ethylenediamine (EDA), diethylenetriamine (DETA), and triethylentetramine (TETA) named as TEPA-NMPs, EDA-NMPs, DETA-NMPs, and TETA-NMPs, respectively (Figure 12.17). Batch adsorption studies were carried out to optimize adsorption conditions, and were found to be highly pH dependent. Adsorption of Cr^{6+} reached equilibrium within 30 min. The data of adsorption kinetics obeyed pseudo-second-order rate mechanism. The adsorption data for Cr^{6+} onto NH_2-NMPs were well fitted to the Langmuir isotherm. The q_m values of NH_2-NMPs for Cr^{6+} were 370.4, 137.0, 149.3, and 204.1 mg g⁻¹ for TEPA-NMPs, EDA-NMPs, DETA-NMPs, and TETA-NMPs, respectively. Thermodynamic parameters such as ΔH^0, ΔS^0, and ΔG^0 suggested that the adsorption processes of Cr^{6+} onto the NH_2-NMPs were endothermic and entropy favored in nature. The adsorption of Cr^{6+} onto the NH_2-NMPs could be related to electrostatic attraction, ion exchange, and coordination interactions. If only electrostatic attraction and ion exchange were used, removal efficiency should decrease gradually as the pH value increases. In the case of coordination interaction, it compensates for the loss of removal efficiency

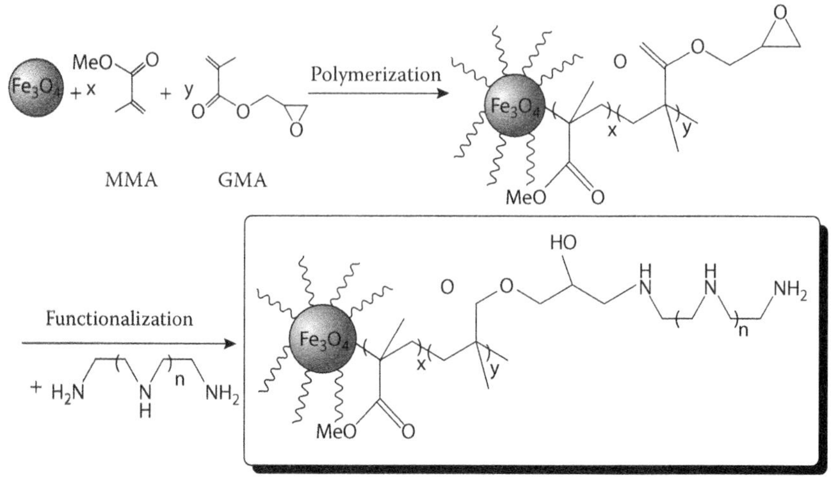

(n = 0, EDA-NMPs; n = 1, DETA-NMPs; n = 2, TETA-NMPs; n = 3, TEPA-NMPs)

FIGURE 12.17 Binding and amino-functionalization procedures of NH_2-NMP nanoparticles. (With kind permission from Springer Science+Business Media: *J. Mater. Sci.*, Preparation and characterization of amino-functionalized nano-Fe_3O_4 magnetic polymer adsorbents for removal of chromium(VI) ions, 45, 2010, 5291–5301, Zhao, Y.-G. et al., figure 1, Copyright 2010.)

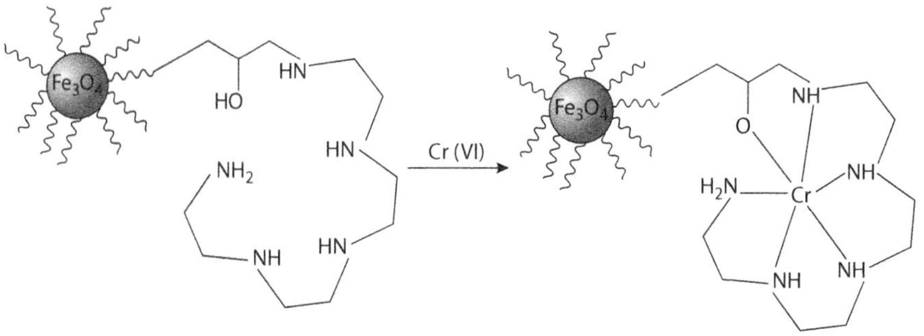

FIGURE 12.18 Presumed coordination interaction process for the removal of Cr^{6+} by TEPA-NMPs. (With kind permission from Springer Science+Business Media: *J. Mater. Sci.*, Preparation and characterization of amino-functionalized nano-Fe_3O_4 magnetic polymer adsorbents for removal of chromium(VI) ions, 45, 2010, 5291–5301, Zhao, Y.-G. et al., figure 1, Copyright 2010.)

resulting from electrostatic attraction and ion exchange with the increasing of the pH value. The presumed process of the coordination interactions for the removal of Cr^{6+} by TEPA-NMPs is illustrated in Figure 12.18.

In earlier studies, the amino-functionalized mesostructured silica containing magnetite (Kim et al. 2003) showed a magnetization value of 10.9 emu g^{-1} and an adsorption capacity of 32 mg g^{-1} for Cu^{2+}. The adsorption capacity of NH_2-functionalized NMPs (Huang and Chen 2009) reached 12.4 mg g^{-1} for Cu^{2+} and 11.2 mg g^{-1} for Cr^{6+}. Hao et al. (2010) developed NMPs by the covalent binding of 1,6-hexanediamine on the surface of Fe_3O_4 nanoparticles for the removal of Cu^{2+} ions from aqueous solution, and the adsorption capacity was 26 mg g^{-1}. More recently, Zhang et al. (2012) prepared Fe_3O_4–SiO_2-poly(1,2-diaminobenzene) submicron particles (FSPs) with a core–shell structure and high saturated magnetization (~60–70 emu g^{-1}) for the removal of As^{3+}, Cu^{2+}, and Cr^{3+} ions from aqueous solution. The sequence of adsorption capacities using the FSPs were As^{3+} (84 ± 5 mg g^{-1}, pH = 6.0) > Cr^{3+} (77 ± 3 mg g^{-1}, pH = 5.3) > Cu^{2+} (65 ± 3 mg g^{-1}, pH = 6.0). The adsorption isotherm fitted the Freundlich model and the adsorption kinetic agreed well with the two-site pseudo-second-order model, which indicated that multilayer adsorption of As^{3+}, Cu^{2+}, and Cr^{3+} ions on FSPs occurred at two sites with different energies of adsorption. The chelating interaction was considered as the main adsorption mechanism. Compared with Fe$_3$O$_4$ particles, the FSPs revealed better stability in acidic as well as basic water systems. The as-prepared materials were chemically stable, with low leaching of Fe (61.7% wt%) and poly(1,2-diaminobenzene) (64.9% wt) in tap water, seawater, and acidic/basic solutions. These metal-loaded FSPs could be easily recovered from aqueous solution using a permanent magnet within 20 s. They could also be easily regenerated with acid. They were also effective in treating metallurgical refinery wastewater. To investigate competitive adsorption, the removal ratio of As^{3+}, Cu^{2+}, and Cr^{3+} ions using FSPs was tested in metallurgical refinery wastewater and river water. In the case of refinery wastewater, with large amounts of coexisting metal ions (Pb^{2+}, Hg^{2+}, Cd^{2+}, Zn^{2+}, Cu^{2+}, Fe^{3+}, pH = 1.1), the removal ratios of As^{3+}, Cu^{2+}, and Cr^{3+} were 97.6%, 98.1%, and 89.7%, respectively. When treating river water (pH = 6.5), the removal ratios of the ions tested were lower than those when treating wastewater. This trend is ascribed to the higher initial concentrations and smaller amounts of adsorbent dosage. A decrease in total organic carbon (TOC) from 2.1 to 1.3 mg L^{-1} was observed after the addition of FSPs into the river water sample, due to the presence of organic matter (e.g., humic acid) in water which could be easily adsorbed onto the small adsorbents. The surface-adsorbed organic matter may induce more functional groups. The metal ions bound strongly to these functional groups, leading to the formation of surface complexes (surface site–metal ion–organic ligand and surface site–organic ligand–metal ion). The saturated magnetization of the as-prepared FSPs was ~60 emu g^{-1}, which was higher than that of ethylenediamine-functionalized magnetic polymers (12.3 emu g^{-1}) (Zhao et al. 2010b), Fe$_3$O$_4$-SiO$_2$-NH$_2$ nanoparticles (36 emu g^{-1}) (Wang et al. 2010), SH-mSi-Fe$_3$O$_4$ (38.4 emu g^{-1}) (Li et al. 2011), and OMS-2/Fe$_3$O$_4$ nanowires (11.3 emu g^{-1}) (Zhang et al. 2011), suggesting that it could be more easily collected from aqueous solution. The magnetic separation could be realized within 20 s. After adsorption, the used FSPs can be washed and reclaimed in 2 M acid, and the adsorption behavior remained relatively stable after five cycles.

Zhang et al. (2015) prepared a series of TEPA-functionalized polymethacrylate–divinylbenzene microbeads as adsorbent for the enhanced removal and selective recovery of Cu^{2+} and Ni^{2+} from a saline solution. The maximum adsorption capacity of 1.21 mmol g^{-1} for Cu^{2+} is superior to commercial multiamine resins such as Purolite S984. Earlier, Shen et al. (2013) studied TEPA-functionalized NMP adsorbents for the removal of Cr^{6+} from aqueous solution in the presence of Cu^{2+}, Ni^{2+}, and Zn^{2+} ions. The maximum adsorption capacities (q_m) of Cr^{6+} were found to be 263, 333, and 354 mg g^{-1} (at 35°C, pH 2.0) when 50 mg L^{-1} of Cu^{2+}, Ni^{2+}, and Zn^{2+}, respectively, coexisted with Cr^{6+} in solution, compared with a capacity of 370.4 mg g^{-1} achieved when only Cr^{6+} was present in the solution. The suppression effect tendency of coexisting ions on the adsorption of Cr^{6+} on TEPA-NMPs was in the order Cu^{2+} > Ni^{2+} > Zn^{2+}. Experimental data on adsorption from a liquid phase fitted the Langmuir equation well with R^2 = 0.9991, while the fit was unsatisfied with the Freundlich model (R^2 = 0.9162) and D-R mode (R^2 = 0.9391). This result showed that the adsorption of Cr^{6+} ions occurs on a homogeneous surface via monolayer adsorption with a q_m of Cr^{6+} at 370.4 mg g^{-1}.

12.3.2.2 APTEs

Wang et al. (2010) prepared amino-functionalized Fe$_3$O$_4$-SiO$_2$ core–shell MNPs using γ-aminopropyltriethoxysilane (APTES) as the silylation agent, which exhibited a magnetization value of 34 emu g^{-1} and higher adsorption affinity for aqueous Cu^{2+}, Pb^{2+}, and Cd^{2+} at ~44.8, ~103.2, and ~39.3 mg g^{-1}, respectively. Ge et al. (2012) evaluated the pH effect (1.0–8.0) on the adsorption of metal ions using a Fe$_3$O$_4$-APTES-acrylic acid (AA)-co-crotonic acid (CA) system. The removal efficiency increased when the pH value was in the range 1.0–4.0, but changed little when pH > 4.0. The pH value of the zero point charge (pH$_{pzc}$) was between 3.0 and 4.0. At low pH levels (pH < pH$_{pzc}$), the surface of the adsorbents presented in carboxyl form and had less adsorption capacity. As the alkalinity of solution increased, the carboxyls turned into carboxylate anions and

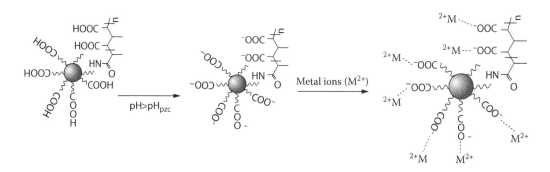

FIGURE 12.19 Possible mechanism for metal ion adsorption by Fe₃O₄-APTES-AA-co-CA. (Reprinted from figure 5, *J. Hazard. Mater.*, 211, Ge, F. et al., Effective removal of heavy metal ions Cd²⁺, Zn²⁺, Pb²⁺, Cu²⁺ from aqueous solution by polymer-modified magnetic nanoparticles. 366–372, Copyright (2012), with permission from Elsevier.)

the adsorption capacity increased gradually until pH > pH$_{pzc}$. After that, the carboxyls completely turned into carboxylate anions with almost no change in adsorption. Considering that the metal ions might precipitate as hydroxide, for example, Pb²⁺ will first precipitate at pH > 5.5 based on the solubility product constant (Ksp) of metal ions and OH⁻. The probable adsorption mechanism is shown in Figure 12.19. The metal ions mainly interacted with the adsorbents by chelation between the ions and the carboxylate anions.

Masoumi et al. (2014) synthesized poly[methyl methacrylate (MMA)-co-maleic anhydride (MA)] by radical polymerization followed by chemical immobilization of APTMS-Fe₃O₄ with the anhydride groups of poly(MMA-co-MA) to form a magnetic nanocomposite (MNC) with the cross-linked shell layer. From adsorption kinetics, the pseudo-second-order rate law, indicating chemisorption, was identified as the rate-limiting step, and agreed well with the Langmuir adsorption model. The maximum adsorption capacities were found to be 90.1, 90.9, 109.9, and 111.1 mg g⁻¹ for Co²⁺, Cr³⁺, Zn²⁺, and Cd²⁺, respectively. MNC synthesis and its adsorption strategies are shown in Figure 12.20. Magnetic Fe₃O₄ nanoparticles were prepared via the chemical coprecipitation method and subsequently modified with APTMS through a silanization reaction with −OH groups present on the NP surface, leading to the formation of amine-terminated APTMS-Fe₃O₄ nanoparticles.

The adsorption behavior was verified under varying pH solutions (2.0–8.0) using HNO₃ or NaOH, adsorbent dosage of 5–30 mg, contact time of 0–100 min, and initial metal ion concentration of 20–100 mg L⁻¹. Adsorption reached the highest level at pH 6.0. At pH 2.0, the adsorbent showed higher absorption for Cr³⁺ ions (51.1%) than for Cd²⁺ ions (33.5%). The relatively moderate adsorption of metal ions at a lower pH level is attributed to the carboxylic groups in the MNC, which act as chelating sites for the cations. The dependence of the metal binding capacity of the amine groups on the concentration of the H⁺ ions in the solution is explained as follows: At lower pH values, the concentration of the H⁺ ions is much higher than of the metal ions and, hence, the H⁺ ions and metal ions compete for the adsorbent surface. As a result, the metal ions will be hindered from reaching the adsorbent binding sites because of repulsive forces arising from the H⁺ ions. With increasing pH levels, more nonprotonated amines are available to bind with metal ions. At higher pH levels, chemical precipitation is possible due to the higher concentration of OH⁻ ions. Maximum removal efficiencies of Cd²⁺ (47.4%), Zn²⁺ (67.5%), Cr³⁺ (76.7%), and Co²⁺ (81.8%) were observed at a pH level of 6 and an adsorbent dosage of 20 mg L⁻¹. Further increases in adsorbent dosage had a lesser effect on metal removal. The increase in the adsorbent dose from 5 to 20 mg resulted in an increase of adsorption efficiency of 10% for Cd²⁺ (37.4%–47.4%), 15% for Co²⁺ (66.8%–81.8%), 18% for Cr³⁺ (58.7%–76.7%), and 19% for Zn²⁺ (48.5%–67.5%). The adsorption efficiency increased considerably with an increase in contact time up to 60 min, and remained constant thereafter. The q_e value reached equilibrium values of 11.8, 16.9, 19.2, and 20.4 mg g⁻¹ for Cd²⁺, Zn²⁺, Cr³⁺, and Co²⁺,

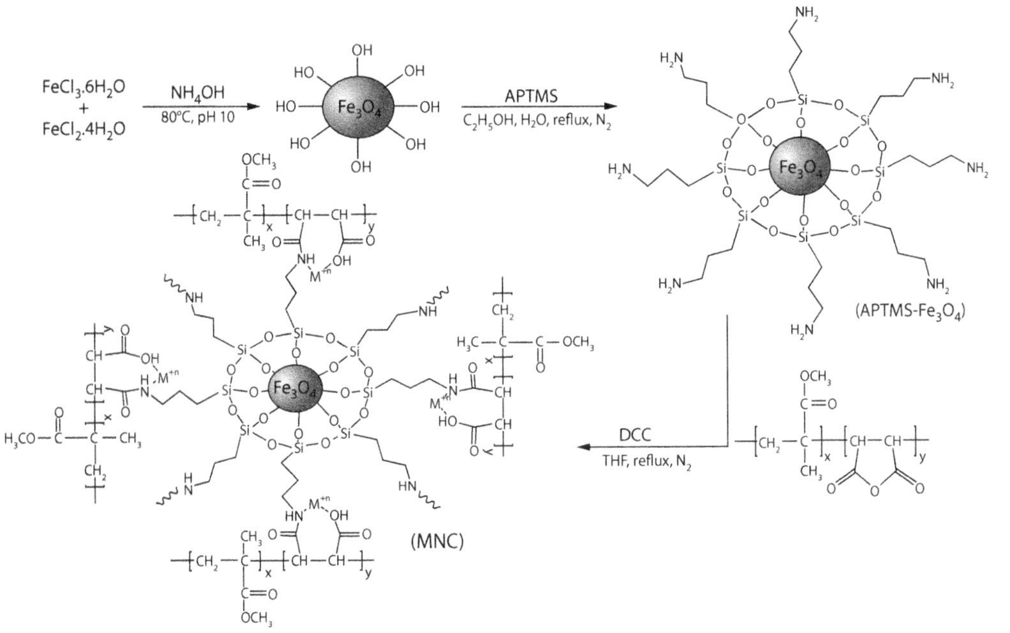

FIGURE 12.20 Synthetic procedures of poly(MMA-co-MA)/APTMS-Fe$_3$O$_4$ MNC. (Reprinted with permission from figure 1, Masoumi, A. et al., 2014, Removal of metal ions from water using poly(MMA-co-MA)/modified-Fe$_3$O$_4$ magnetic nanocomposite: Isotherm and kinetic study, *Ind. Eng. Chem. Res.* 53, 8188–8197. Copyright (2014) American Chemical Society.)

respectively. Preferential adsorption of Co^{2+} and Cr^{3+} ions by the MNC adsorbent was observed, while the Cd^{2+} ions showed the lowest adsorption. This is the opposite of the order of their ionic radius, that is, Cd^{2+} > Zn^{2+} > Co^{2+} ≥ Cr^{3+}. The extraction percentage of the largest metal ions (Cd^{2+}) was 47.4% and for the smallest metal ions (Co^{2+}), this increased to 81.8%. Due to the cross-linking of the shell layer, penetration of the largest ion (Cd^{2+}) into the cross-linked matrix becomes much more difficult and the process becomes diffusion controlled.

In the absence of an external magnetic field, MNC formed highly stable homogeneous dispersions in water, aided by the existence of H-bonding between the –NH$_2$ and –OH groups of MNC and water. However, the MNC was collected easily by applying an external magnetic field (Figure 12.21). Compared with pure Fe$_3$O$_4$, the magnetic strength of MNC probably decreased because of the surface coating, but still, its magnetic strength was sufficient for separation using a conventional magnet. From Table 12.2, the adsorbent capacity values of the MNC (obtained by the Langmuir equation) was much higher than almost all of the reported adsorbents, which indicates that the polyacrylate-APTES-based MNC may be a good candidate for heavy metal ion removal.

12.3.3 POLYVINYL ACETATE-Fe$_3$O$_4$-BASED ADSORBENTS

Tseng et al. (2007) synthesized microsized magnetic polymer adsorbent (MPA) coupled with metal chelating ligands of iminodiacetic acid (IDA) for the removal of Cu^{2+} ions. The superparamagnetic Fe$_3$O$_4$ was prepared via the chemical coprecipitation method and then coated with polyvinyl acetate (PVAc) via suspension polymerization, yielding magnetite-PVAc (M-PVAc). Several sequential procedures, including alcoholysis, epoxide activation, and coupling of IDA were subsequently employed to introduce functional groups onto the surface of the M-PVAc without destroying the magnetite within the particles to yield magnetite-polyvinyl alcohol (M-PVA), magnetite-polyvinyl propenepoxide (M-PVEP), and M-PVAc-IDA. The production yield of M-PVAc was quite high,

(a)   (b)

FIGURE 12.21   (a) Dispersion of poly(MMA-co-MA)/APTMS-Fe$_3$O$_4$ MNC in water and (b) magnetic separation of MNC. (Reprinted with permission from figure 6, Masoumi, A. et al., 2014, Removal of metal ions from water using poly(MMA-co-MA)/modified-Fe$_3$O$_4$ magnetic nanocomposite: Isotherm and kinetic study, *Ind. Eng. Chem. Res.* 53, 8188–8197. Copyright (2014) American Chemical Society.)

TABLE 12.2

Comparison of Maximum Adsorption Capacity of Various Adsorbents

Adsorbents	Adsorbent Capacity (mg g^{-1})			
	Co^{2+}	Cr^{2+}	Zn^{2+}	Cd^{2+}
Poly(MMA-co-MA)/APTMS-Fe$_3$O$_4$	90.09	90.91	109.89	111.11
Nanoalumina modified with 2,4-dinitrophenylhydrazine	—	100.0	—	83.33
Poly[vinylpyridine-ethylene glycol methacrylate (EGMA)-EGDMA]	—	17.38	—	16.5
Poly(MMA-methacryloylamidoglutamic acid)	—			28.2
Poly(EGDMA-n-vinyl imidazole)	—	—	—	69.4
Chitosan	—	—	46.82	89.58
PS-co-MA	—	—	1.95×10^{-3}	—
Fe–Mn binary oxide	32.25	—	—	—
Additive assisted nanostructured goethite	86.6	—	—	29.15
MWCNT-CD polymer	21.44	—	—	—

Source: Extracted with permission from table 2, Masoumi, A. et al., 2014, Removal of metal ions from water using poly(MMA-co-MA)/modified-Fe$_3$O$_4$ magnetic nanocomposite: Isotherm and kinetic study, *Ind. Eng. Chem. Res.* 53, 8188–8197. Copyright (2014), American Chemical Society.

---

with excellent narrow particle size distribution. The synthesized M-PVAc-IDA polymer contained functional groups of metal chelating ligands of IDA, which can chelate metal ions such as Cu^{2+}. The magnetization profile measurement of M-PVAc-IDA featured no magnetic stagnation under normal ambient temperature. As the externally applied magnetic field H was zero, the residual magnetization magnitude (M_r) and coercive force (H_c) were also equal to zero, and exhibited superparamagnetic behavior under normal ambient temperature. The microsized and specified functional groups of metal chelating ligands in M-PVAc-IDA can provide a large S_E and adsorbability of metal ions of the adsorbent, respectively, which are essential to the adsorption. The adsorption capacity is highest at a pH value of 4.5 and the monolayer adsorption capacity is 0.121 mmol g⁻¹. On the other hand, adsorption capacity decreases as the pH value decreases to 1. Moreover, the exhausted M-PVAc-IDA with superparamagnetic properties can be separated from the solution via the applied magnetic force. After removing Cu^{2+} ions at a moderate pH value of 4.5, the exhausted M-PVAc-IDA can be regenerated at a low pH value of 1.

12.3.4 POLYACRYLATE-FE₃O₄-BASED ADSORBENTS

Chen et al. (2006) synthesized magnetic Fe₃O₄–glycidyl methacrylate (GMA)–IDA–styrene–divinyl benzene resin (Figure 12.22c) for the removal of Cu^{2+}, Cd^{2+}, and Pb^{2+} from aqueous solution. The adsorption time required to reach a state of equilibrium was 1 min. The equilibrium adsorption capacities from the single-metal ion solutions were 0.88 mmol g⁻¹ (Cu^{2+}) > 0.81 mmol g⁻¹ (Pb^{2+}) > 0.78 mmol g⁻¹ (Cd^{2+}), respectively. Increasing the concentration (0–0.3 M) of KCl, NaCl, MgCl₂, and CaCl₂ in Cu^{2+} or Pb^{2+} solution affected the adsorption behavior slightly. As the salt concentrations in the Cd^{2+} solution increased, the adsorption capacities of Cd^{2+} decreased in the order: Mg^{2+} > Ca^{2+} > Na^{+} > K^{+}. Within the pH range of 2–5, decreasing the pH of the Cu^{2+} solution did not produce great changes in the equilibrium adsorption capacities. However, significant decrements occurred for the adsorptions of Pb^{2+} or Cd^{2+} when the pH values of the solutions were <3. The competitive adsorption tests verified that this resin had good adsorption selectivity for Cu^{2+} in coexistence of Pb^{2+} and Cd^{2+}. To examine the reusability of the magnetic chelating resin, consecutive adsorption–desorption cycles were repeated 15 times using the same resins. Figure 12.22b shows that only the adsorption capacity decreased by only 5% after 15 cycles of adsorption–desorption operations. Figure 12.22a displays the adsorption–desorption cycles. When the resin–metal ion complexes were immersed in 3 M HCl solution for 60 min at 5°C, high concentrations of H^{+} can replace metal ions adsorbed by GMA–IDA within the resin to form carboxylic acids and quaternary ammonium salts. When these resins were reused, only 50%–60% of initial adsorption capacities

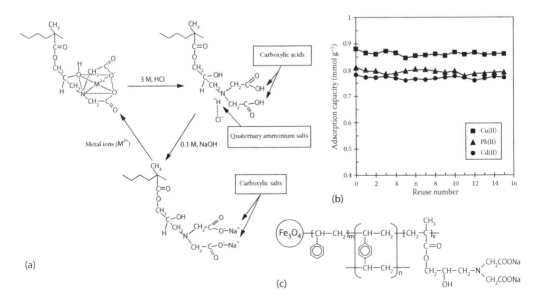

FIGURE 12.22 (a) Resin adsorption–desorption cycles, (b) resin adsorption capacities after repeated adsorption–desorption (pH: 4 at 25°C; metal ion concentration: 3.94 mM; adsorption time: 5 min), and (c) chemical structure of the resin. (Reprinted from scheme 2, figure 6, and scheme 1, respectively, *Sep. Purif. Technol.*, 50, Chen, C.-Y. et al., Adsorptions of heavy metal ions by a magnetic chelating resin containing hydroxy and iminodiacetate groups, 15–21, Copyright (2006), with permission from Elsevier.)

were observed. However, the readsorption capacities could exceed 95% of initial values after these resins were treated with NaOH solution to form carboxylic salts and tertiary amine.

Senel et al. (2008) prepared poly[ethylene glycol dimethacrylate (EGDMA)-vinyl imidazole (VIM)] beads by suspension polymerization in the presence of Fe_3O_4 nanopowder. The specific surface area of the beads was found to be 63.1 m^2 g^{-1}, with a size range of 150–200 μm in diameter and a swelling ratio of 85%. The average Fe_3O_4 content of the resulting beads was 12.4%. The maximum binding capacities of the beads were 32.4 mg g^{-1} for Cu^{2+}, 45.8 mg g^{-1} for Zn^{2+}, 84.2 mg g^{-1} for Cd^{2+}, and 134.5 mg g^{-1} for Pb^{2+}. The affinity order on the mass basis is $Pb^{2+} > Cd^{2+} > Zn^{2+} > Cu^{2+}$. Equilibrium data agreed well with the Langmuir model. The pH value significantly affected the binding capacity of the magnetic beads. The binding capacities of heavy metal ions from synthetic wastewater were 26.2 mg g^{-1} for Cu^{2+}, 33.7 mg g^{-1} for Zn^{2+}, 54.7 mg g^{-1} for Cd^{2+}, and 108.4 mg g^{-1} for Pb^{2+}. The magnetic beads could be regenerated up to about 97% by treating with 0.1 M HNO_3. These features make poly(EGDMA-VIM) beads a potential candidate to support heavy metal removal under a magnetic field.

12.3.5 POLYPYRROLE-FE₃O₄-BASED ADSORBENTS

Wang et al. (2012) synthesized uniform orange-like Fe_3O_4 (that acted as seeds)/polypyrrole (PPy, that acted as pulp and peel) composite microspheres (Figure 12.23a) using the *in situ* interfacial polymerization method under sonication. The composite spheres were treated in an acidic solution to remove residual Fe_3O_4 nanoparticles. The resulting product was not hollow spheres, but solid spherical particles with a porous interior (Figure 12.23b). The magnified PPy microspheres resembled an assembly of segments (Figure 12.23c). The black dots (Arrow I) in the Fe_3O_4/PPy composite microspheres corresponded to the pores (Arrow II) in the PPy microspheres. These results indicated that the residual Fe_3O_4 nanoparticles were embedded in the newly generated PPy microsphere matrix, just as the seeds in the orange segments (Arrow III). A clear image of an orange cross-section is shown in Figure 12.23d for comparison.

A growth mechanism for the Fe_3O_4/PPy composite microspheres was proposed (Figure 12.24): (a) In the early stages, partial pyrrole monomers and H^+ ions were both introduced into the pores/gaps of the Fe_3O_4 microspheres; more H^+ ions enter into the interior of the Fe_3O_4 microspheres due to their smaller size than that of the pyrrole. (b) Subsequently, Fe^{3+} released from the Fe_3O_4 microsphere surface initiated pyrrole monomers to form an outer shell; on the other hand, Fe^{3+} released

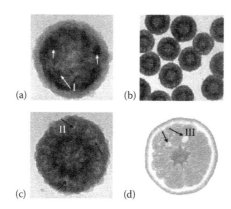

FIGURE 12.23 **(See color insert)** TEM images of (a) single Fe_3O_4/PPy composite microsphere, (b) porous PPy microspheres, (c) single porous PPy microsphere, and (d) digital photograph of a real orange cross-section. (Wang, Y. et al., 2012, Synthesis of orange-like Fe_3O_4/PPy composite microspheres and their excellent Cr(VI) ion removal properties, *J. Mater. Chem.* 22, 9034–9040, figure 2. Reproduced by permission of The Royal Society of Chemistry.)

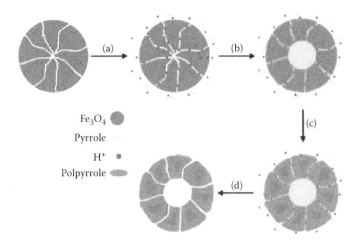

Fe$_3$O$_4$

Pyrrole

H$^+$

Polpyrrole

FIGURE 12.24 **(See color insert)** Synthetic procedure for orange-like Fe$_3$O$_4$/PPy composite microspheres: (a) dispersion of Fe$_3$O$_4$ microspheres in mixed solutions containing pyrrole and acid, (b) release of Fe^{3+} from the surface and the interior of the Fe$_3$O$_4$ microspheres in acidic solution and the polymerization of pyrrole monomers nearby, (c) generation of PPy with decreasing Fe$_3$O$_4$ microspheres, and (d) formation of orange-like Fe$_3$O$_4$/PPy composite microspheres after washing. (Wang, Y. et al., 2012, Synthesis of orange-like Fe$_3$O$_4$/PPy composite microspheres and their excellent Cr(VI) ion removal properties, *J. Mater. Chem.* 22, 9034–9040, figure 6. Reproduced by permission of The Royal Society of Chemistry.)

from the gaps or the interior of the Fe$_3$O$_4$ microspheres initiated pyrrole monomers nearby. Because the particle size within the interior of Fe$_3$O$_4$ microspheres was smaller, often caused by Ostwald ripening, the Fe$_3$O$_4$ microsphere interior was dissolved preferentially, which formed an interior cavity. (c) With further reaction, more and more pyrrole monomers could enter into the gaps/interiors of the Fe$_3$O$_4$ microspheres, and PPy was generated around the template and along the gaps in the template with the continual decrease of Fe$_3$O$_4$. (d) When the reaction stopped, the residual pyrrole monomer and H$^+$ ions in the Fe$_3$O$_4$/PPy composite microspheres were washed repeatedly away with DI water; the final Fe$_3$O$_4$/PPy composite microspheres exhibited an orange-like structure, in which the generated PPy surrounded the residue of the Fe$_3$O$_4$ nanoparticles, just like "PPy" segments of orange with "Fe$_3$O$_4$" seeds, as shown in Figure 12.23. The sonication technique aided in the dispersion of magnetic particles during the reaction and also accelerated the entry of H$^+$ ions into the Fe$_3$O$_4$ microsphere interiors.

Figure 12.25 shows the adsorption isotherm of Cr^{6+} for the as-prepared Fe$_3$O$_4$/PPy composite microspheres. The adsorption data fit with the Langmuir model ($R^2 = 0.99$) rather than the Freundlich model ($R^2 = 0.96$). Compared with the Cr^{6+} adsorption capacity of 139.83 mg g^{-1} at RT using Fe$_3$O$_4$/PPy core–shell nanoparticles (Bhaumik et al. 2011), the strong adsorption capacity of about 209.2 mg g^{-1}, reported by Wang et al. (2012). is due to their porous structure, which provides an accessible diffusion pathway into the interior of the Fe$_3$O$_4$/PPy composite microspheres. Their adsorption capacity is much higher even than those of previous inorganic microspheres such as iron oxides and aluminum oxides (Zhong et al. 2006; Wang et al. 2009).

12.3.6 PEDOT-Fe$_3$O$_4$-Based Adsorbents

Poly(3,4-ethylenedioxythiophene) (PEDOT) is one of the most promising conducting polymers, because of its high conductivity, excellent environmental stability, and simple acid/base doping/dedoping chemistry. In addition, PEDOT contains sulfur, which can endow two unpaired electrons. This thiol-functionalized polymer is readily conjugated with positively charged heavy metal ions, according to the coordination formation. Shin and Jang (2007) reported the simple fabrication of

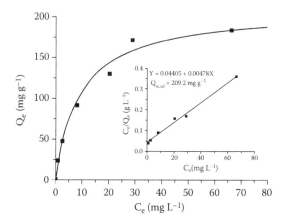

FIGURE 12.25 Cr^{6+} adsorption isotherm of Fe_3O_4/PPy composite microspheres and (inset) C_e/Q_e plotted versus C_e. (Wang, Y. et al., 2012, Synthesis of orange-like Fe_3O_4/PPy composite microspheres and their excellent Cr(VI) ion removal properties, *J. Mater. Chem.* 22, 9034–9040, figure 8. Reproduced by permission of The Royal Society of Chemistry.)

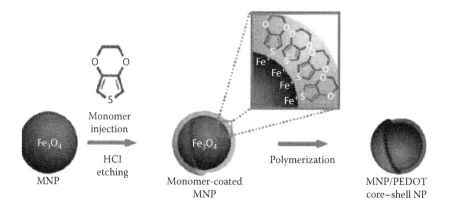

FIGURE 12.26 **(See color insert)** Fabrication procedure of Fe_3O_4–PEDOT NPs by seeded polymerization mediated with acidic etching. (Shin, S. and Jang, J., 2007, Thiol containing polymer encapsulated magnetic nanoparticles as reusable and efficiently separable adsorbent for heavy metal ions, *Chem. Commun.* 41, 4230–4232, scheme 1. Reproduced by permission of The Royal Society of Chemistry.)

thiol-containing polymer-encapsulated magnetic nanoparticles by seeded polymerization mediated with acidic etching. Fe_3O_4–PEDOT core–shell NPs were prepared by inducing ferric cations onto the MNPs with a partial etching process, followed by seeded polymerization. The synthetic procedure for Fe_3O_4–PEDOT NPs is illustrated in Figure 12.26. The adsorption of heavy metal ions progressed by means of surface complex formation between PEDOT shells and heavy metal ions. Uptake saturation occurred within 2 h, when the initial concentration of heavy metal ion was ca. 15 ppm (Figure 12.27a). Based on the weight of PEDOT, the amounts of Ag^+, Hg^{2+}, and Pb^{2+} ion uptake were ca. 28.0, 16.0, and 15.0 mmol g^{-1}, respectively. The maximum adsorption capacity (437 mg g^{-1}) was approximately seven times higher than that of CS-coated Fe_3O_4 NPs. In addition, this value was also high compared with that for thiol-functionalized particles (below 200 mg g^{-1}), due to the high surface area of Fe_3O_4–PEDOT NPs and the abundant unpaired electrons of PEDOT shells. Figure 12.27b displays the uptake of heavy metal ions after 0.5 and 24 h contact time with an excess amount of Fe_3O_4–PEDOT NPs. Up to 95% of heavy metal ions were removed after 24 h, whereas the adsorption rate was observed in the order $Ag^+ > Hg^{2+} > Pb^{2+}$, in accordance with the

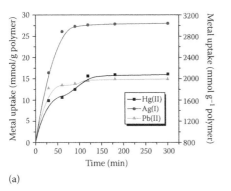
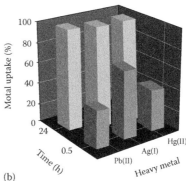

(a)   (b)

FIGURE 12.27   Heavy metal ion uptakes of (a) Fe₃O₄–PEDOT NPs as a function of different contact times, and (b) fixed contact times: 0.5 and 24 h. (Shin, S. and Jang, J., 2007, Thiol containing polymer encapsulated magnetic nanoparticles as reusable and efficiently separable adsorbent for heavy metal ions, *Chem. Commun.* 41, 4230–4232, figure 3. Reproduced by permission of The Royal Society of Chemistry.)

cation radius and interaction enthalpy values. Importantly, Fe₃O₄–PEDOT NPs provide convenient recovery of absorbents in aqueous phase by use of a magnetic field. There was no loss of adsorption capacity and morphological change of Fe₃O₄–PEDOT NPs, even after 10 cycles.

12.3.7   Dendrimer-Fe₃O₄-Based Adsorbents

Poly(amidoamine) (PAMAM) dendrimers are a new class of nanomaterials with 3-D structure, consisting of three basic units: (a) an ethylenediamine core, (b) repeating units, and (c) terminal units. The diameter of PAMAM dendrimers increases roughly by 1 nm/generation. For example, the diameters of G-3 and G-5 PAMAM are 2.9 and 4.5 nm, respectively. The use of PAMAM dendrimers for the adsorptive removal of heavy metals from water and soil has been reported (Diallo et al. 2005; Xu and Zhao 2005). Amine-terminated PAMAM dendrimers exhibit a high binding affinity for metal ions to their surface via coordination to the amine or acid functionality (Crooks et al. 2001). Complexation of metal ions with the dendrimer is pH dependent. The release of metal ions from dendrimers can be readily achieved by the protonation of amine-functional groups at lower pH levels. This feature makes PAMAM dendrimers particularly promising as reusable chelating agents for metal ion separation.

Chou and Lien (2011) developed dendrimer-conjugated magnetic nanoparticles (Gn-MNPs) combining the superior adsorbent of dendrimers with MNPs for effective removal and recovery of Zn²⁺. The adsorption efficiency of Zn²⁺ with Gn-MNPs increases with increases in the pH level. The correlation coefficients (R^2) of the Langmuir and Freundlich models are 0.957 and 0.953, respectively. The maximum adsorption capacity determined by the Langmuir model is 24.3 mg g⁻¹ at pH 7 and 25°C. A synergistic effect between the complexation reaction and the electrostatic interaction may account for the overall performance of Gn-MNPs. The adsorption capacity of G3-MNPs for Zn²⁺ is generally comparable with other nanoadsorbents at pH 7 and 25°C. For example, the Zn²⁺ adsorption capacities of powder-activated carbon, β-FeOOH nanoparticles, MWCNTs, and SWCNTs were 13.4, 27.61, 32.68, and 43.66 mg g⁻¹, respectively (Lu and Chiu 2006; Deliyanni et al. 2007). To determine the reusability of G3-MNPs, adsorption–desorption of Zn²⁺ was carried out in 10 consecutive cycles. Recovery of Zn²⁺ attains an average of >90% in 10 cycles (Figure 12.28a). This suggests that the binding sites on the surface of the adsorbent are reversible. On the contrary, the use of MNPs alone for repetitive adsorption–desorption experiments showed poor performance in terms of Zn²⁺ recovery and regeneration ability of the adsorbent (Figure 12.28b). The average removal efficiency of Zn²⁺ was about 58% and the average recovery of Zn²⁺ was only 38% in eight

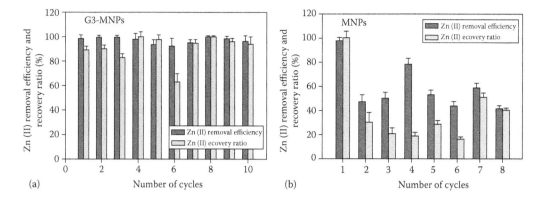

FIGURE 12.28 Adsorption-desorption of Zn^{2+} by (a) G3-MNPs and (b) MNPs (metal loading: 1.0 g/L; Zn^{2+} concentration for each cycle: 0.30 mM; pH: 7). (With kind permission from Springer Science+Business Media: *J. Nanopart. Res.*, Dendrimer-conjugated magnetic nanoparticles for removal of zinc (II) from aqueous solutions, 13, 2011, 2099–2107, Chou, C.-M. and Lien, H.-L., figures 7 and 8, Copyright 2011.)

cycles. This suggests that the interactions between the iron oxide and Zn^{2+} are mainly irreversible. The G3-MNPs used were collected by an external magnet and >75% of the total G3-MNPs can be recovered after 10 consecutive cycles.

12.4 FUTURE TRENDS

Presently, equipment utilized for the magnetic capturing of MNPs is mostly designed for small-scale operations. When considering the future of polymer-coated MNPs, the following issues have to be taken into account: difficulty in controlling particle size distribution, stability of magnetic preparations in various media, large-scale production, long-term storage stability, and cost of sale. Also, the coating methods using fewer chemicals is preferred to transform the overall preparation into a green process. This would ensure less toxicity and open more environmentally friendly applications.

REFERENCES

Babel, S. and T.A. Kurniawan. 2003. Low-cost adsorbents for heavy metals uptake from contaminated water, a review. *J. Hazard. Mater.* 97:219–243.

Badruddoza, A.Z.M., Shawon, Z.B.Z. Tay, W.J.D. Hidajat, K. and M.S. Uddin. 2013. Fe_3O_4/cyclodextrin polymer nanocomposites for selective heavy metals removal from industrial wastewater. *Carbohydr. Polym.* 91:322–332.

Badruddoza, A.Z.M., Tay, A.S.H., Tan, P.Y., Hidajat, K., and M.S. Uddin. 2011. Carboxymethyl-β-cyclodextrin conjugated magnetic nanoparticles as nano-adsorbents for removal of copper ions: Synthesis and adsorption studies. *J. Hazard. Mater.* 185:1177–1186.

Bhaumik, M., Maity, A., Srinivasu, V.V., and M.S. Onyango. 2011. Enhanced removal of Cr(VI) from aqueous solution using polypyrrole/Fe_3O_4 magnetic nanocomposite. *J. Hazard. Mater.* 190:381–390.

Chen, C.-Y., Chiang, C.-L., and P.-C. Huang. 2006. Adsorptions of heavy metal ions by a magnetic chelating resin containing hydroxy and iminodiacetate groups. *Sep. Purif. Technol.* 50:15–21.

Chou, C.-M. and H.-L. Lien. 2011. Dendrimer-conjugated magnetic nanoparticles for removal of zinc (II) from aqueous solutions. *J. Nanopart. Res.* 13:2099–2107.

Crooks, R.M., Zhao, M., Sun, L., Chechik, V., and L.K. Yeung. 2001. Dendrimer-encapsulated metal nanoparticles: Synthesis, characterization, and applications to catalysis. *Acc. Chem. Res.* 31:181–190.

Cumbal, L., Greenleaf, J., Leun, D., and A.K.S. Gupta. 2003. Polymer supported inorganic nanoparticles: Characterization and environmental applications. *React. Funct. Polym.* 54:167–180.

Dabrowski, A., Hubicki, Z., Podkościelny, P., and E. Robens. 2004. Selective removal of the heavy metal ions from waters and industrial wastewaters by ion-exchange method. *Chemosphere* 56:91–106.

Dave, P.N. and L.V. Chopda. 2014. Application of iron oxide nanomaterials for the removal of heavy metals. *J. Nanotechnol.* 2014:1–14.

Deliyanni, E.A., Peleka, E.N., and K.A. Matis. 2007. Removal of zinc ion from water by sorption onto iron-based nanoadsorbent. *J. Hazard. Mater.* 141:176–184.

Diallo, M.S., Christie, S., Swaminathan, P., Johnson, J.H., and W.A. Goddard. 2005. Dendrimer enhanced ultrafiltration. 1. Recovery of Cu(II) from aqueous solutions using PAMAM dendrimers with ethylene diamine core and terminal NH$_2$ groups. *Environ. Sci. Technol.* 39:1366–1377.

Dias, A.M.G.C., Hussain, A., Marcos, A.S., and A.C.A. Roque. 2011. A biotechnological perspective on the application of iron oxide magnetic colloids modified with polysaccharides. *Biotechnol. Adv.* 29:142–155.

Fan, L., Luo, C., Lv, Z., Lu, F., and H. Qiu. 2011. Preparation of magnetic modified chitosan and adsorption of Zn^{2+} from aqueous solutions. *Colloids Surf. B* 88:574–581.

Fang, C. and M. Zhang. 2009. Multifunctional magnetic nanoparticles for medical imaging applications. *J. Mater. Chem.* 19:6258–6266.

Feng, L.Y., Cao, M.H., Ma, X.Y., Zhu, Y.S., and C.W. Hu. 2012. Superparamagnetic high-surface-area Fe$_3$O$_4$ nanoparticles as adsorbents for arsenic removal. *J. Hazard. Mater.* 217:439–446.

Fu, F. and Q. Wang. 2011. Removal of heavy metal ions from wastewaters: A review. *J. Environ. Manage.* 92:407–418.

Fu, F., Xie, L., Tang, B., Wang, A., and S. Jiang. 2012. Application of a novel strategy: Advanced Fenton-chemical precipitation to the treatment of strong stability chelated heavy metal containing wastewater. *Chem. Eng. J.* 189–190:283–287.

Ge, F., Li, M.-M., Ye, H., and B.-X. Zhao. 2012. Effective removal of heavy metal ions Cd^{2+}, Zn^{2+}, Pb^{2+}, Cu^{2+} from aqueous solution by polymer-modified magnetic nanoparticles. *J. Hazard. Mater.* 211:366–372.

Hall, J.L. 2002. Cellular mechanisms for heavy metal detoxification and tolerance. *J. Exp. Bot.* 53:1–11.

Hao, Y.M., Chen, M., and Z.B. Hu. 2010. Effective removal of Cu (II) ions from aqueous solution by amino-functionalized magnetic nanoparticles. *J. Hazard. Mater.* 184:392–399.

Hu, H.B., Wang, Z.H., and L. Pan. 2010. Synthesis of monodisperse Fe$_3$O$_4$@silica core–shell microspheres and their application for removal of heavy metal ions from water. *J. Alloy. Compd.* 492:656–661.

Hu, J., Chen, G., and I.M.C. Lo. 2006. Selective removal of heavy metals from industrial wastewater using maghemite nanoparticle: Performance and mechanisms. *J. Environ. Eng.* 132:709–715.

Huang, S.H. and D.H. Chen. 2009. Rapid removal of heavy metal cations and anions from aqueous solutions by an amino-functionalized magnetic nano-adsorbent. *J. Hazard. Mater.* 163:174–179.

Kim, Y., Lee, B., and J. Yi. 2003. Preparation of functionalized mesostructured silica containing magnetite (MSM) for the removal of copper ions in aqueous solutions and its magnetic separation. *Separ. Sci. Technol.* 38:2533–2548.

Kumar, P.A. and S. Chakraborty. 2009. Fixed-bed column study for hexavalent chromium removal and recovery by short-chain polyaniline synthesized on jute fiber. *J. Hazard. Mater.* 162:1086–1098.

Laurent, S., Forge, D., Port, M. et al. 2008. Magnetic iron oxide nanoparticles: Synthesis, stabilization, vectorization, physicochemical characterizations, and biological applications. *Chem. Rev.* 108:2064–110.

Li, G.L., Zhao, Z.S., Liu, J.Y., and G.B. Jiang. 2011. Effective heavy metal removal from aqueous systems by thiol functionalized magnetic mesoporous silica. *J. Hazard. Mater.* 192:277–283.

Liu, Z., Wang, H., Liu, C., et al. 2012. Magnetic cellulose–chitosan hydrogels prepared from ionic liquids as reusable adsorbent for removal of heavy metal ions. *Chem. Commun.* 48:7350–7352.

Lu, C. and H. Chiu. 2006. Adsorption of zinc(II) from water with purified carbon nanotubes. *Chem. Eng. Sci.* 61:1138–1145.

Lv, L., Xie, Y.H., Liu, G.M., Liu, G., and J. Yu. 2014. Removal of perchlorate from aqueous solution by cross-linked Fe(III)–chitosan complex. *J. Environ. Sci. China* 26:792–800.

Ma, Z.Y., Guan, Y.P., and H.Z. Liu. 2005. Synthesis and characterization of micron-sized monodisperse superparamagnetic polymer particles with amino groups. *J. Polym. Sci. Part A: Polym. Chem.* 43:3433–3439.

Mahlambi, M.M., Malefetse, T.J., Mamba, B.B., and R.W. Krause. 2010. β-Cyclodextrin-ionic liquid polyurethanes for the removal of organic pollutants and heavy metals from water: Synthesis and characterization. *J. Polym. Res.* 17:589–600.

Masoumi, A., Ghaemy, M., and A.N. Bakht. 2014. Removal of metal ions from water using poly(MMA-co-MA)/modified-Fe$_3$O$_4$ magnetic nanocomposite: Isotherm and kinetic study. *Ind. Eng. Chem. Res.* 53: 8188–8197.

McManamon, C., Burke, A.M., Holmes, J.D., and M.A. Morris. 2012. Amine-functionalised SBA-15 of tailored pore size for heavy metal adsorption. *J. Colloid Interf. Sci.* 369:330–337.

Norkus, E. 2009. Metal ion complexes with native cyclodextrins: An overview. *J. Incl. Phenom. Macrocycl. Chem.* 65:237–248.

Ohno, M., Hayashi, H., Suzuki, K., Kose, T., Asada, T., and K. Kawata. 2011. Preparation and evaluation of magnetic carbonaceous materials for pesticide and metal removal. *J. Colloid Interf. Sci.* 359:407–412.

Pang, F.M., Kumar, P., Teng, T.T., Omar, A.K.M., and K.L. Wasewar. 2011. Removal of lead, zinc and iron by coagulation–flocculation. *J. Taiwan Inst. Chem. E.* 42:809–815.

Papelis, C., Roberts, P.V., and J.O. Leekie. 1995. Modeling the rate of cadmium and selenite adsorption on micro- and mesoporous transition aluminas. *Environ. Sci. Technol.* 29:1099–1108.

Pearson, R.G. 1963. Hard and soft acids and bases. *J. Am. Chem. Soc.* 85:3533–3539.

Peng, S., Meng, H., Ouyang, Y., and J. Chang. 2014. Nanoporous magnetic cellulose–chitosan composite microspheres: Preparation, characterization, and application for Cu(II) adsorption. *Ind. Eng. Chem. Res.* 53:2106–2113.

Qiao, R., Yang, C., and M. Gao. 2009. Superparamagnetic iron oxide nanoparticles: From preparations to in vivo MRI applications. *J. Mater. Chem.* 19:6274–6293.

Rastegarzadeh, S., Pourreza, N., Kiasat, A.R., and H. Yahyavi. 2010. Selective solid phase extraction of palladium by adsorption of its 5(p-dimethylaminobenzylidene)rhodanine complex on silica-PEG as a new adsorbent. *Microchim. Acta* 170:135–140.

Ren, Y., Wei, X., and M. Zhang. 2008. Adsorption character for removal Cu(II) by magnetic Cu(II) ion impregnated composite adsorbent. *J. Hazard. Mater.* 158:14–22.

Senel, S., Uzun, L., Kara, A., and A. Denizli. 2008. Heavy metal removal from synthetic solutions with magnetic beads under magnetic field. *J. Macromol. Sci., Pure Appl. Chem.* 45:635–642.

Shen, H., Chen, J., Dai, H., Wang, L., Hu, M., and Q. Xia. 2013. New insights into the sorption and detoxification of chromium(VI) by tetraethylenepentamine functionalized nanosized magnetic polymer adsorbents: Mechanism and pH effect. *Ind. Eng. Chem. Res.* 52:12723–12732.

Shen, Y.F., Tang, J., Nie, Z.H., Wang, Y.D., Ren, Y., and L. Zuo. 2009. Preparation and application of magnetic Fe$_3$O$_4$ nanoparticles for wastewater purification. *Sep. Purif. Technol.* 68:312–319.

Shin, S. and J. Jang. 2007. Thiol containing polymer encapsulated magnetic nanoparticles as reusable and efficiently separable adsorbent for heavy metal ions. *Chem. Commun.* 41:4230–4232.

Song, J., Kong, H., and J. Jang. 2011b. Adsorption of heavy metal ions from aqueous solution by polyrhodanine-encapsulated magnetic nanoparticles. *J. Colloid Interface Sci.* 359:505–511.

Song, J., Oh, H., Kong, H., and J. Jang. 2011a. Polyrhodanine modified anodic aluminum oxide membrane for heavy metal ions removal. *J. Hazard Mater.* 187:311–317.

Tran, H.V., Tran, L.D., and T.N. Nguyen. 2010. Preparation of chitosan/magnetite composite beads and their application for removal of Pb(II) and Ni(II) from aqueous solution. *Mater. Sci. Eng. C* 30:304–310.

Tseng, J.-Y., Chang, C.-Y., Chen, Y.-H., Chang, C.-F., and P.-C. Chiang. 2007. Synthesis of micro-size magnetic polymer adsorbent and its application for the removal of Cu(II) ion. *Colloids Surf. A Physicochem. Eng. Asp.* 295:209–216.

Wang, J., Deng, B.L., Chen, H., Wang, X.R., and J.Z. Zheng. 2009. Removal of aqueous Hg(II) by polyaniline: Sorption characteristics and mechanisms. *Environ. Sci. Technol.* 43:5223–5228.

Wang, J.H., Zheng, S.R., Shao, Y., Liu, J.L., Xu, Z.Y., and D.Q. Zhu. 2010. Amino-functionalized Fe$_3$O$_4$@SiO$_2$ core–shell magnetic nanomaterial as a novel adsorbent for aqueous heavy metals removal. *J. Colloid Interf. Sci.* 349:293–299.

Wang, T., Zhang, L., Li, C., et al. 2015. Synthesis of core–shell magnetic Fe$_3$O$_4$@poly(m-phenylenediamine) particles for chromium reduction and adsorption. *Environ. Sci. Technol.* 49:5654–5662.

Wang, Y., Wang, G., Wang, H., Cai, W., Liang, C., and L. Zhang. 2009. Template-induced synthesis of hierarchical SiO$_2$@γ-AlOOH spheres and their application in Cr(VI) removal. *Nanotechnology* 20:155604.

Wang, Y., Zou, B., Gao, T., Wu, X., Lou, S., and S. Zhou. 2012. Synthesis of orange-like Fe$_3$O$_4$/PPy composite microspheres and their excellent Cr(VI) ion removal properties. *J. Mater. Chem.* 22:9034–9040.

Wen, Y.Z., Ma, J.Q., Chen, J., Shen, C.S., Li, H., and W.P. Liu. 2015. Carbonaceous sulfur-containing chitosan–Fe(III): A novel adsorbent for efficient removal of copper(II) from water. *Chem. Eng. J.* 259:372–380.

Xu, L., Chen, J., Wen, Y., Li, H., Ma, J., and D. Fu. 2015. Fast and effective removal of cadmium ion from water using chitosan encapsulated magnetic Fe$_3$O$_4$ nanoparticles, *Desalin. Water Treat.* 57:8540–8548.

Xu, Y., and D. Zhao. 2005. Removal of copper from contaminated soil by use of poly (amidoamine) dendrimers. *Environ. Sci. Technol.* 39:2369–2375.

Yan, H., Yang, L., Yang, Z., Yang, H., Li, A., and R. Cheng. 2012. Preparation of chitosan/poly(acrylic acid) magnetic composite microspheres and applications in the removal of copper(II) ions from aqueous solutions. *J. Hazard. Mater.* 229–230:371–380.

Yang, W., Ding, P., Zhou, L., Yu, J., Chen, X., and F. Jiao. 2013. Preparation of diamine modified mesoporous silica on multi-walled carbon nanotubes for the adsorption of heavy metals in aqueous solution. *Appl. Surf. Sci.* 282:38–45.

Yu, L., Zou, R., Zhang, Z., et al. 2011. A Zn$_2$GeO$_4$-ethylenediamine hybrid nano ribbon membrane as a recyclable adsorbent for the highly efficient removal of heavy metals from contaminated water. *Chem. Commun.* 47, 10719–10721.

Yuwei, C. and W. Jianlong. 2011. Preparation and characterization of magnetic chitosan nanoparticles and its application for Cu(II) removal. *Chem. Eng. J.* 168:286–292.

Zhang, F., Lan, J., Zhao, Z., Yang, Y., Tan, R., and W. Song. 2012. Removal of heavy metal ions from aqueous solution using Fe$_3$O$_4$-SiO$_2$-poly(1,2-diaminobenzene) core–shell sub-micron particles. *J. Colloid Interface Sci.* 387:205–212.

Zhang, R., Vigneswaran, S., Ngo, H.H., and H. Nguyen. 2006. Magnetic ion exchange (MIEX®) resin as a pre-treatment to a submerged membrane system in the treatment of biologically treated wastewater. *Desalination* 192:296–302.

Zhang, T., Zhang, X.W., Ng, J.W., Yang, H.Y., Liu, J.C., and D.D. Sun. 2011. Fabrication of magnetic cryptomelane-type manganese oxide nanowires for water treatment. *Chem. Commun.* 47:1890–1892.

Zhang, X.-P., Liu, F.-Q., Zhu, C.-Q., et al. 2015. A novel tetraethylenepentamine functionalized polymeric adsorbent for enhanced removal and selective recovery of heavy metal ions from saline solutions. *RSC Adv.* 5:75985–75997.

Zhao, Y.-G., Shen, H.-Y., Pan, S.-D., Hu, M.-Q., and Q.-H. Xia. 2010a. Preparation and characterization of amino-functionalized nano-Fe$_3$O$_4$ magnetic polymer adsorbents for removal of chromium(VI) ions. *J. Mater. Sci.* 45:5291–5301.

Zhao, Y.G., Shen, H.Y., Pan, S.D., and M.Q. Hu. 2010b. Synthesis, characterization and properties of ethylenediamine-functionalized Fe$_3$O$_4$ magnetic polymers for removal of Cr(VI) in wastewater. *J. Hazard. Mater.* 182:295–302.

Zhong, L.S., Hu, J.S., Liang, H.P., Cao, A.M., Song, W.G., and L.J. Wan. 2006. Self-assembled 3D flowerlike iron oxide nanostructures and their application in water treatment. *Adv. Mater.* 18:2426–2431.

Zhu, J.H., Wei, S.Y., Gu, H.B., et al. 2012. One-pot synthesis of magnetic graphene nanocomposites decorated with core@double-shell nanoparticles for fast chromium removal. *Environ. Sci. Technol.* 46:977–985.

13 Nanoscale Materials for the Removal of Arsenic from Wastewater

*Sushmita Banerjee, Puja Rai, Vandani Rawat,
and Ravindra Kumar Gautam*

CONTENTS

ABSTRACT

Arsenic is a toxic metalloid widely distributed in the earth's crust, which has been extensively used in pesticides and antibiotics for agricultural and pharmaceutical purposes. It enters into the aquatic environment through a combination of natural processes such as weathering of rocks, dissolution of minerals, and biological activity, as well as through anthropogenic activities including mining, agriculture, and manufacturing. Thus, water contaminated by arsenic poses a significant threat both to natural organisms and the environment. Most environmental arsenic problems are as a result of mobilization under natural conditions. In natural waters, arsenic is mostly available in inorganic forms, namely, arsenate [As(V)] and arsenite [As(III)]. It is generally recognized that As(III) is more toxic than As(V). The most severe problems associated with arsenic poisoning in drinking water occur in Argentina, Bangladesh, Canada, Chile, India, Japan, Taiwan, and the United States. Among these, Bangladesh and West Bengal, India, are seen as the worst affected regions, where a large number of the population rely on As-contaminated groundwater with concentrations exceeding the World Health Organization (WHO) guideline value. In recent years, the utilization of nanomaterials for the treatment of arsenic-polluted water has been extensively researched and the results exhibit a high arsenic removal efficiency. This chapter intends to provide in-depth information about the technical feasibility of various nanoscale materials employed for the removal of As from aqueous solutions and groundwater. Furthermore, the effectiveness of various nanoscale materials under different physicochemical process parameters and their comparative adsorption capacity toward the adsorption of As is also presented.

Keywords: Arsenic, Toxicity, Nanometal oxides, Nanomagnetic materials, Bimetallic oxides, Layered double hydroxides, Carbon nanotubes, Graphene, Metal organic framework

13.1 INTRODUCTION

Arsenic (As) is regarded as one of the most highly toxic elements of the periodic table (Mondal et al. 2006). It is widely distributed in the earth's crust and it ranks twentieth in crustal abundance (Mandal and Suzuki 2002). It is ubiquitously present in air, water, groundwater, soil, rock, mineral deposits, and other media in variable concentrations. The frequent detection of As in the environment may be due to the combined effect of natural and anthropogenic processes. The natural processes include weathering of As-bearing rocks, leaching from mineral deposits, especially sulfide deposits such as arsenides, sulfides and sulfosalts, volcanic eruptions, and biological activities, while anthropogenic activities that are responsible for As release comprise the combustion of coal and its by-products, mining and smelting, the use of inorganic and organic arsenic compounds as pesticides, herbicides, and silvicides in various agricultural- and forest- related activities, the manufacturing of plastics, paints, lacquers, and so on. Figure 13.1 depicts various possible sources of arsenic in the environment.

Chemically, As mainly exists in four oxidation states, namely, arsenic (As^0), arsine (As^{-III}), arsenite [As(III)], and arsenate [As(V)]. However, in the aqueous state, As is available as oxyanions, namely, arsenite (AsO_3^{3-}) and arsenate (AsO_4^{3-}), with corresponding oxidation states [As(III)] and [As(V)], respectively (Choong et al. 2007). It has been shown that As(III) is 60 times more toxic than As(V) because of its high mobility and solubility (Kundu et al. 2004). It has also been reported that pH and Eh (redox potential) significantly influence the mobility of the arsenic species. Further bioavailability of As is primarily a function of several factors such as the chemical state of the arsenic, the pH of the aqueous media, the presence of hydrated manganese and iron oxides, clay minerals, and competing ions and organic matter (Martinson and Reddy 2009). It has been established that, under aerobic conditions, As(V) is the most common species, while an anoxic environment promotes the formation of As(III) species. Therefore, As(V) is anticipated to be thermodynamically more stable than As(III) in surface aquifers (Zhao et al. 2010). However, in subsurface water resources, both species of arsenic predominate, as both species undergo various redox processes and, thus, their oxidation states oscillate between As(III) and As(V). Thus, the occurrence of arsenic in a particular chemical state at a certain specific time phase is determined by the prevailing chemical conditions and biological activities that may occur during that period of time. Figure 13.2 illustrates the prevalence of aqueous arsenic in different chemical states at different pH and Eh values.

From Figure 13.1, it can be seen that arsenic exists in different protonated forms as a function of pH and redox potential. As for the As(III) system, it includes uncharged arsenite ($H_3AsO_3^0$) and

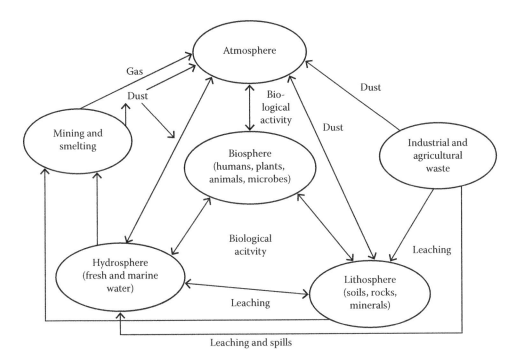

FIGURE 13.1 **(See color insert)** Arsenic input from various natural and anthropogenic sources. (Redrawn from Mudhoo et al. 2011.)

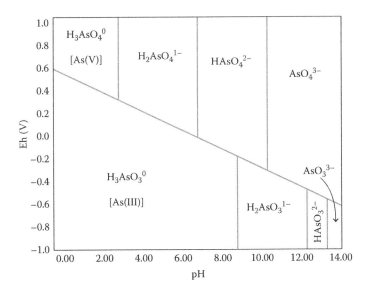

FIGURE 13.2 Speciation of inorganic arsenic species in aqueous media under different pH-Eh conditions at 25°C and 1 bar total pressure. (Redrawn from Garcia et al. 2012.)

other anionic derivatives such as $H_2AsO_3^{1-}$, $HAsO_3^{2-}$, and AsO_3^{3-} that predominate mostly under reduced conditions at pH < 9.2, 9.2, 12.7, and 13.4, respectively; while, under oxidizing conditions, the As(V) system, such as arsenate ($H_3AsO_4^0$), is present in highly acidic conditions at pH < 2.1, and its dissociated forms such as $H_2AsO_4^-$, $HAsO_4^{2-}$, and AsO_4^{3-} exist at pH values of 2.1, 6.7, and 11.2, respectively (Gupta and Chen 1978; Garcia et al. 2012). The dissociation of arsenite (Equation 13.1) and arsenate (Equation 13.2) into various deprotonated forms under differing pH conditions is summarized as follows:

$$H_3AsO_3 \xrightarrow[-H^+;pK_a=9.2]{} H_2AsO_3^- \xrightarrow[-H^+;pK_a=12.7]{} HAsO_3^{2-} \xrightarrow[-H^+;pK_a=13.4]{} AsO_3^{3-} \quad (13.1)$$

$$H_3AsO_4 \xrightarrow[-H^+;pK_a=2.1]{} H_2AsO_4^- \xrightarrow[-H^+;pK_a=6.7]{} HAsO_4^{2-} \xrightarrow[-H^+;pK_a=11.2]{} AsO_4^{3-} \quad (13.2)$$

From these equations, it can be presumed that for surface and subsurface waters whose pH value falls within the range of 6.5 to 8.5, the dominating arsenic species are H_3AsO_3, $H_2AsO_4^{-1}$, and $HAsO_4^{2-}$ (Abejon and Garea 2015). In the environment, arsenic is also present in various organic forms such as methylarsonic acid, dimethylarsinic acid, dimethylarsinoyl ribosides, trimethylarsine oxide, arsenobetaine, arsenosugars, and arsenolipids (Mondal et al. 2006).

13.2 ARSENIC TOXICITY

As mentioned earlier, arsenic is the one of the most toxic elements; its toxicity is so deadly that this element is often referred to as the "king of poisons." However, centuries ago this very substance was extensively used for healing. It was reportedly one of the most popular ingredients of traditional Chinese medicine. In addition, in the Western world it was used in the treatment of some severe ailments such as trypanosomiasis and syphilis. Moreover, its use was also reported during the Victorian era, when women used it to whiten their skin and in other cosmetic applications. Nevertheless, if we peep into history, the incidence of arsenic poisoning is one of the most preferred murder weapons. Arsenic poisoning was extensively reported in Europe from the time of the Roman Empire in 117 AD through the Middle Ages (410 AD) to the Renaissance period of 1350 AD. However, its potent toxic behavior was recognized earlier, in around 300 BC by Hippocrates, but symptoms of poisoning were quite difficult to detect as they appeared similar to cholera; as a result of the lack of proper medication, the victim failed to survive. It was not until almost the mid-nineteenth century that an English chemist, James Marsh, successfully developed a technique that could easily diagnose arsenic toxicity. But, before this, many notable personalities such as Pope Alexander VI, George III, Napoleon Bonaparte, Charles Francis Hall, and a few others lost their lives due to severe poisoning. Though these were incidences of intentional arsenic poisoning, according to the data of the World Health Organization (WHO), in 2012 (George et al. 2014) nearly 200 million people worldwide suffered due to unintentional arsenic poisoning. The most common source that paves the way for arsenic to enter into the human body is through drinking water. Drinking water that is primarily extracted from aquifers near arsenic-rich strata, or those that receive runoff from nearby agricultural land or industries, are extremely vulnerable to arsenic contamination. Arsenic exhibits acute, chronic, and lethal health effects in humans. However, all these effects depend on the dosage of arsenic ingested by the individual. According to Ellenhorn et al. (1997), the lethal dose of arsenic toxicity in an adult human is estimated to be in the range of 1–3 mg As/kg. Acute arsenic toxicity usually occurs when arsenic is ingested in large quantities with low exposure time, while a long exposure to low levels (<10 µg/L) of arsenic causes chronic toxicity. However, the level of toxicological effects is found to vary greatly with changes in the chemical characteristics of the elemental species (Mondal et al. 2006). Therefore, in the case of arsenic, its inorganic forms exhibit more toxicological effects than its organic forms (Mondal et al. 2013). It was reported that, in the biological system during the metabolism of arsenic, several intermediate organic arsenic metabolites such as arsenocholine, arsenobetaine, and arsenosugars were synthesized and existed in a benign state without causing any harm to the organisms (Saha et al. 1999). Thus, from this perspective, it is imperative to determine the chemical state of arsenic present in the background and within the biological system, which further facilitates better understanding of the risk due to exposure to arsenic compounds.

A case of arsenic toxicity was first reported in Taiwan by Tseng et al. (1977). The symptoms of arsenic toxicity include muscular pain, weakness, severe nausea, vomiting, abdominal pain, diarrhea, and red rashes on the skin. Collapse of the circulatory system and incorrect functioning of the kidneys, with minimal urine output, was also reported. In some cases, psychological problems such as hallucinations, phobia, confusion, and restlessness have been observed in many patients (Saha et al. 1999). However, severe toxicity results in seizures, convulsion, coma, and ultimately, death. Moreover, a positive correlation between arsenic exposure and cancer has been investigated by many researchers; therefore, it is classified as a Group I carcinogen (Khan et al. 2003; Ungureanu et al. 2015; Jadav et al. 2015). Chronic arsenic toxicity causes skin lesions, hyperpigmentation, and respiratory, cardiovascular, hepatic, and gastrointestinal problems, and impairs hormonal activity and the nervous system (Kapaj et al. 2006). Diseases such as peripheral neuropathy, black foot disease, and arsenicosis have reportedly affected large numbers of the population who inhabit regions where the incidence of arsenic toxicity is prevalent.

13.3 GLOBALLY AFFECTED AREAS WITH ARSENIC AND ENVIRONMENTAL STANDARDS

The problem of arsenic contamination in drinking water has been reported in more than 70 countries such as Argentina (Bhattacharya et al. 2005), Australia (O'Shea et al. 2007), Bangladesh (van Halem et al. 2010), Cambodia (Amrose et al. 2013), Canada (Wang et al. 2006), Chile (Romero et al. 2003), China (Lado et al. 2013), Finland (Parviainen et al. 2015), Ghana (Asante et al. 2007), Hungary (Rowland et al. 2011), India (Ghosh 2013), Japan (Yoshizuka et al. 2010), Korea (Ahn 2012), Mexico (Camacho et al. 2011a), Nepal (Pokhrel et al. 2009), Pakistan (Malik et al. 2009), Romania (Rowland et al. 2011), Serbia (Tubic et al. 2010), Taiwan (Liang et al. 2013), Thailand (Cho et al. 2011), the United States (Barringer et al. 2011; Yang et al. 2014), and Vietnam (Nguyen et al. 2009). Thus, due to the implications of high levels of arsenic, the European Union (Council of the European Communities 1976) and the U.S. Environmental Protection Agency (EPA 1982) considered arsenic as a pollutant of priority interest. Furthermore, to abate the problems associated with arsenic, environmental authorities imposed stringent regulations on the presence of arsenic in water. In 1993, WHO, and in 1996, the National Health and Medical Research Committee (NHMRC), issued guidelines for drinking water that recommended that the maximum permissible level of arsenic in drinking water should not exceed 10 µg/L and 7 µg/L, respectively. In 2003, the European Commission decided to lower the permissible limit of arsenic from 50 to 10 µg/L. Countries such as India, Japan, Taiwan, the United States, and Vietnam have adopted 10 µg/L as the maximum permissible level for arsenic, while countries such as China, Bangladesh, and some of the nations of South America have retained the higher concentration of 50 µg/L (Chakraborti et al. 2010; Reddy and Roth 2013; Jadhav et al. 2015).

Thus, the widespread prevalence of arsenic in surface and underground water and its adverse toxic effects has raised global concern toward the urgent removal of arsenic from aqueous environments. So far, several removal techniques such as oxidation, coagulation, precipitation, membrane separation, ion exchange, and adsorption have been developed to treat arsenic-contaminated water. Adsorption-based removal is regarded as the most promising and highly acceptable method among other available methods due to its simple operation, low cost, lack of sludge generation, and great scope for adsorbent regeneration and reuse (Gautam et al. 2015; Banerjee et al. 2015; Rai et al. 2015). However, some setbacks were also associated with this technique; for instance, it is a time-consuming and slow process; most of the adsorbents are not versatile, so the process cannot be used for all types of contaminants; not all adsorbents are commercially acceptable due to poor removal efficiency; and the column-based performance of adsorbents at the industrial level are still ignored. In recent years, nanoengineered materials have gained significant momentum in pollution-related studies over their role in water and wastewater treatment, which is highly commendable. Nanoparticles can serve as potential adsorbents due to their unique properties such as small size, high surface area, high reactivity, excellent catalytic performance, being easily separable, and the presence of a large number of active sites that promote effective interaction with different pollutants (Ali 2012). Due to these beneficial characteristics of nanosized materials, researchers have employed nanoscale adsorbents for the decontamination of arsenic-polluted waters. Nanoadsorbents exhibit rapid and remarkable arsenic scavenging competence, along with a high grade of decontamination efficiency.

In this chapter, we provide an outline of existing arsenic removal techniques. Moreover, the aim of this chapter is limited to providing an overview related to the use of nanoscale materials in the treatment of arsenic-contaminated water. Special emphasis has been laid on the latest findings and recent developments in the field of nanosized materials.

13.4 TREATMENT TECHNIQUES FOR REMOVAL OF ARSENIC FROM WATER/WASTEWATER

The stringent environmental regulations and hazardous effects on health of arsenic have stimulated various researchers, scientific workers, and government agencies to ensure its immediate removal from water/wastewater. Different treatment techniques, including both conventional and advanced, have been extensively used for the removal of As from aqueous media, under both laboratory and field conditions. The most commonly employed techniques for the elimination of As from contaminated water include oxidation, coagulation–flocculation, ion exchange, membrane processes, and adsorption. In the next section, the performance of these techniques are briefly described, along with their merits and shortcomings.

13.4.1 OXIDATION

In natural water, As occurs in the two oxidation states +3 (arsenite) and +5 (arsenate). It has been demonstrated that the removal of arsenite is more tedious than that of arsenate, as most of the treatment processes are more effective at removing the latter. Therefore, for the removal of arsenite, the treatment process includes the conversion of arsenite into arsenate by oxidation followed by its subsequent elimination (Bissen and Fimmel 2003). However, this conventional oxidation-based treatment does not ensure removal of the arsenic species from water; this method is a typical pretreatment process, followed by removal through another process, most often adsorption or coagulation (Lee et al. 2003). The oxidation of arsenite can be carried out in various ways, such as in the presence of air or pure oxygen, in the presence of oxidants such as sodium hypochlorite, chlorine dioxide, potassium permanganate, or hydrogen peroxide, or by an oxidant generating photochemical oxidation via the ultraviolet (UV)/H_2O_2, UV/iron, and TiO_2/UV systems. Lee et al. (2003) investigated arsenite oxidation by Fe(VI); arsenite was successfully oxidized to arsenate by ferrate using chemicals with the stoichiometry of 3:2 (arsenite:ferrate). It was also proposed that oxygen transfer was the basic mechanism that initiates the oxidation process. Sorlini and Gialdini (2010) explored the efficacy of the oxidation of arsenite by employing four conventional oxidants, namely, chlorine dioxide, sodium hypochlorite, potassium permanganate, and monochloramine. The results demonstrated that the best yield was obtained with potassium permanganate, with efficiency up to 100% and oxidation reaction completed within 7 s, followed by hypochlorite, chlorine dioxide, and monochloramine with the lowest yield. Likewise, Hu et al. (2012b) prepared a water treatment system that exhibited promising potential for the oxidation of arsenite and its subsequent removal by coagulation. The oxidation of As(III) was carried out by means of active chlorine using sodium hyplochlorite (NaOCl). The stoichiometric rate was maintained at 0.99 mg Cl_2/mg As(III). Recently, an advanced oxidative treatment system was proposed by Tong et al. (2014). The process involved induced reductive dissolution of As-rich Fe(III) oxyhydroxides in the absence of oxygen. The study was based on the hypothesis that arsenic in anaerobic conditions can be effectively immobilized by enhancing the Eh of the iron-rich groundwater. An electrode system was employed to accomplish the oxidation reaction. An Fe anode and mixed metal oxide (MMO) cathode liberated Fe(II) and O_2, respectively, while OH^- was produced from the MMO cathode. The arsenite was oxidized due to the presence of O_2 and, meanwhile, Fe(II) precipitated as Fe(III) hydroxide, on which arsenite or arsenate can be adsorbed.

The advantageous aspects of this technique involve simple operation, easy handling, the possibility of *in situ* treatment, cost-effectiveness, and the capability of treating large volumes of wastewater. The drawbacks include the generation of a large volume of sludge and carcinogenic by-products and the presence of foreign substances that may reduce removal efficiency, as this is a pretreatment process that requires an additional treatment process.

13.4.2 Coagulation/Flocculation/Precipitation

Coagulation and flocculation are the most promising and widely used arsenic removal techniques in developing nations (Sancha 2006). The coagulation process can lower the pollution load from the milligram per liter level to the microgram per liter level. The process includes all reaction mechanisms that result in the formation of small flocs followed by large aggregates. The overall process comprises coagulant formation, destabilization of chemical particles, and physical interparticle contact. Coagulation involves the removal of colloidal and settleable particles of sizes between 0.001 and 100 microns, and greater than 100 microns, respectively. Coagulation-based arsenic removal involves the conversion of soluble arsenic into insoluble large aggregates or precipitates, as the coagulants added strongly reduce the absolute values of the zeta potentials of the particles (Song et al. 2006). Thus, the well-developed aggregates can be eventually separated from the system by means of sedimentation or filtration. Well-known coagulants used for arsenic removal are aluminum salts such as alum, polyaluminum chloride, aluminum chloride, aluminum sulfate, and ferric salts such as ferric chloride and ferric sulfate (Baskan and Pala 2010; Hu et al. 2012a). In addition to iron and aluminum compounds, manganese, calcium, and magnesium compounds can also be used as effective coagulants for removing arsenic from water in a neutral-pH medium. The effective pH range for arsenic removal has been reported as 5–7 with aluminum ions, and 5–8 with ferric ions (Song et al. 2006). Results demonstrated that iron salts are more efficient in arsenic removal than aluminum salts. Sometimes, effective coagulation did not occur; in such cases, a coagulant aid, in the form of an organic polymer, was added to promote the coagulation process (Wang et al. 1978). The coagulation process was greatly influenced by coagulant type and dose, the oxidation state of arsenic, pH, contact time, agitation speed, and the presence of competing ions. Pallier et al. (2010) reported that a higher coagulant dose could remove As(III) more efficiently without an oxidation step. Arsenic removal by coagulation was investigated by various researchers (Song et al. 2006; Pallier et al. 2010; Baskan and Pala 2010; Hu et al. 2012a). Song et al. (2006) reported the removal of arsenic from highly arsenic-contaminated water from acid mine drainage in Mexico through an enhanced coagulation process using ferric ions and coarse calcite. The results of the experiment suggested that coagulates produced from the ferric ions were very fine and could easily be filtered through a microfiltrating membrane. While undergoing treatment with coarse calcite, arsenic coagulates were coated onto the calcite surfaces and thus settled at the bottom of the reactor. The settled coagulates were easily separated by conventional filtration. The process led to the removal of more than 99% of arsenic from contaminated water. The effect of organic matter in arsenic removal by coagulation was studied by Pallier et al. (2010). It was reported that arsenite removal depended more on the coagulant dose and the availability of reactive sites on hydroxide surfaces than on the pH level and the presence of organic matter, while arsenate removal was influenced by the pH level, zeta potential, and the presence of organic matter. Baskan and Pala (2010) investigated that at low initial arsenate concentrations, the highest arsenate removal efficiency was obtained at high aluminum sulfate doses, while at high initial arsenate concentrations, the highest arsenate removal efficiency was achieved with low doses of aluminum sulfate.

The advantages of the coagulation process are that the process is simple and economical, it can be effectively used over a wide pH range, it is applicable to the treatment of large volumes of water, and it has high arsenate removal efficiency. The disadvantages are that the treatment process requires high amounts of coagulant dose, the generation of sludge creates environmental concerns, and an additional treatment step is required for the separation of coagulates from the reactor.

13.4.3 Ion Exchange

Ion exchange is also considered to be an attractive technology for the elimination of arsenate from water and wastewater. It is a physicochemical process that involves the exchange of anions present on the solid resin phase with targeted anionic contaminants in the aqueous phase. Thus, in the ion-exchange process, solid resin plays a leading role. The solid resins are typically organic molecules with three-dimensional network structures that electrostatically attach to large numbers of ionizable functional groups, with which the targeted anions are easily replaced. For the removal of anionic species, strong-base anion resins are commonly used, due to their strong exchange affinity. Inorganic resins are also reportedly used as ion exchangers and are considered to be more advantageous than strong-base anion resins, due to their high resistance against attacks by acids, alkalis, and oxidants (Suzuki et al. 2000). In the case of arsenite species, their removal is not possible using anion-exchange resin, as arsenite usually exists as a neutral species; thus, it is not exchanged. Therefore, the pretreatment of arsenite into arsenate is a requisite step (Mondal et al. 2013) However, oxyanionic species of arsenate such as $H_2AsO_4^-$, $HAsO_4^{2-}$, and AsO_3^{4-} are exchanged with high efficiency when using anionic-charged functional resin groups, thus producing high-quality effluents with low arsenate concentration (Singh et al. 2015). Arsenate anions can be easily eliminated through the use of strong-base anion resins, available either in the form of chlorides or hydroxides (An et al. 2010; Donia et al. 2011). The efficacy of the removal process largely depends on the solution pH, competitive anions such as nitrates, sulfates, and phosphates, the type of resin employed, alkalinity, and influent concentration. Barakat and Shah (2013) reported the exchange of arsenate and arsenite ions from an aqueous solution through the ion-exchange process by using a strong-base chloride to form anion-exchange resin spectra/gel. The results suggested that arsenite was weakly adsorbed, whereas arsenate was strongly retained on the resin. The removal efficiency of arsenate was 99.2%.

The merits of the ion-exchange technique are that the removal process is independent of the solution pH and influent arsenic concentration, and removal efficacy is moderately high. The disadvantages are that the direct removal of arsenite is not possible and a pretreatment is required, removal efficiency is greatly reduced due to interfering anions, clogging of the resin frequently occurs, and large volumes of toxic brine are released during resin regeneration.

13.4.4 Membrane Separation

In recent years, membrane-based techniques have gained significant attention due to their high removal performance and reliability in the removal of arsenic from water/wastewater. The removal of arsenic through membrane-based filtration techniques works on the principle of differences in pressure. Membranes are typically composed of synthetic material with billions of small pores that act as a selective barrier that allows certain materials to pass through the membrane while others are rejected. Thus, the movement of certain specific constituents across the membrane is regulated by a driving force, such as the pressure difference between the two sides of the membrane (Uddin et al. 2007). In general, membrane separation processes are classified into five categories, namely, microfiltration (MF), ultrafiltration (UF), nanofiltration (NF), electrodialysis (ED), and reverse osmosis (RO) (Jadhav et al. 2015). The separation accomplished through this technique relies on the pore size of the membrane; for MF and UF, separation is achieved through mechanical sieving, while for NF and RO, capillary action or solute diffusion are responsible for the separation (Shih et al. 2005). The advantages of this technique include deep removal of arsenic species, high removal efficiency for arsenate ions along with additional contaminants, and no scope for the generation of secondary wastes. The disadvantages are the process's low removal efficiency for arsenite species, that it requires high energy input, often demands a pretreatment step, and has high operational costs.

13.4.4.1 Microfiltration

The technique is useful for the separation of particles from aqueous media whose size is in the range of 0.1 to 10 microns. The arsenic removal efficacy depends greatly on the size distribution of the arsenic particles present in the fluid mixture. However, the technique is not suitable for the removal of dissolved and colloidal arsenic species, as the available pores are too large to screen them. Thus, an MF membrane is highly applicable to the removal of the particulate form of arsenic, but it rarely exists in particulate form in the natural aquatic environment. Therefore, the removal efficiency of arsenic can be enhanced by increasing the size of the arsenic-bearing species by employing techniques such as coagulation and flocculation (Ghurye et al. 2004). For instance, Han et al. (2002) investigated arsenic removal from drinking water by flocculation followed by MF. Flocculation was performed by using ferric chloride and ferric sulfate; moreover, flocculation efficiency was enhanced by the addition of small amounts of cationic polymeric flocculation aids. Arsenic removal by flocculation and MF depends on the interaction between arsenic and ferric complexes and also on the rejection of arsenic-bearing flocs by the membrane. It was also investigated whether flocculation followed by MF was better for arsenic removal than flocculation–sedimentation. The efficacy of the process also depends on the solution pH and the presence of interfering ions. Maximum removal was observed when using mixed esters of cellulose acetate and a cellulose nitrate MF membrane with a pore size of 0.22 micron, as compared with a membrane with a 1.2 micron pore size. The process was reported as economical, due to lower energy input as compared with other membrane-based removal techniques.

Ghosh et al. (2011) also reported arsenic removal using a similar technique. An arsenic species from the feed solution with a concentration of 200 μg/L, in the presence of fluoride and iron contaminants with arsenic content of 8.7 μg/L, was first coagulated electrically and then filtered using a ceramic membrane. However, the efficiency of MF for arsenic removal is doubtful as, due to the large pore size, arsenic species of small particle size can easily escape through the pores.

13.4.4.2 Ultrafiltration

In this membrane technique, the filtration mechanism is driven by low pressure and follows size-based exclusion. The UF membrane allows separation of molecules of sizes less than 10 Å. The UF membrane bears pores of sizes in the range of 100–10,000 nm, in which macromolecules, colloids, and solutes with molecular weights ranging from 300 to 5,00000 Da are easily retained. UF is not considered a favorable removal technique due to its large pore size; thus, it fails to screen small-sized contaminant species. UF assisted by electric repulsion demonstrates greater removal efficiency in contrast to pore-dependent sieving by UF only. The filtration-based removal of As was reportedly influenced by the presence of divalent anions and natural organic matter. Arsenic rejection is more highly reduced in the presence of divalent ions than monovalent ones. It was also observed in some cases that in the presence of divalent cations, arsenate ions start to interact with these solutes, resulting in poor filtration of the arsenate. Moreover, in the presence of organic matter, the performance of the UF membrane improves and higher arsenate rejection is reported, which is probably due to complexation of organic matter with divalent cations. To improve the arsenic removal efficiency through UF membranes, Gecol et al. (2004) investigated micellar-assisted enhanced UF. Through this technique, arsenic was removed in the form of micelles as, prior to treatment, arsenic in the form of arsenate and arsenite was allowed to interact with the cationic surfactant and formed micelles, which can be easily filtered through a UF membrane. Iqbal et al. (2007) also reported the removal of arsenic from groundwater by micellar-enhanced UF. The study suggests that the removal efficiency of arsenic also depended on the type of surfactant used. The study was conducted using four types of surfactant, namely, hexadecylpyridinium chloride (CPC), hexadecyltrimethylammonium bromide (CTAB), benzalkonium chloride (BC), and octadecylamine acetate (ODA). The highest removal of 96% was achieved when CPC was used, followed by CTAB with a removal efficiency of 94%; however, the lowest removal of 57% was reported with BC, due to a higher critical micelle concentration. Another attempt to enhance the removal performance of UF membranes was carried out by Lohokare et al. (2008), who investigated the applicability of polyacrylonitrile (PAN)-based negatively charged UF membranes to the removal of arsenic from aqueous solutions. It was reported that the surface hydrolysis of PAN-modified UF membranes by NaOH led to a reduction in pore size, due to the formation of carboxylate groups on the membrane wall as well as within the pore surface wall. The PAN-modified membrane demonstrated excellent potential for the removal of arsenate ions and the removal mechanism followed the Donnan exclusion principle. In the removal process, the size exclusion mechanism contributes little, while electrostatic repulsion plays an important role.

13.4.4.3 Nanofiltration

The removal technique of employing an NF membrane is another of the best possible removal technologies, especially for small water treatment units (Sato et al. 2002). The technique offers high-grade removal of dissolved arsenic from contaminated water that contains a very low amount of suspended solids (Figoli et al. 2010). The low-pressure-based operation of the NF process is beneficial to energy consumption. The technique is advantageous with respect to other membrane techniques, in terms of higher selectivity and increased water flux at a much lower operating pressure (Vrijenhoek and Waypa 2000). The technique is capable of eliminating contaminants of sizes less than 1 nm and molecular weights less than 1000 Da. NF membranes are generally applicable to the separation of large and multivalent ions from monovalent ones. Hence, the NF process is competent to preserve the mineral content of the treated water (Uddin et al. 2007). NF membranes are also termed as *loose RO membranes*, as they provide relatively higher water fluxes. The membranes bear charged surfaces, which play a prominent role in the separation of molecules. The NF membranes usually possess random and negatively charged ions at neutral and alkaline pH levels, while at acidic pH levels, the membranes lose their charge. The removal of anions is due to the combined effect of solute diffusion and ion exclusion. Urase et al. (1998) examined the efficiency of negatively charged NF membranes at different pH levels for the removal of arsenate, arsenite, and dimethylarsinic acid (DMMA) from groundwater. It was reported that a change in the solution pH caused a change in the electric charge of arsenic species. Thus, in a variable pH environment, the rejection rate of arsenic also varied. The results suggested that for DMAA and arsenate, the rejection rate was as high as nearly 99% and 89%, respectively, regardless of the variable pH conditions. For arsenite, the rejection rate increased from 50% to 89% with an increase of the pH value from 3 to 10. The reason for the change in the rejection rate with changes in the pH level reveals that, for charged solutes, rejection was higher as compared with noncharged ones. Sato et al. (2002) investigated whether arsenic removal by NF membranes such as ES-10, NTR-729HF, and NTR-7250 was more effective in treating arsenate and arsenite as compared with rapid sand filtration with interchlorination. Maximum removal of arsenite and arsenate of 80% and 97%, respectively, was observed at a pressure of 1.2 MPa. Uddin et al. (2007) reported that the pH level, initial arsenic concentration, and temperature were the decisive factors that determined arsenic concentration in the treated effluents.

13.4.4.4 Reverse Osmosis

This process has been declared to be the most technically superior removal method by the EPA. This technique facilitates deep arsenic removal, and effluents produced after treatment usually contain low levels of arsenic concentration, that is to say, below the MCL limit. This is one of the most favored treatment systems of drinking water contaminated with arsenic species. The method is effective for the removal of species of sizes less than 1 nm and offers very high rejection of low molecular weight compounds. The major disadvantage of this technique is that the rejection rate is very low for arsenite ions. Several researchers have investigated the efficiency of the RO technique for arsenic removal. Ning (2002) reported that arsenic in common high oxidation states is very effectively removed by the RO process. Kang et al. (2000) investigated the removal of arsenic from drinking water using ES-10 and NTR-729HF RO membranes. It was reported that, for the removal of neutral species, the radius of the particle plays a significant role, while ionic solutes show no removal dependence on the particle radius. Moreover, a higher removal rate was achieved for arsenate ions than for arsenite. Removal efficiency is also enhanced with alkaline pH conditions. Thus, pH control is considered an important criterion that ensures greater arsenate removal efficiency. Akin et al. (2011) investigated the removal of arsenite and arsenate from water via SWHR membrane and BW-30 membrane. In the study, it was demonstrated that not only the pH level, but also the concentration of the feed water and operating pressure determined the rate of rejection. For both membranes, arsenate removal was higher at pH values above 4.0, while pH values above 9.1 were suitable for arsenite removal. Moreover, a high operating pressure favors the rejection of both anions.

13.5 ADSORPTION

Adsorption is a well-known water treatment technique that has been employed over centuries and has remained to the present day as a preferred purification technique (Ali 2012). Among other treatment techniques, adsorption is widely used for dearsenication of water and wastewater. It is considered to be one of the proven decontamination techniques that offers economical, highly efficient, and environmentally benign treatment (Banerjee et al. 2015). In the adsorptive removal of contaminants, the adsorbent plays a prominent role in the removal process. The physical and chemical characteristics of the adsorbent, such as surface area, porosity, size distribution, density, and surface charge greatly influence the sorption process (Yazdanbakhsh et al. 2011). Moreover, the adsorptive performance of the adsorbent is also strongly dependent on the chemical state of the adsorbate species in which it exists at the time of treatment. Thus, knowledge of aqueous chemistry and speciation of arsenic is essential to understand what would further facilitate the achievement of high arsenic removal efficiency. Considerable work has been reported on arsenic removal by adsorption using different kinds of adsorbent materials. Arsenic removal by conventional adsorbents such as activated carbon, alumina, minerals, and ores has been extensively explored by researchers. But, the major setbacks posed by these adsorbents include low removal efficiency, high equilibrium time, and the problem of regeneration. The adsorption characteristics of these adsorbents are depicted in Table 13.1.

The adsorption characteristics illustrated in Table 13.1 reflect that, in most of the cases, adsorbents take a longer contact time to equilibrate, ranging from hours to days; for instance, activated alumina grain (Lin and Wu 2001) attains equilibrium after 2400 min, hematite (Pajany et al. 2009) equilibrates at 1440 min, and clinoptilolite (Camacho et al. 2011a,b) exhibits a high equilibrium period of 2580 min. Hence, adsorbents that exhibit slow adsorption kinetics are generally not acceptable at the commercial level, as the treatment is considered not to be economically beneficial. Thus, to overcome the challenges offered by conventional adsorbents, the research focus has shifted toward exploring the prospect of nanoscale material in the scavenging of arsenic ions present in wastewater. Nanoscale materials possess high competence for the rapid removal of contaminants from the liquid phase within a few minutes of contact time with a relatively small dose, which makes their application effective and economical (McDonald et al. 2015). To date, different classes of nanomaterials, such as various metal oxides of transition metals, bimetallic oxides, layered double hydroxides, magnetic nanoparticles, carbon-based nanoparticles, and metal organic frameworks were extensively used in the removal of arsenic-bearing wastewater. In the next section, a brief outline of these nanomaterials is presented; in addition, a list representing the adsorption characteristics of various nanoscale materials is depicted in Table 13.2.

13.5.1 Different Nanoadsorbents Used for Arsenic Removal

13.5.1.1 Nanometal Oxides of Different Transition Metals

Iron-based metal oxides have been one of the extensively explored adsorbent materials by researchers, due to their low cost, high stability, environmental affability, and strong affinity for arsenic

TABLE 13.1

Adsorption Characteristics of Various Conventional Adsorbents for Removal of Arsenic Ions from Wastewater

Arsenic Species	Adsorbent	Equilibrium Time	pH, Temperature (K)	Adsorbent Dose; Adsorbate Conc. (mg/L)	Adsorption Capacity/ Removal Efficiency	References
As(V)	Oat hull-based activated carbon (AC)	100 min	5; 297 K	15 mg/L; 25–200 µg/L	3.09 mg As/g	Chuang et al. 2005
As(III) and As(V)	Activated carbon	—	As(III) 12; As(V) 3	1.0 g/10 mL; 20–150 ppm	As(III) 95% at 10 ppm; As(V) 84% at 10 ppm	Ansari and Sadegh 2007
As(III)	Activated carbon	360 min	3.56–3.70; 298 K	0.2 g/100 mL 5–10 mg/L	58% at 298 K	Wu et al. 2008
As(V)	Beet pulp AC	1440	9.1–9.4; room temp.	0.05 g/100 mL; 100–1000 µg/L	691 µg/g	Lodeiro et al. 2013
As(III) and As(V)	Coal-based mesoporous AC	As(III) 60 min; As(V) 30 min	6; 280 K	0–400 mg/L; 0.1–0.5 mg/L	As(III) 1.49 mg/g; As(V) 1.76 mg/g	Li et al. 2014b
As(V)	Apricot stone-activated carbon	100 min	9.0; 303 K	0.1 g/100 mL; 75 mg/L	26.3 mg/g at 293 K; 27 mg/g at 303 K	Hassan et al. 2014
As(III) and As(V)	Activated alumina grains	2400 min	As(III) 6.9; As(V) 5.2; 298 K	0.1–0.5 g/L; 0.02–12 mg total As/L	As(III) 3.48 mg/g; As(V) 15.9 mg/g	Lin and Wu 2001
As(III) and As(V)	Mesoporous Alumina	2200 min	5.0; room temp.	0.1 g/20 mL; 0.1–20 mmol total As/L	As(III) 0.63 mmol/g; As(V) 1.62 mmol/g	Kim et al. 2004
As(III)	Activated alumina	360 min	7.6; 298 K	1.0 g/100 mL; 0.5–1.5 mg/L	0.1803 mg/g	Singh and Pant 2004
As(III) and As(V)	Alkali-treated laterite (6 N HCl)	180 min	6.5; 305 K	0.5 g/L; 1000 µg/L	As(III) 8.0 mg/g; As(V) 24.1 mg/g	Maiti et al. 2010
As(III) and As(V)	Activated alumina	—	7; room temp.	100–5000 µg/L	As(III) 0.005 g/g; As(V) 0.008 g/g	Lescano et al. 2015
As(III) and As(V)	Red mud (bauxite)	As(III) 45 min; As(V) 90 min	As(III) 9.5, As(V) 1.1–3.2; 298 K	20 g/L; 2.5–30 mg/L	As(III) 8.86 µg/g; As(V) 6.86 µg/g	Altundogan et al. 2000
As(III) and As(V)	Acid treated red mud	As(III) 45 min; As(V) 90 min	As(III) 5.8–7.5, As(V) 1.8–3.5; 298 K; 343 K	20 g/L; 125–1500 µg/L;	As(III) 11.8 µmol/g; As(V) 17.71 µmol/g	Altundogan et al. 2002
As(V)	FeCl$_3$ modified red mud	1440 min	6.0; room temp.	100 mg/L; 1 mg/L	68.5 mg/g	Zhang et al. 2008
As(III) and As(V)	Ferruginous manganese ore	30 min	2.0–8.0; room temp.	0.2 g/100 mL; As(III) 0.12 mg/L As(V) 0.19 mg/L	As(III) 0.53 mg/g; As(V) 15.38 mg/g	Chakravarty et al. 2002
As(V)	Goethite	120 min	5.0; 302 K	100 mg/250 mL; 10 mg/L	4.7 mg/g at pH 5.0	Lakshmipathiraj et al. 2006

As(V)	Granular ferric hydroxide	30 min	6.5; 313 K	250 mg/L;100 µg/L	4.57 mg/L	Banerjee et al. 2008
As(V)	Hematite	1440 min	7.0; room temp.	0.2 g/50 mL; 500 µg/L	>80% As(V) removal observed at pH 7.0	Pajany et al. 2009
As(V)	Iron-coated zeolite	30 min	3–10; room temp.	3 g/30 mL.; 2 mg/L	0.68 mg/g	Jeon et al. 2009
As(V)	Hematite	180 min	5.0; room temp.	0.5 g/100 mL; 5.0 mg/L	0.206 mg/g	Aredes et al. 2013
As(V)	Magnetite	180 min	5.0; room temp.	10 g/100 mL; 5.0 mg/L	0.0495 mg/g	Aredes et al. 2013
As(V)	Goethite	180 min	5.0; room temp.	10 g/100 mL; 5.0 mg/L	0.0495 mg/g	Aredes et al. 2013
As(V)	Laterite	180 min	5.0; room temp.	10 g/100 mL; 5.0 mg/L	0.0495 mg/g	Aredes et al. 2013
As(V)	Siderite	48 h	2.0; 318 K	10 g/L 1.0–20.0 mg/L	2.904 mg/g	Zhao and Guo et al. 2014
As(V)	Kenyaite	360 min	2.0; 298 K	3.0 g/L; 25.0 mg/L	5.45 mmol/g	Guerra et al. 2009
As(III) and As(V)	Titanium dioxide	As(III) 240 min; As(V) 63 min	As(III) 8.5; As(V) 4.5	1.0 g/L; 300 µg/L	As(III) 32.4 mg/g; As(V) 41.4 mg/g	Bang et al. 2005
As(III)	Titanium dioxide	120 min	9.4; room temp.	0.05 g/L;1–15 µmol/L	49.2 µmol/g	Liu et al. 2008
As(III) and As(V)	Fe-exchanged Clinoptilolite	30 min	As(III) 6.0; As(V) 3.0	1.0 g/50 mL.; 0.5 mg/L As	As(III) 100 mg/kg; As(V) 50 mg/Kg	Li et al. 2011
As(V)	Fe-modified Clinoptilolite	60 min	7.0; 293 K	500 mg/50 mL; 100 µg/L	9.2 µg/g	Baskan and Pala 2011
As(V)	Clinoptilolite	2580	6.0; 298 K	1.0 g/100 mL; 0.5 to 50 µg/L	0.0079 µg/g	Camacho et al. 2011b
As(V)	Manganese oxide-coated clinoptilolite	2580	4.0–6.0; 298 K	1.0 g/100 mL; 0.5 to 50 µg/L	0.0338 µg/g	Camacho et al. 2011b
As(V)	Manganese modified-clinoptilolite	80 min	10.0; 311 K	1.0 g/100 mL; 0.1–3.0 mg/L	150.9 µg/g	Massoudinejad et al. 2015

TABLE 13.2

Adsorption Characteristics of Some Nanoscale Adsorbents for Removal of Arsenic Species from Aqueous Solutions

Type of Arsenic Ion Treated	Type of Nanomaterial	Nanomaterial	Size and Surface Area	pH, Temp (K)	Equilibrium Time	Adsorbent Dose (g/L) and Adsorbate Conc. (mg/L) or (mol/L)	Adsorption Capacity (mg/g or μg/g) or Removal Efficiency (%)	References
As(III) and As(V)	Metal oxide	Zerovalent iron	>90 nm; 24.4 m²/g	3.0–9.0; 298 K	As(III) 30 min; As(V) 30 min	As(III) 2.0 mg/L; As(V) 20.0 mg/L	As(III) 90%; As(V) 96%	Giasuddin et al. 2005
As(III)	Metal oxide	Zerovalent iron	2–10 nm; 64 m²/g	7.0; room temp.	—	12 g; 500 μg/L flow rate; 1.8 mL/min	50 cm column; 100% of 0.5 mg/L of As(III) was removed for 90 days	Kanel et al. 2007
As(III)	Metal oxide	Iron oxide	20–100 nm; 39 m²/g	5.0; room temp.	60 min	100 mg/100 mL; 0.25 μg/mL	96% at pH 4.5–7.5	De et al. 2009
As(III)	Metal oxide	Iron oxide hydroxide nanoflower	Average size 20 nm; 6.577 m²/g	7.0; 298 K	100 min	1.0 g/L; 200 μg/L, 500 μg/L	475 μg/g	Raul et al. 2014
As(III)	Metal oxide	Hierarchical copper oxide	Each crystallite < 10 nm; 87 m²/g	4.0; 298 K		0.03 g/20 mL; 11.5 mg/L	5.7 mg/g	Cao et al. 2007
As(III) and As(V)	Metal oxide	Copper oxide	12–18 nm; 85 m²/g	As(III) 10.0; As(V) 7.8; room temp.	30 min	2.0 g/L; 0.9 mg/L	As(III) 26.9; As(V) 22.6	Martinson and Reddy 2009
As(III)	Metal oxide	Copper oxide	52.11 m²/g	8.0; 323 K	100 min	1.0 g/L; 100–1000 μg/L	1086.2 μg/g	Goswami et al. 2012
As(III)	Metal oxide	Hierarchical copper oxide	Each crystallite < 20 nm, 119 m²/g	7.0; 298 K	120 min	0.4 g/L; 1.0 mg/L	12.9 mg/g	Yu et al. 2012
As(III) and As(V)	Metal oxide	Nest-like MgO	<100 nm; 32.96 m²/g	7.0; 298 K	As(III) 200 min; As(V) 100 min	0.1–0.5 g/L; As(III) 4.639 mg/L, As(V) 7.189 mg/g	As(III) 643.84 mg/g; As(V) 378.79 mg/g	Yu et al. 2011

Target	Category	Material	Size/surface area	pH; temperature	Time	Conditions	Capacity/removal	Reference
As(III) and As(V)	Metal oxide	Zirconium oxide	Crystallite size 7 nm; 98 m²g	7.1; room temp.	—	As(III) 0.005–0.06 g/L; 0.212 mg/L; As(V) 0.01–0.1 g/L; 0.355 mg/L	As(III) 16.5 mg/g; As(V) 13.56 mg/g	Cui et al. 2013
As(V)	Metal oxide	Zinc aluminate	Crystallite 5–7 nm; 145 m²/g	7.0; room temp.	180 min	0.05 g/30 mL; 10–200 µg/L	86% removal	Kumari and Bhaumik 2015
As(III)	Metal oxide	γ-alumina	30–70 nm; 497 m²/g	6.0; 298 K	60 min	0.1 g/100 mL–10 µg/L	79.0% removal	Patra et al. 2012
As(V)	Functionalized metal oxide	Mesoporous SBA-15 (Al10SBA-15)	450 m²/g	6.6; 298 K	100 min	0.2 g/50 mL; 44.7 mg/L	19.77 mg/g	He et al. 2015
As(V)	Functionalized metal oxide	Nanoalumina dispersed in chitosan-grafted polyacrylamide	—	7.2; 303 K	360 min	25–80 mg/L	4.23 mg/g	Saha and Sarkar 2012
As(III) and As(V)	Magnetic material	γ-Fe₂O₃ nanoparticles	7–12 nm; 168.73 m²/g	3.0; 323 K	30 min	0.08 g/100 mL; 100 mg/L	As(III) 74.83 mg/g; As(V) 105.25 mg/g	Lin et al. 2012
As(V)	Magnetic material	Magnetic nanomaterial synthesized from red mud	6–14 nm	2.0; 338 K	30 min	8 g/L; 10–1000 µg/L	400 µg/g	Akin et al. 2012
As(III) and As(V)	Functionalized magnetic material	Ascorbic acid-coated Fe₃O₄ nanoparticles	Average size 10 nm; 179 m²/g	As(III) 5.0; As(V)2.0; 300 K	180 min	60 mg/L; 0.1 mg/L	As(III) 46.6 mg/g; As(V) 16.56 mg/g	Feng et al. 2012
As(III) and As(V)	Functionalized magnetic material	Mg-doped α-Fe₂O₃	2–7 nm, 438.2 m²/g	7.0; 298 K	250 min	0.02 g/L; As(III) 0.097 mg/L; As(V) 0.101 mg/L	As(III) 127.4 mg/g; As(V) 83.2 mg/g	Tang et al. 2013
As(V)	Bimetallic oxides	Fe-Al nanoparticles	131 m²/g	2.0; 303 K	90 min	100 mg/50 mL; 6.5 mg/L	54.55 mg/g	Basu et al. 2012
As(III)	Bimetallic oxides	Mn-Al nanoparticles	44.73 m²/g	7.0; room temp.	600 min	0.3 g/100 mL; 1–90 mg/L	142.2 mg/g	Wu et al. 2012
As(V)	Bimetallic oxides	Y-Mn nanoparticles	1–20 µm	10; 298 K	1500 min	0.01 g/100 mL; 1–100 mg/L	279.9 mg/g	Yu et al. 2015
As(V)	Layered double hydroxides	Mg-Fe LDH	50–200 nm; 90.2 m²/g	12; room temp.	50 min	0.1 g/750 mL; 20 µg/L	480 µg/g at pH 12; 1 ppm As(V)	Park and Kim 2011 (Continued)

TABLE 13.2 (CONTINUED)
Adsorption Characteristics of Some Nanoscale Adsorbents for Removal of Arsenic Species from Aqueous Solutions

Type of Arsenic Ion Treated	Type of Nanomaterial	Nanomaterial	Size and Surface Area	pH, Temp (K)	Equilibrium Time	Adsorbent Dose (g/L) and Adsorbate Conc. (mg/L) or (mol/L)	Adsorption Capacity or Removal Efficiency	References
As(V)	Carbon nanoparticles	Zr immobilized carbon nanoparticles	Carbon nanoparticles 50–60 nm	2.0–3.0; room temp.	75% of adsorption within 10 min 600 min	1.0 g/L; 200 mg/L	110 mg/g	Mahanta and Chen 2013
As(III) and As(V)	CNTs	Magnetic activated CNTs	20–30 nm; 662.1 m²/g	—	10 min	—	As(III) 8.13 mg/g; As(V) 9.74 mg/g	Ma 2013
As(III) and As(V)	Carbon iron composite	Graphene in Fe_3O_4 nanocomposite	Fe_3O_4 average size 8 nm; 53 m²/g	7.0; 296 K	—	10 mg/50 mL; 1–10 mg/L	As(III) 8.67 mg/g; As(V) 61.73 mg/g	Paul et al. 2015
As(V)	Metal organic framework	Iron and 1,3,5-benzenetricarboxylic metal organic framework	—	4.0; 298 K	10 min	5.0 g/L; 5 mg/L	12.287 mg/g	Zhu et al. 2012
As(V)	Metal organic framework	Zeolitic imidazolate framework-8	—	4.0; 298 K	1500 min	40 mg/500 mL	76.5 mg/g	Li et al. 2014a
As(V)	Metal organic framework	MIL-53(Fe)	5–8 nm; 14 m²/g	5.0; 298 K	5 mg/L: 90 min; 10–15 mg/L: 120 min	1.0 g/L; 5 mg/L	21.27 mg/L	Vu et al. 2015
As(V)	Metal organic framework	MOF-808	150–200 nm	<7.0; room temp.	50 min	10 mg/50 mL; 5 mg/L	24.83 mg/g	Li et al. 2015

species (Tang et al. 2013). The use of nanoscale iron oxides for arsenic removal is widely reported (Kanel et al. 2007; De et al. 2009; Luther et al. 2012; Raul et al. 2014). Among nanoiron particles, zerovalent iron (NZVI) is regarded as a promising material, as it removes both arsenite and arsenate species simultaneously from aqueous media, without the need for any preoxidation step. De et al. (2009) effortlessly synthesized iron oxide nanoparticles from a precursor of Fe3+ species dispersed in PVA molecules, followed by precipitation using an ammonia solution. The synthesized particles were revealed to be of nanoscale size and rhombohedral structure, as given in Figure 13.3a, and the particle size distribution histogram (Figure 13.3b) demonstrates that the synthesized particles have a wide size distribution of 20–100 nm, while the average size was estimated to be 45 nm.

The mechanism of arsenic removal is presumably due to adsorption through the formation of surface complexes followed by coprecipitation of arsenic with Fe(II) and Fe(III) oxides/hydroxides that form *in situ* during the oxidation of ZVI (Ling and Zhang 2014). The probable reaction, proposed by Kanel et al. (2005), has been given in Equations 13.3 and 13.4, and represents oxidation of ZVI by water and O_2 that results in the formation of ferrous ions. Further, these ferrous ions, depending on the redox condition and a suitable pH level, react to produce magnetite (Fe_3O_4), ferrous hydroxide Fe $(OH)_2$, and ferric hydroxide $Fe(OH)_3$, represented by Equations 13.5, 13.6, and 13.8, respectively.

$$Fe^0 + 2H_2O \rightarrow 2Fe^{2+} + H_2 + 2OH^- \tag{13.3}$$

$$Fe^0 + O_2 + 2H_2O \rightarrow 2Fe^{2+} + 4OH^- \tag{13.4}$$

$$6Fe^{2+} + O_2 + 6H_2O \rightarrow 2Fe_3O_4(s) + 12H^+ \tag{13.5}$$

$$Fe^{2+} + 2OH^- \rightarrow Fe(OH)_2(s) \tag{13.6}$$

$$6Fe(OH)_2(s) + O_2 \rightarrow 2Fe_3O_4(s) + 6H_2O \tag{13.7}$$

$$Fe_3O_4(s) + O_2(aq) + 18H_2O \leftrightarrow 12Fe(OH)_3(s) \tag{13.8}$$

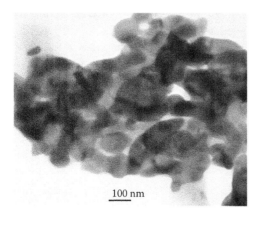

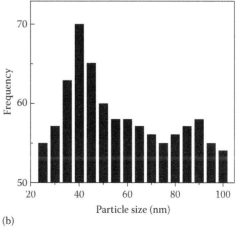

(a)   (b)

FIGURE 13.3 (a) HRTEM image and (b) histogram showing particle size distribution of iron nanoparticles. (From De, D. et al., *J. Environ. Sci. Health* A, 44, 152–162, 2009.)

Kanel et al. (2005) synthesized NZVI material by the reduction method, using NaBH$_4$ solution. The particles exist in the nanorange (1–120 nm), which was confirmed by an atomic force microscopy (AFM) study. The adsorption of arsenite exhibits rapid kinetics, as removal of more than 80% was observed within 7 min, and 99.9% removal was achieved in 60 min. The maximum adsorption capacity of 3.5 mg of As(III)/g of NZVI was estimated at 298 K. Tanboonchuy et al. (2011) investigated the effect of operational parameters on the adsorption of arsenate and arsenite via NZVI. The adsorption reaction of arsenite was found to reach equilibrium within 10, 15, and 30 min for pH values 4, 7, and 9, respectively, and for arsenate, the equilibrium condition was attained at 7, 20, and 40 min, respectively, for the corresponding pH values. At a pH of 4, a removal percentage of 81% was obtained within 10 min in the case of arsenite, and 99% removal was observed within 7 min of contact time for arsenate. The removal efficiency for both arsenic species reportedly declined as the pH value increased from 4.0 to 9.0. The initial arsenic concentration also influences the adsorption of arsenic species, and the amount of uptake of arsenic increases with an increase in the initial arsenic concentration. Dissolved oxygen was also investigated as an influencing factor; arsenite removal was greater than 80% after 10 min of reaction under oxygenated conditions, while only 52% was removed under deoxygenated conditions. With As(V), 99% and 76% were removed after 7 min of reaction under aerobic and deoxygenated conditions, respectively. Arsenic removal was anticipated to be enhanced in the presence of oxygen, due to the formation of inner- and outer-sphere arsenic complexation with iron hydroxides. The feasibility of arsenic removal by NZVI in the presence of humic acid was investigated by Giasuddin et al. (2007). The results suggested that humic acid exhibits a strong competitive effect with arsenic as well as other groundwater pollutants. The adsorption of arsenite and arsenate in the presence of humic acid reduced from 100% to 43% and 68%, respectively. However, removal of arsenic with iron oxide particles was also extensively reported. Luther et al. (2011) synthesized nanophase Fe$_3$O$_4$ and Fe$_2$O$_3$ particles for the removal of arsenite and arsenate ions from aqueous solutions. Such removal was found to be pH dependent; maximum removal was observed in the pH range 6–9. Outside this range, arsenic removal efficiency declined significantly. The presence of interfering ions (sulfate, carbonate, phosphate) considerably affected the removal performance of nanoparticles for arsenic species; in the case of arsenite ions, the interfering ions completely inhibited their binding with Fe$_3$O$_4$. Kanel et al. (2007) reported that the aggregation tendency of iron nanoparticles during the oxidative reaction of Fe(III) oxide/hydroxide limits the removal performance of nanoparticles, which can be controlled by surface modification, using stabilizers such as polyacrylic acid, starch, and oil. Kanel et al. (2007) modified iron nanoparticles with a nonionic surfactant (polyoxyethylene sorbitan monolaurate) and employed this in the removal of arsenite through the column mode. It was found that 100% of As(III) in 0.5 mg/L solution (flow rate 1.8 mL/min) was removed by modified iron nanoparticles. Recently, Raul et al. (2014) investigated the feasibility of iron oxide/hydroxide nanoparticles in the removal of arsenite ions. The adsorbent media was highly efficient and capable of removing arsenite from 300 mg/L to less than 10 mg/L, demonstrating a high removal capacity of 475 µg/g.

The performance of copper nanoparticles (CuO) in the removal of arsenic ions has been investigated by several researchers (Cao et al. 2007; Martinson and Reddy 2009; Goswami et al. 2012; Yu et al. 2012; McDonald et al. 2015). CuO was considered as an effective arsenic adsorbent, as it did not require particular pH levels or preoxidation of arsenite ions; moreover, it performed well in the presence of competing anions (Martinson and Reddy 2009). Adsorption of arsenite and arsenate by CuO nanoparticles was investigated by Martinson and Reddy (2009). The nanoparticles exhibited rapid adsorption within a few minutes. Maximum adsorption of >80% was reported for arsenite at pH 10, while arsenate removal was not very pH dependent, as removal remained >95% for the pH range 6–10. The presence of sulfate and silicate in the water showed no effect on the adsorption of arsenate, but only slightly inhibited the adsorption of arsenite; however, high concentrations of phosphate (>0.2 mM) reduced the adsorption of arsenic onto the CuO nanoparticles. The removal mechanism of CuO nanoparticles is interpreted as the oxidation of arsenite species into arsenate, followed by its adsorption over the surface of CuO nanoparticles. A similar removal mechanism was also suggested by McDonald et al. (2015). Goswami et al. (2012) also researched the adsorption characteristics of CuO nanoparticles for the removal of arsenic. It was found that arsenic adsorption by CuO was strongly dependent on the initial arsenic concentration, adsorbent dose, pH, competing ions, and temperature, while stirring speed had no influence on arsenic removal efficiency. CuO revealed a high adsorption capacity of 1086.2 µg/g.

Zinc oxide nanoparticles have also reportedly been used in the removal of arsenic ions. However, this metal oxide is the least investigated in the removal of arsenic ions, as compared with iron and copper oxide. Yang et al. (2011) synthesized ZnO microtubes via self-assembly of ZnO nanoparticles. The results suggested that the self-assembled ZnO microtubes were extremely efficient in the removal of arsenite from natural water in neutral pH conditions. The nanomaterials demonstrated a high arsenite uptake capability of 10 mg/g, even at a low arsenite concentration of 0.1 mg/L. Kumari and Bhaumik (2015) investigated the removal of arsenite species from wastewater using mesoporous zinc aluminate ($ZnAlO_4$) nanoparticles with high removal efficiency.

The use of γ-alumina nanoparticles in the removal of arsenic anions was reported by several researchers (Patra et al. 2012; Darban et al. 2013). However, the efficiency of γ-alumina nanoparticles in the removal of arsenic species is not very commendable, due to its irregular pore structure, low surface area, and slow adsorption rate; thus, bare alumina nanoparticles are rarely investigated in removal studies.

13.5.1.2 Functionalized Metal Oxides

The principal logic behind the application of functionalized metal oxides concerns the enhancement of removal efficiency of certain pollutants by these metal oxides, through the tailoring of the surface property of the adsorbent with certain chemicals. For instance, Patra et al. (2012) synthesized self-assembled mesoporous γ-alumina nanoparticles, using sodium salicylate as a template. The removal efficiency of the mesoporous Al_2O_3 nanoparticles exhibited a high removal efficiency of 73.7% for arsenate ions, as compared with bare alumina nanoparticles with a low removal efficiency of 52.6%. Saha and Sarkar (2012) also reported enhanced removal efficiency for arsenic ions by synthesizing alumina nanoparticles in chitosan-graft-polyacrylamide (CTSgPA). Singh et al. (2013) investigated selective removal of arsenic using acetate functionalized zinc oxide nanoparticles. The ZnO nanoparticles were synthesized using zinc acetate precursors. The presence of acetate groups over the surface of the ZnO nanoparticles offered host sites that efficiently bound arsenic ions through the formation of arsenic–acetate complexes. He et al. investigated the high uptake of arsenate ions using alumina-functionalized highly ordered mesoporous SBA-15 (Al_x–SBA-15). The adsorbent material was synthesized through modification of the SBA-15 surface with alumina nanoparticles. Al_{10}-SBA-15 exhibits high arsenate removal in a relatively wide pH range (2.0–8.2), and treated effluents bear an arsenate concentration of <2.235 mg/L.

13.5.1.3 Magnetic Nanomaterials

It has been demonstrated that the separation of nonmagnetic nanometal oxides after effluent treatment is a highly cumbersome task, as inefficient separation could result in their dispersion into the aqueous environment, which would further increase treatment costs; moreover, it could create a potentially damaging threat to natural organisms, as well as the environment. This problem can be dealt with by employing magnetic adsorbents in water treatment plants. The technique facilitates easy separation of used nanoadsorbents from treated water bodies by application of magnetic fields or by using simple hand magnets. Lin et al. (2012) reported the removal of arsenic contaminants with magnetic γ-Fe_2O_3 nanoparticles. The adsorbent material was synthesized easily by the coprecipitation method. The removal of arsenite and arsenate was very rapid and a steady state was reached within 30 min of contact time. The adsorption was found to be highly temperature dependent; the maximum uptake amounts of 74.83 mg/g for arsenite and 105.25 mg/g for arsenate were observed at 323 K. The saturated magnetic γ-Fe_2O_3 nanoparticles were reportedly recovered easily with the assistance of a magnetic field of strength greater than 0.35 T. Akin et al. (2012)

synthesized magnetic nanoparticles with waste red mud, and studied its application in the removal of arsenate ions from groundwater samples. The material exhibited high magnetic properties, with an estimated high magnetization saturation value of 55.3 emu/g. The maximum arsenate removal of 82% was seen at a pH level of 2.0. The magnetic nanoparticles were capable of bringing down the initial arsenite and arsenate concentrations of 1570 and 280 µg/L, respectively, to 15.3 µg/L from natural groundwater samples. Kilianova et al. (2013) reported the economical synthesis of ultrafine iron(III) oxide nanoparticles with a narrow size distribution of 3–8 nm. The material was further tested for its potential application in the removal of arsenate ions from aqueous media. The results indicated that the magnetic Fe_2O_3 nanoparticles possessed a strong magnetic response with a maximum magnetization value of 39.6 A m^2/kg. It was also reported that a high surface area and the pattern of arrangement of the nanoparticles into a mesoporous structure due to magnetic interaction played a significant role in the adsorption of arsenate ions.

13.5.1.4 Functionalized Magnetic Nanoparticles

Researchers have considered that bare magnetic nanoparticles are extremely difficult to recycle due to their high susceptibility to oxidation when exposed to the atmosphere due to their small size. Thus, to overcome this obstacle, the surfaces of the magnetic nanoparticles are functionalized, either by doping with some other elements, such as the doping of Mg into ultrafine α-Fe_2O_3 nanocrystallites (Tang et al. 2013), or by using some stabilizers or capping agents such as oleic acid ligands (Yavuz et al. 2006) or ascorbic acid (Feng et al. 2012). The ascorbic acid–coated magnetic nanoparticles synthesized by Feng et al. (2012) exhibited high dispersibility and stability; moreover, this efficiently inhibited the leaching of Fe from Fe_2O_3. The functionalized nanoparticles revealed a high surface area of 179 m^2/g and demonstrated superparamagnetic properties at room temperature with a saturation magnetization of 40 emu/g. The material exhibited a rapid removal of kinetics within 30 min, approaching a steady state. The maximum removal of arsenic species was reported at neutral pH conditions. Saiz et al. (2013) proficiently functionalized SiO_2/Fe_3O_4 magnetic nanoparticles with aminopropyl groups for the efficient adsorption of arsenic anions. Low temperatures favored the uptake of arsenate species. The adsorption capacities were reported as 14.7 mg/As(III) g and 121 mg/As(V) g. Superparamagnetic magnesium ferrite nanoadsorbents synthesized by Tang et al. (2013) by the doping of Mg^{2+} into α-Fe_2O_3 revealed an extremely high surface area of 438.2 m^2/g and represented superparamagnetic behavior with a high saturation magnetization of 32.9 emu/g. Arsenic adsorption was due to the formation of inner-sphere complexes; moreover, the presence of abundant hydroxyl groups on the surface of the magnetic nanoparticles enhanced removal performance. The adsorption capacity was reported as 127.4 mg/g for arsenite and 83.24 mg/g for arsenate species. Easy magnetic separation of these nanoparticles from treated water bodies was also proposed.

13.5.1.5 Bimetallic Oxides

In recent years, considerable efforts have been made toward the development of MMOs to improve the scavenging efficiency of adsorbent material for arsenic species from waste streams. Moreover, it was anticipated that the incorporation of other metal ions into nanostructured iron oxide would amend the physical characteristics of the host material such as porosity, surface area, surface charge, and crystallinity, which would further augment adsorption efficiency for undesirable solutes (Basu et al. 2012). Zhang et al. (2005) investigated Fe-Ce bimetallic oxide for the removal of arsenate ions from groundwater. The adsorption test revealed that the bimetallic oxide exhibited better sorption capacity as compared with its reference Ce and Fe oxides. The mechanism of adsorption was investigated by an XPS (X-ray photoelectron spectroscopy) study, and it was proposed that the presence of high concentrations of Fe-OH groups would enhance the amount of uptake of the arsenate species as arsenate anions were replaced with the OH group of Fe-OH. Additionally, the presence of Ce promoted breakage of the magnetite structure through oxidation of Fe(II), and stimulated Fe atoms to capture more hydroxyl ions. The adsorption of arsenate ions was as a result of surface complexation with the hydroxyl group of Fe-OH. Mahmood et al. (2012) also investigated the sorption of arsenate ions from aqueous media on to the binary mixed Fe-Si oxide, synthesized in the ratio of 3:1. The adsorption affinity of arsenate toward various metal oxides/hydroxides was considered in decreasing order as mixed oxide > iron hydroxide > iron oxide (nanoparticles) > iron oxide > silica. In the presence of a background electrolyte, the adsorption of arsenate ion increased; it was suggested that the anions that were adsorbed through inner-sphere complexes showed little or no sensitivity to ionic strength. The influence of thermal treatment on bimetallic oxides for arsenate removal was also investigated, and it was found that, with an increase in calcination temperature, adsorption capacity was remarkably decreased due to the transformation of the amorphous phase to a crystalline phase. Wu et al. (2012) reported a new class of nonferric bimetallic Mn-Al oxides as adsorbents for the removal of arsenite ions. The MMO performed a dual process, which included the oxidation of arsenite to arsenate and its subsequent adsorption over the surface of the Mn-Al oxide. The solution pH significantly influenced the sorption process; the oxidation of arsenite and adsorption of arsenate were inhibited by pH increases over a range of 4–10. The maximum adsorption capacity was reported as 142.19 mg/g for arsenite and 99.7 mg/g for arsenate in a neutral-pH environment. Recently, Y-Mn binary composites were investigated by Yu et al. (2015). The presence of interfering ions such as fluoride, sulfate, bicarbonate, phosphate, and humic acid had a low influence on arsenate adsorption. The metal composite exhibited a very high capacity of 279.9 mg/g for arsenate adsorption.

13.5.1.6 Layered Double Hydroxides

Layered double hydroxides (LDH) are anionic-class clay compounds, and are regarded as potentially good adsorbents for a diverse number of anionic pollutants, as they offer a large interlayer surface to accommodate contaminant species, with the additional advantage that they can be easily regenerated. LDHs are one of the preferred adsorbent materials that were extensively investigated for the liquid-based removal of arsenic ions. Yang et al. (2005) investigated the removal of trace levels of arsenic from an aqueous solution via calcined and uncalcined Mg/Al-CO$_3$ LDH. Both calcined and uncalcined LDH exhibited poor removal for arsenite ions. In the case of arsenate ions, the calcined LDH showed better adsorption capacity than the uncalcined one. With the calcined LDH, the arsenate adsorption reached equilibrium within 60 min of contact time, while with the uncalcined LDH, the arsenate reached saturation after 400 min of contact time. The study on the effect of competitive ions for arsenate ions using calcined LDH revealed that the NO_3^- ions exert no influence on the adsorption performance, but the SO_4^{2-}, CO_3^{2-}, and HPO_4^{2-} ions have a great influence on arsenate adsorption. Turk et al. (2009) studied the sorptive removal of arsenate ions using an Mg/Fe-CO$_3$ LDH. More than 90% of removal was observed in 30 min of contact time. The results revealed that the Mg/Fe LDH has high arsenate removal efficiency, and can efficiently lower arsenate concentration from an initial value of 330 µg/L to <10 µg/L. Guo et al. (2012) investigated adsorption characteristics of Cu/Mg/Fe/La for the adsorption of arsenate ions. The uptake of arsenate ions increased with an increase of the lanthanum content of the LDH. The adsorbent material exhibited a high surface area of >100 m^2/g. The maximum arsenate uptake of 72 mg/g was observed at an acidic pH value of 3.0. The adsorption temperature also played a significant role in removal efficiency, which increased considerably from 26 to 57 mg/g as the temperature increased from 288 K to 323 K. The ion-exchange process was proposed as the principal mechanism for the uptake of arsenate ions by LDH. A recent investigation of arsenate adsorption was reported by Kameda et al. (2015), using Fe^{2+}-doped Mg/Al LDH. It reported that this showed better a removal performance of arsenate ions than simple Mg/Al LDH. The equilibrium condition for arsenate removal was achieved quickly, within 30 min. It was further suggested that Fe-doped Mg/Al LDH could easily take up arsenate ion from the aqueous phase by anion exchange with Cl$^-$ ions.

13.5.1.7 Carbon-Based Nanomaterials

Carbon nanoparticles are blessed with several functional groups on their surface, which can be easily tailored according to requirements. In addition, many intriguing and unique physical and chemical properties of carbon nanoparticles render them potential pollutant-scavenging materials. Mahanta and Chen (2013) reported that carbonaceous materials that are synthesized at high temperature achieve enhanced porosity but their surface functional groups may be eliminated. This problem can be managed through functionalization of the carbon nanoparticles. Mahanta and Chen (2013) investigated Zr(IV) ion-functionalized carbon nanoparticles for the removal of arsenate ions from water. The functionalization of carbon nanoparticles was carried out by chemical immobilization of Zr(IV) ions to the hydroxyl group of the carbon. The modified material exhibited a rapid removal performance; it was demonstrated that 70%–75% of its final adsorption was achieved within the first 10 min, much faster than any other adsorbent; moreover, no significant effects were detected on the uptake in the presence of competing anions such as fluoride, phosphate, and nitrate, as well as humic acid, but the presence of silicate ions significantly impeded the sorption process. Thus, the material can be effectively used for the treatment of arsenic-contaminated surface- and groundwater.

However, among carbon nanomaterials, carbon nanotubes (CNTs) are highly studied adsorbent materials for the removal of arsenic anions. CNTs are most desirable candidates for this purpose in the current scientific research, due to their novel properties such as high aspect ratio and good thermal, electrical, and mechanical properties (Mishra and Ramaprabhu 2011). CNTs comprise a large number of adsorption sites for various types of contaminants, as well as good support for other adsorbent materials. In addition, their commendable surface properties—a hollow and layered nanostructure—make them a versatile adsorbent material. Tawabini et al. (2011) investigated modified CNTs for the removal of arsenic from aqueous solutions. It was also recommended that the adsorption efficiency of CNTs may be enhanced by considering certain factors such as mode of synthesis, purification and modification processes, pH, CNT dosage, contact time, and mixing rate. The oxidation and purification of CNTs have a great impact on adsorption efficiency. Tawabini et al. (2011) modified CNTs with the COOH group and by impregnating iron nanoparticles to improve arsenic removal performance. The multiwalled CNT was synthesized by the chemical vapor deposition method. The synthesized material revealed (Figure 13.4a and b) a hollow, tubular structure with uniform diameter distribution and no structural deformity.

Further, the synthesized CNTs were modified by carboxylic functional groups and iron nanoparticles. The performance of unmodified and modified CNTs was investigated and it was found that unmodified CNTs showed zero arsenic removal, while COOH-CNT exhibited a small removal percentage of only 10.7% at pH 5, while the highest removal of 77.45% was obtained by Fe-CNT at pH 8 (Figure 13.5).

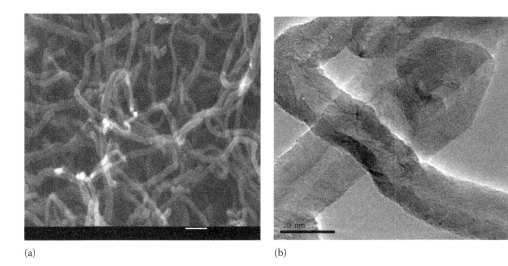

(a)   (b)

FIGURE 13.4 (a) Scanning electron microscope (SEM) image and (b) transmission electron microscope (TEM) image of carbon nanotube (From Tawabini, B.S. et al., *J. Environ. Sci. Health A*, 46, 215–223, 2011.)

Another interesting arsenic removal approach was investigated by Mishra and Ramaprabhu (2011). They reported the simultaneous adsorption of functionalized multiwalled carbon nanotubes (MWNTs) for sodium and arsenic from aqueous solution by employing a new type of carbon fabric–supported functionalized MWNT (f-MWNT)-based supercapacitor. The material exhibited high arsenate and sodium removals of 75% and 77%, respectively; while, in the case of sodium arsenite-containing water, there was removal of 51% and 52% of As and Na, respectively, with 20 repeated cycles and 100 mg of f-MWNTs loaded at each electrode, using an initial arsenic concentration of 200 ppm (Figure 13.6). Removal of nearly equal percentages of As and Na in each case

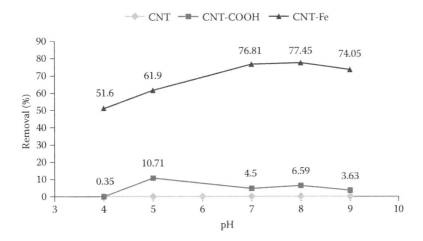

FIGURE 13.5 Removal of arsenic in different pH conditions using unmodified and modified CNTs. (From Tawabini, B.S. et al., *J. Environ. Sci. Health A*, 46, 215–223, 2011.)

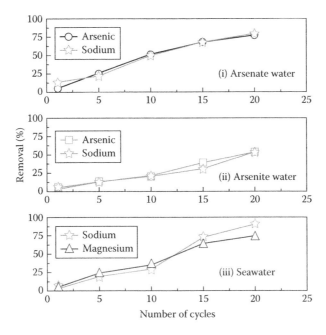

FIGURE 13.6 **(See color insert)** Simultaneous removal performance of Na and As: (i) sodium arsenate-containing water and (ii) sodium arsenite-containing water; (iii) removal efficiency of sodium and magnesium (desalination) from seawater (From Mishra, A.K. and Ramaprabhu, S., *J. Exp. Nanosci.*, 7, 85–97, 2011.)

suggested that the removal of each metallic contaminant was not affected, even in the presence of other impurities.

Similarly, others also reported arsenic removal performance by CNTs, such as Ma et al. (2013), who described magnetic activated CNTs for the removal of arsenic, and Liu et al. (2014), who reported titanium dioxide–coated CNTs for the efficient sorption of arsenic species from water.

Recently, another important class of carbon-based nanostructure, popularly known as graphene, emerged as a new generation of adsorbent that offers high-grade treatment efficiency. A few treatment studies using graphene-based materials make an entry into the exhaustive list of adsorbent used for arsenic remediation. Vadahanambi et al. (2013) described a three-dimensional graphene–carbon nanotube–iron oxide nanostructure composite for the efficient removal of arsenic from contaminated water. Enhanced arsenic removal was witnessed due to the combined effect of each component of the nanocomposite. Paul et al. (2015) reported that a graphene-Fe_3O_4 nanocomposite eliminated the negative effect of humic acid present in natural water that further enhanced the uptake of arsenic species.

13.5.1.8 Metal Organic Frameworks

Metal organic frameworks (MOFs) are an advanced class of adsorbent material that offers high surface area and high porosity, and display outstanding removal potential for undesirable materials from contaminated water. These are a new class of hybrid materials assembled with metal cations and organic linkers. They have received a great deal of attention in recent years due to their unique properties, including large surface areas, crystalline open structures, tunable pore sizes, and functionality (Vu et al. 2015). MOFs have the ability to adjust their pore openings according to the size of the guest species and exhibit a high level of flexibility. Zhu et al. (2012) reported an iron–1,3,5-benzenetricarboxylic (Fe-BTC) metal organic framework. Arsenate adsorption occurred swiftly and the equilibrium value was reached within 10 min. Within the pH range of 2–10, removal efficiency remained higher than 96%. The removal mechanism of Fe-BTC was also interpreted, and it was proposed that the arsenate ions adsorbed only onto the interior of the Fe-BTC polymer but not onto the outer surface. Li et al. (2014) investigated the removal of trace (10^{-9} mg/L) amounts of arsenate from water by using zeolitic imidazolate framework-8 (ZIF-8). The material displayed a high removal capacity of 76.5 mg/g at a low initial arsenate concentration of 9.8 µg/L. However, the presence of PO_4^{3-} ions seriously impaired removal efficiency. The arsenate removal mechanism was predicted as the formation of a large number of surface active sites in the form of Zn-OH through the dissociative adsorption of water followed by complexation of metal hydroxides with the arsenate ions. Recently, Wang et al. (2015) reported the high efficiency of zirconium metal organic framework UiO-66 (constructed with $Zr_6O_4(OH)_4$ clusters and terephthalate [1,4-benzenedicarboxylate (BDC)] linkers) for arsenic decontamination from water. The adsorbent material UiO-66 performed superbly in the pH range of 1–10, and achieved the high arsenate uptake capacity of 303 mg/g at pH 2. The high removal of arsenic was attributed to the high surface area (569.6 m²/g) and the availability of a large number of active sites. Arsenic species were captured by metal hydroxide $Zr_6O_4(OH)_4$ through the formation of a Zr–O–As complex. A few studies also reported the removal of arsenic by MOFs, such as Vu et al. (2015), who investigated arsenic removal by MIL-53 (Fe); and Li et al. (2015), who used $Zr_6O_4(OH)_4(BTC)_2(HCOO)_6$ (MOF-808) nanoparticles for arsenate removal.

13.6 CONCLUSIONS

A brief overview on arsenic remediation from water/wastewater using various conventional and advanced nanoparticle-based adsorbents has been presented. All the conventional adsorbents such as alumina, iron hydroxides, aluminum hydroxides, granular ferric hydroxide, ferrihydrite, hematite, laterite, magnetite, siderite, and TiO_2 exhibited slow removal kinetics with insignificant removal efficiency due to poor adsorption capacity; moreover, due to their small surface area and particle instability, their performance in continuous flow systems was always questionable. The nanoparticles demonstrated relatively better arsenic removal efficiency. However, the efficacy of the arsenic adsorption process relies on the physical characteristics of the adsorbent species and the chemical form of the adsorbate species. In addition, operational parameters such as pH, temperature, adsorbent dose, and competing ions also play a significant role in the sorption process. However, the research findings established that nanoparticles exhibit serious toxic effects, and their biological outcome has still not been thoroughly assessed. Therefore, before employing nanomaterials in the decontamination of arsenic-polluted water, a risk assessment of the adsorbent is critically required.

ACKNOWLEDGMENTS

SB and RKG are thankful to the Council of Scientific and Industrial Research (CSIR) and the University Grant Commission (UGC), New Delhi for providing financial assistance as a senior research fellowship (SRF).

REFERENCES

Abejon, R., A. Garea, A bibliometric analysis of research on arsenic in drinking water during the 1992–2012 period: An outlook to treatment alternatives for arsenic removal, *J. Water Process Eng.* 6 (2015) 105–119.

Ahn, J.S., Geochemical occurrences of arsenic and fluoride in bedrock groundwater: A case study in Geumsan County, Korea, *Environ. Geochem. Health* 34 (2012) 43–54.

Akin, I., G. Arslan, A. Tor, Y. Cengeloglu, M. Ersoz, Removal of arsenate [As(V)] and arsenite [As(III)] from water by SWHR and BW-30 reverse osmosis, *Desalination* 281 (2011) 88–92.

Akina, I., G. Arslana, A. Torb, M. Ersoza, Y. Cengeloglu, Arsenic(V) removal from underground water by magnetic nanoparticles synthesized from waste red mud, *J. Hazard. Mater.* 235–236 (2012) 62–68.

Ali, I., New generation adsorbents for water treatment, *Chem. Rev.* 112 (2012) 5073–5091.

Altundogan, H.S., S. Altundogan, F. Tumen, M. Bildik, Arsenic removal from aqueous solutions by adsorption on red mud, *Waste Manage.* 20 (2000) 761–767.

Altundogan, H.S., S. Altundogan, F. Tumen, M. Bildik, Arsenic adsorption from aqueous solutions by activated red mud, *Waste Manage.* 22 (2002) 357–363.

Amrose, S., A. Gadgil, V. Srinivasan, K. Kowolik, M. Muller, J. Huang, R. Kostecki, Arsenic removal from groundwater using iron electrocoagulation: Effect of charge dosage rate, *J. Environ. Sci. Health A* 48 (2013) 1019–1030.

An, B., Z. Fu, Z. Xiong, D. Zhao, A.K. SenGupta, Synthesis and characterization of a new class of polymeric ligand exchangers for selective removal of arsenate from drinking water, *React. Funct. Polym.* 70 (2010) 497–507.

Ansari, R., M. Sadegh, Application of activated carbon for removal of arsenic ions from aqueous solutions, *E-J. Chemistry* 4 (2007) 103–108.

Aredes, S., B. Klein, M. Pawlik, The removal of arsenic from water using natural iron oxide minerals, *J. Clean. Product.* 60 (2013) 71–76.

Asante, K.A., T. Agusa, A. Subramanian, O.D. Ansa-Asare, C.A. Biney, S. Tanabe, Contamination status of arsenic and other trace elements in drinking water and residents from Tarkwa, a historic mining township in Ghana, *Chemosphere* 66 (2007) 1513–1522.

Banerjee, K., G.L. Amy, M. Prevost, S. Nour, M. Jekel, P.M. Gallagher, C.D. Blumenschein, Kinetic and thermodynamic aspects of adsorption of arsenic onto granular ferric hydroxide (GFH), *Water Res.* 42 (2008) 3371–3378.

Banerjee, S., R.K. Gautam, A. Jaiswal, M.C. Chattopadhyaya, Y.C. Sharma, Rapid scavenging of methylene blue dye from aqueous solutions by adsorption on nanoalumina, *RSC Adv.* 5 (2015) 14425–14440.

Bang, S., M. Patel, L. Lippincott, X. Meng, Removal of arsenic from groundwater by granular titanium dioxide adsorbent, *Chemosphere* 60 (2005) 389–397.

Barakat, M.A., S.I. Shah, Utilization of anion exchange resin spectra/gel for separation of arsenic from water, *Arab. J. Chemistry* 6 (2013) 307–311.

Barringer, J.L., Z. Szabo, T.P. Wilson, J.L. Bonin, T. Kratzer, K. Cenno, T. Romagna, M. Alebus, B. Hirst, Distribution and seasonal dynamics of arsenic in a shallow lake in northwestern New Jersey, U.S.A., *Environ. Geochem. Health* 33 (2011) 1–22.

Baskan, M.B., A. Pala, A statistical experiment design approach for arsenic removal by coagulation process using aluminum sulfate, *Desalination* 254 (2010) 42–48.

Baskan, M.B., A. Pala, Removal of arsenic from drinking water using modified natural zeolite, *Desalination* 281 (2011) 396–403.

Basu, T., K. Gupta, U.C. Ghosh, Performances of As(V) adsorption of calcined (250°C) synthetic iron(III)–aluminum(III) mixed oxide in the presence of some groundwater occurring ions, *Chem. Eng. J.* 183 (2012) 303–314.

Bhattacharya, P., M. Claesson, J. Bundschuh, O. Sracek, J. Fagerberg, G. Jacks, R.A. Martin, A.R. Storniolo, J.M. Thir, Distribution and mobility of arsenic in the Rio Dulce Alluvial aquifers in Santiago del Estero Province, Argentina, *Sci. Total Environ.* 358 (2005) 97–120.

Bissen, M., F.H. Frimmel, Arsenic: A review. Part II: Oxidation of arsenic and its removal in water treatment, *Acta Hydrochim. Hydrobiol.* 31 (2003) 97–107.

Camacho, L.M., M. Gutierrez, M.T. Alarcon-Herrera, M.L. Villalba, S. Deng, Occurrence and treatment of arsenic in groundwater and soil in northern Mexico and southwestern USA, *Chemosphere* 83 (2011b) 211–225.

Camacho, L.M., R.R. Parra, S. Deng, Arsenic removal from groundwater by MnO2-modified natural clinoptilolite zeolite: Effects of pH and initial feed concentration, *J. Hazard. Mater.* 189 (2011a) 286–293.

Cao, A., J.D. Monnell, C. Matranga, J. Mi. Wu, L.L. Cao, D. Gao, Hierarchical nanostructured copper oxide and its application in arsenic removal. *J. Phys. Chem. C* 111 (2007) 18624–18628.

Chakraborti, D., M.M. Rahman, B. Das, M. Murrill, S. Dey, S.C. Mukherjee, Status of groundwater arsenic contamination in Bangladesh: A 14-year study report, *Water Res.* 44 (2010) 5789–5802.

Chakravarty, S., V. Dureja, G. Bhattacharyya, S. Maity, S. Bhattacharjee, Removal of arsenic from groundwater using low cost ferruginous manganese ore, *Water Res.* 36 (2002) 625–632.

Cho, K.H., S. Sthiannopkao, Y.A. Pachepsky, K.W. Kim, J.H. Kim, Prediction of contamination potential of groundwater arsenic in Cambodia, Laos, and Thailand using artificial neural network, *Water Res.* 45 (2011) 5535–5544.

Choong, T., T.G. Chuah, Y. Robiah, F.L. Gregory Koay, I. Azni, Arsenic toxicity, health hazards and removal techniques from water: An overview, *Desalination* 217 (2007) 139–166.

Chuang, C.L., M. Fan, M. Xu, R.C. Brown, S. Sung, B. Saha, C.P. Huang, Adsorption of arsenic(V) by activated carbon prepared from oat hulls, *Chemosphere* 61 (2005) 478–483.

Council of the European Communities. Council directive 76/464/EEC of 4 May 1976 on pollution caused by certain dangerous substances discharged into the aquatic environment of the community. CELEX-EUR Off. J. L 129, 23–29, (1976) 18 May.

Cui, H., Y. Su, Q. Li, S. Gao, J.K. Shang, Exceptional arsenic (III,V) removal performance of highly porous, nanostructured ZrO_2 spheres for fixed bed reactors and the full-scale system modeling, *Water Res.* 47 (2013) 6258–6268.

Darban, A.K., Y. Kianinia, E. Taheri-Nassaj, Synthesis of nano-alumina powder from impure kaolin and its application for arsenite removal from aqueous solutions, *J. Environ. Health Sci. Eng.* 11 (2013) 1–11.

De, D., S.M. Mandal, J. Bhattacharya, S. Ram, S.K. Roy, Iron oxide nanoparticle-assisted arsenic removal from aqueous system, *J. Environ. Sci. Health A* 44 (2009) 152–162.

Donia, A.M., A.A. Atia, D.A. Mabrouk, Fast kinetic and efficient removal of As(V) from aqueous solution using anion exchange resins, *J. Hazard. Mater.* 191 (2011) 1–7.

Ellenhorn, M.J., S. Schonwald, G. Ordog, J. Wasserberger, *Medical Toxicology: Diagnosis and Treatment of Human Poisoning*, Williams and Wilkins, Baltimore, MD, 1997.

EPA, 1982. Code of Federal Regulations, Title 40: Protection of Environment, Part 23, Appendix A-List of 126 Priority Pollutants. Available at: http://www.epa. gov/region1/npdes/permits/generic/prioritypollutants.pdf [accessed November 16, 2015].

European Commission, Directive 98/83/EC, related with drinking water quality intended for human consumption. Brussels, Belgium, (1998).

Feng, L., M. Cao, X. Ma, Y. Zhu, C. Hu, Superparamagnetic high-surface-area Fe_3O_4 nanoparticles as adsorbents for arsenic removal, *J. Hazard. Mater.* 217–218 (2012) 439–446.

Figoli, A., A. Cassano, A. Criscuoli, M.S.I. Mozumder, M.T. Uddin, M.A. Islam, E. Drioli, Influence of operating parameters on the arsenic removal by nanofiltration, *Water Res.* 44 (2010) 97–104.

Gallegos-Garcia, M., K. Ramírez-Muñiz, S. Song, Arsenic removal from water by adsorption using iron oxide minerals as adsorbents: A review, *Miner. Process. Extr. Metall. Rev.* 33 (2012) 301–315.

Gautam, R.K., P.K. Gautam, S. Banerjee, S. Soni, S.K. Singh, M.C. Chattopadhyaya, Removal of Ni(II) by magnetic nanoparticles, *J. Mol. Liq.* 204 (2015) 60–69.

Gecol, H., E. Ergican, A. Fuchs, Molecular level separation of arsenic(V) from water using cationic surfactant micelles and ultrafiltration membrane, *J. Membr. Sci.* 241 (2004) 105–119.

George, C.M., L. Sima, M.H.J. Arias, J. Mihalic, L.Z. Cabrera, D. Danz, W. Checkley, R.H. Gilman, Arsenic exposure in drinking water: An unrecognized health threat in Peru, *Bull. World Health Organ.* 92 (2014) 565–572.

Ghosh, A., Evaluation of chronic arsenic poisoning due to consumption of contaminated ground water in West Bengal, India, *Int. J. Prev. Med.* 4 (2013) 976–979.

Ghosh, D., C.R. Medhi, M.K. Purkait, 2011. Treatment of fluoride, iron and arsenic containing drinking water by electrocoagulation followed by microfiltration. In: *Proceedings of 12th International Conference on Environmental Science and Technology*, Rhodes, Greece.

Ghurye, G., D. Clifford, A. Tripp, Iron coagulation and direct microfiltration to remove arsenic from groundwater, *J. Am. Water Works Ass.* 96 (2004) 143–152.

Giasuddin, A.B.M., S.R. Kanel, H. Choi, Adsorption of humic acid onto nanoscale zerovalent iron and its effect on arsenic removal, *Environ. Sci. Technol.* 41 (2007) 2022–2027.

Goswami, A., P.K. Raul, M.K. Purkait, Arsenic adsorption using copper (II) oxide nanoparticles, *Chem. Eng. Res. Des.* 90 (2012) 1387–1396.

Guerraa, D.L., C. Airoldi, R.R. Viana, Adsorption of arsenic(V) into modified lamellar Kenyaite, *J. Hazard. Mater.* 163 (2009) 1391–1396.

Guo, Y., Z. Zhu, Y. Qiu, J. Zhao, Adsorption of arsenate on Cu/Mg/Fe/La layered double hydroxide from aqueous solutions, *J. Hazard. Mater.* 239–240 (2012) 279–288.

Gupta, S.K., K.Y. Chen, Arsenic removal by adsorption, *J. Water Pollut. Contr. Fed.* 50(3) (1978) 493–506.

Hassan, A.F., A.M.A. Mohsen, H. Elhadidy, Adsorption of arsenic by activated carbon, calcium alginate and their composite beads, *Int. J. Biol. Macromol.* 68 (2014) 125–130.

He, S., C. Han, H. Wang, W. Zhu, S. He, D. He, Y. Luo, Uptake of arsenic(V) using alumina functionalized highly ordered mesoporous SBA-15 (Al$_x$SBA-15) as an effective adsorbent, *J. Chem. Eng. Data* 60 (2015) 1300–1310.

Hu, C., H. Liu, G. Chen, W.A. Jefferson, J. Qu, As(III) oxidation by active chlorine and subsequent removal of As(V) by Al13 polymer coagulation using a novel dual function reagent, *Environ. Sci. Technol.* 46 (2012) 6776–6782.

Hu, C., H. Liu, G. Chen, J. Qu, Effect of aluminum speciation on arsenic removal during coagulation process, *Sep. Purif. Technol.* 86 (2012) 35–40.

Iqbal, J., H.J. Kim, J.S. Yang, K. Baek, J.W. Yang, Removal of arsenic from groundwater by micellar-enhanced ultrafiltration (MEUF), *Chemosphere* 66 (2007) 970–976.

Jadhav, S.V., Bringas, G.D. Yadav, V.K. Rathod, I. Ortiz, K.V. Marathe, Arsenic and fluoride contaminated groundwaters: A review of current technologies for contaminants removal, *J. Environ. Manage.* 162 (2015) 306–325.

Jeon, C.S., K. Baek, J.K. Park, Y.K. Oh, S.D. Lee, Adsorption characteristics of As(V) on iron-coated zeolite, *J. Hazard. Mater.* 163 (2009) 804–808.

Kameda, T., E. Kondo, T. Yoshioka, Equilibrium and kinetics studies on As(V) and Sb(V) removal by Fe^{2+} doped Mg-Al layered double hydroxides, *J. Environ. Manage.* 151 (2015) 303–309.

Kanel, S.R., B. Manning, L. Charlet, H. Choi, Removal of arsenic(III) from groundwater by nanoscale zerovalent iron, *Environ. Sci. Technol.* 39 (2005) 1291–1298.

Kanel, S.R., D. Nepal, B. Manning, H. Choi, Transport of surface-modified iron nanoparticle in porous media and application to arsenic(III) remediation, *J. Nanopart. Res.* 9 (2007) 725–735.

Kang, M., M. Kawasaki, S. Tamada, T. Kamei, Y. Magara, Effect of pH on the removal of arsenic and antimony using reverse osmosis membranes, *Desalination* 131 (2000) 293–298.

Kapaj, S., H. Peterson, K. Liber, P. Bhattacharya, Human health effects from chronic arsenic poisoning: A review, *J. Environ. Sci. Health A* 41 (2006) 2399–2428.

Kilianova, M., R. Prucek, J. Filip, J. Kolarik, L. Kvitek, A. Panacek, J. Tucek, R. Zboril, Remarkable efficiency of ultrafine superparamagnetic iron(III) oxide nanoparticles toward arsenate removal from aqueous environment, *Chemosphere* 93 (2013) 2690–2697.

Kim, Y., C. Kim, I. Choi, S. Rengaraj, J. Yi, Arsenic removal using mesoporous alumina prepared via a templating method, *Environ. Sci. Technol.* 38 (2004) 924–931.

Kumari, V., A. Bhaumik, Mesoporous ZnAl$_2$O$_4$: An efficient adsorbent for the removal of arsenic from contaminated water, *Dalton Trans.* 44 (2015) 11843–11851.

Kundu, S., S.S. Kavalakatt, A. Pal, A.K. Ghosh, M. Mandal, T. Pal, Removal of arsenic using hardened paste of Portland cement: Batch adsorption and column study, *Water Res.* 38 (2004) 3780–3790.

Lado, L.R., G.F. Sun, M. Berg, Q. Zhang, H.B. Xue, Q.M. Zheng, C.A. Johnson, Ground water arsenic contamination throughout China, *Science* 341 (2013) 866–868.

Lakshmipathiraj, P., B.R.V. Narasimhan, S. Prabhakar, G.B. Raju, Adsorption of arsenate on synthetic goethite from aqueous solutions, *J. Hazard. Mater.* B136 (2006) 281–287.

Lee, Y., I.H. Um, J. Yoon, Arsenic(III) oxidation by iron(VI) (ferrate) and subsequent removal of arsenic(V) by iron(III) coagulation, *Environ. Sci. Technol.* 37 (2003) 5750–5756.

Lescano, M.R., C. Passalía, C.S. Zalazar, R.J. Brandi, Arsenic sorption onto titanium dioxide, granular ferric hydroxide and activated alumina: Batch and dynamic studies, *J. Environ. Sci. Health A* 50 (2015) 424–431.

Li, J., Y. Wu, Z. Li, B. Zhang, M. Zhu, X. Hu, Y. Zhang, F. Li, Zeolitic imidazolate framework-8 with high efficiency in trace arsenate adsorption and removal from water, *J. Phys. Chem. C* 118 (2014a) 27382–27387.

Li, W.G., X.J. Gong, K. Wang, X.R. Zhang, W.B. Fan, Adsorption characteristics of arsenic from micro-polluted water by an innovative coal-based mesoporous activated carbon, *Bioresor. Technol.* 165 (2014b) 166–173.

Li, Z., J.S. Jean, W.T. Jiang, P.H. Chang, C.J. Chen, L. Liao, Removal of arsenic from water using Fe-exchanged natural zeolite, *J. Hazard. Mater.* 187 (2011) 318–323.

Li, Z.Q., J.C. Yang, K.W. Sui, N. Yin, Facile synthesis of metal-organic framework MOF-808 for arsenic removal, *Mater. Lett.* 160 (2015) 412–414.

Liang, C.P., C.S. Jang, J.S. Chen, S.W. Wang, J.J. Lee, C.W. Liu, Probabilistic health risk assessment for ingestion of seafood farmed in arsenic contaminated groundwater in Taiwan, Environ. *Geochem. Health* 35 (2013) 455–464.

Lin, S., D. Lu, Z. Liu, Removal of arsenic contaminants with magnetic γ-Fe_2O_3 nanoparticles, *Chem. Eng. J.* 211–212 (2012) 46–52.

Lin, T.F., J.K. Wu, Adsorption of arsenite and arsenate within activated alumina grains: Equilibrium and kinetics, *Water Res.* 35 (2001) 2049–2057.

Ling, L., W. Zhang, Sequestration of arsenate in zero-valent iron nanoparticles: Visualization of intraparticle reactions at angstrom resolution, *Environ. Sci. Technol. Lett.* 1 (2014) 305–309.

Liu, G., X. Zhang, J.W. Talley, C.R. Neal, H. Wang, Effect of NOM on arsenic adsorption by TiO_2 in simulated As(III)-contaminated raw waters, *Water Res.* 42 (2008) 2309–2319.

Liu, H., K. Zuo, C.D. Vecitis, Titanium dioxide-coated carbon nanotube network filter for rapid and effective arsenic sorption, *Environ. Sci. Technol.* 48 (2014) 13871–13879.

Lodeiro, P., S.M. Kwan, J.T. Perez, L.F. González, C. Gérente, Y. Andrès, G. McKay, Novel Fe loaded activated carbons with tailored properties for As(V) removal: Adsorption study correlated with carbon surface chemistry, *Chem. Eng. J.* 215–216 (2013) 105–112.

Lohokare, H.R., M.R. Muthu, G.P. Agarwal, U.K. Kharul, Effective arsenic removal using polyacrylonitrile-based ultrafiltration (UF) membrane, *J. Membrane Sci.* 320 (2008) 159–166.

Luther, S., N. Borgfeld, J. Kim, J.G. Parsons, Removal of arsenic from aqueous solution: A study of the effects of pH and interfering ions using iron oxide nanomaterials, *Microchem. J.* 101 (2012) 30–36.

Ma, J., Z. Zhu, B. Chen, M. Yang, H. Zhou, C. Li, F. Yu, J. Chen, One-pot, large-scale synthesis of magnetic activated carbon nanotubes and their applications for arsenic removal, *J. Mater. Chem. A* 1 (2013) 4662–4666.

Mahanta, N., J.P. Chen, A novel route to the engineering of zirconium immobilized nano-scale carbon for arsenate removal from water, *J. Mater. Chem. A* 1 (2013) 8636–8644.

Mahmood, T., S.U. Din, A. Naeem, S. Mustafa, M. Waseem, M. Hamayun, Adsorption of arsenate from aqueous solution on binary mixed oxide, of iron and silicon, *Chem. Eng. J.* 192 (2012) 90–98.

Maiti, A., J.K. Basu, S. De, Development of a treated laterite for arsenic adsorption: Effects of treatment parameters, *Ind. Eng. Chem. Res.* 49 (2010) 4873–4886.

Malik, A.H., Z.M. Khan, Q. Mahmood, S. Nasreen, Z.A. Bhatti, Perspectives of low cost arsenic remediation of drinking water in Pakistan and other countries, *J. Hazard. Mater.* 168 (2009) 1–12.

Mandal, B.K., K.T. Suzuki, Arsenic round the world: A review, *Talanta* 58 (2002) 201–235.

Martinson, C.A., K.J. Reddy, Adsorption of arsenic(III) and arsenic(V) by cupric oxide nanoparticles, *J. Colloid Interface Sci.* 336 (2009) 406–411.

Massoudinejad, M., A. Asadi, M. Vosoughi, M. Gholami, B. Kakavandi, M.A. Karami, A comprehensive study (kinetic, thermodynamic and equilibrium) of arsenic (V) adsorption using $KMnO_4$ modified clinoptilolite, *Korean J. Chem. Eng.* 32 (2015) 2078–2086.

McDonald, K.J., B. Reynolds, K.J. Reddy, Intrinsic properties of cupric oxide nanoparticles enable effective filtration of arsenic from water, *Sci. Rep.* 5 (2015) 11110.

Mishra, A.K., S. Ramaprabhu, The role of functionalised multiwalled carbon nanotubes based supercapacitor for arsenic removal and desalination of sea water, *J. Exp. Nanosci.* 7 (2012) 85–97.

Mondal, P., S. Bhowmick, D. Chatterjee, A. Figoli, B.V. der Bruggen, Remediation of inorganic arsenic in groundwater for safe water supply: A critical assessment of technological solutions, *Chemosphere* 92 (2013) 157–170.

Mudhoo, A., S.K. Sharma, V.K. Garg, C.H. Tseng, Arsenic: An overview of applications, health, and environmental concerns and removal processes, *Crit. Rev. Environ. Sci. Technol.* 41 (2011) 435–519.

Nguyen, V.A., S. Bang, P.H. Viet, K.W. Kim, Contamination of groundwater and risk assessment for arsenic exposure in Ha Nam province, Vietnam, *Environ. Inter.* 35 (2009) 466–472.

Ning, R.Y., Arsenic removal by reverse osmosis, *Desalination* 143 (2002) 237–241.

O'Shea, B., J. Jankowski, J. Sammut, The source of naturally occurring arsenic in a coastal sand aquifer of eastern Australia, *Sci. Total Environ.* 379 (2007) 151–166.

Pajany, Y.M., C. Hurel, N. Marmier, M. Romeo, Arsenic adsorption onto hematite and goethite, *C. R. Chimie* 12 (2009) 876–881.

Park, J.Y., J.H. Kim, Characterization of adsorbed arsenate on amorphous and nano crystalline MgFe-layered double hydroxides, *J. Nanopart. Res.* 13 (2011) 887–894.

Parviainen, A., K. Loukola-Ruskeeniemi, T. Tarvainen, T. Hatakka, P. Härmä, B. Backman, T. Ketola, et al., Arsenic in bedrock, soil and groundwater: The first arsenic guidelines for aggregate production established in Finland, *Earth Sci. Rev.* 150 (2015) 709–723.

Patra, A.K., A. Dutta, A. Bhaumik, Self-assembled mesoporous γ-Al_2O_3 spherical nanoparticles and their efficiency for the removal of arsenic from water, *J. Hazard. Mater.* 201–202 (2012) 170–177.

Paul, B., V. Parashar, A. Mishra, Graphene in the Fe_3O_4 nano-composite switching the negative influence of humic acid coating into an enhancing effect in the removal of arsenic from water, *Environ. Sci.: Water Res. Technol.* 1 (2015) 77–83.

Pokhrel, D., B.S. Bhandari, T. Viraraghavan, Arsenic contamination of groundwater in the Terai region of Nepal: An overview of health concerns and treatment options, *Environ. Inter.* 35 (2009) 157–161.

Rai, P., R.K. Gautam, S. Banerjee, V. Rawat, M.C. Chattopadhyaya, Synthesis and characterization of a novel $SnFe_2O_4$@activated carbon magnetic nanocomposite and its effectiveness in the removal of crystal violet from aqueous solution, *J. Environ. Chem. Eng.* 3 (2015) 2281–2291.

Raul, P.K., R.R. Devi, I.M. Umlong, A.J. Thakur, S. Banerjee, V. Veer, Iron oxide hydroxide nanoflower assisted removal of arsenic from water, *Mater. Res. Bull.* 49 (2014) 360–368.

Reddy, K.J., T.R. Roth, Arsenic removal from natural groundwater using cupric oxide, *Natl. Ground. Assoc.* 51 (2013) 83–91.

Romero, L., H. Alonso, P. Campano, L. Fanfani, R. Cidu, C. Dadea, T. Keegan, I. Thornton, M. Farago, Arsenic enrichment in waters and sediments of the Rio Loa (Second Region, Chile), *Appl. Geochem.* 18 (2003) 1399–1416.

Rowland, H.A.L., E.O. Omoregie, R. Millot, C. Jimenez, J. Mertens, C. Baciu, S.J. Hug, M. Berg, Geochemistry and arsenic behaviour in groundwater resources of the Pannonian Basin (Hungary and Romania), *Appl. Geochem.* 26 (2011) 1–17.

Saha, S., P. Sarkar, Arsenic remediation from drinking water by synthesized nano-alumina dispersed in chitosan-grafted polyacrylamide, *J. Hazard. Mater.* 227–228 (2012) 68–78.

Saiz, J., E. Bringas, I. Ortiz, Functionalized magnetic nanoparticles as new adsorption materials for arsenic removal from polluted waters, *J. Chem. Technol. Biotechnol.* 89 (2014) 909–918.

Sancha, A.M., Review of coagulation technology for removal of arsenic: Case of Chile, *J. Health Popul. Nutr.* 24 (2006) 267–272.

Sato, Y., M. Kang, T. Kamei, Y. Magara, Performance of nanofiltration for arsenic removal, *Water Res.* 36 (2002) 3371–3377.

Shih, M.C., An overview of arsenic removal by pressure driven membrane processes, *Desalination* 172 (2005) 85–97.

Singh, N., S.P. Singh, V. Gupta, H.K. Yadav, T. Ahuja, S.S. Tripathy, Rashmi, A process for the selective removal of arsenic from contaminated water using acetate functionalized zinc oxide nanomaterials, *Environ. Progress Sustain. Energy* 32 (2013) 1023–1029.

Singh, R., S. Singh, P. Parihar, V.P. Singh, S.M. Prasad, Arsenic contamination, consequences and remediation techniques: A review, *Ecotoxicol. Environ. Safety* 112 (2015) 247–270.

Singh, T.S., K.K. Pant, Equilibrium, kinetics and thermodynamic studies for adsorption of As(III) on activated alumina, *Sep. Purif. Technol.* 36 (2004) 139–147.

Song, S., A.L. Valdivieso, D.J.H. Campos, C. Peng, M.G.M. Fernandez, I.R. Soto, Arsenic removal from high-arsenic water by enhanced coagulation with ferric ions and coarse calcite, *Water Res.* 40 (2006) 364–372.

Sorlini, S., F. Gialdini, Conventional oxidation treatments for the removal of arsenic with chlorine dioxide, hypochlorite, potassium permanganate and monochloramine, *Water Res.* 44 (2010) 5653–5659.

Tang, W., Y. Su, Q. Li, S. Gao, J.K. Shang, Superparamagnetic magnesium ferrite nanoadsorbent for effective arsenic (III, V) removal and easy magnetic separation, *Water Res.* 47 (2013) 3624–3634.

Tawabini, B.S., S.F. Al-Khaldi, M.M. Khaled, M.A. Atieh, Removal of arsenic from water by iron oxide nanoparticles impregnated on carbon nanotubes, *J. Environ. Sci. Health A* 46 (2011) 215–223.

Tong, M., S. Yuan, P. Zhang, P. Liao, A.N. Alshawabkeh, X. Xie, Y. Wang, Electrochemically induced oxidative precipitation of Fe(II) for As(III) oxidation and removal in synthetic groundwater, *Environ. Sci. Technol.* 48 (2014) 5145–5153.

Tseng, W.P. Effect of dose-response relationships of skin cancer and Blackfoot disease with arsenic, *Environ. Health Perspect.* 19 (1977) 109–119.

Tubic, A., J. Agbaba, E. Dalmacija, I. Ivancev-Tumnas, M. Damlacija, Removal of arsenic and natural organic matter from groundwater using ferric and alum salts: A case study of central Banat region (Serbia), *J. Environ. Sci. Health A* 45 (2010) 363–369.

Turk, T., I. Alp, H. Deveci, Adsorption of As(V) from water using Mg–Fe-based hydrotalcite (FeHT), *J. Hazard. Mater.* 171 (2009) 665–670.

Uddin, M.T., M.S.I. Mozumder, A. Figoli, M.A. Islam, E. Drioli, Arsenic removal by conventional and membrane technology: An overview, *Indian J. Chem. Technol.* 14 (2007) 441–450.

Ungureanu, G., S. Santos, R. Boaventura, C. Botelho, Arsenic and antimony in water and wastewater: Overview of removal techniques with special reference to latest advances in adsorption, *J. Environ. Manage.* 151 (2015) 326–342.

Urase, T., J. Oh, K. Yamamoto, Effect of pH on rejection of different species of arsenic by nanofiltration, *Desalination* 117 (1998) 11–18.

Vadahanambi, S., S.H. Lee, W.J. Kim, I.K. Oh, Arsenic removal from contaminated water using three-dimensional graphene-carbon nanotube-iron oxide nanostructures, *Environ. Sci. Technol.* 47 (2013) 10510–10517.

van Halem, D., et al., Subsurface iron and arsenic removal: Low-cost technology for community-based water supply in Bangladesh, *Water Sci. Technol.* 62 (2010) 2702–2709.

Vrijenhoek, E.M., J.J. Waypa, Arsenic removal from drinking water by a "loose" nanofiltration membrane, *Desalination* 130 (2000) 265–277.

Vu, T.A., G.H. Le, C.D. Dao, L.Q. Dang, K.T. Nguyen, Q.K. Nguyen, P.T. Dang, H.T.K. Tran, Q.T. Duong, T.V. Nguyen, G.D. Lee, Arsenic removal from aqueous solutions by adsorption using novel MIL-53(Fe) as a highly efficient adsorbent, *RSC Adv.* 5 (2015) 5261–5268.

Wang, C., X. Liu, J.P. Chen, K. Li, Superior removal of arsenic from water with zirconium metal-organic framework UiO-66, *Sci. Rep.* 5 (2015) 16613.

Wang, L.K., M.H. Wang, J.F. Kao, Application and determination of organic polymers, *Water Air Soil Pollut.* 9 (1978) 337–348.

Wang, S., C.N. Mulligan, Occurrence of arsenic contamination in Canada: Sources, behavior and distribution, *Sci. Total Environ.* 366 (2006) 701–721.

Wu, Y., X. Ma, M. Feng, M. Liu, Behavior of chromium and arsenic on activated carbon, *J. Hazard. Mater.* 159 (2008) 380–384.

Wua, K., T. Liu, W. Xue, X. Wang, Arsenic(III) oxidation/adsorption behaviors on a new bimetal adsorbent of Mn-oxide-doped Al oxide, *Chem. Eng. J.* 192 (2012) 343–349.

Yang, L., Z. Shahrivari, P.K.T. Liu, M. Sahimi, T.T. Tsotsis, Removal of trace levels of arsenic and selenium from aqueous solutions by calcined and uncalcined layered double hydroxides (LDH), *Ind. Eng. Chem. Res.* 44 (2005) 6804–6815.

Yang, N., L.H.E. Winkel, K.H. Johannesson, Predicting geogenic arsenic contamination in shallow groundwater of South Louisiana, United States, *Environ. Sci. Technol.* 48 (2014) 5660–5666.

Yang, W., Q. Li, S. Gao, J.K. Shang, High efficient As(III) removal by self-assembled zinc oxide micro-tubes synthesized by a simple precipitation process, *J. Mater. Sci.* 46 (2011) 5851–5858.

Yavuz, C.T., J.T. Mayo, W.W. Yu, A. Prakash, J.C. Falkner, S.J. Yean, L.L. Cong, et al., Low-field magnetic separation of monodisperse Fe_3O_4 nanocrystals, *Science* 314 (2006) 964–967.

Yazdanbakhsh, M., H. Tavakkoli, S.M. Hosseini, Characterization and evaluation catalytic efficiency of $La_{0.5}Ca_{0.5}NiO_3$ nanopowders in removal of reactive blue 5 from aqueous solution, *Desalination* 281 (2011) 388–395.

Yoshizuka, K., S. Nishihama, H. Sato, Analytical survey of arsenic in geothermal waters from sites in Kyushu, Japan, and a method for removing arsenic using magnetite, *Environ. Geochem. Health* 32 (2010) 297–302.

Yu, X.Y., T. Luo, Y. Jia, Y.X. Zhang, J.H. Liu, X.J. Huang, Porous hierarchically micro-/nanostructured MgO: Morphology control and their excellent performance in As(III) and As(V) removal, *J. Phys. Chem. C* 115 (2011) 22242–22250.

Yu, X.Y., R.X. Xu, C. Gao, T. Luo, Y. Jia, J.H. Liu, X.J. Huang, Novel 3D hierarchical cotton-candy-like CuO: Surfactant-free solvothermal synthesis and application in As(III) removal, *ACS Appl. Mater. Interfaces* 4 (2012) 1954–1962.

Yu, Y., L. Yu, J.P. Chen, Introduction of an yttrium–manganese binary composite that has extremely high adsorption capacity for arsenate uptake in different water conditions, *Ind. Eng. Chem. Res.* 54 (2015) 3000–3008.

Zhang, S., C. Liu, Z. Luan, X. Peng, H. Rena, J. Wang, Arsenate removal from aqueous solutions using modified red mud, *J. Hazard. Mater.* 152 (2008) 486–492.

Zhang, Y., M. Yang, X.M. Dou, H. He, D.S. Wang, Arsenate adsorption on an Fe-Ce bimetal oxide adsorbent: Role of surface properties, *Environ. Sci. Technol.* 39 (2005) 7246–7253.

Zhao, F.J., S.P. McGrath, A.A. Meharg, Arsenic as a food chain contaminant: Mechanism of plant uptake and metabolism and mitigation strategies, *Annu. Rev. Plant Biol.* 61 (2010) 535–559.

Zhao, K., H. Guo, Behavior and mechanism of arsenate adsorption on activated natural siderite: Evidences from FTIR and XANES analysis, *Environ. Sci. Pollut. Res.* 21 (2014) 1944–1953.

Zhu, B.J., X.Y. Yu, Y. Jia, F.M. Peng, B. Sun, M.Y Zhang, T. Luo, J.H. Liu, X.J. Huang, Iron and 1,3,5-benzenetricarboxylic metal–organic coordination polymers prepared by solvothermal method and their application in efficient As(V) removal from aqueous solutions, *J. Phys. Chem. C* 116 (2012) 8601–8607.

14 Metal Organic Framework-Based Adsorbents in Water Treatment

Heecheul Kim, Dhevagoti Manjula Dhevi,
Kap Jin Kim, Arun Anand Prabu, and Xubiao Luo

CONTENTS

ABSTRACT

Metal organic frameworks (MOFs) are a new class of crystalline porous materials composed of metal oxide units joined by organic linkages with ion-exchange, gas sorption, or catalytic properties. Although earlier studies on MOFs were focused on their structural diversities and application in gas storage, in recent times, MOFs have generated a steadily growing interest worldwide in the treatment of wastewater generated from industries. In this chapter, we have explored the usage of MOFs as dye and heavy metal adsorbents, their various challenges, and future prospects. Compared with conventional and new adsorbents such as carbon materials, biomass, magnetic nanoparticles, and chelating polymers, MOFs exhibit better characteristics such as high surface area and capacity, thermal stability ($T_d > 350°C$), and excellent affinity/selectivity for dyes/heavy metals. However, MOFs suffer from their low hydrolytic stability. Many researchers have focused on the synthesis of hierarchically mesostructured and hydrolytically more stable MOFs using mineralizing and structure-directing agents. The synthesized materials were found to be capable of adsorption and removal of ionic dyes, as well as heavy metals, from contaminated water. The adsorption kinetics and equilibrium dye adsorption followed the pseudo-second-order and Langmuir models, respectively. Overall, this chapter will be useful to both academic and industrial researchers and technologists to identify suitable MOF-based adsorbents for efficient and selective remediation of dyes and heavy metals from industrial effluents.

14.1 INTRODUCTION

With the rapid increase in global industrial activities, biologically toxic and carcinogenic compounds and metal ions are being constantly released into the water environment. On the other hand, toxic metal concentration in drinking water sources also affects human health. For instance, arsenic (As) is one of the most toxic and carcinogenic chemical elements, and is regarded as the first priority issue of toxic substances by the World Health Organization (WHO). Its practical and effective removal from groundwater remains an important and intractable challenge in water treatment. Adsorption has always been the choice of treatment for the recovery/reuse of heavy metals in industrial effluents. Though zeolites, activated carbon, and magnetic (Fe_2O_3) nanoparticles (MNPs) are known to be more versatile, there is a growing need for more efficient, economical, and specific adsorbents. For instance, MNPs are widely used as adsorbents in the removal of heavy metals from industrial effluents, but one disadvantage is their aggregation (Tollefson 2007), which can be overcome in two main ways: one method is by loading functional nanoparticles (NPs) on common porous adsorbents such as activated carbon, porous alumina, zeolite, diatomite, and so on, which can partly resolve these problems by sacrificing adsorption capacities. The other method is to design and synthesize micro-nanohierarchically structured adsorbents, such as metal organic frameworks (MOFs), which make a compromise between high adsorption capacity and NP stability.

Among the new class of heavy metal adsorbents, MOFs are microporous organic coordination polymers (CPs) assembled through the coordination between metal ions or metal clusters and organic ligands. These crystalline/porous compounds have strong metal–ligand interactions, and their crystalline structure is made up of extended 3-D networks of small discrete clusters/ions connected by multidentate organic linkers. Compared with NPs, metal–organic CPs have at least two advantages in adsorption applications: one is that the CPs are tailored nanoporous host materials, on the basis of the self-assembly of metal ions linked together by specific polyatomic organic bridging ligands. The most striking difference from common materials and NPs is probably the total lack of nonaccessible bulk volume in coordination polymer structures (Figure 14.1). The other advantage is that the CPs possess high thermal and mechanical stability, which avoids the aggregation problem that NP materials suffer.

Compared with other templates, MOFs show distinct advantages such as well-ordered crystalline structure, high porosity, large surface area, and tunable pore size. Research in the field of MOFs has focused on achieving a basic understanding of the interactions occurring between the framework and the adsorbed/reacting molecules. Development of *in situ* spectral and diffraction methods has also helped in the understanding of MOF formation and their working mechanisms involving molecular binding sites in the frameworks. Further, analysis of adsorption, diffusion, and the reaction of guests inside MOFs, structural changes of the molecules, and dynamic processes in the frameworks have also been studied by many researchers (Bansal et al. 2013; Hei et al. 2014). MOFs designed using suitable building blocks are characterized by custom-made pore systems, and specific interactions of molecules inside the framework help in the realization of applications ranging from gas storage and sensors, separation of molecular species inside MOFs, heavy metal adsorbents, to name a few. The usage of MOFs in gas-phase separation has been extensively studied, while there is relatively little research on the liquid-phase area. For use as sensor materials, a change in their physical properties is used for the detection of molecules. For metal adsorption, having specific active sites in the framework or the pores is an important factor. There are only a few reports on the application of MOFs to remove organic contamination from aqueous solution, including benzothiophene, methyl orange (MO), and methylene blue (MB) (Haque et al. 2010, 2011; Khan et al. 2011). For these reasons, MOFs are regarded as promising materials in the selectively

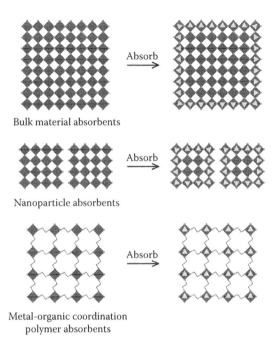

Bulk material absorbents

Nanoparticle absorbents

Metal-organic coordination
polymer absorbents

FIGURE 14.1 **(See color insert)** Schematic illustration of as-proposed new strategy for efficient adsorbent. (Reprinted with permission from scheme 1, Zhu, B.J. et al., 2012, Iron 1,3,5-benzenetricarboxylic metal–organic coordination polymers prepared by solvothermal method and their application in effcient As(V) removal from aqueous solutions, *J. Phys. Chem. C* 116, 8601–8607. Copyright (2012) American Chemical Society.)

adsorptive removal of hazardous materials from wastewater, though their weak stability in aqueous solution is a limiting factor.

14.2 CATION EXCHANGE IN MOFs

Cation exchange is an emerging synthetic route for the modification of the secondary building units (SBUs) of MOFs, and was only first demonstrated with MOFs in 2007 (Dinca and Long 2007). This technique has been used extensively to enhance the properties of nanocrystals and molecules. Cation exchange is the partial or complete substitution of a metal ion at the site of another. Geometric flexibility and the ability of metal sites to interact with the solvent are requisites for cation exchange (Figure 14.2). In these materials, the exchange occurs at the inorganic clusters, often called the *metal nodes* or SBUs. Although these clusters are integral to the MOF structure, the metal ions can be replaced, sometimes entirely and in a matter of hours, without compromising the structure. Ringwood's rule states that ions with similar electronegativity replace each other (Ringwood 1955), and ions with lower value will be exchanged more because they will form bonds with greater ionic character. The metal ion does not readily leave the cluster as a dissociated cation. Instead, solvent molecules may associate stepwise to the exiting metal ion as it remains partially bound to the cluster. Since cation exchange occurs in "paddle-wheel" structures with either a solvent or 4,4'-bipyridine at the axial position of the metal site, the clusters must be flexible enough to accommodate the inserted metal ions. An alternative model suggests that the MOF ligands dynamically dissociate from metal sites in the presence of coordinating solvents and thereby enable cation exchange. The ability of coordination-saturated metal sites to exchange when surrounded by a weak field of carboxylates, but not bridging O^{2-} ligands, suggests that cation exchange may become a predictable tool by quantifying the interaction of the SBU with the metal ions. Installing cations with open coordination sites and open-shell electronic structures enhances the adsorption interaction between the SBU and the

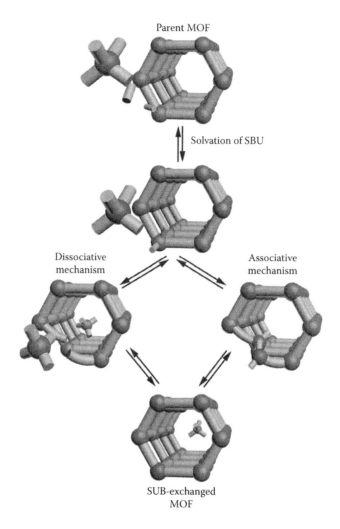

FIGURE 14.2 **(See color insert)** Simplified mechanistic pathways for cation exchange at MOF SBUs. Green and red spheres represent exiting and entering metal ions, respectively. Organic linkers are shown in gray and solvent is depicted in yellow. (Brozek, C.K. and Dinca, M., 2014, Cation exchange at the secondary building units of metal–organic frameworks, *Chem. Soc. Rev.* 43, 5456–5467, scheme 1. Reproduced by permission of The Royal Society of Chemistry.)

guest molecule. Being able to substitute specific metal ions into a predefined environment is a level of control uncommon to solid-state synthetic chemistry, and cation exchange in the SBUs of MOFs is already unlocking materials with unprecedented properties (Brozek and Dinca 2014).

14.3 MESOPOROUS MOFs

One limiting factor against MOFs as a primary source for porous materials is the scarcity of mesoporous MOFs. So far, the majority of all reported porous MOFs are microporous. Only a few mesoporous MOFs containing channel or cavity sizes in the +2 nm realm have been reported. For instance, Eddaoudi et al. (2002) prepared a 3-D mesoporous MOF with channel sizes of 28.8 Å using a linear linker (1,1′:4′,1″-terphenyl-4,4″-dicarboxylate) and an SBU [$Zn_4O(CO_2)_6$]. However, no further characterization of N_2-sorption was reported, due to the structural instability of the MOF after the removal of guest molecules. Augmentation of channels or cavity size in MOFs to the mesoporous range still poses a great challenge. Though ligand extension is a favorable strategy, open MOFs constructed from large ligands tend to disintegrate after the removal of guest molecules. Inhibition of framework interpenetration is another difficulty faced while obtaining mesoporous MOFs containing elongated ligands. Interpenetration occurs because it adds stability by drastically reducing the sizes of channels or cavities. The major drawback of interpenetration is the obstruction of large molecules from entering the internal pores, thus limiting the application potential of MOFs. Therefore, the strategy for the preparation of mesoporous MOFs should focus on ligand extension, while preventing interpenetration, yet maximizing framework stability.

14.3.1 Preparation of Adsorbents

To understand the methodologies involved in the preparation of MOF-based adsorbents, the preparation of a representative material, Materials of Institut Lavoisier (MIL)-101 (Luo et al. 2015) is as follows: MIL-101 was synthesized under solvothermal synthesis conditions using hydrofluoric acid (HF) as a mineralization agent. Amounts of 820 mg of terephthalic acid (TPA) and 2.0 g of $Cr(NO_3)_3 \cdot 9H_2O$ were dissolved in 35 mL of ultrapure water containing 100 μL of HF (40%) under ultrasonic stirring for 30 min. The solution was then transferred to an autoclave and heated at 220°C for 8 h. After cooling to room temperature (RT) (25°C), the resultant solid green-colored products were collected via centrifugation. To remove the unreacted TPA, the as-synthesized MIL-101 was further purified by two-step processes with ultrapure water, dimethylformamide (DMF), and hot ethanol. The precipitate was washed by ultrapure water (100 mL) and DMF (100 mL) three times, using ultrasonic stirring for 10 min followed by centrifugation (15 min, 8000 rpm). A solvothermal treatment was then performed using ethanol (50 mL, 100°C, 20 h). The resulting material was recovered by filtration and finally dried overnight at 150°C under air atmosphere.

MIL-100(Fe) was synthesized (Liu et al. 2014) as follows: a reaction mixture of 0.416 g Fe°, 1.032 g 1,3,5-benzene tricarboxylic or trimesic acid (H_3BTC), 300 μL HF, 280 μL HNO_3, and 30 mL H_2O was placed in a Teflon-lined autoclave, held for 12 h at 150°C, and finally cooled to RT. The light-orange solid product was recovered by filtration, washed with deionized (DI) water and dried at 80°C. The synthesized MIL-100(Fe) was placed in a ceramic boat and calcined in an air furnace with a heating rate of 10°C min^{-1} from RT to 350°C, 550°C, and 750°C (maintained for 2 h), followed by natural cooling to RT.

Another representative adsorbent material (iron-1,3,5-benzenetricarboxylic [Fe−BTC]) is prepared (Zhu et al. 2012) as follows: The reactants $FeCl_3 \cdot 6H_2O$ and H_3BTC (1:1 molar ratio) were dissolved in 5 mL DMF, and heated at 150°C for 24 h in a Teflon-lined stainless-steel autoclave with a total volume of 23 mL. After the workup, a homogeneous red Fe−BTC polymer was formed. The as-synthesized polymer was washed with DMF and ethanol several times, centrifuged and then dried at 60°C in a vacuum for 6 h to obtain the dried product.

14.3.2 Functionalization of Adsorbents

To improve the stability of MOF-based adsorbents and to increase the adsorptive sites to enable stronger interaction with metal ions, functionalization of the adsorbents is a must. In one such study (Luo et al. 2015), ethylenediamine (ED)-modified MIL-101 was formed through coordination bonding of unsaturated Cr metal centers in MIL-101 with the −NH$_2$ group in ED by a postsynthetic modification technique. Cr trimers in mesoporous cages of MIL-101 possess terminal water, and are removable from the framework after vacuum treatment at 150°C for 12 h (Figure 14.3a). This, in turn, can generate open metal sites or coordinate-unsaturated sites (CUSs) as Lewis acid sites in the structure, which are usable for surface modification. Thus, ED, when used as the common functional group, can be easily coordinated on the CUSs of MIL-101. The amino groups modified on the pore surface of MIL-101 afford chelating binding sites for the adsorption of Pb^{2+} ions, which coordinate well with nitrogen atoms of amino groups.

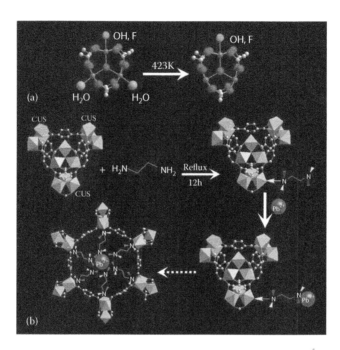

FIGURE 14.3 **(See color insert)** (a) Generation of coordinated–unsaturated sites from Cr trimers in MIL-101 after vacuum treatment at 423 K for 12 h, and (b) adsorption principle of amino-functionalized MIL-101 for Pb²⁺ ions. (Reprinted with permission from figure 1, Luo, X. et al., 2015, Adsorptive removal of Pb(II) ions from aqueous samples with amino-functionalization of metal–organic frameworks MIL-101(Cr), *J. Chem. Eng. Data* 60, 1732–1743. Copyright (2015) American Chemical Society.)

Typically, 1.0 g of as-synthesized MIL-101 sample was dehydrated at 150°C in a vacuum oven for 12 h to generate CUSs (Figure 14.3a) and then suspended in 75 mL of anhydrous toluene. After that, an appropriate amount of 15.0 mol L^{-1} ED solution (2 mmol, 136 μL for sample A; 5 mmol, 340 μL for sample B; 10 mmol, 680 μL for sample C) was introduced, and the mixture was further refluxed for 12 h (Figure 14.3b). The final sample was collected by filtration and washing with excess ethanol and dried overnight at RT (25°C). In the process of preparing ED-MIL-101, the material was optimized through the constant adjustment of the grafting amount of ED. Depending on the amount of ED (2, 5, and 10 mmol), the product exhibits different performances on the Brunauer–Emmett–Teller (BET) surface area, morphology, and adsorption capacity. Finally, it was determined that 5 mmol grafted on MIL-101 is the optimal grafting amount of ED.

14.3.3 SOLUTION PREPARATION FOR ADSORPTION STUDIES

In most cases, batch adsorption experiments were conducted at 25°C for 6 h, using the necessary adsorbents, in a 100 mL stoppered conical flask containing 20 mL of test solution to obtain equilibrium data. Hei et al. (2014) prepared solutions containing different concentrations of As³⁺ and As⁵⁺ using As_2O_3 and $NaH_2AsO_4 \cdot 12H_2O$ as the sources of heavy metal ions, respectively. Luo et al. (2015) mixed 20 mg of the adsorbents with 20 mL of Pb²⁺ solutions of different pH and initial concentrations. The pH of the aqueous solution was adjusted by adding 0.1 mol L^{-1} HNO_3 and 0.1 mol L^{-1} NaOH, and the concentration of Pb²⁺ ions was 50 mg L^{-1}.

14.3.4 ADSORPTION KINETIC EXPERIMENTS

An amount of 200 mg of adsorbent was added into 200 mL of a 500 mg L^{-1} metal ion solution at RT (25°C) with constant stirring. Samples were taken out at certain intervals, and the adsorbent and liquid were separated by centrifugation. The concentration of metal ions in the remaining solution were analyzed. The amount of metal ions bound on the adsorbent was calculated as follows:

$$Q_e = \frac{C_0 V - C_e V}{m} \tag{14.1}$$

where:

Q_e (mg g^{-1})	represents the adsorption capacity
C_0 (mg L^{-1})	is the initial concentration of metal ions
C_e (mg L^{-1})	is the equilibrium concentration of metal ions
m (g)	is the mass of adsorbents
V (L)	is the volume of the metal ion solution

Adsorption selectivity experiments were carried out as follows: Selectivity of the adsorbent toward the metal ions was valued by competitive adsorption in the presence of various competitive metal ions. For example, a mixture of divalent Pb, Cu, Zn, Co, and Ni ions were added to DI water to obtain an initial concentration of 30 mg L^{-1} for each ion. After a competitive adsorption equilibrium was reached for 6 h, the concentrations of the metal ions in the remaining samples were detected using an atomic absorption spectrometer (AAS). The following equations were used to evaluate the selectivity of adsorbent.

Static distribution coefficient

$$K_D = \frac{Q_e}{C_e} \tag{14.2}$$

Selectivity coefficient

$$\alpha = \frac{K_{D1}}{K_{D2}} \tag{14.3}$$

where:

K_{D1}	represents the distribution coefficients of Pb^{2+} ions
K_{D2}	represents the distribution coefficients of Cu^{2+}, Zn^{2+}, Co^{2+}, and Ni^{2+} ions
α	is the selectivity coefficient

14.3.5 CHARACTERIZATION TECHNIQUES

Different characterization techniques could be employed to analyze the structural changes in the MOFs. Elemental (C, H, and N) and inductively coupled plasma (ICP) analyses, AAS, TGA, x-ray powder diffraction (XRD), scanning electron microscopy (SEM), transmission electron microscopy (TEM), x-ray photoelectron spectroscopy (XPS), and BET were some of the essential characterization methods used. XRD patterns were obtained on a diffractometer using Cu K α radiation ($\lambda = 0.1541$ nm) in a scanning range of 3° to 50° at a scanning rate of 1°min^{-1}. Powder samples were dispersed on low-background quartz disks for analyses. Simulation of the powder x-ray diffraction (PXRD) spectra was carried out by using the single crystal data and diffraction-crystal module of the Mercury program. N$_2$ adsorption isotherms were obtained at −196°C. The samples were outgassed under vacuum at 150°C for 10 min prior to the adsorption measurements. The surface electronic states of the synthesized samples were investigated by XPS using Al K α radiation. The XPS data were internally calibrated fixing the binding energy of C 1s at 284.6 eV. TGA was performed under N_2 atmosphere with a heating rate of 2°C min^{-1}. The BET surface area measurement was performed with N_2 adsorption–desorption isotherms at liquid N_2 temperature (–196°C) after the samples were dehydrated under vacuum at 110°C for 12 h. AAS was used for the determination of metal ion concentrations.

14.4 TYPES OF MOFs

MOFs with diverse architectures and morphologies have been recognized as promising precursors/ templates in the remediation of heavy metals/dyes from effluents. The preparation of nanostructured MOFs can be classified into top-down or bottom-up methods and can be easily prepared by the one-pot synthesis method. This section deals with the synthesis of selected MOFs, their structural characteristics, and adsorption properties, such as adsorption equilibrium, kinetics, thermodynamics, and adsorption mechanism.

14.4.1 MAGNETIC NP-BASED MOFs

Although MOFs have several practical applications in pollutant removal, there are still some disadvantages to MOFs, and the most important is separating them from liquid media. Magnetic separation based on superparamagnetic Fe-NPs is widely used in wastewater treatment mainly because of its convenience, economy, and efficiency. Thus, the combination of MNPs and MOFs could overcome the disadvantage of MOFs, and cannot only be easily separated from liquid media, but also exhibit high specific surface areas. MOFs have been recognized as promising precursors/templates to develop porous iron oxides. However, simple and controlled synthesis of γ-Fe$_2$O$_3$ MNPs from the thermolysis of Fe-based MOFs is still rarely reported. Xu et al. (2010) reported the fabrication of spindle-like mesoporous α-Fe$_2$O$_3$ of <20 nm size using MOF MIL-88(Fe) as a template. Many researchers have investigated the efficiency of Fe$_2$O$_3$ NPs as promising adsorbents for the removal of As from aqueous solution (Zhang et al. 2003; Guo and Chen 2005; Iesan et al. 2008). The interaction between As and Fe$_2$O$_3$ is very strong and irreversible, especially due to their large surface-to-volume ratio resulting from their small sizes (Kanel et al. 2005). Heavy metal ion adsorption by metal oxide is likely, due to the combination of static electrical attraction between oxides and heavy metal ions and ion exchange in aqueous solution.

Zhu et al. (2012) synthesized Fe–BTC-based MOFs via a simple solvothermal method for the removal of As. The as-synthesized Fe–BTC polymers exhibit gel behavior, which is stable in common organic solvents or in water. The effect of contact time on the removal of As^{5+} by Fe–BTC, Fe$_2$O$_3$ NPs, and bulk counterparts is depicted in Figure 14.4a. For Fe–BTC polymer and bulk Fe$_2$O$_3$ powders, a rather fast adsorption of the As^{5+} occurred and then reached equilibrium value within 10 min. For the Fe$_2$O$_3$ NPs, the equilibrium was reached within 60 min. To analyze the adsorption rate of As^{5+} on the three adsorbents, the pseudo-second-order rate equation was evaluated based on the experimental data.

$$\frac{t}{q_t} = \frac{1}{k_2 q_e^2} + \frac{1}{q_e} t \tag{14.4}$$

where:
  q_e and q_t are the amount of As^{5+} adsorbed at equilibrium and at time t, respectively
    k_2 represents the pseudo-second-order rate constant of adsorption [g.(mg min)$^{-1}$]

The linear plot feature of t/q_t versus t (shown in Figure 14.4b) was achieved, and the k_2 values were calculated from the slopes. The correlation coefficients of the pseudo-second-order rate model for the linear plots are very close to 1, which suggests that kinetic adsorption of the three adsorbents

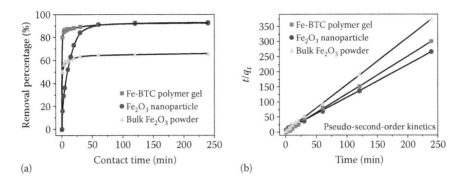

(a)   (b)

FIGURE 14.4 (a) Effect of contact time on absorption rate, and (b) pseudo-second-order kinetic plots for As^{5+} adsorption by Fe–BTC polymer gel, Fe$_2$O$_3$ nanoparticles, and bulky Fe$_2$O$_3$ powders (pH = 4, m/V = 5.0 g L^{-1}, T = 25°C). (Reprinted with permission from figure 4, Zhu, B.J. et al., 2012, Iron 1,3,5-benzenetricarboxylic metal–organic coordination polymers prepared by solvothermal method and their application in efficient As(V) removal from aqueous solutions, *J. Phys. Chem. C* 116, 8601–8607. Copyright (2012) American Chemical Society.)

can be described by the pseudo-second-order rate equation. The pH value is another important factor that controls As speciation. H$_2$AsO$_4^-$ dominates at low pH values, which are defined as less than about pH 6.9. At high pH values, HAsO$_4^{2-}$ is dominant. H$_3$AsO$_4$ and AsO$_4^{3-}$ may be present in strong acid (<pH 2.3) or base (>pH 11) conditions, respectively. To determine the optimum pH for the adsorption of As^{5+} on the Fe–BTC polymer, the uptake of As^{5+} as a function of pH was studied. The removal efficiency of As^{5+} by Fe–BTC was above 96% in the range of pH 2–10, with the maximum removal efficiency of 98.2% at pH 4. Increasing the pH to 12 decreased the removal efficiency of As^{5+} to 35.8%. The as-prepared Fe–BTC polymer was unstable in strong base conditions and dissolved gradually. As^{5+} was preferably adsorbed by the Fe–BTC polymer in a wide pH range. Although optimum As^{5+} removal existed in acidic conditions, high removal performance was still found near neutral pH values. In general, the pH value of natural water is in the range of 6–8.5. So, a pH preadjustment is not needed for contaminated water when the Fe–BTC polymer is applied to the removal of As^{5+}.

Figure 14.5a shows the adsorption isotherms of As^{5+} (at 25°C) on the as-obtained Fe–BTC polymer, Fe$_2$O$_3$ NPs with a size of 50 nm, and bulk Fe$_2$O$_3$ powders with a size of ca. 2 μm. The

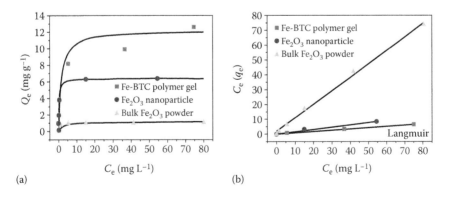

(a)   (b)

FIGURE 14.5 (a) Adsorption isotherms, and (b) linearized Langmuir isotherms for As^{5+} adsorption by Fe–BTC gel, Fe$_2$O$_3$ nanoparticles, and bulk Fe$_2$O$_3$ powders (pH = 4, m/V = 5.0 g L^{-1}, T = 25°C). (Reprinted with permission from figure 6, Zhu, B.J. et al., 2012, Iron 1,3,5-benzenetricarboxylic metal–organic coordination polymers prepared by solvothermal method and their application in efficient As(V) removal from aqueous solutions, *J. Phys. Chem. C* 116, 8601–8607. Copyright (2012) American Chemical Society.)

Langmuir isotherm assumes a surface with homogeneous binding sites, equivalent adsorption energies, and no interaction between adsorbed species. Therefore, the adsorption saturates, and no further adsorption occurs. The Freundlich isotherm is based on an exponential distribution of adsorption sites and energies, and is derived from the multilayer adsorption model and adsorption onto heterogeneous surfaces. Thermodynamic analysis indicates that the adsorption is spontaneous. The adsorption isotherms can be well described by the Langmuir equation, and the Q_m values of Fe−BTC, Fe_2O_3 NPs, and bulk counterparts are 12.29, 6.37, and 1.10 mg g^{-1}, respectively. Maximum adsorption capacity related to the mass of iron that the adsorbent contains is represented as Q_m', and its values for Fe−BTC, Fe_2O_3 NPs, and bulk Fe_2O_3 are 57.71, 9.10, and 1.57 mg g^{-1}, which indicate that the effective adsorption sites in the Fe−BTC polymers are nearly 6.5 times that of Fe_2O_3 NPs with a diameter of 50 nm and 36.8 times that of commercial Fe_2O_3 powders with sizes of ca. 2 μm. The curve fitting results of the Langmuir model are shown in Figure 14.5b. From the correlation coefficients, it can be seen that the adsorption data of Fe−BTC, Fe_2O_3 NPs, and Fe_2O_3 bulk powders fit the Langmuir isotherm model better than the Freundlich model. The preparation of Fe−BTC-based MOFs can be considered as a new method to conquer the dilemma between the excellent properties of the nanoscale effect and the aggregation of small-sized particles in the adsorption application of NPs.

Hei et al. (2014) synthesized MIL-100(Fe) crystals by a simple low-temperature (<100°C) synthesis route and used them as the template for metal oxide. MIL-100(Fe) is a novel mesoporous Fe^{3+} carboxylate crystal MOF with large BET specific surface area and pore volume, and a significant amount of accessible Lewis acid metal sites on dehydration. Using MIL-100(Fe) as the template, porous γ-Fe_2O_3 NPs were fabricated by simple thermolysis of MIL-100(Fe) powder via a two-step calcination treatment: (a) under N_2 gas flow (1°C min^{-1}) in a tube furnace heated to 450°C and maintained at that temperature for 2 h, and (b) the resultant black powder was placed in a box furnace, heated to 450°C (1°C min^{-1}) and maintained at that temperature for 2 h in air. Compared with the ordinary γ-Fe_2O_3 materials obtained by the solvothermal process and subsequent calcination, this strategy is simple, inexpensive, and scalable. This material has a relatively large specific surface area of 123.5 m^2 g^{-1}, which is triple that of mixed Fe_2O_3 prepared from Prussian Blue (Zhang et al. 2013). Such a high surface area value is presumed to benefit from the two-step calcination of the template MIL-100(Fe). On the other hand, the pore size distribution reveals that most of the pores are focused on 10 nm, indicating the effective formation of γ-Fe_2O_3 mesopores. The large specific surface area, mesoporous nature, and high pore volume assist γ-Fe_2O_3 to serve as an efficient adsorbent for heavy metal ion removal in water treatment. From the adsorption studies, it can be seen that the first 30 min correspond to a rapid adsorption stage for both As^{3+} and As^{5+}. The residual concentration of As^{3+} was <40 mg L^{-1} and that of As^{5+} was <20 mg L^{-1} after this fast stage. Thereafter, the adsorption rates decreased and reached equilibrium. The Langmuir model ($R^2 > 0.98$), representing monolayer adsorption, was applied to fit the experimental data for both As^{3+} and As^{5+}. The maximal adsorption capacity of the γ-Fe_2O_3 NPs obtained is around 62.9 mg g^{-1} for As^{3+} and 90.6 mg g^{-1} for As^{5+}. The removal capacity values of the γ-Fe_2O_3 obtained were measured without any pH value adjustment and, thus, the removal capacities achieved here are more indicative of the potential of this material in practical applications.

Liu et al. (2014) synthesized hematite nanorods and MNPs via direct thermolysis of an iron-based MOF material, MIL-100(Fe), and used them as absorbents for the removal of As^{5+}. MIL-100(Fe) possesses two sets of mesoporous cages that are accessible through microporous windows and corresponding large Langmuir surface areas. Fe Lewis acid sites can be produced by the removal of the two terminal H_2O molecules of iron octahedral and the partial departure of anions through vacuum activation. Thus, MIL-100(Fe) is confirmed to have a large amount of Fe Lewis acid sites in its pores. These structural characters indicate that this material has a significant advantage for sorption. Calcination temperature and atmosphere tend to affect pore sizes and types of the obtained MOFs. SEM images (Figure 14.6a–c) show that this bulk solidity structure is composed of clustered porous α-Fe_2O_3 nanorods. The average lengths of the nanorods obtained at

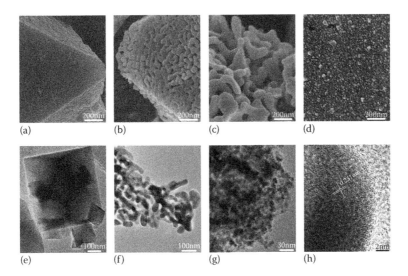

(a)   (b)   (c)   (d)

(e)   (f)   (g)   (h)

FIGURE 14.6 SEM images of α-Fe$_2$O$_3$ calcined at (a) 350°C, (b) 550°C, (c) 750°C, and (d) Fe$_3$O$_4$; TEM images of obtained (e) MIL-100(Fe), (f and h) α-Fe$_2$O$_3$ calcined at 550°C, and (g) Fe$_3$O$_4$. (Reprinted from figure 2, *Mater. Lett.*, 132, Liu, Z.-M. et al., Novel hematite nanorods and magnetite nanoparticles prepared from MIL-100(Fe) template for the removal of As(V), 8–10, Copyright (2014), with permission from Elsevier.)

350°C, 550°C, and 750°C are 50, 150, and 200 nm, respectively. Moreover, as the calcination temperature increased, α-Fe$_2$O$_3$ nanorods became larger as a result of crystal growth. An individual α-Fe$_2$O$_3$ nanorod can be clearly seen in Figure 14.6f with a mean length of 150 nm and diameter of about 30 nm. The size of Fe$_3$O$_4$ NPs (Figure 14.6d) is found to be in the range of 10–20 nm (Figure 14.6g). The octahedron-like morphology of the MIL-100(Fe) is confirmed by TEM analysis (Figure 14.6e). A TEM image of an individual rod shows clear and regular lattice fringes (Figure 14.6h) with interplanar spacing of 0.251 nm, which fits well to that of the (110) plane of α-Fe$_2$O$_3$. The α-Fe$_2$O$_3$ and Fe$_3$O$_4$ had relatively high BET surface areas (92.8 and 150.9 cm^2 g^{-1}). Figure 14.7 shows the As^{5+} adsorption capacity of α-Fe$_2$O$_3$ (at 25°C, pH 3) calcined at 350°C, 550°C, and 750°C as 94.9, 74.1, and 70.5 mg g^{-1}, respectively, compared with Fe$_3$O$_4$ (80.5 mg g^{-1}). It can be explained that the adsorption capacities have a relationship with the surface areas, pore volumes, and pore sizes. More importantly, the adsorption capacities increase with the increase of the pore volumes in this situation. This approach may provide reference for the synthesis of other metal oxide NPs.

Taghizadeh et al. (2013) prepared a novel magnetic MOF from dithizone-modified Fe$_3$O$_4$ NPs and a Cu-BTC-based MOF and used them in the preconcentration of Cd^{2+}, Pb^{2+}, Ni^{2+} and Zn^{2+} ions. The parameters affecting preconcentration were optimized by a Box–Behnken design through response surface methodology. Three variables (extraction time, amount of magnetic sorbent, and pH value) were selected as the main factors affecting adsorption, while four variables (type, volume, concentration of the eluent, and desorption time) were selected for desorption in the optimization study. Following preconcentration and elution, the ions were quantified by a flame atomic absorption spectrometer (FAAS). The limits of detection are 0.12, 0.39, 0.98, and 1.2 ng mL^{-1} for Cd^{2+}, Zn^{2+}, Ni^{2+}, and Pb^{2+} ions, respectively. The relative standard deviations were <4.5% for five separate batch determinations of 50 ng mL^{-1} of Cd^{2+}, Zn^{2+}, Ni^{2+} and Pb^{2+} ions. The adsorption capacities (in mg g^{-1}) of this new MOF are 188 for Cd^{2+}, 104 for Pb^{2+}, 98 for Ni^{2+}, and 206 for Zn^{2+}. The MOF has a higher capacity than the Fe$_3$O$_4$/dithizone conjugate, and was successfully applied to the rapid extraction of trace quantities of heavy metal ions in fish, sediment, soil, and water samples. Figure 14.8 shows the schematic diagram for the synthesis of magnetic MOF-DHz nanocomposites.

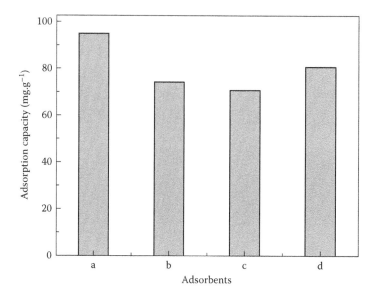

FIGURE 14.7 As^{5+} adsorption capacity of α-Fe$_2$O$_3$ calcined at (a) 350°C, (b) 550°C, (c) 750°C, and (d) As^{5+} adsorption capacity of Fe$_3$O$_4$. (Reprinted from figure 4, *Mater. Lett.*, 132, Liu, Z.-M. et al., Novel hematite nanorods and magnetite nanoparticles prepared from MIL-100(Fe) template for the removal of As(V), 8–10, Copyright (2014), with permission from Elsevier.)

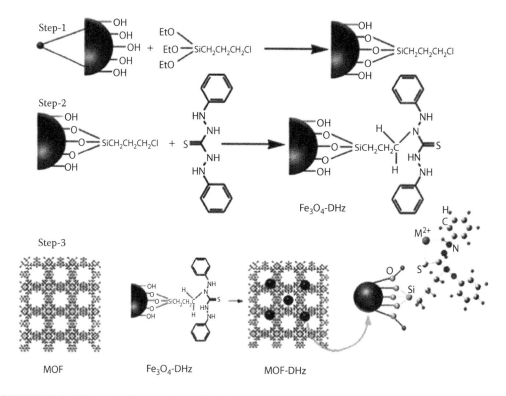

FIGURE 14.8 Schematic diagram for synthesis of magnetic MOF-DHz nanocomposite. (With kind permission from Springer Science+Business Media: *Microchim. Acta*, A novel magnetic metal organic framework nanocomposite for extraction and preconcentration of heavy metal ions, and its optimization via experimental design methodology, 180, 2013, 1073–1084, Taghizadeh, M. et al., graphical abstract.)

14.4.2 AMINO-FUNCTIONALIZED MOFs

Jhung et al. (2007) synthesized MIL-101 by a microwave method and reported a larger capacity for MIL-101 than SBA-15, HZSM-5 zeolite, and activated carbon in the adsorption of benzene. Zhu et al. (2012) reported six times more As adsorption capacity using Fe–BTC than Fe_2O_3 NPs, and 36 times more than commercial iron oxide powders. Ke et al. (2011) reported remarkably strong adsorption affinity and high adsorption capacity of Hg^{2+} ions using thiol-functionalized $[Cu_3(BTC)_2]_n$ MOFs. Zou et al. (2013) used Hong Kong University of Science and Technology (HKUST)-1-MW-$H_3PW_{12}O_{40}$ to achieve a remarkably high adsorption affinity and adsorption capacity of metal ions, but this cannot selectively remove a specific metal ion, which is important in metal ion recycling.

Fang et al. (2010) used $Zn_4O(CO_2)_6$ as an SBU and two extended ligands containing amino functional groups to form two different isostructural mesoporous MOFs (porous coordination network, PCN-100 and 101) with cavities up to 2.7 nm, while preventing interpenetration by ligand design and stabilizing the open framework through hydrogen bonding. In the PCN-100 framework, eight 4,4′,4″-s-triazine-1,3,5-triyltri-p-aminobenzoate (TATAB) ligands and six $Zn_4O(CO_2)_6$ clusters as SBUs formed a mesoporous cavity with an internal diameter of approximately 2.7 nm (the measured distance between opposite atoms); the size of the windows is about 1.32×1.82 nm. These mesoporous cavities are further interconnected through TATAB ligands to generate a 3-D noninterpenetrating extended open network, in which the vacant space is filled with 17 N,N-diethylformamide guest molecules and three H_2O guest molecules per formula unit (Figure 14.9). Removal of the guest molecules reveals that the effective free volume of PCN-100 is 80.8% of the crystal volume (16462.7 Å of the 20379.0 Å unit cell volume). Each $Zn_4O(CO_2)_6$ SBU and TATAB ligand can be defined as a 6-connected and a 3-connected node, respectively. Based on this simplification, PCN-100 can be described as a 3,6-connected 3-D network with the Schläfli symbol $(6^{12}.8^3)(6^3)_2$, which corresponds to a pyrite topology. N_2 sorption isotherms of both the MOFs showed typical type IV behavior, indicating their mesoporous nature. PCN-100 was employed to capture heavy metal ions (Cd^{2+} and Hg^{2+}) by constructing complexes within the pores with a possible coordination mode similar to that found in aminopyridinato complexes. Capture of metal ions becomes particularly valuable when coordination of heavy metals such as Cd and Hg is possible (Figure 14.10). The metal capture reaction was performed in a DMF solution of different metal salts $[Co(NO_3)_2, Cd(NO_3)_2,$ or $HgCl_2]$ and a sample of PCN-100. A direct indication that a reaction between PCN-100 and Co^{2+} ions took place was the change in solid color from colorless to red. A concomitant change of the metal content of PCN-100 was observed by ICP corresponding to the impregnation of about 1.61 Co^{2+}, 1.63 Cd^{2+}, or 1.38 Hg^{2+} per formula, respectively. Similar experiments were performed on the isostructural PCN-101, which lacks the aminopyridinato-type chelating coordination environment.

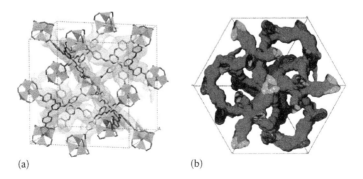

(a)   (b)

FIGURE 14.9 **(See color insert)** (a) View of packing of PCN-100, and (b) view of channels and free volume of PCN-100. (Reprinted with permission from figure 3, Fang, Q.-R. et al., 2010, Functional mesoporous metal-organic frameworks for the capture of heavy metal ions and size-selective catalysis, *Inorg. Chem.* 49, 11637–11642. Copyright (2010) American Chemical Society.)

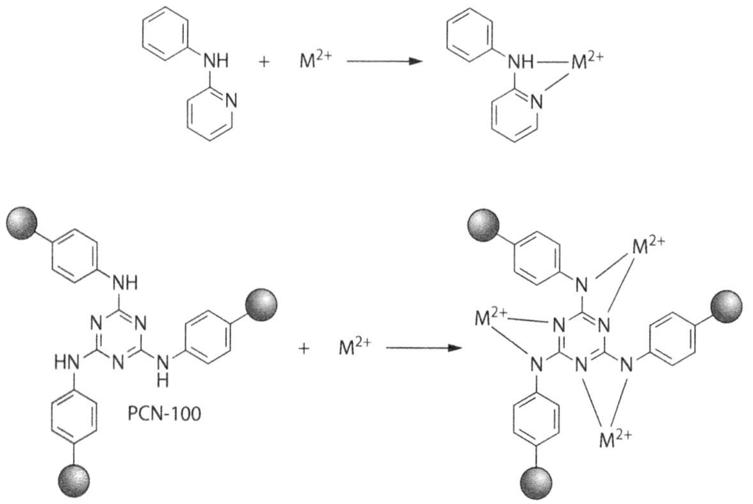

FIGURE 14.10 Possible chelating coordination mode for metal ions in PCN-100. (Reprinted with permission from scheme 2, Fang Q.-R. et al., 2010, Functional mesoporous metal-organic frameworks for the capture of heavy metal ions and size-selective catalysis, *Inorg. Chem.* 49, 11637–11642. Copyright (2010) American Chemical Society.)

The ICP results showed that there was only 0.59 Co^{2+}, 0.69 Cd^{2+}, or 0.61 Hg^{2+} contained in the framework of PCN-101 per formula unit. From these results, it can be deduced that an additional 0.51 Co^{2+} ions were captured due to a chelating effect in PCN-100. Similarly, the ICP measurements revealed that there were an additional 0.47 Cd^{2+} or 0.39 Hg^{2+} ions captured due to this chelating effect. These interesting results suggest that PCN-100 and similar MOFs can be applied in the elimination of heavy metal ions from waste liquid.

Inspired by the fact that the adsorption capacity of Pb^{2+} ions using amino groups modified-silica-based mesoporous materials is apparently higher than that of normal silica-based materials, Luo et al. (2015) revealed a simple strategy to fabricate amine-grafted MOFs (ED-MIL-101) as a novel adsorbent for selective removal of Pb^{2+} ions from aqueous samples. The influence of the pH value on the adsorption curves of Pb^{2+} is shown in Figure 14.11a. It can be observed that the adsorption capacity of MIL-101 and ED-MIL-101 (5 mmol) for Pb^{2+} ions increases with the solution pH increasing from 2 to 6. The rate of increase in the adsorption capacity of ED-MIL-101 (5 mmol) is much higher than that of MIL-101 for Pb^{2+} ions. This is because some of the free carboxyl groups on the MIL-101 can easily lose hydrogen ions and provide the chelating binding sites for Pb^{2+} ions. However, for ED-MIL-101, the amino groups modified on the pore surface of MIL-101 also afford the chelating binding sites for the adsorption of Pb^{2+} ions, except for the original free carboxyl groups, which results in a much higher rate of increase in adsorption capacity of ED-MIL-101 (5 mmol) than of MIL-101. The results verified further that there are more chelating binding sites in ED-MIL-101 (5 mmol) than in MIL-101. When the solution pH < 4, a large amount of H^+ can coordinate with the ionized carboxyl of MIL-101 and the amino groups of ED-MIL-101. The adsorption capacity of MIL-101 and ED-MIL-101 (5 mmol) is very low as the pH value of the solution increases, indicating a decrease in the concentration of H^+, which is due to a large amount of H^+ competing with Pb^{2+} ions and the protonation of the amino group moieties. The adsorption of Pb^{2+} ions on MIL-101 and ED-MIL-101 (5 mmol) increased substantially when the pH value rose from 4 to 6. The main reason is that the protonation of carboxyl groups and amino groups are relatively weakened and most of them chelate Pb^{2+} ions. When the solution pH > 6, the adsorption capacity of MIL-101 and ED-MIL-101 (5 mmol) decreases, due to the predominance of hydrolysis products of Pb^{2+} ions over the adsorption of MIL-101 and ED-MIL-101 (5 mmol) for Pb^{2+} ions. Therefore, the pH of the solution should be adjusted to 6 to obtain optimal adsorption efficiency.

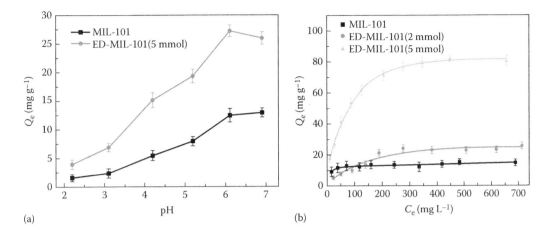

FIGURE 14.11 Adsorption capacity of MIL-101 and EDMIL-101 for Pb²⁺ ions: (a) Effect of pH on adsorption capacity, and (b) adsorption isotherms. (Reprinted with permission from figures 8 and 9, respectively, Luo, X. et al., 2015, Adsorptive removal of Pb(II) ions from aqueous samples with amino-functionalization of metal–organic frameworks MIL-101(Cr), *J. Chem. Eng. Data* 60, 1732–1743. Copyright (2015) American Chemical Society.)

Figure 14.11b shows the adsorption isotherms of Pb²⁺ ions at pH 6. Because the structures of ED-MIL-101 (10 mmol) had been destroyed, MIL-101, ED-MIL-101 (2 mmol) and ED-MIL-101 (5 mmol) were chosen for Pb²⁺ adsorption. It can be seen that the equilibrium adsorption capacities of ED-MIL-101 (2 mmol) and ED-MIL-101 (5 mmol) for Pb²⁺ increased rapidly in a low concentration range, then increased slowly, and finally achieved a maximum equilibrium adsorption capacity. The maximum adsorption capacities of MIL-101, ED-MIL-101 (2 mmol), and ED-MIL-101 (5 mmol) are 15.8, 25.6, and 81.1 mg g⁻¹, respectively, indicating that ED-MIL-101 (5 mmol) has five times' stronger affinity and more adsorption sites for Pb²⁺ ions than MIL-101. Langmuir and Freundlich models were used to fit the adsorption process to describe the Pb²⁺ ion distribution between the liquid and adsorbent phase and achieve the maximum adsorption uptake. The Langmuir and Freundlich isotherms are represented by

$$\frac{C_e}{Q_e} = \frac{C_e}{Q_m} + \frac{1}{K_L Q_m} \tag{14.5}$$

$$\log Q_e = \log K_F + \frac{1}{n} \log C_e \tag{14.6}$$

where:

K_L (L mg⁻¹) is the Langmuir constant that represents the affinity between solute and adsorbent

n and K_F {mg^[1−(1/n)] L^(1/n) g⁻¹} are Freundlich constants related to the adsorption capacity and intensity, respectively

The isotherms of Pb²⁺ ion adsorption on MIL-101 and ED-MIL-101 can be well described by the Langmuir isotherm equation, and the adsorption kinetics of Pb²⁺ ions can fit well with the pseudo-second-order model.

For the adsorption selectivity studies, Ni²⁺, Co²⁺, Zn²⁺, and Cu²⁺ were chosen as competitor ions due to the same charge as Pb²⁺ ions. The radii of Pb²⁺, Ni²⁺, Co²⁺, Zn²⁺, and Cu²⁺ ions are 0.119, 0.069, 0.075, 0.074, and 0.073 nm, respectively. The adsorption capacities of Pb²⁺, Ni²⁺, Co²⁺, Zn²⁺,

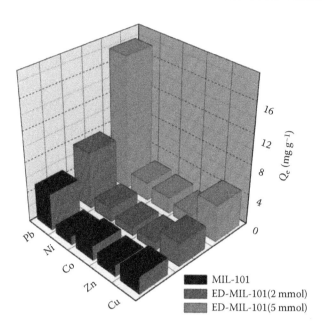

FIGURE 14.12   (See color insert) Selectivity adsorption capacity of MIL-101 and ED-MIL-101 for Pb^{2+} ions in the presence of coions. (Reprinted with permission from figure 14 from Luo, X. et al., 2015, Adsorptive removal of Pb(II) ions from aqueous samples with amino-functionalization of metal–organic frameworks MIL-101(Cr), *J. Chem. Eng. Data* 60, 1732–1743. Copyright (2015) American Chemical Society.)

and Cu^{2+} ions by MIL-101 and ED-MIL-101 are shown in Figure 14.12. It can be observed that the ED-MIL-101 exhibits higher adsorption capacities for Pb^{2+}, Cu^{2+}, and Ni^{2+} than the other ions. This phenomenon can be explained by the complexation constants of EDTA with metal ions (18.8 for Cu^{2+}, 18.6 for Ni^{2+}, 18.0 for Pb^{2+}, 16.5 for Zn^{2+}, and 16.3 for Co^{2+}). Although the complexation constant of amines with Pb^{2+} is slightly smaller than that of Cu^{2+} and Ni^{2+}, the radius of Pb^{2+} is much larger than that of Cu^{2+} or Ni^{2+} and, therefore, the special pore structure of MIL-101 with the appropriate ED grafted groups can coordinate with Pb^{2+} ions more effectively, which leads to a higher adsorption capacity for Pb^{2+} than for the other ions. The selectivity coefficients of ED-MIL-101 (5 mmol) for Pb^{2+} ions with respect to Cu^{2+}, Zn^{2+}, Co^{2+}, and Ni^{2+} are about 6.9, 24.0, 15.7, and 14.5, respectively, which is higher than that of MIL-101 and ED-MIL-101 (2 mmol). This can indicate a higher adsorptive removal ability of ED-MIL-101 (5 mmol) for Pb^{2+} ions with respect to the competing ions. These results imply that the special pore structure of MIL-101 with the appropriate ED grafted groups and specific binding sites for Pb^{2+} ions have been formed in ED-MIL-101, and the $-NH_2$ groups, which are coordinated with metal ions in ED-MIL-101, are unable to match well with Ni^{2+}, Co^{2+}, Zn^{2+}, and Cu^{2+} ions, resulting in a high recognition ability and high selectivity of ED-MIL-101 for Pb^{2+} ions. Moreover, ED-MIL-101 achieves almost 97.2% removal efficiency for Pb^{2+} ions. These results indicate that ED-MIL-101 has great potential in selectively removing Pb^{2+} ions from the water environment.

Abbasi et al. (2015) hydrothermally synthesized a new TATAB-based MOF $\{[Co_2(TATAB)(OH)(H_2O)_2]\cdot H_2O\cdot 0.6O\}_n$ in nanoscale using ultrasound radiation, and studied its removal efficiency of Al^{3+}, Pb^{2+}, Cd^{2+}, Hg^{2+} and Fe^{3+} ions. Figure 14.13a shows the image of MOF crystals before and after capture of the ions, which change from purple after each metal ion adsorption. With an increasing pH value from 2 to 6 and with an exposure time of 60 min, the removal percentage of Fe^{3+} and Al^{3+} metals increased to the highest adsorption level (Figure 14.13b). Repeating the same procedure in the presence of nanostructured TATAB-MOF gave better adsorption results compared with those of bulk TATAB-MOF. The effect of exposure time on the adsorption of these ions showed that the removal percentage of metal ions improved with increased exposure time (Figure 14.13c). For Fe^{3+}

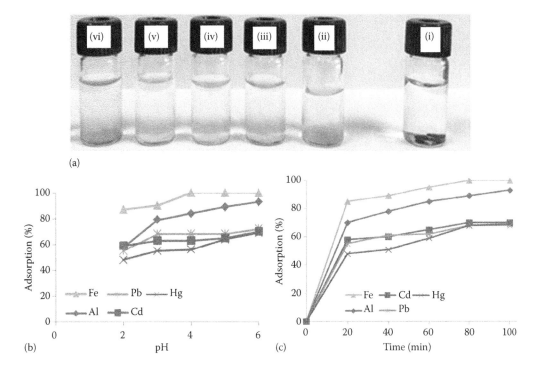

FIGURE 14.13 **(See color insert)** (a) Photographs of (i) TATAB-based MOF, and (ii) adsorbed heavy metal ions: (ii) Al^{+3}, (iii) Hg^{+2}, (iv) Cd^{+2}, (v) Pb^{2+}, and (vi) Fe^{3+}, (b) effect of pH, and (c) effect of exposure time on the adsorption of heavy metal ions by TATAB-MOF nanostructures. (Reprinted from figures 5, 7, and 8, respectively, *Inorg. Chim. Acta*, 430, Abbasi, A. et al., A new 3D cobalt (II) metal–organic framework nanostructure for heavy metal adsorption, 261–267, Copyright (2015), with permission from Elsevier.)

ions, the maximum removal percentage was achieved after 80 min, while for Al^{3+}, Cd^{2+}, Hg^{2+}, and Pb^{2+}, the highest percentage was obtained after 100 min. It seems that TATAB-MOF shows a removal capacity in the order of $Fe^{3+} > Al^{3+} > Cd^{2+} \geq Hg^{2+} \geq Pb^{2+}$. The results showed that the prepared compounds are able to adsorb the metal ions efficiently under RT conditions.

Li et al. (2015a) fabricated MOFs based on amino sulfonic-Cu-(4,4′-bipy)$_2$ (ASC) by the solvo-thermal method for the removal of perchlorate from aqueous solution. The removal rate of perchlorate increased from 72.8% to 98.3% with the adsorbent dosage increasing from 0.5 to 3.0 g L^{-1}, whereas its adsorption capacity decreased steeply from 145.8 to 25.3 mg g^{-1} (Figure 14.14a). When the ASC dosage exceeded 2.0 g L^{-1}, both removal rate and adsorption capacity reached equilibrium, and was hence selected as the optimum dosage. The maximum sorption amount of perchlorate reached 133.5 mg g^{-1} at pH 7, and the perchlorate could be removed effectively over a broad range of pH values (from 2 to 11) at RT. From Figure 14.14b, it can be seen that the adsorption capacity of ASC increased rapidly in the initial 10 h due to the huge number of available vacant surface sites of ASC in the initial stages of the adsorption. The Q_m values for ASC at 10°C, 30°C, and 50°C were found to be 48.4, 45.7, and 40.8 mg g^{-1}, respectively. The values of $q_{e,exp}$ decreased with increasing temperature, because the surface tension between ASC and solution might be weakened rapidly with a temperature increase. The kinetics and equilibrium adsorption data could fit well with the pseudo-first-order kinetic model and Langmuir model, respectively. The proposed adsorption mechanism was the coeffect of electrostatic force and ion exchange between the perchlorate ions and MOF material, which improved the adsorption capacity greatly (Figure 14.15). In the Fourier transform infrared spectroscopy (FTIR) spectrum, a peak of 1494 cm^{-1}, which is supposed to represent the stretching vibration of N–H in −NH$_2$ groups, disappeared after adsorption process, which indicated that the −NH$_2$ groups might be surrounded by perchlorate ions due to its positive charge.

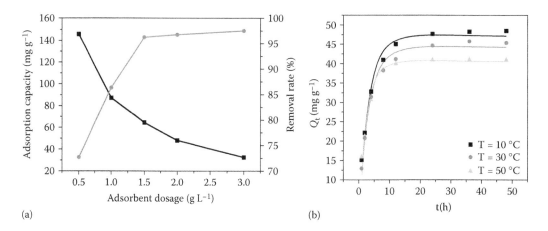

(a)   (b)

FIGURE 14.14 (a) Effect of adsorbent dose on removal rate and adsorption capacity (at 30°C), and (b) plots of pseudo-first-order kinetics model of perchlorate ion adsorption onto ASC (adsorbent dose = 2 g L⁻¹) at three different temperatures (10°C, 30°C, and 50°C). Initial concentration: 100 mg L⁻¹; contact time = 48 h, and pH 7 ± 0.2, maintained for both (a) and (b). (Reprinted from figures 6 and 8, *Chem. Eng. J.*, 281, Li, T. et al., Perchlorate removal from aqueous solution with a novel cationic metal–organic frameworks based on amino sulfonic acid ligand linking with Cu-4,4′-bipyridyl chains, 1008–1016, Copyright (2015), with permission from Elsevier.)

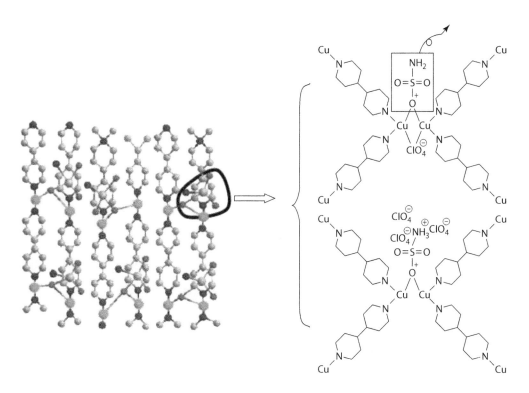

FIGURE 14.15 Potential mechanisms of the perchlorate ion adsorption onto ASC. (Reprinted from figure 11, *Chem. Eng. J.*, 281, Li, T. et al., Perchlorate removal from aqueous solution with a novel cationic metal–organic frameworks based on amino sulfonic acid ligand linking with Cu-4,4′-bipyridyl chains, 1008–1016, Copyright (2015), with permission from Elsevier.)

Moreover, the experiment data matched well with the Langmuir isotherm model, which also demonstrated that electrostatic adsorption, as a monolayer surface adsorption, was the main one.

14.4.3 HEAVY METAL REMOVAL USING OTHER MOFS

Li et al. (2015b) synthesized MOF-808 [$Zr_6O_4(OH)_4(BTC)_2(HCOO)_6$] with a household microwave oven under varying irradiation times (5–30 min). The obtained octahedral nanocrystals were in the range of 150–200 nm, much smaller than those obtained under conventional hydrothermal conditions. They also exhibited high superacidity, making them an excellent adsorbent for As removal. When the initial concentration of As^{5+} was 5 ppm, the adsorption capacity was calculated to be 24.8 mg g^{-1}. This value is much higher than other porous adsorbents such as hybrid silica and aluminum oxide. This phenomenon could be explained in that the surface of the MOF-808 NPs became positively charged at low pH levels, which would enhance the electrostatic interaction of adsorbent and As^{5+}. The adsorption rate of As^{5+}, with an initial concentration of 5 ppm using MOF-808 irradiated for 10 min, is shown in Figure 14.16a. Adsorption occurred rapidly during the early stage of adsorption (about 0–10 min), and approximately 95% of As^{5+} could be removed in 30 min. The correlation coefficient of the pseudo-second-order rate model for the linear plots was very close to 1, and the calculated q_e value was nearly equal to the experimentally obtained value, which suggests that the As adsorption process meets the pseudo-second-order kinetics. The stability and regeneration ability of the adsorbent is very important for practical applications. Typically, the adsorbent was washed with 20 mL of 0.5 M Na_2SO_4 solution three times to remove the adsorbed As^{5+} ions. As shown in Figure 14.16b, the adsorbents still maintained about 82.1% of removal efficiency even after five cycles, which indicates that MOF-808 NPs can serve as a regenerable As^{5+} adsorbent.

Maleki et al. (2015) measured Cr adsorption at various pH values after equilibration with a Cu–BTC-based MOF at 25°C (Figure 14.17a). The adsorbed amounts increased with increasing pH values from 2 to 8. Moreover, adsorption of Cr at lower pH levels (2–6) did not cause significant changes in the equilibrium pH. At lower pH levels, the surface charge of Cu–BTC may become positively charged, thus making H$^+$ ions compete with metal ions, causing a decrease in the amount of metal adsorbed. At higher pH levels, the metal contains chromate anion, which is exchanged to have a displacement reaction with NaOH to produce an NaCl aqueous solution, which may decrease the adsorption of Cr on Cu–BTC. Maximum removal without structural degradation of the adsorbent was achieved at pH 7. The influence of varying the initial metal concentration (10, 20, 30, and 40 mg L^{-1}) was assessed, and the other parameters were kept constant at optimum values (Figure 14.17b). It is obvious that the higher the initial metal concentration, the lower the percentage

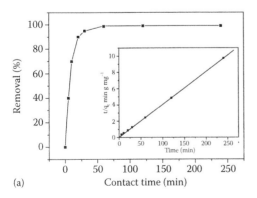
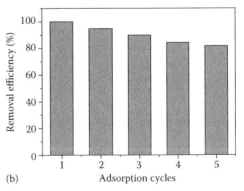

(a)   (b)

FIGURE 14.16 (a) Effect of contact time on adsorption rate; inset shows pseudo second-order kinetic plots for As removal by MOF-808 nanoparticles, and (b) adsorption–desorption cycles. (Reprinted from figure 3, *Mater. Lett.*, 160, Li, Z.-Q. et al., Facile synthesis of metal-organic framework MOF-808 for arsenic removal, 412–414, Copyright (2015), with permission from Elsevier.)

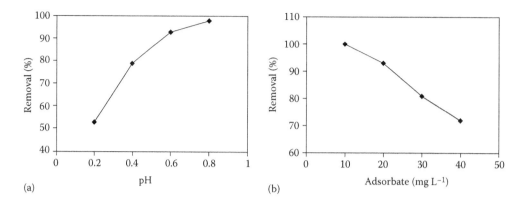

(a)   (b)

FIGURE 14.17 (a) Effect of pH on adsorptive removal of MB (A) and equilibrium pH (B) after adsorption of MB ($m = 5.0$ mg, $V = 10$ mL, $T = 25°C$), and (b) Effect of initial metal on Cr adsorption by Cu-BTC (pH = 7, $C_0 = 20$ mg L^{-1}, $T = 25°C$). (Reprinted from figures 5 and 7, respectively, *J. Ind. Eng. Chem.*, 28, Maleki, A. et al., Adsorption of hexavalent chromium by metal organic frameworks from aqueous solution, 211–216, Copyright (2015), with permission from Elsevier.)

of metal adsorbed. The adsorption values are acceptable under a Cr concentration of 20 mg L^{-1}. In addition, the material can be reused at least three times after washing with ethanol.

Bakhtiari and Azizian (2015) synthesized zinc-based MOF-5 for the adsorption of Cu ions from aqueous solution. Initially, the numbers of available sites for adsorption of Cu^{2+} are greater than close to equilibrium, and hence the uptake of Cu ions into MOF-5 is rapid for the first 30 min (Figure 14.18). The pseudo-first-order kinetic model describes the adsorption of Cu ions onto MOF-5, which indicates that the rate constant of this system is time dependent, due to the presence of different active sites for adsorption (heterogeneous surface) and, therefore, the changing of favorable active sites of adsorption over time. Figure 14.19a shows the equilibrium adsorption data at 25°C. According to the correlation coefficient r^2 and root mean square (RMS) error values, the Langmuir–Freundlich model, which represents the adsorbent with heterogeneous surface, that is, with a different distribution of adsorption energies, describes the experimental data better than other models. The Q_m values of Cu ions using MOF-5 is about 290 mg g^{-1}, which is higher than that obtained using 100.0 mg g^{-1} polyaniline-g-chitosan beads (Igberase et al. 2014), 67.2 mg g^{-1} PNIPAM-Co-AA hydrogels (Chen et al. 2013), and 103.5 mg g^{-1} keratin/PA6 blend nanofibers 90% (Aluigi et al. 2011) for the removal of Cu ions. The results of fit of the equilibrium data with the Langmuir–Freundlich isotherm and kinetic data with the fractal-like pseudo-first-order model are

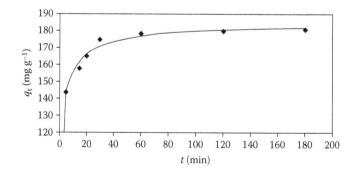

FIGURE 14.18 Effect of contact time on adsorption of copper ions (300 ppm) by MOF-5. The symbols are the experimental data and the line indicates the values predicted by the FL-PFO model. (Reprinted from figure 5, *J. Mol. Liq.*, 206, Bakhtiari, N. and Azizian, S. Adsorption of copper ion from aqueous solution by anoporous MOF-5: A kinetic and equilibrium study, 114–118, Copyright (2015), with permission from Elsevier.)

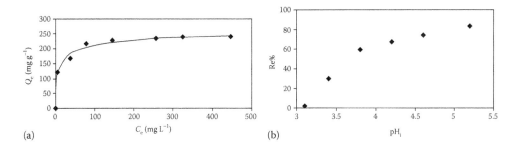

(a)

(b)

FIGURE 14.19 (a) Adsorption isotherm of Cu ions by MOF-5. The symbols are experimental values and the line represents the values predicted by the FL isotherm, and (b) effect of initial solution pH on the removal percentage of Cu ions by MOF-5. (Reprinted from figures 6 and 9, *J. Mol. Liq.*, 206, Bakhtiari, N. and Azizian, S. Adsorption of copper ion from aqueous solution by anoporous MOF-5: A kinetic and equilibrium study, 114–118, Copyright (2015), with permission from Elsevier.)

in agreement with each other because both of them indicate that the surface of the adsorbent is heterogeneous; that is, there are different active sites for adsorption. The removal percentage of Cu ions was studied within the pH range of 3.0–5.2 (Figure 14.19b). This range was selected because Cu ions precipitate as $Cu(OH)_2$ at $pH \geq 6$. The removal efficiency increases significantly by increasing the pH value of the solution, that is, decreasing the H^+ ion concentration, which leads to decreased repulsion between Cu^{2+} and the positive surface. Also, the removal percentage of Cu ions increase by decreasing the H^+ ions, which compete with Cu^{2+} ions for the adsorption sites.

Wu et al. (2015) synthesized novel thiol-functionalized $[Cu_4O(BDC)]_n$-based MOFs (BDC: terephthalic acid), and the thiol-modified MOFs showed remarkable and selective adsorption for four heavy metals (Hg^{2+}, Cr^{6+}, Pb^{2+}, Cd^{2+}; especially for Hg^{2+}). Figure 14.20a shows the Hg^{2+} adsorption equilibrium isotherms of the original and modified MOFs. The adsorption for Hg^{2+} by the modified MOFs reached saturation when the concentration of Hg^{2+} reached 0.6 mg mL^{-1}. In contrast, the unmodified MOF showed almost unchanged Hg^{2+} adsorption with increasing concentration of Hg^{2+}. Obviously, the maximum adsorption amount of the modified MOFs for Hg^{2+} (235.3 mg g^{-1}) was much greater than that of the original MOFs (22.0 mg g^{-1}). The adsorption for Hg^{2+} by the modified

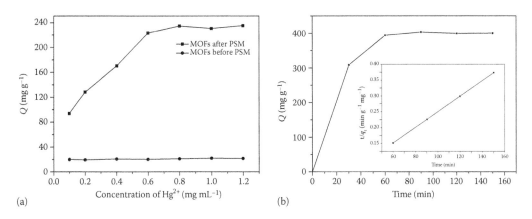

(a)

(b)

FIGURE 14.20 (a) Adsorption curve of Hg^{2+} using the as-synthesized $[Cu_4O(BDC)]_n$ before and after PSM (adsorbent amount = 10 mg), and (b) adsorption kinetics of thiol-modified $[Cu_4O(BDC)]_n$ for Hg^{2+}; inset: pseudo-second-order kinetic plot for the adsorption of Hg^{2+} by the modified MOFs (Hg^{2+} concentration = 1.0 mg mL^{-1}, adsorbent amount = 10 mg). (Reprinted from figures 5 and 6, *Micropor. Mesopor. Mat.*, 210, Wu, Y. et al., Postsynthetic modification of copper terephthalate metal-organic frameworks and their new application in preparation of samples containing heavy metal ions, 110–115, Copyright (2015), with permission from Elsevier.)

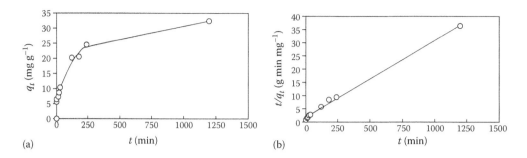

(a) t (min)   (b) t (min)

FIGURE 14.21 (a) Effect of immersion time on the adsorption of Cu ions (50 ppm) by MOF-NC. Symbol plots represent the experimental data and solid line shows the values predicted by the mixed surface reaction and diffusion-controlled model, and (b) plots of t/q_t versus t. (Reprinted from figure 2, *Micropor. Mesopor. Mat.*, 217, Bakhtiari, N. et al., Study on adsorption of copper ion from aqueous solution by MOF-derived nanoporous carbon, 173–177, Copyright (2015), with permission from Elsevier.)

MOFs was better described by the Freundlich model. The modified MOFs achieved saturated Hg^{2+} adsorption within 90 min (Figure 14.20b). Moreover, the q_e value (400 mg g^{-1}) calculated for the pseudo-second-order kinetic model also agreed with the experimental data (405.6 mg g^{-1}), suggesting that Hg^{2+} adsorption followed that model. The recycling ability of the modified MOF was also measured using EDTA as a desorbent. It could be reused three times at most, due to structural collapse as a result of the addition of EDTA. EDTA competitively combined with Cu^{2+} in the structure of thiol-functionalized $[Cu_4O(BDC)]_n$, resulting in the collapse of the crystals.

Bakhtiari et al. (2015) prepared a nanoporous carbon (NC)-derived MOF with high surface area and graphitized wall by direct carbonization of the zeolitic imidazolate framework for the removal of Cu ions. Cu ions were adsorbed rapidly onto the MOF-NC at an early stage of the contact period and then equilibrated after a few minutes (Figure 14.21a). The q_e value obtained by the mixed surface reaction and diffusion-controlled model was 33.4 mg g^{-1}, which is very close to the experimental value (32.5 mg g^{-1}). From Figure 14.21b, it can be observed that both the diffusion into the pores and the adsorption of the Cu ions on the active sites of the MOF-NC surface greatly affected the rate of Cu removal from aqueous solution. The removal percentages (Figure 14.22) of Cu ions by MOF-NC after 0.5 and 4 h are at least five times higher than other commercial activated carbons. Figure 14.23a shows that the removal percentage increases with increasing temperature, suggesting

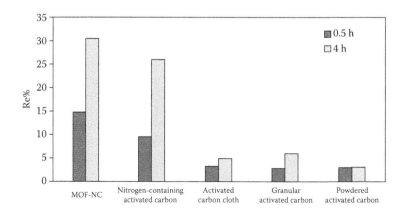

FIGURE 14.22 Comparative removal percentage of Cu ions by MOF-NC and other activated carbons after 0.5 and 4 h. (Reprinted from figure 4, *Micropor. Mesopor. Mat.*, 217, Bakhtiari, N. et al., Study on adsorption of copper ion from aqueous solution by MOF-derived nanoporous carbon, 173–177, Copyright (2015), with permission from Elsevier.)

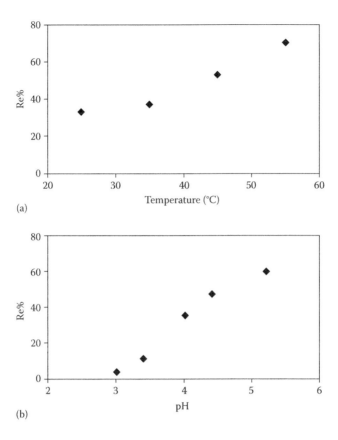

FIGURE 14.23 (a) Effect of temperature on the removal percentage of Cu ions (100 ppm) by MOF-NC, and (b) effect of initial solution pH on the removal percentage of Cu ions by MOF-NC. (Reprinted from figure 5, *Micropor. Mesopor. Mat.*, 217, Bakhtiari, N. et al., Study on adsorption of copper ion from aqueous solution by MOF-derived nanoporous carbon, 173–177, Copyright (2015), with permission from Elsevier.)

an endothermic adsorption behavior ($\Delta H_{ad} > 0$). By considering $\Delta G_{ad} < 0$ ($\Delta G_{ad} = \Delta H_{ad} - T\Delta S_{ad}$), it can be found that the entropy $\Delta S_{ad} > 0$. This increase may be attributed to the release of hydrated water molecules when the Cu ions bound to the surface. The effect of the pH level on the removal percentage of Cu ions was studied in the pH range from 3.0 to 5.4 (Cu ions precipitate as $Cu(OH)_2$ at pH≥ 6) (Figure 14.23b). The pH value corresponding to the point of zero charge (pH$_{pzc}$) is the pH of the solution when the net charge of the adsorbent surface is zero, and is an important adsorbent characteristic. At pH$>$pH$_{pzc}$, the surface charge is negative, while at pH$<$pH$_{pzc}$, the surface charge is positive. The adsorption of Cu ions depends on the pH of the solution, as indicated from the decrease in removal percentage with decreasing pH. In highly acidic solution, the competition between $H^+_{(aq)}$ and $Cu^{2+}_{(aq)}$ probably decreases the adsorption capability of Cu ions from aqueous solution and the carbon surface active sites, such as the hydroxyl and carboxyl groups, tend to be protonated.

14.4.4 Dye Removal Using MOFs

Organic dyes and pigments are becoming ubiquitous sources of environmental pollution because of their carcinogenic and mutagenic effects to certain forms of aquatic life, and also human life. MB is one of the most widely used dyes in textile industry. However, a high dose of MB could induce methemoglobinemia in the body, which tends to be poisonous and harmful to humans. Unfortunately, MB in wastewater is difficult to degrade on account of its complex aromatic structure. Moreover, owing to the development of dye technology, the dye materials could resist common oxidizing agents and are also very stable to light. Quite a few researchers have attempted to overcome this problem using MOFs for the removal of dyes from the effluents.

Haque et al. (2010) have utilized chemically modified MIL-101(Cr) for the adsorptive removal of MO from water, with an adsorption capacity of 194 mg g^{-1}, which is 16 times more than activated carbon. Huang et al. (2012) reported the usage of MIL-101(Cr) for the removal of xylenol orange from aqueous solution. Hierarchically mesostructured MIL-101 MOFs were synthesized under solvothermal synthesis conditions using a cationic surfactant (cetyltrimethylammonium bromide) as a supramolecular template. Pore size distribution analyses of the as-synthesized MOF samples revealed well-defined trimodal distributions, showing simultaneous existence of meso- and macropore channel systems. They exhibited remarkably accelerated adsorption kinetics for dye removal in comparison with bulk MIL-101 crystals.

Haque et al. (2011) used both anionic dye (MO) and a cationic dye (MB) over MOF-235 [Fe$_3$O(terephthalate)$_3$(DMF)$_3$][FeCl$_4$], and studied its removal efficiency. MOF-235 is readily available at low temperature and is composed of nontoxic Fe$_3$O clusters. MOF-235 frameworks have +1 charge per formula unit, which is balanced with an [FeCl$_4$]$^-$ ion. The adsorptive removal of MO and MB is interesting, because both anionic and cationic dyes are adsorbed in the liquid phase, even though MOF-235 is regarded as a nonporous material as evidenced from the negligible adsorption of N$_2$ over MOF-235 at low temperature. As shown in Figure 14.24a, the adsorbed quantity of MO and MB over MOF-235 is much higher compared with activated carbon. Figure 14.24b shows the plots of the pseudo-second-order kinetics of MO and MB adsorption over MOF-235 and activated carbon at initial dye concentration (C_i) of 30 ppm. Adsorption kinetic constants for MO adsorption over MOF-235 are larger than the constants over activated carbon. However, the kinetic constants for MB over MOF-235 are nearly similar to those of activated carbon. The kinetic constants of MOF-235 and activated carbon increase slightly with increasing initial dye concentrations, showing rapid adsorption in the presence of dyes in high concentration. Adsorption kinetics data are in accordance with the results of the pseudo-second-order kinetic model. Adsorption isotherms obtained after adsorption for 12 h are shown in Figure 14.25a. The amount of dyes adsorbed over MOF-235 is higher than that over activated carbon. The Q_0 value of MO over MOF-235 is greater than that over activated carbon by around 43 times (Figure 14.25b). In the case of MB, Q_0 over MOF-235 is greater than activated carbon by about seven times. The adsorption mechanism may be explained as an electrostatic interaction between the dyes and adsorbents (Figure 14.26). MO and

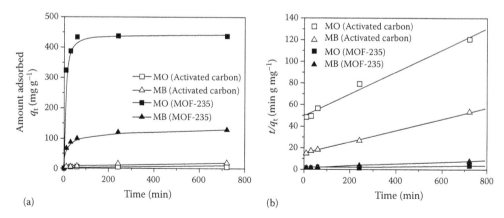

(a)   (b)

FIGURE 14.24 (a) Effect of contact time on MO and MB adsorption over MOF-235 and activated carbon, and (b) plots of pseudo-second-order kinetics of MO and MB adsorption over MOF-235 and activated carbon (C_i: 30 ppm). (Reprinted from figure 1, *J. Hazard. Mater.*, 185, Haque, E. et al., Adsorptive removal of methyl orange and methylene blue from aqueous solution with a metal-organic framework material, iron terephthalate (MOF-235), 507–511, Copyright (2011), with permission from Elsevier.)

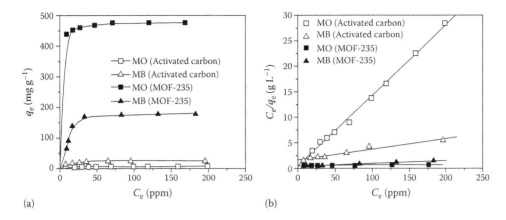

FIGURE 14.25 (a) Adsorption isotherms for MO and MB adsorption over MOF-235 and activated carbon, and (b) Langmuir plots of the isotherms of (a). (Reprinted from figure 2, *J. Hazard. Mater.*, 185, Haque, E. et al., Adsorptive removal of methyl orange and methylene blue from aqueous solution with a metal-organic framework material, iron terephthalate (MOF-235), 507–511, Copyright (2011), with permission from Elsevier.)

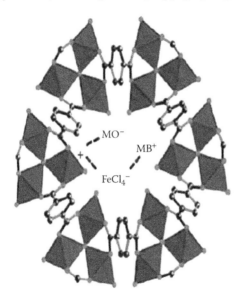

FIGURE 14.26 **(See color insert)** Proposed electrostatic interaction between dyes and adsorbents. (Reprinted from scheme 2, *J. Hazard. Mater.*, 185, Haque, E. et al., Adsorptive removal of methyl orange and methylene blue from aqueous solution with a metal-organic framework material, iron terephthalate (MOF-235), 507–511, Copyright (2011), with permission from Elsevier.)

MB usually exist in negative and positive forms, and there will be an electrostatic interaction with the MOF adsorbent having positively charged frameworks and negatively charged charge-balancing anions, respectively. However, other mechanisms such as π–π interactions between benzene rings of MOF-235 and dyes cannot be ruled out. Important mechanisms for the adsorption of hazardous materials over MOFs are summarized in Figure 14.27. The mechanisms of interactions for selective adsorptions include electrostatic interaction, acid–base interaction, hydrogen bonding, π–π stacking/interaction, and hydrophobic interaction (Hasan and Jhung 2015).

Xu et al. (2015) described a simple method for the preparation of MOF-based MNPs consisting of Cu^{2+}–BTC for rapid magnetic solid-phase extraction of Congo red (CR) dye from aqueous solution. Magnetization of the MOF and solid-phase extraction of CR was simultaneously accomplished

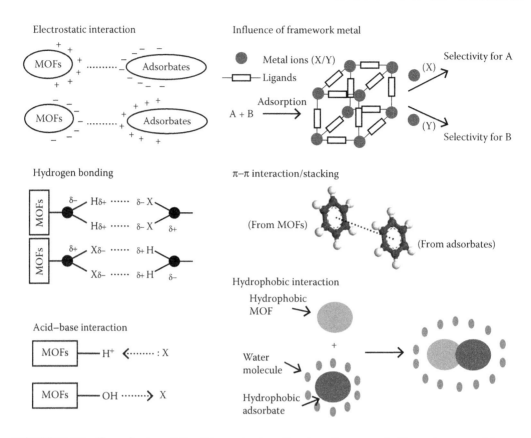

FIGURE 14.27 **(See color insert)** Possible mechanisms for adsorptive removal of hazardous materials over MOFs. (Reprinted from figure 9, *J. Hazard. Mater.*, 283, Hasan, Z. and Jhung, S.H. Removal of hazardous organics from water using metal-organic frameworks MOFs): Plausible mechanisms for selective adsorptions, 329–339, Copyright (2015), with permission from Elsevier.)

by ultrasonication of MOF and silica-coated magnetite microparticles in solution. Under optimized conditions, the MOF hybrid exhibited a fast adsorption rate and high removal efficiency (>97%) toward CR even after 15 times reuse, which also featured a high adsorption capacity of 97.7% and 92.5% for cationic MB and crystal violet, respectively. Desorption of CR from the MOF was realized by washing it with ethanol and water. This MOF material is considered to be a promising new adsorbent for use in wastewater treatment and in analytical preconcentration.

Tan et al. (2015) modulated the average particle size of MIL-100(Fe) from 100 nm to 20 μm by changing the valence state of iron in iron sources and the amount of HF during hydrothermal reactions. The products were denoted as MIL-100(Fe)-x-yF ($x = 1$–4, representing the molar ratios of Fe^{II}:Fe^{III} of 0:4, 1:3, 2:2, and 4:0, respectively, and $y = 0$, 1, or 3, representing the molar ratio of HF:Fe). From adsorption isotherms (Figure 14.28a), the adsorbed amount reaches a plateau at a higher equilibrium solution concentration, and the plateau rises with increases in the particle size. The highest adsorption capacity for MB, at 1105 mg g^{-1}, belongs to the sample synthesized with maximum HF content (MIL-100(Fe)-4-3F), which is much higher than previous reports concerning MOFs. The sorption capacity of MIL-100(Fe) for MB dye increased with particle size. The high sorption capacity could be attributed to high surface area and suitable pore structure. Additionally, the particle size determined the adsorption rate. The smaller the particle size, the higher the adsorption rate. The adsorption isotherms fit adequately with the Langmuir model and the kinetic data followed the pseudo-second-order model. As shown in Figure 14.28b, MIL-100(Fe)-4-0F exhibits 89% adsorption in the first hour, whereas this value is about 41% for MIL-100(Fe)-4-3F. Therefore, the

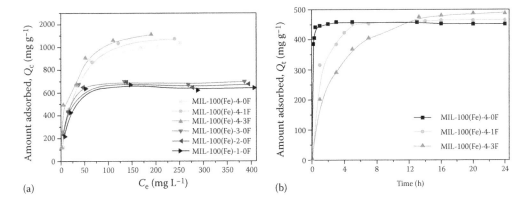

(a)

(b)

FIGURE 14.28 (a) Adsorption isotherms, and (b) time-resolved adsorption capacity of MB on MIL-100(Fe) with different particle sizes and amounts of HF, respectively, at 35°C, 24 h. (Reprinted from figures 6 and 7, *Chem. Eng. J.*, 281, Tan, F. et al., Facile synthesis of size-controlled MIL-100(Fe) with excellent adsorption capacity for methylene blue, 360–367, Copyright (2015), with permission from Elsevier.)

smaller particle size of MIL-100(Fe)-4-0F may be responsible for its faster adsorption. Although the adsorption rate of the large particles is limited, the condensation of dye molecules within the mesopores could improve adsorption capacity significantly. As shown in Figure 14.29, the adsorption ability of MIL-100(Fe)-4-3F remained at 403 mg g^{-1} even after three repeated cycles, decreasing slightly from the initial adsorption amount of 489 mg g^{-1}. Thus, MIL-100(Fe)-4-3F displayed great removal performance of MB from wastewater and can be regenerated by washing with ethanol.

Shao et al. (2016) fabricated Fe$_3$O$_4$ nanospheres by the hydrothermal method, and further modified them using poly(acrylic acid) (PAA) via distillation–precipitation polymerization. Core–shell MOFs were prepared using Fe$_3$O$_4$-PAA NPs as templates through a step-by-step self-assembly strategy (Figure 14.30) as follows: Fe$_3$O$_4$-PAA nanospheres were dispersed alternately in an ethanolic solution of FeCl$_3$·6H$_2$O for 15 min and then in an ethanolic solution of benzene-1,3,5-tricarboxylic acid for 30 min at 25°C. Between each step, the products formed were collected using a magnet and washed several times with ethanol. Fe$_3$O$_4$-MIL-100(Fe) core–shell nanospheres with different shell thicknesses were obtained on repeated cyclic growth. TEM images confirmed the nearly sphere-shaped magnetic particles (average diameter 361 ± 24 nm) and also the coating of individual magnetite particles with a uniform PAA shell of 30 nm thickness. Magnetization

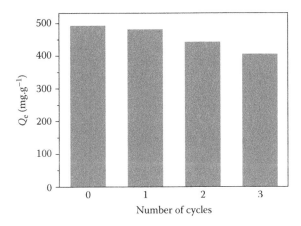

FIGURE 14.29 Recyclability of 20 mg MIL-100(Fe)-4–3F for adsorptive removal of 50 mL MB (200 mg L^{-1}) at 35°C, 24 h. (Reprinted from figure 10, Tan, F. et al., Facile synthesis of size-controlled MIL-100(Fe) with excellent adsorption capacity for methylene blue, 360–367, Copyright (2015), with permission from Elsevier.)

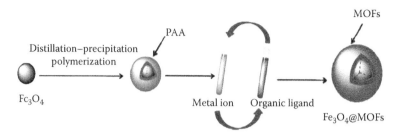

FIGURE 14.30 Schematic illustration of fabrication process of Fe_3O_4-MIL-100(Fe). (Reprinted from figure 1, *Chem. Eng. J.*, 283, Shao, Y. et al., Magnetic responsive metal–organic frameworks nanosphere with core–shell structure for highly efficient removal of methylene blue, 1127–1136, Copyright (2016), with permission from Elsevier.)

saturation values of Fe_3O_4 and Fe_3O_4-MIL-100(Fe) obtained after 20 and 40 cycles at RT were measured to be 64.7, 40.6, and 31.9 emu g^{-1}, respectively, revealing the strong magnetic properties of the MOFs' nanospheres (Figure 14.31). The MNPs were attracted toward the magnet within 5 s, directly demonstrating the convenient separation of the core–shell nanospheres from liquids using an external magnetic field. The effect of contact time on the adsorption of MB onto Fe_3O_4-MIL-100(Fe) composites is shown in Figure 14.32a. Adsorption capacity was found to be rapidly increased in the first 2 h and reached equilibrium after that, implying that the magnetic composite possessed both high adsorption capacity and high removal efficiency toward pollutants in water. Pure MIL-100(Fe) exhibited a higher adsorption capacity of MB than its composite. Due to the presence of the Fe_3O_4 core in Fe_3O_4-MIL-100(Fe) composite, the proportion of MOF is smaller in the composite than in pure MIL-100(Fe), which, in turn, helps in the easier separation of the adsorbent from the liquid medium for reuse. Pseudo-first-order and second-order kinetic models were employed to illustrate the adsorption mechanism (Figure 14.32b): the pseudo-second-order kinetic curves gave a good fit for the experimental kinetic data ($R^2 > 0.998$), indicating chemical adsorption. The pseudo-first-order models failed to describe the experimental data as indicated by lower R^2 values. The Q_m value of MB was calculated to be 73.8 mg g^{-1} at 25°C, which is higher than other magnetic MOFs (20.2 mg g^{-1}) (Zheng et al. 2014). Most importantly, the surface functional group

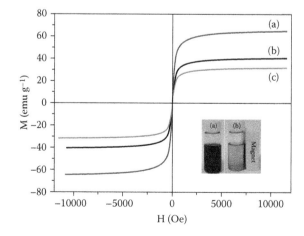

FIGURE 14.31 **(See color insert)** Magnetic hysteresis loops of (a) Fe_3O_4 and (b) Fe_3O_4-MIL-100(Fe) obtained after 20, and (c) 40 assembly cycles. Inset demonstrates Fe_3O_4-MIL-100(Fe) (a) well dispersed in water and (b) separated easily from water by a magnet. (Reprinted from figure 7, *Chem. Eng. J.*, 283, Shao, Y. et al., Magnetic responsive metal–organic frameworks nanosphere with core–shell structure for highly efficient removal of methylene blue, 1127–1136, Copyright (2016), with permission from Elsevier.)

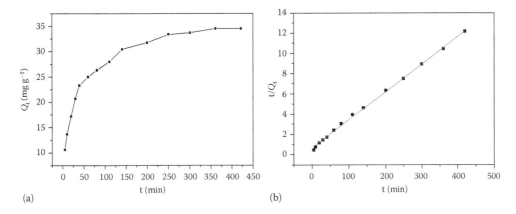

FIGURE 14.32 Kinetic adsorption plots of Fe_3O_4-MIL-100(Fe) for MB: (a) plot of removal amount (Q_t) versus time t, and (b) plot of pseudo-second-order kinetics for the adsorption process of the hybrid. Initial concentration of dye solution was 60 mg L^{-1} and Fe_3O_4-MIL-100(Fe) was 1 g L^{-1}. (Reprinted from figure 9, *Chem. Eng. J.*, 283, Shao, Y. et al., Magnetic responsive metal–organic frameworks nanosphere with core–shell structure for highly efficient removal of methylene blue, 1127–1136, Copyright (2016), with permission from Elsevier.)

could also easily be changed by alternating the organic ligands in the assembly process based on different adsorbates.

14.5 FUTURE TRENDS

The MOF synthesis field is still relatively new and has been recognized to have evolved from coordination and solid-state/zeolite chemistry. In the last few years, there is clearly an increasing trend in the number of works dealing with MOFs for heavy metal/dye adsorption, as well as the separation and recovery of MOFs as discussed in detail in this chapter. As more research data is generated in this field, our focus would be toward tackling the challenges that may arise in the future: (a) emerging micropollutants from the pharmaceutical/textile industries and other industrial residuals are likely to pose new challenges to the adsorbents, and hence, MOFs capable of tackling those pollutants need to be identified/developed; (b) as sensitivity and selectivity arise from the functionalization of the MOFs, more attention should be paid to the preparation of new and effectively modified materials with improved superadsorbent characteristics; (c) in most cases, the adsorption studies were based on laboratory-scale batch preparation of effluents rather than real-time industrial effluents. Hence, the testing conditions need to involve all the parameters usually present in the real-time analysis cases, which, in turn, could increase the efficiency of the developed MOFs in real-time use; and (d) although, performance-wise, MOFs are far better than conventional adsorbents such as clay and zeolites, the fabrication/regeneration costs of MOFs are still on the high side, which need to be reduced to bring down the operating costs.

REFERENCES

Abbasi, A., Moradpour, T., and K.V. Hecke. 2015. A new 3D cobalt (II) metal–organic framework nanostructure for heavy metal adsorption. *Inorg. Chim. Acta* 430:261–267.

Aluigi, A., Tonetti, C., Vineis, C., Tonin, C., and G. Mazzuchetti. 2011. Adsorption of copper(II) ions by keratin/PA6 blend nanofibres. *Eur. Polym. J.* 47:1756–1764.

Bakhtiari, N. and S. Azizian. 2015. Adsorption of copper ion from aqueous solution by anoporous MOF-5: A kinetic and equilibrium study. *J. Mol. Liq.* 206:114–118.

Bakhtiari, N., Azizian, S., Alshehri, S.M., Torad, N.L., Malgras, V., and Y. Yamauchi. 2015. Study on adsorption of copper ion from aqueous solution by MOF-derived nanoporous carbon. *Micropor. Mesopor. Mat.* 217:173–177.

Bansal, P., Bharadwaj, L.M., Deep, A., and P. Kaushik. 2013. Zn based metal organic framework as adsorbent material for mecoprop. *Res. J. Recent Sci.* 2:84–86.

Brozek C.K. and M. Dinca. 2014. Cation exchange at the secondary building units of metal–organic frameworks. *Chem. Soc. Rev.* 43:5456–5467.

Chen, J.J., Ahmad, A.L., and B.S. Ooi. 2013. Poly (N-isopropylacrylamide-co-acrylic acid) hydrogels for copper ion adsorption: Equilibrium isotherms, kinetic and thermodynamic studies. *Environ. Chem. Eng.* 1:339–348.

Dinca, M. and J.R. Long. 2007. High-enthalpy hydrogen adsorption in cation-exchanged variants of the microporous metal–organic framework $Mn_3[(Mn_4Cl)_3(BTT)_8(CH_3OH)_{10}]_2$ *J. Am. Chem. Soc.* 129:11172–11176.

Eddaoudi, M., Kim, J., Rosi, N., et al. 2002. Systematic design of pore size and functionality in isoreticular mofs and their application in methane storage. *Science* 295:469–472.

Fang, Q.-R., Yuan, D.-Q., Sculley, J., Li, J.-R., Han, Z.-B., and H.-C. Zhou. 2010. Functional mesoporous metal-organic frameworks for the capture of heavy metal ions and size-selective catalysis. *Inorg. Chem.* 49:11637–11642.

Guo, X. and F. Chen. 2005. Removal of arsenic by bead cellulose loaded with iron oxyhydroxide from groundwater. *Environ. Sci. Technol.* 39:6808–6818.

Haque, E., Jun, J.W., and S.H. Jhung. 2011. Adsorptive removal of methyl orange and methylene blue from aqueous solution with a metal-organic framework material, iron terephthalate (MOF-235). *J. Hazard. Mater.* 185:507–511.

Haque, E., Lee, J.E., Jang, I.T., et al. 2010. Adsorptive removal of methyl orange from aqueous solution with metal–organic frameworks, porous chromium–benzenedicarboxylates. *J. Hazard. Mater.* 181:535–542.

Hasan, Z., and S.H. Jhung. 2015. Removal of hazardous organics from water using metal-organic frameworks (MOFs): Plausible mechanisms for selective adsorptions. *J. Hazard. Mater.* 283:329–339.

Hei, S., Jin, Y., and F. Zhang. 2014. Fabrication of γ-Fe_2O_3 nanoparticles by solid-state thermolysis of a metal-organic framework, MIL-100(Fe), for heavy metal ions removal. *J. Chem.* 2014–2019.

Huang, X.-X., Qiu, L.-G., Zhang, W., et al. 2012. Hierarchically mesostructured MIL-101 metal–organic frameworks: Supramolecular template-directed synthesis and accelerated adsorption kinetics for dye removal. *Cryst. Eng. Comm.* 14:1613–1617.

Iesan, C.M., Capat, C., Ruta, F., and I. Udrea. 2008. Evaluation of a novel hybrid inorganic/organic polymer type material in the arsenic removal process from drinking water. *Water Res.* 42:4327–4333.

Igberase, E., Osifo, P., and A. Ofomaja. 2014. The adsorption of copper (II) ions by polyaniline graft chitosan beads from 3 aqueous solution: Equilibrium, kinetic and desorption studies. *Environ. Chem. Eng.* 2:362–369.

Jhung, S.H., Lee, J.H., Yoon, J.W., Serre, C., Ferey, G., and J.S. Chang. 2007. Microwave synthesis of chromium terephthalate MIL-101 and its benzene sorption ability. *Adv. Mater.* 19:121–124.

Kanel, S.R., Manning, B., Charlet, L., and H. Choi. 2005. Removal of arsenic(III) from groundwater by nanoscale zero-valent iron. *Environ. Sci. Technol.* 39:1291–1298.

Ke, F., Qiu, L.G., Yuan, Y.P., et al. 2011. Thiol-functionalization of metal–organic framework by a facile coordination-based postsynthetic strategy and enhanced removal of Hg^{2+} from water. *J. Hazard. Mater.* 196:36–43.

Khan, N.A., Jun, J.W., Jeong, J.H., and S.H. Jhung. 2011. Remarkable adsorptive performance of a metal–organic framework, vanadium-benzenedicarboxylate (MIL-47), for benzothiophene. *Chem. Commun.* 47:1306–1308.

Li, T., Yang, Z., Zhang, X., Zhu, N., and X. Niu. 2015a. Perchlorate removal from aqueous solution with a novel cationic metal–organic frameworks based on amino sulfonic acid ligand linking with Cu-4,4′-bipyridyl chains. *Chem. Eng. J.* 281:1008–1016.

Li, Z.-Q., Yang, J.-C., Sui, K.-W., and N. Yin. 2015b. Facile synthesis of metal-organic framework MOF-808 for arsenic removal. *Mater. Lett.* 160:412–414.

Liu, Z.-M., Wu, S.-H., Jia, S.-Y., et al. 2014. Novel hematite nanorods and magnetite nanoparticles prepared from MIL-100(Fe) template for the removal of As(V). *Mater. Lett.* 132:8–10.

Luo, X., Ding, L., and J. Luo. 2015. Adsorptive removal of Pb(II) ions from aqueous samples with amino-functionalization of metal–organic frameworks MIL-101(Cr). *J. Chem. Eng. Data* 60:1732–1743.

Maleki, A., Hayati, B., Naghizadeh, M., and S.W. Joo. 2015. Adsorption of hexavalent chromium by metal organic frameworks from aqueous solution. *J. Ind. Eng. Chem.* 28:211–216.

Ringwood, A.E. 1955. The principles governing trace element distributions during magnetic crystallization. *Geochim. Cosmochim. Acta* 7:189–202.

Shao, Y., Zhou, L., Bao, C., Ma, J., Liu, M., and F. Wang. 2016. Magnetic responsive metal–organic frameworks nanosphere with core–shell structure for highly efficient removal of methylene blue. *Chem. Eng. J.* 283:1127–1136.

Taghizadeh, M., Asgharinezhad, A.A., Pooladi, M., Barzin, M., Abbaszadeh, A., and A. Tadjarodi. 2013. A novel magnetic metal organic framework nanocomposite for extraction and preconcentration of heavy metal ions, and its optimization via experimental design methodology. *Microchim. Acta* 180:1073–1084.

Tan, F., Liu, M., Li, K., et al. 2015. Facile synthesis of size-controlled MIL-100(Fe) with excellent adsorption capacity for methylene blue. *Chem. Eng. J.* 281:360–367.

Tollefson, J. 2007. Worth its weight in platinum. *Nature* 450:334–335.

Wu, Y., Xu, G., Liu, W., et al. 2015. Postsynthetic modification of copper terephthalate metal-organic frameworks and their new application in preparation of samples containing heavy metal ions. *Micropor. Mesopor. Mat.* 210:110–115.

Xu, X., Cao, R., Jeong, S., and J. Cho. 2010. Spindle-like mesoporous α-Fe_2O_3 anode material prepared from MOF template for high rate lithium batteries. *Nano Lett.* 12:4988–4991.

Xu, Y., Jin, J., Li, X., Han, Y., Meng, H., Song, C., and X. Zhang. 2015. Magnetization of a Cu(II)-1,3,5-benzenetricarboxylate metal-organic framework for efficient solid-phase extraction of Congo Red. *Microchim. Acta* 182:2313–2320.

Zhang, L., Wu, H.B., Xu, R., and X.W. Lou. 2013. Porous Fe_2O_3 nanocubes derived from MOFs for highly reversible lithium. *Cryst. Eng. Comm.* 15:9332–9335.

Zhang, Y., Yang, M., and X. Huang. 2003. Arsenic(V) removal with a Ce(IV)-doped iron oxide adsorbent. *Chemosphere* 51:945–962.

Zheng, J., Cheng, C., Fang, W.-J., et al. 2014. Surfactant-free synthesis of a Fe_3O_4@ZIF-8 core–shell heterostructure for adsorption of methylene blue. *Cryst. Eng. Comm.* 16:3960–3964.

Zhu, B.J., Yu, X.Y., Jia, Y., et al. 2012. Iron 1,3,5-benzenetricarboxylic metal–organic coordination polymers prepared by solvothermal method and their application in effcient As(V) removal from aqueous solutions. *J. Phys. Chem. C* 116:8601–8607.

Zou, F., Yu, R., Li, R., and W. Li. 2013. Microwave-assisted synthesis of HKUST-1 and functionalized HKUST-1-@$H_3PW_{12}O_{40}$: Selective adsorption of heavy metal ions in water analyzed with synchrotron radiation. *Chem. Phys. Chem.* 14:2825–2832.

15 Environmental Fate and Ecotoxicity of Engineered Nanoparticles
Current Trends and Future Perspective

Anamika Kushwaha, Radha Rani, and Vishnu Agarwal

CONTENTS

ABSTRACT

Nanotechnology research is currently an area of intense scientific interest, because of a wide variety of potential applications due to unique size-dependent properties. These properties make nanoparticles superior and indispensable, as they show unusual physical and chemical properties such as conductivity, heat transfer, melting temperature, optical properties, and magnetization. Due to the wide production and applications of engineered nanoparticles, their release into the environment has significantly increased. The behavior of nanoparticles and their effects on both biotic and abiotic components of the ecosystems are not yet well established. Moreover, the process and rate of degradation of these nanoparticles are unknown, thereby leading to their accumulation in the environment. Therefore, nanoparticles have major impacts on ecosystems and the environment. Some of the engineered nanoparticles, such as silver nanoparticles, have a known antimicrobial effect, and their presence may lead to the killing of pathogenic as well as beneficial microorganisms in the environment, which may affect various environmental processes such as biogeochemical cycles. Nanoparticles also tend to accumulate in marine and aquatic environments, thus affecting aquatic microorganisms and macroorganisms. Therefore, in this chapter, the uses of nanoparticles, their entry into the ecosystem, and their impact on the organisms present there are discussed.

Keywords: Engineered nanoparticles, Ecotoxicity, Environment, Fate, Microorganisms

15.1   INTRODUCTION

Nanotechnology is an area where research niches are growing fast and attract substantial funding from both the public and private sectors. Nanoparticles (NPs) are particles with at least one dimension less than 100 nm. According to Lux Research, a New York–based independent intelligence and technology research and advisory firm, investments in the nanotechnology industry grew from $13 billion in 2004 to $50 billion in 2006. It has been estimated that this amount will reach $1 trillion by 2015 and that the industry will employ about two million workers (Nel 2006; Roco and Bainbridge 2005) and output is projected to grow to over half a million tons by 2020 (Robichaud et al. 2009; Stensberg et al. 2011). According to BCC Research (2012) the total market for NPs in biotechnology, drug discovery, and development was valued at $17.5 billion in 2011 and is predicted to reach $53.5 billion in 2017. Thus, a large number of materials and products made of NPs may come into contact with the environment, either during production, transport, use, or when they end up as waste.

Nanotechnology is a collective term that implies the capacity to work with materials (surface structures, pores, particles, etc.) at a nanometer scale. Thus, engineered nanoparticles (ENPs) find application in a wide range of areas, from energy (production, catalysis, storage), materials (lubricants, abrasives, paints, tires, sportsware), electronics (chips, screens), optics, remediation (pollution absorption, water filtering, disinfection), to food (additives, packaging), cosmetics (skin lotions and sunscreens), and medicine (diagnostics, drug delivery). Currently, nanotechnology has been applied in the development or production of a range of products, such as nanoporous membranes to filter water (to remove microbes, pollutants, or salts); nanoetched computer chips to reduce the size and energy of microprocessors; silver particle coatings in refrigerators, tennis shoes, Band-Aids, and so on to kill bacteria and reduce odor problems.

The increased production of nanobased products will lead to their greater release into air, water, and soil (Nowack and Bucheli 2007), where their fate and behavior are largely unknown. Nanomaterial wastes are released into the environment either during their manufacturing processes or from operating or disposing of nanodevices. The unique properties of ENPs, such as a large surface area, an abundant reactive site on the surface due to the presence of a large number of atoms on the exterior, and mobility could probably lead to unexpected health or environmental hazards (Maynard 2006; Wiesner et al. 2006). Therefore, organisms such as algae, plants, and fungi that interact strongly with their immediate environment are expected to be affected by their exposure to ENPs. Thus, ENPs not only affect humans, but also the whole ecosystem. In this chapter, we examine the release and transport of ENPs into the environment and the toxic effects of NPs on different organisms.

15.2 CLASSIFICATION AND OCCURRENCE OF NANOPARTICLES

Since time immemorial, organisms and the environment have been exposed to natural NPs like volcanic dust, ash, combustion by-products (e.g., carbon black, soot), organic matter like humic and fulvic acids, proteins, peptides, and colloidal inorganic species present in natural water and soil systems (Figure 15.1) (Buffle 2006). In contrast to nascent and incidental NPs, ENPs are produced by processing materials at the nanoscale.

15.2.1 NATURAL NPs

NPs are not a human invention, but have existed since life began on earth. During evolution, all forms of life have been exposed to some types of NP and they have developed some mechanisms to tolerate their presence (Buffle 2006). There are several mechanisms that create NPs in the environment; these may be either geological or biological. Geological mechanisms include physicochemical weathering, authigenesis/neoformation (e.g., in soils), and volcanic activity. These geological processes typically produce inorganic NPs. Biological mechanisms typically produce organic

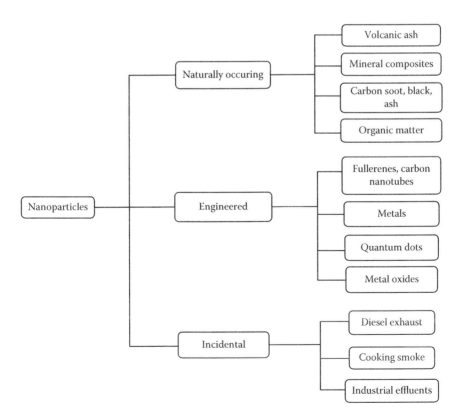

FIGURE 15.1 Types of nanoparticles that can be released into the environment.

nanomolecules, although some organisms can produce mineral granules in cells. Common natural NPs are soil colloids, which consist of silicate clay minerals, iron or aluminum oxides/hydroxides, or humic organic matter, including black carbon. NPs may also be produced during a volcanic eruption and found in glacial ice cores some 10,000 years old (Murr et al. 2004). There is an evidence of a natural NP formation in the sediments at the Cretaceous–Tertiary (K–T) boundary (Verma et al. 2002). The atmospheric dust produced per year globally accounts for about one billion metric tons (Kellogg and Griffin 2006), and even with a fraction of this being ultrafine particles, this would equate to millions of tons of natural NPs.

15.2.2 Engineered NPs

ENPs are classified based on their chemical composition, size, or morphology characteristics. A more detailed classification distinguishes between

- Fullerenes (buckminsterfullerenes, carbon nanotubes [CNTs], etc.)
- Metal ENPs (e.g., elemental gold, silver, iron, and platinum)
- Oxides (e.g., TiO_2, ZnO, Fe_2O_3, Fe_3O_4, SiO_2, MgO, and Al_2O_3)
- Complex compounds (e.g., cobalt-zinc iron oxide)
- Quantum dots (or q-dots) (e.g., cadmium-selenide [CdSe])
- Organic polymers (dendrimers, polystyrene, etc.)

15.2.2.1 Fullerenes

These are newly discovered carbon allotropes (Gr. *allos*, other, and *tropos*, manner) made up of pure carbon. Graphite and diamond are other well-known carbon allotropes. The simplest fullerene, C_{60}, resembles a stitching pattern on a ball made up of 60 carbon atoms arranged as 12 pentagons and 20 hexagons. These are well known as buckminsterfullerenes or "buckyballs," named after the architect Richard Buckminster Fuller. Fullerenes are hydrophobic in nature and soluble in organic solvents, such as toluene. They have wide applications as organic photovoltaics, antioxidants, catalysts, polymers, in water purification, biohazard protective agents, and in various medical and pharmaceutical applications (Yadav and Kumar 2008). Fullerene nanoparticles also exist in C_{70}, C_{74}, C_{76}, C_{78}, and so on. Fullerenes form tubes with spherical ends when grown by vapor deposition, and thus they are called *carbon nanotubes*. When fullerenes contain atoms, ions, or small clusters inside their spherical structure, they are called *endofullerenes* (e.g., M@C82, where M is a metal that is enclosed by a fullerene consisting of 82 carbon atoms). The surface of a fullerene may be modified by covalently attaching functional groups to a carbon atom of the fullerene. A common functionalization is to attach OH groups, which make the molecule more hydrophilic (Brant et al. 2007).

CNTs are long, thin cylinders of carbon. They were discovered in 1991 by Sumio Iijima, and consist of rolled-up graphene sheets that may or may not be capped at the ends by a fullerene half-sphere. Arc discharge or laser ablation were first used to produce CNTs, but nowadays these methods have been replaced by low temperature chemical vapor deposition (CVD) techniques (<800°C). The starting material consists of graphite or is made from CH_4 or other carbon-containing gases. CNTs include single-, multi-, and double-walled nanotubes (SWNTs, MWNTs, and DWNTs). CNTs possess excellent extraordinary electrical conductivity, heat conductivity, and mechanical properties. They are probably the best electron field emitters known, largely because of their high length-to-diameter ratios. Due to their extraordinary properties, they are widely used in field emitters/emission, conductive or reinforced plastics, and molecular electronics: CNT-based nonvolatile RAM, CNT-based transistors, energy storage, CNT-based fibers and fabrics, and CNT-based ceramics and biomedical applications. Due to the enormous demand for carbon-based NPs in the market, especially in the electronics and polymer sectors, demand is expected to reach $1.096 billion by 2015 (Garland 2009).

15.2.2.2 Metal ENPs

Metal ENPs are made by manipulating heavy metals such as gold, iron, platinum, and silver. They possess specific properties based on their shape, size, and dissolution medium. Recently, green synthesis methods have been widely used to make silver and iron NPs (Ramteke et al. 2012; Sahu and Soni 2013). These metal ENPs have a wide application; for example, colloidal gold has been used for the treatment of rheumatoid arthritis in an animal model (Tsai et al. 2007), as a drug carrier (Gibson et al. 2007), and as an agent for detecting tumors (Qian et al. 2008); and it has also been used as a contrast agent for biological probes such as antibodies, nucleic acids, glycans, and receptors (Horisberger and Rosset 1977). Silver NPs are used in medicine as a disinfectant, an antiseptic, in surgical masks, and in wound dressings that have antibacterial activity (Chopra 2007); and in many textiles, keyboards, cosmetics, water purifier appliances, plastics, and biomedical devices (Li et al. 2010). Iron nanoparticles are utilized in magnetic recording media and tapes, as a catalyst in Fischer–Tropsch synthesis, in drug delivery, magnetic resonance imaging (MRI), and the treatment of hyperthermia (Huber 2005); and for the remediation of industrial sites (Zhang 2003). Platinum NPs exhibit antioxidant properties, but what applications they are to be put to is as yet undeciphered; however, they have been found to increase roundworm longevity (Kim et al. 2008).

15.2.2.3 Metal Oxides

Commercially available metal oxide nanomaterials include TiO_2, ZnO, Fe_2O_3, Fe_3O_4, SiO_2, MgO, and Al_2O_3. These nanomaterials have applications in catalyst devices, environmental remediation, and different commercial products such as cosmetics, sunscreens, textiles, paints, varnishes, and household appliances (Rodríguez and Fernández-García 2007). Metal NPs such as MgO, TiO_2, CaO, and BaO are used as scrubber material for gaseous pollutants (e.g., CO_2, NO_x, SO_x) in the chemical industry (Forzatti 2000). Iron NPs such as FeO (iron oxide), Fe_3O_4 (magnetite), α-Fe_2O_3 (hematite), and γ-Fe_2O_3 (maghemite) occur naturally in bacteria, insects, weathered soils, rocks, the natural atmosphere, and polluted aerosols (Cornell and Schwertmann 1996). Magnetite has been approved by regulatory authorities for use as an antibacterial agent.

15.2.2.4 Quantum Dots

Quantum dots (QDs) or Q-dots are fluorescent semiconductor nanocrystals (~2–100 nm) having electronic properties between bulk semiconductors and discrete molecules. Such nanomaterials include CdTe, CdSe/ZnS, PbSe, CdHg, and InP. Due to their small size, they can emit light (i.e., photons) with a specific wavelength that is limited by quantum confinement and determined by particle size and composition (i.e., the resulting band gap energy). These properties make QDs a candidate for applications in biomedical imaging; for example, to localize tumor cells, to target specific cell membrane receptors (Alivisatos 2004; Chen et al. 2002; Lidke et al. 2004), and to visualize cellular biomolecules such as peroxisomes (Colton et al. 2004) and DNA (Dubertret et al. 2002).

15.2.2.5 Organic Polymers

A series of organic ENPs have been developed that have found application in pharmaceutical products (targeted drug delivery). These organic ENPs include liposomes, micelles, and dendrimers. These ENPs are used to deliver drugs to the target tissues. Starch polymers are also the basis for a type of organic ENP that is used, for example, in biodegradable plastic films, sometimes combined with other ENPs such as nanoclays, to modify permeability properties.

15.3 RELEASE PATHWAYS OF ENGINEERED NANOPARTICLES INTO THE ENVIRONMENT

NPs are released into the atmospheric environment either intentionally (point/stationary sources) or accidentally (nonpoint/mobile sources) (Figure 15.2). The point or stationary sources include the production of NPs in industry and wastewater treatment plants. Wastewater treatment plants are

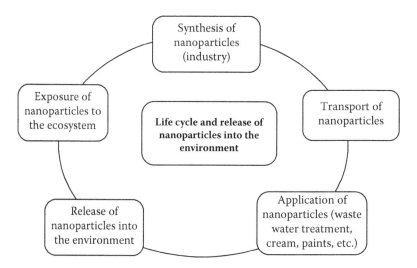

FIGURE 15.2 Life cycle and release of nanoparticles into the environment.

the major source of the release of NPs into the environment. During wastewater treatment, NPs interact with various inorganic and organic pollutants and form complexes, or new compounds (Pandey and Kumar 1990); and thus, these NPs interact with other environmental constituents after passing through treatment procedures. In the effluents from U.S., European, and Swiss wastewater treatment plants, high concentrations of TiO_2, ZnO, CNT, Ag, and fullerenes have been found (Gottschalk et al. 2009). Nonpoint or mobile sources include paints, varnishes, cosmetics, and cleaning agents (Biswas and Wu 2005), during the production and transportation of nanomaterial-containing products. Some nanomaterials are released into the environment intentionally for the remediation of groundwater and wastewater (Nowack and Bucheli 2011). NPs that are released into aquatic systems are either runoff from land or industrial and household wastewater effluents. The major source of aquatic contamination by NPs is the use of metal-based NPs for the remediation of water (Defra 2007).

15.4 FATE AND TRANSPORT OF ENGINEERED NANOPARTICLES IN THE ENVIRONMENT

After release of ENPs into environments such as natural waters, sediments, or soils, their behavior is likely to be complex and involve several processes. The physicochemical properties such as pH, ionic strength, and presence of organic matter of environmental media affect the mobility of ENPs. The effect of various environmental media on the transport of ENPs is described in Sections 15.4.1 through 15.4.3.

15.4.1 AIR

The transport of ENPs in air is similar to that of fluids (Mädler and Friedlander 2007). In the gaseous atmosphere, the transport of nanoparticles is essentially by Brownian diffusion, due to which particles disperse into the air. Dispersion of particles may also occur from mechanical mixing during industrial processes. ENPs can also be transported in air via agglomeration. During Brownian motion, individual particles collide with each other and agglomerate, leading to increases in size (Bandyopadhyaya et al. 2004). The physiochemical characteristics of ENPs also affect their transport in air. Particles with a diameter of ≤100 nm have greater efficiency to remain suspended in air for longer periods of time and are capable of diffusion (Aitken et al. 2004). Particle size has an inverse relationship with the diffusion rate, whereas it is directly proportional to gravitational settling. Particles of <80 nm tend to be short lived and to agglomerate, whereas particles of >2000 nm settle due to gravitational force. Particles of intermediate size, between >80 and <2000 nm, are able to persist in the atmosphere for longer periods of time.

15.4.2 WATER

The transport, behavior, and fate of ENPs in aquatic ecosystems are given by colloid chemistry and the movement of colloids in aquatic systems. In water, hydrophobic colloids are insoluble and thus they are stabilized by their electrokinetic properties. The magnitude of charge is responsible for the stability of colloids and is referred to as the *zeta potential*. The zeta potential is a repelling force that protects cells from coalescing due to intermolecular or interparticle forces (Van der Waal's forces). This happens when forces of attraction overcome forces of repulsion (Sawyer and McCarty 1967). Guzman et al. (2006) described the effect of the pH on TiO_2 NPs in an aqueous environment. They found that when the pH reaches the zero point, charged NPs have the tendency to agglomerate. The agglomerated NPs may be clogged within pores and thus their mobility decreases, inducing sedimentation.

Natural organic matter (NOM; fulvic acid, humic acids, and polysaccharides) affects the transport of ENPs in aqueous systems. The adsorption of ENPs to NOM changes the surface charge and density charge of ENPs, and thus affects their mobility in water (Ghosh et al. 2008; Guzman et al. 2006; Hyung and Fortner 2007). The electrophoretic mobility of NPs (TiO_2, ZnO and CeO_2) in groundwater, lakes, rivers, and sea water is greatly influenced by particle size, the presence of NOM, and the ionic strength of the transport medium (Keller et al. 2010).

15.4.3 SOIL

ENPs aggregate and deposit in porous structures of soil ecosystems. The fate and transport of NPs in soil depends on their size, surface characteristics, and matrix constituents (Darlington et al. 2009). Guzman et al. (2006) described three mechanisms for the transport of NPs in porous media; that is to say, particle interception with media, gravitational sedimentation, and diffusion. The transport of different ENPs (namely, CNTs, AgNPs, TiO_2, ZnO, SWNTs, and QDs) was studied on unsaturated soil or sand columns (Tian et al. 2010; Milani et al. 2010; Solovitch et al. 2010; Navarro et al. 2011). The results showed the aggregation of NPs through the column. The fate of ENPs in soil is similar to their fate in other media. The fate of NPs in soil systems is greatly affected by the transformation mechanism. For example, the high surface area of metallic ENPs favors their sorption into soil particles, which renders them immobile. Alternately, nanomaterials easily insert themselves into smaller spaces of soil particles or travel larger distances before becoming trapped in the soil matrix.

15.5 ECOTOXICITY OF ENGINEERED NANOPARTICLES

Due to the development of nanotechnologies, the production of nanobased products has increased, which has introduced large amounts of manufactured NPs into the environment. Thus, to protect human health and wildlife from the toxic effects of a broad range of nanomaterials, a number of studies have been conducted on the assessment of the toxicity of the NPs commonly used in industry (Yang and Watts 2005).

A convenient method to measure toxicity and ecotoxicity is the measurement of oxidative stress in terms of reactive oxygen species (ROS) generation. The cells combat this oxidative stress by an increasing number of protective responses that can easily be measured as enzymatic or by genetic expression responses (Kovochich et al. 2005). Sayes et al. (2004) conducted *in vitro* studies and found that the presence of NPs such as TiO_2 and fullerenes initiated the formation of ROS; but, on the other hand, some authors found that NPs such as fullerenes protected against oxidative stress

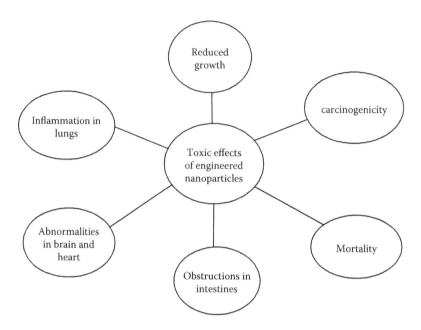

FIGURE 15.3 Toxic effects of engineered nanoparticles in different organisms.

(Daroczi et al. 2006). The toxic effects of ENPs in different organisms have been summarized in Figure 15.3. The organisms tested to date include microbes such as bacteria, protozoans, invertebrates, nematodes, earthworms, fish, and mammals. The toxic effects of some ENPs in different organisms have been presented in Table 15.1.

15.5.1 MICROBES

15.5.1.1 Bacteria

Lyon et al. (2005) used *Escherichia coli* and *Bacillus subtilis* as model organisms to test the toxicity of C_{60} fullerenes. They found that the minimum inhibitory concentration (MIC) of nano C_{60} fullerenes for *E. coli* is much less (0.5–1.0 mg L^{-1}) than for *B. subtilis* (1.5–3.0 mg L^{-1}). On comparing the toxicity of bulk materials such as carboxyfullerenes and benzene, they showed less toxicity than C_{60} fullerenes. Fortner et al. (2005) also used *E. coli* and *B. subtilis* as a test organism and results showed reduced growth and respiration at concentrations of <0.4 and 4 mg L^{-1}. Other studies have also shown the toxic effects of C_{60}, even at low concentration, on bacterial membrane composition and fluidity (Fang et al. 2007). In the presence of C_{60}, *Pseudomonas putida* showed a decrease in unsaturated fatty acids and an increase in cyclopropane fatty acids, possibly to combat oxidative stress. Fourier transform infrared spectroscopy (FTIR) data showed a slight increase in phase transition temperature and membrane fluidity of bacterial cells. These changes could be due to conformational changes in acyl chains at a growth-inhibiting concentration (0.5 mg L^{-1}) of C_{60}. In *B. subtilis*, a lower dose of C_{60} (0.01 mg L^{-1}) resulted in an increase in levels of branched fatty acids, whereas a higher dose (0.75 mg L^{-1}) resulted in an increase in monounsaturated fatty acids. The increase in fatty acids could be due to the interaction of C_{60} with the lipid fraction of bacterial cell membranes and other constituents to generate lipid peroxidation (Fang et al. 2007). These reports show the physiological adaptation of bacteria toward response to ENPs.

   E. coli K12 cells were used as a model organism to test the toxicity of SWNTs. Results indicated a substantial loss in viability of treated cells within 60 min, compared with controls. Moreover, there was an increase in the average percentage of viability loss with time. Thus, SWNTs exhibit a strong antibacterial activity (Kang et al. 2007).

TABLE 15.1

Toxic Effects of Engineered Nanoparticles in Different Organisms

Nanoparticle	Organism Studied	Toxic Effects	Refs.
C_{60} fullerenes	*Escherichia coli* and *Bacillus subtilis*	Reduced growth and respiration	Fortner et al. 2005
	Pseudomonas putida	Decrease in unsaturated fatty acids and increase in cyclopropane fatty acids	Fang et al. 2007
	B. subtilis	Increase in levels of branched fatty acids and monounsaturated fatty acids	Fang et al. 2007
	Lumbricus rubellus	Reduced growth	Van der Ploeg et al. 2011
	Pimephales promelas	Lipid peroxidation in brain, depletion of GSH levels of gills	Oberdörster 2004
SWNTs	*E. coli* K12	Loss in viability	Kang et al. 2007
	Oncorhynchus mykiss	Decrease in thiobarbituric acid-reactive substance (TBARS) in gills, brain, and liver; increase in total glutathione levels in gills and liver	Smith et al. 2007
	Danio rerio	Delay in hatching	Cheng et al. 2007
MWNTs	*Stylonychia mytilus*	Damage to macronucleus and external membranes of cells	Zhu et al. 2006
DWNTs	*Eisenia veneta*	Reduced growth, impaired cocoon production	Scott-Fordsmand et al. 2008
	Danio rerio	Delay in hatching	Cheng et al. 2007
Silver NPs	*E. coli*	Deactivation of membrane-bound enzymes, destruction of membrane permeability	Li et al. 2010
	Caenorhabditis elegans	Decrease in survival rate and reproduction	Roh et al. 2009
	Oryzias latipes	Edema and abnormalities in the heart, fins, brain, spine, and eyes	Wu et al. 2010
TiO_2, SiO_2 and ZnO NPs	*B. subtilis* and *E. coli*	Highest antibacterial activity observed in SiO_2, followed by TiO_2, and lowest in ZnO	Adams et al. 2006
ZnO NPs	*E. coli*	Disorganization of membrane	Brayner et al. 2006
CeO_2 NPs	*E. coli*	Oxidative stress	Thill et al. 2006
	Anabaena (CPB4337) and *Pseudokirchneriella subcaptitata*	Cellular damage and membrane damage	Palomares et al. 2011
SiO_2 NPs	*Scenedesmus obliquus*	Decrease in chlorophyll content	Wei et al. 2010
TiO_2 NPs	*Desmodesmus subspicatus*	Growth reduction	Hund-Rinke and Simon 2006
	Daphnia magna and *Thamnocephalus platyurus*	Acute toxicity	Baun et al. (2008)
	Caenorhabditis elegans	Reduced growth and reproduction	Wang et al. 2009
	Lumbricus rubellus	DNA damage	Hu et al. 2010

Li et al. (2010) conducted a study to determine the toxic effects of silver NPs on the growth of *E. coli*. Results revealed that silver NPs deactivate membrane-bound enzymes, resulting in leakage of reducing sugars and proteins, signifying that silver NPs have the ability to destroy membrane permeability. This is due to the potential of Ag^+ ions to interrupt the proton gradient across the cell membrane, leading to cell death.

Adams et al. (2006) studied the effect of TiO_2, SiO_2, and ZnO water suspensions on *B. subtilis* and *E. coli*. *B. subtilis* was found to be more susceptible to these NPs than *E. coli*. The highest antibacterial activity was observed in SiO_2, followed by TiO_2, with the lowest in ZnO. Brayner et al. (2006) reported disorganization of the membrane in *E. coli* when treated with ZnO NPs (1.4–14 nm). ZnO NPs showed 100% inhibition of bacterial growth at a concentration of $10^{-2} - 3.0 \times 10^{-3}$ M, but at a lower concentration (1.5×10^{-3} and 10^{-3} M), bacterial colonies were seen. Transmission electron microscopy (TEM) revealed cellular internalization and increased membrane permeability for Zn^{2+} ions. Thill et al. (2006) elicited the cytotoxicity caused by CeO_2 NPs (7 nm) on *E. coli*. TEM, x-ray absorption spectroscopy, and sorption isotherm data revealed the presence of NPs in the bacterial membrane, which later causes oxidative stress. Antibacterial activity of QDs has been investigated in *B. subtilis*, *E. coli*, and *P. aeruginosa* in acidic (pH 2) or alkaline (pH 12) environments (Mahendra and Zhu 2008). The degradation of the surface coating of QDs releases cadmium (Cd) and selenium (Se) ions, which are toxic to the organisms.

The studies described were conducted on different bacterial species, different types of ENPs, and under varying conditions. Such varying conditions affect the interaction between bacteria and NPs. For example, hydrophobic carbon–based NPs damage DNA, lipid fractions of cell membranes, and other lipid components of cells; but in natural environments, due to their interaction with the soil, they do not elicit toxicity, as they become unavailable for interaction with biological materials. On the other hand, metal oxide NPs generate ROS and oxidative stress once inside the cell. These NPs pass through the ion channels and replace the cations at their site of action. Metal oxide NPs interact with thiols, carboxylates, phosphates, hydroxyls, nitrates, and amines present in cellular constituents. Therefore, more research is needed to understand the nature and behavior of ENPs.

15.5.1.2 Algae

The toxicity of SiO_2 NPs (diameter 10–20 nm) was investigated in water alga *Scenedesmus obliquus*. When SiO_2 NPs were exposed to *S. obliquus* at moderate to high concentrations (50, 100, and 200 mg L^{-1}), *S. obliquus* showed a decrease in chlorophyll content after 96 h of exposure (Wei et al. 2010). In another study, Miao et al. (2010) reported the accumulation of nanosilver in freshwater alga *Ochromonas danica*. The toxic effect of CeO_2 NPs has been studied on a self-luminescent cyanobacterial recombinant strain of *Anabaena* (CPB4337) and on the green alga *Pseudokirchneriella subcapitata*. CeO_2 NPs caused cellular damage and membrane damage, which is due to the direct contact of CeO_2 NPs and organism cells (Rodea-Palomares et al. 2011).

15.5.1.3 Protozoa

The exposure of MWNTs to unicellular protozoan *Stylonychia mytilus* resulted in their absorption, and they subsequently passed on after cell division (Zhu et al. 2006). As the concentration of MWNTs increased, growth was inhibited. The results of fluorescence and electron microscopy revealed damage to the macronucleus and external membranes of cells caused by MWNTs. The authors concluded that the toxic effects of MWNTs on the macronucleus, micronucleus, and cell membrane results in damage to mitochondria. A similar study has been performed on unicellular protozoan *Tetrahymena pyriformis*; it showed growth stimulation on exposure to MWNTs–peptone conjugates in a proteose peptone yeast extract medium (PPY), whereas growth inhibition was shown in the presence of filtered pond water. The toxic effect of MWNTs was observed by measurement of malondialdehyde (MDA) and superoxide dismutase (SOD) (Zhu et al. 2006).

15.5.2 ANIMALS

15.5.2.1 Invertebrates

15.5.2.1.1 Crustaceans

Crustaceans have the ability to sequester toxic metals in the hepatopancreas and other tissues. The common crustaceans found in freshwater lakes and ponds are *Daphnia magna*. Oberdörster et al. (2006) reported a decrease in reproduction in *D. magna* when exposed to fullerenes and, when exposed to higher concentrations of fullerenes, the mortality rate increased. Baun et al. (2008) also reported the toxic effect of TiO_2 in *D. magna* and *Thamnocephalus platyurus*. When *D. magna* cells were exposed to a 20 mg L^{-1} concentration of TiO_2, *D. magna* showed a 60% toxic effect, whereas *T. platyurus* showed marginal toxic effects. Blinova et al. (2010) investigated the toxic effect of metal oxide NPs (ZnO and CuO) on *D. magna*, *T. platyurus*, and *Tetrahymena thermophila* in both natural and artificial waters. The toxicity of ZnO NPs was lower than that of CuO NPs and bulk CuO. *D. magna* and *Vibrio fischeri* were tested for the toxic effects of TiO_2 (6 nm), Al (100 nm), ALEX (aluminum explosive), NPs coated with Al_2O_3, L-ALEX-NPs coated with carboxylate groups, and boron NPs (10–20 nm) (Strigul et al. 2009). TiO_2 and L-ALEX showed low toxicity to *D. magna*, whereas nanoaluminum was highly toxic. Boron NPs displayed slightly more toxicity to *D. magna* than *V. fischeri*. Templeton et al. (2006) performed experiments on estuarine copepod *Amphiascus tenuiremis* for SWNTs. The copepod ingested purified SWNTs and showed no significant effects on mortality, development, or reproduction.

15.5.2.1.2 Nematodes

The most well-known soil nematode is *Caenorhabditis elegans*. Most of the researchers evaluated the toxicity of NPs on *C. elegans*. Roh et al. (2009) evaluated the toxicity of silver NPs on *C. elegans* and found a decrease in survival rate and reproduction due to oxidative stress. Wang et al. (2009) compared the toxicity of both the nano- and bulk forms of ZnO, Al_2O_3, and TiO_2 in *C. elegans*. Both the nano- and bulk forms of Al_2O_3 and TiO_2 significantly affected the growth and reproduction of *C. elegans*. The toxic effect of CeO_2 and TiO_2 NPs was related to the cyp35a2 gene expression. This gene is related to the growth, fertility, and survival in *C. elegans* (Roh et al. 2010).

15.5.2.1.3 Earthworms

Van der Ploeg et al. (2011) exposed earthworms (*Lumbricus rubellus*) to C_{60} (doses: 0, 15.4, and 154 mg kg^{-1} in soil). After 4 weeks, dose-dependent effects were observed in juveniles and adults. The juvenile was found to be more sensitive to C_{60} NPs. Scott-Fordsmand et al. (2008) studied the toxic effect of DWCNTs and C_{60} on *Eisenia veneta* at concentration levels of 0, 50, 100, 300, and 495 mg kg^{-1}. No significant effect was observed at low concentrations of C_{60}, but at high concentrations, a 20% reduction in growth was observed. DWCNTs impaired cocoon production and reproduction at a concentration of 37 mg kg^{-1}. Exposure of earthworms to TiO_2 and ZnO NPs resulted in DNA damage when doses were greater than 1.0 mg kg^{-1} (Hu et al. 2010). Coleman et al. (2010) performed experiments to test the toxic effect of Al_2O_3 NPs on earthworms. They found that when a concentration of 10,000 mg kg^{-1} was used, earthworm mortality resulted.

15.5.2.2 Fish

Fish gills are sensitive to NPs. Oberdörster (2004) reported the depletion of glutathione (GSH) levels of gills and lipid peroxidation in the brain after 48 h of exposure to C_{60} on *Pimephales promelas*. Fe_2O_3 NPs cause harm to the gills, delays in embryo hatching, malformation in some zebra fish embryos and larvae, and ultimately mortality (Chen et al. 2012).

When rainbow trout (*Oncorhynchus mykiss*) were exposed to SWNTs, there was a significant decrease in thiobarbituric acid-reactive substances (TBARS) in gills, brain, and liver, whereas a significant increase in total glutathione levels in gills (28%) and liver (18%) was observed. Similarly, when TiO_2 NPs were exposed, there was a sublethal effect on fish (Smith et al. 2007). Cheng et al. (2007) studied the toxic effects of unprocessed single-walled and double-walled CNTs to zebrafish (*Danio rerio*). There was a significant delay in hatching in zebrafish embryos when exposed to SWCNTs and DWCNTs at concentrations greater than 120 and 240 mg L^{-1}, respectively. The delay in hatching is due to the presence of cobalt and nickel catalysts used in the production of SWCNTs that remained even after purification.

Exposure of Japanese medaka (*Oryzias latipes*) to silver NPs resulted in edema and abnormalities in the heart, fins, brain, spine, and eyes (Wu et al. 2010). Similar abnormalities were reported in zebrafish when tested against starch and bovine serum albumin (BSA)-capped AgNPs

(Asharani et al. 2008). In addition to acute toxicity studies, other investigations have been undertaken on expression levels of stress-related genes. Chae et al. (2009) conducted a study on Japanese medaka (*O. latipes*) to evaluate the toxicity of silver NPs. In this study, heat shock protein-70 (HSP 70), p53, cytochrome P450 1A (CYP1A), and transferring genes were selected as stress markers. The silver NPs caused cellular damage, DNA damage, carcinogenicity, and oxidative stress. Pham et al. (2012) conducted a similar study and indicated that AgNPs are potential inducers of metal detoxification and oxidative/inflammatory stress.

15.5.2.3 Mammals

The toxicity of NPs has been investigated in rats, mice, and guinea pigs. Roursgaard et al. (2008) studied the toxicity of fullerol NPs and quartz in mice. Quartz induced greater inflammation than fullerol in mice. Handy and Shaw (2007) studied the toxicity of CNTs (doses: 0.1–12.5 mg kg^{-1}), ultrafine TiO$_2$ NPs (doses: 0.5, 2.0, 10 mg L^{-1}), ultrafine cadmium particles (70 µg L^{-1}), and metal oxide particles (1–5 mg) in rats. There was significant damage to lungs, inflammation, and fibrotic responses. The toxicity of polyalkyl-sulfonated C$_{60}$ has been reported in rats, due to its accumulation in the liver, spleen, and kidneys (Chen et al. 1998). Muller et al. (2005) studied the toxicity of MWNTs (doses: 0.5, 2, or 5 mg) in Sprague Dawley rats. Results indicated the persistence of MWNTs in lung tissues causing inflammation and fibrosis. Warheit et al. (2004) conducted a similar study in rats with SWNTs (doses: 1–5 mg kg^{-1}) and found cell injury and transient inflammation. Wang et al. (2006) studied gastrointestinal dosing in mice, with either nanozinc oxide or larger particles. Due to the accumulation of these NPs, this resulted in severe symptoms such as lethargy, vomiting, diarrhea, and death from intestinal obstruction by aggregated nanozinc oxide.

The effects of ENPs on human beings have yet to be studied. But, it is believed that exposure to these ENPs result in the formation of ROS and oxidative stress leading to respiratory and cardiovascular inflammation (Xia et al. 2009). For example, exposure of workers to manganese during mining in central India revealed a genetic polymorphism in cytochrome P450 2D6 (CYP2D6*2), glutathione-S-transferase M1 (GSTM1), and NAD(P)H quinine oxidoreductase 1 (NQO1) genes (Vinayagamoorthy et al. 2010).

15.5.3 Plants

Plants are particularly important for the study of econanotoxicity, as they interact with air, soil, and water, all of which contain ENPs. In addition, plants are the major source of transfer of ENPs to the food chain, as they are consumed by lower trophic-level organisms, animals, and people. Yang and Watts (2005) first conducted nanotoxicity studies on plants. They studied the effect of Al$_2$O$_3$ NPs on five plants (*Zea mays, Cucumis sativus, Glycine max, Brassica oleracea*, and *Daucus carota*). Exposure to aluminum NPs at a 2 mg mL^{-1} concentration after 24 h resulted in stunted root growth. CNTs were tested for their toxicity in tomato seeds (10–40 µg mL^{-1}) and it was found that the NPs penetrated the seed coat, resulting in an increased germination rate (Khodakovskaya et al. 2009). Wang et al. (2011) reported the effects of magnetite NPs (Fe$_3$O$_4$) on ryegrass (*Lolium perenne*) and pumpkin (*Cucurbita mixta* cv. white cushaw). These NPs induced more oxidative stress than bulk materials.

Lee et al. (2008) studied the toxicity of insoluble copper oxide (CuO) NPs to *Triticum aestivum* (wheat) and *Phaseolus radiates* (mung beans). The authors reported the aggregation of CuO inside the plant cell and found that wheat was more tolerant to CuO NPs than mung beans. Li et al. (2010) studied the effect of ZnO, Al$_2$O$_3$, SiO$_2$, and Fe$_3$O$_4$ NPs on *C. pepo* and found that ZnO NPs were more toxic than others. However, Stampoulis et al. (2009) did not find any significant toxicity effects of NPs on *C. pepo*. Lee et al. (2012) reported a decrease in the growth and germination of two edible plants (*P. radiatus* and *Sorghum bicolor*) with an increase in the concentration of silver NPs.

Lin and Xing (2007) studied the effect of MWCNTs, aluminum, alumina, zinc, and zinc oxide on six different plant species: radish, rape, ryegrass, lettuce, corn, and cucumber. Zinc showed an inhibition of seed germination in ryegrass; ZnO NPs showed an inhibition of seed germination in corn, whereas the pattern of root growth differed from plant to plant and type of NPs used. ZnO NPs showed reduced biomass and vacuolated epidermal or cortical cells in ryegrass (*L. perenne*) when exposed to concentrations of 10, 20, 50, 100, 200, and 1000 mg L^{-1} (Lin and Xing 2008).

15.6 ENVIRONMENTAL RISK ASSESSMENT OF NANOPARTICLES

The first step in the risk assessment of NPs is the identification of hazards and exposure routes for humans. Traditionally the risk assessment is divided into four steps:

- Hazard assessment
- Dose–response assessment
- Environmental exposure assessment
- Risk characterization

15.6.1 HAZARD ASSESSMENT

In the assessment of hazard, the capability of NPs to cause harm is evaluated. Hazards can be divided into toxicity and ecotoxicity; they can be measured by identifying physiological, genetic, or functional effects, either acute or chronic. The negative effect on the environment should also be assessed, such as the impact on atmospheric/stratospheric processes, stability of soil, and effects on bioavailability of mineral nutrients (Joner et al. 2008).

15.6.2 DOSE–RESPONSE ASSESSMENT

To establish dose–response relationships, mathematical models or experiments in the laboratory are required. However, dose–response may be less relevant for NPs, as doses based on mass concentration and preparation of NPs result in differences in surface reactivity and thus toxicity (Joner et al. 2008).

15.6.3 EXPOSURE ASSESSMENT

Exposure to NPs begins with their production. Thus, information about the steps involved in the production, purification, functionalization, conditioning, packing, and transport is necessary. Knowledge about their exposure to the environment is important, to have empirical data or procedures to identify their persistence and mobility in soil, air, and water (Joner et al. 2008).

15.6.4 RISK CHARACTERIZATION

This is the last step in the risk assessment procedure, in which data from the hazard identification, dose–response, and exposure steps are studied together to determine and communicate the actual likelihood of risk to exposed populations.

15.7 FUTURE PROSPECTS

The incessant need for research is related to the environmental impact of NPs to establish the degree of mobility and bioavailability of NPs in the environment. These parameters will help to resolve whether NPs can be taken up and cause harm to various organisms including plants, as NPs obtain entry into the food chain through drinking water and food. Research on the impact of NPs on the environment has started, but there are methodological obstacles such as its detection, characterization, and the tracing of which NPs should be examined; the points to be considered first and foremost regarding the small size of NPs and the complexity of the environments where their impact should be studied.

To study the behavior of NPs in the environment, the following points should be considered:

1. Mobility of NPs in soils, sediments, and waste
2. Adsorption/desorption of NPs to biological components of soils, sediments, and water

To study the ecotoxicity of NPs, the following points should be considered:

1. Selection of suitable terrestrial/aquatic organisms to judge the ecotoxicity of NPs
2. Establishment of dose–response relationships and toxicokinetics

ACKNOWLEDGMENT

The authors are grateful to TEQIP-II and MNNIT, Allahabad for financial support for this research.

REFERENCES

Adams, L. K., D. Y. Lyon, and P. J. J. Alvarez. 2006. Comparative eco-toxicity of Nanoscale TiO_2, SiO_2, and ZnO water suspensions. *Water Research* 40: 3527–32.

Alivisatos, P. 2004. The use of nanocrystals in biological detection. *Nature Biotechnology* 22(1): 47–52.

Asharani, P. V., Y. L. Wu, Z. Gong, and S. Valiyaveettil. 2008. Toxicity of silver nanoparticles in Zebrafish models. *Nanotechnology* 19: 1–8.

Aitken, R. J., K. S. Creely, and C. L. Tran. 2004. Nanoparticles: An occupational hygiene review. Research Report 274. Prepared by the Institute of Occupational Medicine for the Health and Safety Executive, North Riccarton, Edinburgh.

Bandyopadhyaya, R., A. A. Lall, and S. K. Friedlander. 2004. Aerosol dynamics and the synthesis of fine solid particles. *Powder Technology* 139: 193–99.

Baun, A., N. B. Hartmann, K. Grieger, and K. Ole Kusk. 2008. Ecotoxicity of engineered nanoparticles to aquatic invertebrates: A brief review and recommendations for future toxicity testing. *Ecotoxicology* 17(5): 387–95.

BCC Research. 2012. Market research reports and technical publications. Product catalog. Available from: ///C:/Users/pk_naoghare/Downloads/2012_BCC_Catalog%20(1).pdf.

Biswas, P., and C. Y. Wu. 2005. Journal of the air & waste management nanoparticles and the environment. *Nanoparticles and the Environment* 55: 37–41.

Blinova, I., A. Ivask, M. Heinlaan, M. Mortimer, and A. Kahru. 2010. Ecotoxicity of nanoparticles of CuO and ZnO in natural water. *Environmental Pollution* 158(1): 41–47.

Brant, J. A., J. Labille, C. Ogilvie, and M. Wiesner. 2007. Fullerol cluster formation in aqueous solutions: Implications for environmental release. *Journal of Colloid and Interface Science* 314: 281–88.

Brayner, R., R. Ferrari-Iliou, N. Brivois, S. Djediat, M. F. Benedetti, and F. Fievet. 2006. Toxicological impact studies based on *Escherichia coli* bacteria in ultrafine ZnO nanoparticles colloidal medium. *Nano Letters* 6(4): 866–70.

Buffle, J. 2006. The key role of environmental colloids/nanoparticles for the sustainability of life. *Environmental Chemistry* 3: 155–58.

Chae, Y. J., C. H. Pham, J. Lee, E. Bae, J. Yi, and M. B. Gu. 2009. Evaluation of the toxic impact of silver nanoparticles on Japanese medaka (*Oryzias latipes*). *Aquatic Toxicology* 94(4): 320–27.

Chan, W. C. W., D. J. Maxwell, X. Gao, R. E. Bailey, M. Han, and S. Nie. 2002. Luminescent quantum dots for multiplexed biological detection and imaging. *Current Opinion in Biotechnology* 13: 40–46.

Chen, H. H. C., Y. Chi, T. H. Ueng, S. Chen, B. J. Chen, and K. J. Huang. 1998. Acute and subacute toxicity study of water-soluble polyalkylsulfonated C_{60} in rats. *Toxicologic Pathology* 26(1): 143–51.

Chen, X., X. Zhu, R. Li, H. Yao, Z. Lu, and X. Yang. 2012. Photosynthetic toxicity and oxidative damage induced by Nano-Fe_3O_4 on *Chlorella vulgaris* in aquatic environment. *Journal of Ecology* 2(1): 21–28.

Cheng, J., E. F. Lahaut, and S. H. Cheng. 2007. Effect of carbon nanotubes on developing zebrafish (*Danio rerio*) embryos. *Environmental Toxicology and Chemistry* 26(4): 708–16.

Chopra, I. 2007. The increasing use of silver-based products as antimicrobial agents: A useful development or a cause for concern? *Journal of Antimicrobial Chemotherapy* 59(4): 587–90.

Coleman, J. G., D. R. Johnson, J. K. Stanley, A. J. Bednar, C. A. Weiss, R. E. Boyd, and J. A. Steevens. 2010. Assessing the fate and effects of nano aluminum oxide in the terrestrial earthworm, *Eisenia fetida*. *Environmental Toxicology and Chemistry* 29(7): 1575–80.

Colton, H. M., J. G. Falls, H. Ni, P. Kwanyuen, D. Creech, E. McNeil, W. M. Casey, G. Hamilton, and N. F. Cariello. 2004. Visualization and quantitation of peroxisomes using fluorescent nanocrystals: Treatment of rats and monkeys with fibrates and detection in the liver. *Toxicological Sciences* 80(1): 183–92.

Cornell, R. M., and U. Schwertmann. 1996. *The Iron Oxides: Structure, Properties, Reactions, Occurrences and Uses*. Wiley, Weinheim.

Darlington, T., A. M. Neigh, M. T. Spencer, O. T. N. Guyen, and S. J. Oldenburg. 2009. Nanoparticle characteristics affecting environmental fate and transport through soil. *Environmental Toxicology and Chemistry* 28(6): 1191–99.

Daroczi, B., G. Kari, M. F. Mcaleer, J. C. Wolf, U. Rodeck, and A. P. Dicker. 2006. *In vivo* radioprotection by the fullerene nanoparticle DF-1 as assessed in a zebrafish model. *Clinical Cancer Research* 12(23): 7086–92.

Defra. 2007. Characterizing the potential risks posed by engineered nanoparticles: A second UK government research report. Department for Environment, Food and Rural Affairs, London 90. http://www.defra.gov.uk/environment/nanotech/research/reports/index.htm.

Dubertret, B., P. Skourides, and D. J. Norris. 2002. *In vivo* imaging of quantum dots encapsulated in phospholipid micelles. *Science* 298: 1759.

Fang, J., D. Y. Lyon, M. R. Wiesner, J. Dong, and P. J. J. Alvarez. 2007. Effect of a fullerene water suspension on bacterial phospholipids and membrane phase behavior. *Environmental Science and Technology* 41(7): 2636–42.

Fortner, J. D., D. Y. Lyon, C. M. Sayes, A. M. Boyd, J. C. Falkner, and E. M. Hotze. 2005. C_{60} in water: Nanocrystal formation and microbial response. *Environmental Science & Technology* 39(11): 4307–16.

Forzatti, P. 2000. Environmental catalysis for stationary applications. *Catalysis Today* 62(1): 51–65.

Garland, A. 2009. The global market for carbon nanotubes to 2015: A realistic market assessment. Nanoposts. http://www.nanoposts.com/index.php?mod nanotubes.

Ghosh, S., H. Mashayekhi, B. Pan, P. Bhowmik, and B. Xing. 2008. Colloidal behavior of aluminum oxide nanoparticles as affected by pH and natural organic matter. *Langmuir* 40(24): 12385–91.

Gibson, J. D., B. P. Khanal, and E. R. Zubarev. 2007. Paclitaxel-functionalized gold nanoparticles. *Journal of the American Chemical Society* 129(37): 11653–61.

Gottschalk, F., T. Sonderer, R. W. Scholz, and B. Nowack. 2009. Modeled environmental concentrations of engineered nanomaterials (TiO_2, ZnO, Ag, CNT, fullerenes) for different regions. *Environmental Science and Technology* 43(24): 9216–22.

Guzman, K. A. D., M. P. Finnegan, and J. F. Banfield. 2006. Influence of surface potential on aggregation and transport of titania nanoparticles. *Environmental Science & Technology* 40(24): 7688–93.

Handy, R. D., and B. J. Shaw. 2007. Toxic effects of nanoparticles and nanomaterials: Implications for public health, risk assessment and the public perception of nanotechnology. *Health, Risk & Society* 9(2): 125–44.

Horisberger, M., and J. Rosset. 1977. Colloidal gold, a useful marker for transmission and scanning electron microscopy. *The Journal of Histochemistry and Cytochemistry* 25(4): 295–305, http://www.epa.gov/osa/pdfs/nanotech/epa-nanotechnology-whitepaper-0207.pdf.

Hu, C. W., M. Li, Y. B. Cui, D. S. Li, J. Chen, and L. Y. Yang. 2010. Soil biology & biochemistry toxicological effects of TiO_2 and ZnO nanoparticles in soil on earthworm *Eisenia fetida*. *Soil Biology and Biochemistry* 42(4): 586–91.

Huber, D. L. 2005. Synthesis, properties, and applications of iron nanoparticles. *Small* 5: 482–501.

Hund-Rinke, K., and M. Simon. 2006. Ecotoxic effect of photocatalytic active nanoparticles (TiO_2) on algae and daphnids. *Environmental Science and Pollution Research International* 13(4): 1–8.

Hyung, H., and J. D. Fortner. 2007. Natural organic matter stabilizes carbon nanotubes in the aqueous phase. *Environmental Science & Technology* 41(1): 179–84.

Kang, S., M. Pinault, L. D. Pfefferle, and M. Elimelech. 2007. Single-walled carbon nanotubes exhibit strong antimicrobial activity. *Langmuir* 23(17): 8670–73.

Keller, A. A., H. Wang, D. Zhou, H. S. Lenihan, G. Cherr, B. J. Cardinale, R. Miller, and Z. Ji. 2010. Stability and aggregation of metal oxide nanoparticles in natural aqueous matrices. *Environmental Science and Technology* 44(6): 1962–67.

Kellogg, C. A., and D. W. Griffin. 2006. Aerobiology and the global transport of desert dust. *Trends in Ecology and Evolution* 21(11): 638–44.

Kim, J., M. Takahashi, T. Shimizu, T. Shirasawa, M. Kajita, A. Kanayama, and Y. Miyamoto. 2008. Effects of a potent antioxidant, platinum nanoparticle, on the lifespan of *Caenorhabditis elegans*. *Mechanisms of Ageing and Development* 129(6): 322–31.

Kovochich, M., T. Xia, J. Xu, J. I. Yeh, and A. E. Nel, 2005. Principles and procedures to assess nanoparticles. *Environmental Science and Technology* 39(5): 1250–56.

Lee, W. M., Y. J. An, H. Yoon, and H. S. Kweon. 2008. Toxicity and bioavailability of copper nanoparticles to the terrestrial plants mung bean (*Phaseolus radiatus*) and wheat (*Triticum aestivum*): Plant agar test for water-insoluble nanoparticles. *Environmental Toxicology and Chemistry* 27(9): 1915–21.

Lee, W. M., J. I. Kwak, and Y. J. An. 2012. Chemosphere effect of silver nanoparticles in crop plants *Phaseolus radiatus* and *Sorghum bicolor*: Media effect on phytotoxicity. *Chemosphere* 86(5): 491–99.

Li, W. R., X. B. Xie, and Q. S. Shi. 2010. Antibacterial activity and mechanism of silver nanoparticles on *Escherichia coli*. *Applied Microbiology and Biotechnology* 85(4): 1115–22.

Lidke, D. S., P. Nagy, R. Heintzmann, D. J. Arndt-Jovin, J. N. Post, H. E. Grecco, E. A. Jares-Erijman, and T. M. Jovin. 2004. Quantum dot ligands provide new insights into erbB/HER receptor-mediated signal transduction. *Nature Biotechnology* 22(2): 198–203.

Lin, D., and B. Xing. 2007. Phytotoxicity of nanoparticles: Inhibition of seed germination and root growth. *Environmental Pollution* 150(2): 243–50.

Lin, D., and B. Xing. 2008. Root uptake and phytotoxicity of ZnO nanoparticles. *Environmental Science & Toxicology* 42(15): 5580–85.

Lyon, D. Y., J. D. Fortner, C. M. Sayes, V. L. Colvin, and J. B. Hughes. 2005. Bacterial cell association and antimicrobial activity of a C_{60} water suspension. *Environmental Toxicology & Chemistry* 24(11): 2757–62.

Mädler, L., and S. K. Friedlander. 2007. Transport of nanoparticles in gases: Overview and recent advances. *Aerosol and Air Quality Research* 7(3): 304–42.

Mahendra, S., and H. Zhu. 2008. Quantum dot weathering results in microbial toxicity. *Environmental Science & Technology* 42(24): 9424–30.

Maynard, A. D. 2006. Safe handling of nanotechnology. *Nature* 444: 267–69.

Miao, A. J., Z. Luo, C. S. Chen, W. C. Chin, and P. H. Santschi. 2010. Intracellular uptake: A possible mechanism for silver engineered nanoparticle toxicity to a freshwater alga *Ochromonas danica*. *PloS One* 5(12): 6–13.

Milani, N., M. J. McLaughlin, G. M. Hettiaratchchi, D. G. Beak, J. K. Kirby, and S. Stacey. 2010. Fate of nanoparticulate zinc oxide fertilisers in soil: solubility, diffusion and solid phase speciation. In *Soil Solutions for a Changing World: 19th World Congress of Soil Science, Brisbane*, Australia, pp. 1–6.

Muller, J., N. Moreau, P. Misson, M. Delos, M. Arras, A. Fonseca, J. B. Nagy, and D. Lison. 2005. Respiratory toxicity of multi-wall carbon nanotubes. *Toxicology and Applied Pharmacology* 207: 221–31.

Murr, L. E., E. V. Esquivel, J. J. Bang, G. De Rosa, and J. L. Gardea-torresdey. 2004. Chemistry and nanoparticulate compositions of a 10,000 year-old ice core melt water. *Water Research* 38: 4282–96.

Navarro, D. A., S. Banerjee, D. F. Watson, and D. S. Aga. 2011. Differences in soil mobility and degradability between water-dispersible CdSe and CdSe/ZnS quantum dots. *Environmental Science & Technology* 45(15): 6343–49.

Nel, A. 2006. Toxic potential of materials at the nanolevel. *Science* 311: 622.

Nowack, B., and T. D. Bucheli. 2007. Occurrence, behavior and effects of nanoparticles in the environment. *Environmental Pollution* 150(1): 5–22.

Oberdörster, E. 2004. Manufactured nanomaterials (fullerenes, C_{60}) induce oxidative stress in the brain of juvenile largemouth bass. *Environmental Health Perspectives* 1058(10): 1058–62.

Oberdörster, E., S. Zhu, and T. Michelle Blickley. 2006. Ecotoxicology of carbon-based engineered nanoparticles: Effects of fullerene (C_{60}) on aquatic organisms. *Carbon* 44: 1112–20.

Pandey, R. A., and A. Kumar. 1990. Heavy metals in coal carbonization wastewater and their complexation: Some observations. *Water, Air and Soil Pollution* 50: 31–38.

Pham, C. H., J. Yi, and M. B. Gu. 2012. Biomarker gene response in male Medaka (*Oryzias latipes*) chronically exposed to silver nanoparticle. *Ecotoxicology and Environmental Safety* 78: 239–45.

Qian, X., X. H. Peng, D. O. Ansari, Q. Yin-Goen, G. Z. Chen, D. M. Shin, L. Yang, A. N. Young, M. D. Wang, and S. Nie. 2008. *In vivo* tumor targeting and spectroscopic detection with surface-enhanced Raman nanoparticle tags. *Nature Biotechnology* 26(1): 83–90.

Raam, J. B., and P. J. J. Alvarez. 2010. Developmental phytotoxicity of metal oxide nanoparticles to *Arabidopsis thaliana*. *Environmental Toxicology and Chemistry* 29(3): 669–75.

Ramteke, C., T. Chakrabarti, B. K. Sarangi, and R. A. Pandey. 2012. Synthesis of silver nanoparticles from the aqueous extract of leaves of *Ocimum sanctum* for enhanced antibacterial activity. *Journal of Chemistry* 2013: 7.

Robichaud, C. O., A. L. I. Emre Uyar, M. R. Darby, L. G. Zucker, and M. R. Wiesner. 2009. Policy analysis estimates of upper bounds and trends in nano-TiO_2 production as a basis for exposure assessment. *Environmental Science & Technology* 43(12): 4227–33.

Roco, M. C., and W. S. Bainbridge. 2005. Societal implications of nanoscience and nanotechnology: Maximizing human benefit. *Journal of Nanoparticle Research* 7(1): 1–13.

Rodea-Palomares, I., K. Boltes, F. Ferna, J. Santiago, and R. Rosal. 2011. Physicochemical characterization and ecotoxicological assessment of CeO_2 nanoparticles using two aquatic microorganisms. *Toxicological Sciences* 119 (1): 135–45.

Rodríguez, J. A., and M. Fernández-García. 2007. *Synthesis, Properties and Applications of Oxide Nanoparticles*. Wiley, Totowa, NJ, pp. 335–51.

Roh, J. Y., and S. J. Sim. 2009. Ecotoxicity of silver nanoparticles on the soil nematode *Caenorhabditis elegans* using functional ecotoxicogenomics. *Environmental Science & Toxicology* 43(10): 3933–40.

Roh, J. Y., Y. K. Park, K. Park, and J. Choi. 2010. Ecotoxicological investigation of CeO_2 and TiO_2 nanoparticles on the soil nematode *Caenorhabditis elegans* using gene expression, growth, fertility, and survival as endpoints. *Environmental Toxicology and Pharmacology* 29: 167–72.

Roursgaard, M., S. S. Poulsen, C. L. Kepley, M. Hammer, G. D. Nielsen, and S. T. Larsen. 2008. Polyhydroxylated C60 fullerene (fullerenol) attenuates neutrophilic lung inflammation in mice. *Basic and Clinical Pharmacology and Toxicology* 103(4): 386–88.

Sahu, N., D. Soni, B. Chandrashekhar, B. K. Sarangi, D. Satpute, and R. A. Pandey. 2012. Synthesis and characterization of silver nanoparticles using Cynodon dactylon leaves and assessment of their antibacterial activity. *Bioprocess and Biosystems Engineering* 36(7): 999–1004.

Sawyer, C. N., and P. L. McCarty. 1967. *Chemistry for Sanitary Engineers*. 2nd edn. McGraw-Hill Series in Sanitary Science and Water Resources Engineering, McGraw-Hill, Toronto.

Sayes, C. M., J. D. Fortner, W. Guo, D. Lyon, A. M. Boyd, K. D. Ausman, et al. 2004. The differential cytotoxicity of water-soluble fullerenes. *Nano Letters* 4(10): 1881–7.

Scott-Fordsmand, J. J., P. H. Krogh, M. Schaefer, and A. Johansen. 2008. Ecotoxicology and environmental safety the toxicity testing of double-walled nanotubes-contaminated food to *Eisenia veneta* earthworms. *Ecotoxicology and Environmental Safety* 71: 616–19.

Smith, C. J., B. J. Shaw, and R. D. Handy. 2007. Toxicity of single walled carbon nanotubes to rainbow trout (*Oncorhynchus mykiss*): Respiratory toxicity, organ pathologies, and other physiological effects. *Aquatic Toxicology* 82: 94–109.

Solovitch, N., J. Labille, J. Rose, P. Chaurand, D. Borschneck, M. R. Wiesner, and J. Y. Bottero. 2010. Concurrent aggregation and deposition of TiO_2 nanoparticles in a sandy porous media. *Environmental Science and Technology* 44(13): 4897–902.

Stensberg, M. C., Q. Wei, E. S. McLamore, D. M. Porterfield, A. Wei, and M. S. Sepúlveda. 2011. Toxicological studies on silver nanoparticles: challenges and opportunities in assessment, monitoring and imaging. *Nanomedicine* 6(5): 879–98.

Stampoulis, D., S. K. Sinha, and J. C. White. 2009. Assay-dependent phytotoxicity of nanoparticles to plants. *Environmental Science & Toxicology* 43(24): 9473–79.

Strigul, N., L. Vaccari, C. Galdun, M. Wazne, X. Liu, C. Christodoulatos, and K. Jasinkiewicz. 2009. Acute toxicity of boron, titanium dioxide, and aluminum nanoparticles to *Daphnia magna* and *Vibrio fischeri*. *Desalination* 248(1–3): 771–82.

Templeton, R. C., P. Lee Ferguson, K. M. Washburn, and W. A. Scrivens. 2006. Life-cycle effects of single-walled carbon nanotubes (SWNTs) on an estuarine meiobenthic copepod. *Environmental Science & Technology* 40(23): 7387–93.

Thill, A., O. Zeyons, O. Spalla, F. Chauvat, J. Rose, M. Auffan, and A. M. Flank. 2006. Cytotoxicity of CeO_2 nanoparticles for *Escherichia coli*: physico-chemical insight of the cytotoxicity mechanism. *Environmental Science & Technology* 40: 6151–56.

Tian, Y., B. Gao, and K. J. Ziegler. 2010. Transport of engineered nanoparticles in saturated porous media. *Journal of Nanoparticle Research* 12: 2371–80.

Tsai, C. Y., A. L. Shiau, S. Y. Chen, Y. H. Chen, P. C. Cheng, M. Y. Chang, D. H. Chen, C. H. Chou, C. R. Wang, and C. L. Wu. 2007. Amelioration of collagen-induced arthritis in rats by nanogold. *Arthritis & Rheumatology* 56(2): 544–54.

Van der Ploeg, M. J. C., J. M. Baveco, A. Van Der Hout, R. Bakker, I. M. C. M. Rietjens, and N. W. Van Den Brink. 2011. Effects of C_{60} nanoparticle exposure on earthworms (*Lumbricus rubellus*) and implications for population dynamics. *Environmental Pollution* 159(1): 198–203.

Verma, H. C., C. Upadhyay, A. Tripathi, R. P. Tripathi, and N. Bhandari. 2002. Thermal decomposition pattern and particle size estimation of iron minerals associated with the Cretaceous-Tertiary boundary at Gubbio. *Meteoritics and Planetary Science* 37(7): 901–10.

Vinayagamoorthy, N., K. Krishnamurthi, S. S. Devi, P. K. Naoghare, R. Biswas, A. R. Biswas, S. Pramanik, A. R. Shende, and T. Chakrabarti. 2010. Chemosphere genetic polymorphism of CYP2D6*2C-T 2850, GSTM1, NQO1 genes and their correlation with biomarkers in manganese miners of central India. *Chemosphere* 81(10): 1286–91.

Wang, B., W. Y. Feng, T. C. Wang, G. Jia, M. Wang, J. W. Shi, F. Zhang, Y. L. Zhao, and Z. F. Chai. 2006. Acute toxicity of nano- and micro-scale zinc powder in healthy adult mice. *Toxicology Letters* 161: 115–23.

Wang, H., R. L. Wick, and B. Xing. 2009. Toxicity of nanoparticulate and bulk ZnO, Al_2O_3 and TiO_2 to the nematode *Caenorhabditis elegans*. *Environmental Pollution* 157: 1171–77.

Wang, H., X. Kou, Z. Pei, J. Q. Xiao, X. Shan, and B. Xing. 2011. Physiological effects of magnetite (Fe_3O_4) nanoparticles on perennial ryegrass (*Lolium perenne L*) and pumpkin (*Cucurbita mixta*) plants. *Nanotoxicology* 5(1): 30–42.

Warheit, D. B., B. R. Laurence, K. L. Reed, D. H. Roach, G. A. M. Reynolds, and T. R. Webb. 2004. Comparative pulmonary toxicity assessment of single-wall carbon nanotubes in rats. *Toxicological Sciences* 125: 117–25.

Wei, C., Y. Zhang, J. Guo, B. Han, X. Yang, and J. Yuan. 2010. Effects of silica nanoparticles on growth and photosynthetic pigment contents of *Scenedesmus obliquus*. *Journal of Environmental Sciences* 22(1): 155–60.

Wiesner, M. R., G. V. Lowry, P. Alvarez, D. Dionysiou, and P. Biswas. 2006. Assessing the risks of manufactured nanomaterials. *Environmental Science and Technology* 40: 4336–45.

Wu, Y., Q. Zhou, H. Li, W. Liu, T. Wang, and G. Jiang. 2010. Effects of silver nanoparticles on the development and histopathology biomarkers of Japanese medaka (*Oryzias latipes*) using the partial-life test. *Aquatic Toxicology* 100(2): 160–67.

Xia, T., N. Li, and A. E. Nel. 2009. Potential health impact of nanoparticles. *Annual Review of Public Health* 30: 137–50.

Yadav, B. C., and R. Kumar. 2008. Structure, properties and applications of fullerenes. *International Journal of Nanotechnology and Application* 1(1): 15–24.

Yang, L., and D. J. Watts. 2005. Particle surface characteristics may play an important role in phytotoxicity of alumina nanoparticles. *Toxicology Letters* 158: 122–32.

Zhang, W. X. 2003. Nanoscale iron particles for environmental remediation: an overview. *Journal of Nanoparticle Research* 5: 323–32.

Zhu, Y., Q. Zhao, Y. Li, X. Cai, and W. Li. 2006. The interaction and toxicity of multi-walled carbon nanotubes with Stylonychia mytilus. *Journal of Nanoscience and Nanotechnology* 6(5):1357–64.

Index

A

ACCs, *see* Activated carbon cloths (ACCs)
ACF, *see* Activated carbon fibers (ACF)
Acid Brown-14, 266
Activated carbon cloths (ACCs), 33
Activated carbon fiber (ACF), 80, 90, 91, 119, 121
Active oxygen, 71
Adsorption, 24, 185, 214, 231, 233, 257–264, 285, 287; *see also* Sorption
    heavy metal removal using, 292–294
        magnetic separation technology, 293
        MNPs-based adsorbents, 293–294
    isotherms and kinetics, 260–262, 263–264, 378
    kinetic experiments, 360–361
    nanoadsorbents used for arsenic removal, 333–346
        bimetallic oxides, 342–343
        carbon-based nanomaterials, 343–346
        functionalized magnetic nanoparticles, 342
        functionalized metal oxides, 341
        layered double hydroxides, 343
        magnetic nanomaterials, 341–342
        metal organic frameworks, 346
        nanometal oxides of transition metals, 333–341
    nanomaterial, 228–230
        effects of, functional groups on, 228–229
        effects of water chemistry on, 229–230
    solution preparation for, 360
    of water pollutant species on LDHs, 257–260
Advanced oxidation processes (AOPs), 4, 5, 66, 109–127
    heterogeneous Fenton process, 112–118
    electro-Fenton process, 119–127
Aflatoxin-detoxifizyme (APTZ), 11
AHPS, *see* 4-amino-3-hydroxy-2-p-tolylazo-naphthalene-1-sulfonic acid (AHPS)
Algae, 396
Alginate beads, 124, 126
Alizarin Red, 80, 81
Alphazurine A (AZA), 76
American Association of Textile Chemists and Colorists, 62
Amine-Fe$_3$O$_4$-based adsorbents, 307–311
    APTEs, 309–311
    tetraethylenepentamine and its related, 307–309
Amino-functionalized MOFs, 367–373
Amino sulfonic-Cu-(4,4′-bipy)$_2$ (ASC), 371
Amphiascus tenuiremis, 397
Animals, 396–398
    fish, 397–398
    invertebrates, 396–397
        crustaceans, 396–397
        earthworms, 397
        nematodes, 397
    mammals, 398
Anodic oxidation, 39, 40–41, 43, 69, 70–81
    direct electrolysis, 70

    electrogenerated hydroxyl radicals, 70–81
        comparison of anode materials, 80–81
        dissociative adsorption of water, 71
        dye degradation, 73–80
        electrolytic discharge of water, 71–73
Antimicrobial property, 172–173
AOPs, *see* Advanced oxidation processes (AOPs)
APTES, *see* γ-Aminopropyltriethoxysilane (APTES)
APTMS, *see* γ-Aminopropyltrimethoxysilane (APTMS)
APTZ, *see* Aflatoxin-detoxifizyme (APTZ)
Array biosensors, 10–11
Arsenate, 328
Arsenic (As), 135, 324–347
    adsorption, 333–346
        nanoadsorbents used for, removal
    globally affected areas with, and environmental standards, 327–328
    toxicity, 326–327
    treatment techniques for removal of, 328–333
        coagulation and flocculation, 329–330
        ion exchange, 330
        membrane separation, 330–333
        oxidation, 328–329
        precipitation, 329–330
Arsenite, 328
ASC, *see* Amino sulfonic-Cu-(4,4′-bipy)$_2$ (ASC)
Auxochrome, 62
AZA, *see* Alphazurine A (AZA)
Azo dyes, 62, 66
Azure B, 91, 96

B

Bacillus subtilis, 235, 236, 238, 394, 395
Bacteria, 394–396
Ballistic conduction, 208
Basic Red 46, 74
Basic Yellow, 74, 97
BDD, *see* Boron-doped diamond (BDD)
Beer, 74
BET, *see* Brunauer-Emmett-Teller (BET) surface area
Bimetallic oxides, 342–343
Biofouling, 172
Biogranulation method, 270
Biological mechanisms, 389
Biological treatments, 66
Biopolymeric nanofibers, 173
Biosensors, 9, 10
Bipolar electrode, 41
Bipolaron, 158
BIS, *see* Bureau of Indian Standards (BIS)
Boron-doped diamond (BDD), 72, 76–79, 81, 85, 91
Bottom-up approach, 180
Brillas, Enric, 89
Brucite sheets, 246

Printed and bound by CPI Group (UK) Ltd, Croydon, CR0 4YY

01/11/2024

01782603-0008